Lecture Notes in Computer Science 9019

Commenced Publication in 1973
Founding and Former Series Editors:
Gerhard Goos, Juris Hartmanis, and Jan van Leeuwen

More information about this series at http://www.springer.com/series/7407

António Gaspar-Cunha · Carlos Henggeler Antunes
Carlos Coello Coello (Eds.)

Evolutionary Multi-Criterion Optimization

8th International Conference, EMO 2015
Guimarães, Portugal, March 29 – April 1, 2015
Proceedings, Part II

 Springer

Editors
António Gaspar-Cunha
Institute for Polymers and Composites/I3N
University of Minho
Guimarães
Portugal

Carlos Coello Coello
CINVESTAV-IPN
Depto. de Computacíon
Col. San Pedro Zacatenco
Mexico

Carlos Henggeler Antunes
Dept. of Electrical and Computer Engg.
University of Coimbra
Coimbra
Portugal

ISSN 0302-9743
Lecture Notes in Computer Science
ISBN 978-3-319-15891-4
DOI 10.1007/978-3-319-15892-1

ISSN 1611-3349   (electronic)

ISBN 978-3-319-15892-1   (eBook)

Library of Congress Control Number: 2015932656

LNCS Sublibrary: SL1 – Theoretical Computer Science and General Issues

Springer Cham Heidelberg New York Dordrecht London

Springer International Publishing AG Switzerland is part of Springer Science+Business Media
(www.springer.com)

Preface

EMO is a biennial international conference series devoted to the theory and practice of evolutionary multi-criterion optimization.

The first EMO took place in 2001 in Zürich (Switzerland), with later conferences taking place in Faro (Portugal) in 2003, Guanajuato (Mexico) in 2005, Matsushima-Sendai (Japan) in 2007, Nantes (France) in 2009, Ouro Preto (Brazil) in 2011, and Sheffield (UK) in 2013. The proceedings of this series of conferences have been published as a volume in Lecture Notes in Computer Science (LNCS), respectively, in volumes 1993, 2632, 3410, 4403, 5467, 6576, and 7811.

The 8th International Conference on Evolutionary Multi-Criterion Optimization (EMO 2015) took place in Guimarães, Portugal, from March 29 to April 1, 2015. The event was organized by the University of Minho. Following the success of the two previous EMO conferences, a special track was offered aiming to foster further cooperation between the EMO and the multiple criteria decision making (MCDM). Also, a special track on real-world applications (RWA) was endorsed.

EMO 2015 received 90 full-length papers, which were submitted to a rigorous single-blind peer-review process, with a minimum of three referees per paper. Following this process, a total of 68 papers were accepted for presentation and publication in this volume, from which 40 were chosen for oral and 24 for poster presentation. The selected papers were distributed through the different tracks as follows: 46 main track, 6 MCDM track, and 16 RWA track.

The conference benefitted from the presentations of plenary speakers on research subjects fundamental to the EMO field: Thomas Stützle, from the IRIDIA laboratory of Université libre de Bruxelles (ULB), Belgium; Murat Köksalan, from the Industrial Engineering Department of Middle East Technical University, Ankara, Turkey; Luís Santos, from the University of São Paulo and Embraer, Brazil; Carlos Fonseca, from the University of Coimbra, Portugal.

From the beginning, this conference provided significant advances in relevant subjects of evolutionary multi-criteria optimization. This event aimed to continue these type of developments, being the papers presented focused on: theoretical aspects, algorithms development, many-objectives optimization, robustness and optimization under uncertainty, performance indicators, multiple criteria decision making, and real-world applications.

Finally, we would express our gratitude to the plenary speakers for accepting our invitation, to all the authors who submitted their work, to the members of the International Program Committee for their hard work, to the members of the Organizing Committee, particularly Lino Costa, and to the Track Chairs Kaisa Miettinen, Salvatore Greco, and

Robin Purshouse. We would like to acknowledge the support of the School of Engineering of the University of Minho. We would also like to thank Alfred Hofmann and Anna Kramer at Springer for their support in publishing these proceedings.

March 2015

António Gaspar-Cunha
Carlos Henggeler Antunes
Carlos Coello Coello

Organization

Committees

General Chairs

António Gaspar-Cunha	University of Minho, Portugal
Carlos Henggeler Antunes	University of Coimbra, Portugal
Carlos Coello Coello	CINVESTAV-IPN, Mexico

MCDM Track Chairs

Kaisa Miettinen	University of Jyväskylä, Finland
Salvatore Greco	University of Catania, Italy

Real-world Applications Track Chairs

Robin Purshouse	University of Sheffield, UK
Carlos Henggeler Antunes	University of Coimbra, Portugal

EMO Steering Committee

Carlos Coello Coello	CINVESTAV-IPN, Mexico
David Corne	Heriot-Watt University, UK
Kalyanmoy Deb	Michigan State University, USA
Peter Fleming	University of Sheffield, UK
Carlos M. Fonseca	University of Coimbra, Portugal
Hisao Ishibuchi	Osaka Prefecture University, Japan
Joshua Knowles	University of Manchester, UK
Kaisa Miettinen	University of Jyväskylä, Finland
J. David Schaffer	Binghamton University, USA
Lothar Thiele	ETH Zürich, Switzerland
Eckart Zitzler	PH Bern, Switzerland

Local Organization Committee

António Gaspar-Cunha	University of Minho, Portugal
Pedro Oliveira	University of Porto, Portugal
Lino Costa	University of Minho, Portugal
M. Fernanda P. Costa	University of Minho, Portugal
Isabel Espírito Santo	University of Minho, Portugal
A. Ismael F. Vaz	University of Minho, Portugal
Ana Maria A.C. Rocha	University of Minho, Portugal

Program Committee

Ajith Abraham	MIR Labs, USA
Adiel Almeida	Federal University of Pernambuco, Brazil
Maria João Alves	University of Coimbra, Portugal
Helio Barbosa	Laboratório Nacional de Computação Científica, Brazil
Matthieu Basseur	LERIA Lab., France
Juergen Branke	University of Warwick, UK
Dimo Brockhoff	INRIA Lille - Nord Europe, France
Marco Chiarandini	University of Southern Denmark, Denmark
Sung-Bae Cho	Yonsei University, South Korea
Leandro Coelho	Pontifical Catholic University of Parana, Brazil
Salvatore Corrente	University of Catania, Italy
M. Fernanda P. Costa	University of Minho, Portugal
Clarisse Dhaenens	University of Lille 1, France
Yves De Smet	Université libre de Bruxelles, Belgium
Alexandre Delbem	Institute of Mathematical and Computer Sciences, Brazil
Michael Doumpos	Technical University of Crete, Greece
Michael Emmerich	Leiden University, Netherlands
Andries Engelbrecht	University of Pretoria, South Africa
Isabel Espírito Santo	University of Minho, Portugal
Jonathan Fieldsend	University of Exeter, UK
José Figueira	CEG-IST, Portugal
Peter Fleming	University of Sheffield, UK
António Gaspar-Cunha	University of Minho, Portugal
Martin Josef Geiger	Helmut-Schmidt-Universität, Germany
Ioannis Giagkiozis	University of Sheffield, UK
Christian Grimme	University of Münster, Germany
Walter Gutjahr	University of Vienna, Austria
Francisco Herrera	University of Granada, Spain
Jin-Kao Hao	University of Angers, France
Evan Hughes	White Horse Radar Limited, UK
Masahiro Inuiguchi	Osaka University, Japan
Hisao Ishibuchi	Osaka Prefecture University, Japan
Yaochu Jin	University of Surrey, UK
Laetitia Jourdan	University of Lille 1, France

Daniel Vanderpooten	Paris Dauphine University, France
Fernando J. Von Zuben	University of Campinas, Brazil
Tobias Wagner	Technische Universität Dortmund, Germany
Elizabeth Wanner	CEFET-MG, Brazil
Farouk Yalaoui	University of Technology of Troyes, France
Gary Yen	Oklahoma State University, USA
Qingfu Zhang	University of Essex, UK
Andre de Carvalho	University of São Paulo, Brazil
Alexis Tsoukias	CNRS, France

Support Institution

School of Engineering, University of Minho

Contents – Part II

Mulit-Criterion Decision Making (MCDM)

Real World Applications

Contents – Part I

Many-Objectives Optimization, Performance and Robustness

Evolutionary Many-Objective Optimization Based on Kuhn-Munkres' Algorithm

José A. Molinet Berenguer and Carlos A. Coello Coello[(⊠)]

Computer Science Department, CINVESTAV-IPN,
Av. IPN 2508. San Pedro Zacatenco, 07300 México D.F., Mexico
`jmolinet@computacion.cs.cinvestav.mx`, `ccoello@cs.cinvestav.mx`

Abstract. In this paper, we propose a new multi-objective evolutionary algorithm (MOEA), which transforms a multi-objective optimization problem into a linear assignment problem using a set of weight vectors uniformly scattered. Our approach adopts uniform design to obtain the set of weights and Kuhn-Munkres' (Hungarian) algorithm to solve the assignment problem. Differential evolution is used as our search engine, giving rise to the so-called Hungarian Differential Evolution algorithm (HDE). Our proposed approach is compared with respect to a MOEA based on decomposition (MOEA/D) and with respect to an indicator-based
MOEA (the *S metric selection Evolutionary Multi-Objective Algorithm*, SMS- EMOA) using several test problems (taken from the specialized literature) having from two to ten objective functions. Our preliminary experimental results indicate that our proposed HDE outperforms MOEA/D and is competitive with respect to SMS-EMOA, but at a significantly lower computational cost.

Keywords: Many-objective optimization · Multi-Objective Evolutioanry Algorithms · Kuhn-Munkres algorithm

1 Introduction

A large number of problems that arise in academic and industrial areas have several conflicting objectives that need to be optimized simultaneously [7]; they are called multi-objective optimization problems (MOPs). The most commonly adopted notion of optimum in multi-objective optimization is Pareto optimality, which refers to finding the best possible trade-offs among the objectives of a multi-objective problem. These trade-off solutions constitute the so-called Pareto optimal set. The image of the Pareto optimal set is called the Pareto front. Among the different techniques available to solve MOPs, multi-objective evolutionary algorithms (MOEAs) have become very popular, mainly because of their flexibility and ease of use. Modern MOEAs normally aim at producing, in

C.A. Coello Coello—The second author acknowledges support from CONACyT project no. 221551.

A. Gaspar-Cunha et al. (Eds.): EMO 2015, Part II, LNCS 9019, pp. 3–17, 2015.
DOI: 10.1007/978-3-319-15892-1_1

a single run, several different solutions, which are as close as possible to the true Pareto front [7]. For several years, MOEAs adopted a selection mechanism based on Pareto optimality. However, in recent years, it was found that Pareto-based MOEAs cannot properly differentiate individuals when dealing with problems having four or more objectives (the so-called many-objective optimization problems [13]). This has motivated the development of alternative selection schemes from which the use of performance indicators has been (until now) the most popular choice [26]. When using indicator-based selection, the idea is to identify the solutions that contribute the most to the improvement of the performance indicator adopted in the selection mechanism.

From the several performance indicators currently available, the hypervolume [24] has become the most popular choice for implementing indicator-based MOEAs, mainly because of its good theoretical properties [5]. The hypervolume is the only unary indicator that is known to be Pareto compliant and it has been proved that its maximization is equivalent to finding the Pareto optimal set [11]. However, the main disadvantage of adopting this indicator is that the best algorithms known to compute the hypervolume have a computational cost which grows exponentially on the number of objectives [4]. Although some researchers have proposed schemes to approximate the hypervolume contributions at an affordable computational cost (see for example [1]), the performance of such approaches seems to degrade very quickly in high dimensionality at the expense of reducing their computational cost. This has motivated the development of other selection schemes based on different performance indicators (see for example [6]).

On the other hand, MOEAs based on decomposition have also become popular in recent years. Perhaps, MOEA/D is the most popular MOEA based on decomposition. This algorithm decomposes the MOP into N scalar optimization subproblems and it solves these subproblems simultaneously using an evolutionary algorithm. MOEA/D has shown to be a good alternative to solve MOPs with low or high dimensionality (regarding objective function space). However, MOEA/D has two important disadvantages. The first is that it generates a new solution from a unique neighborhood, i.e., the new solution cannot be generated from individuals of different neighborhoods. And, the second is that a new solution with a high fitness can replace several solutions, and then, the population can lose diversity, see Figure 1. Li and Zhang proposed in [16] a variant of MOEA/D and they called it "MOEA/D-DE". This proposal allows that a new individual will be generated from individuals of different neighborhoods. Also, it restricts the number of solutions that can be replaced by the same individual. However, both proposals MOEA/D and MOEA/D-DE generate a new solution, and then, they look in which subproblem the new solution is better than the current solution but they do not consider the case where the solution which was replaced could improve the solution of another subproblem, i.e, both algorithms assign the best individual to each subproblem in an independent way, without considering the best assignment globally. Figure 2 shows the assignment made by MOEA/D and MOEA/D-DE and Figure 3 shows the global optimal assignment.

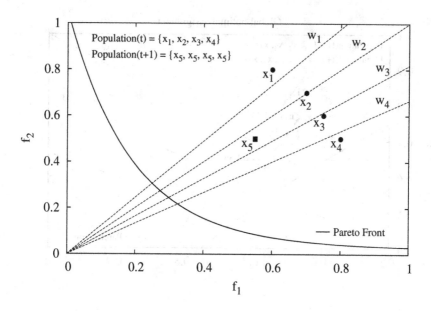

Fig. 1. Disadvantage of MOEA/D when it replaces the solutions. For each weight vector w_i, $i = 1, 2, 3, 4$, the solution x_5 has the highest utility value. Therefore, x_5 is the best solution of the four subproblems.

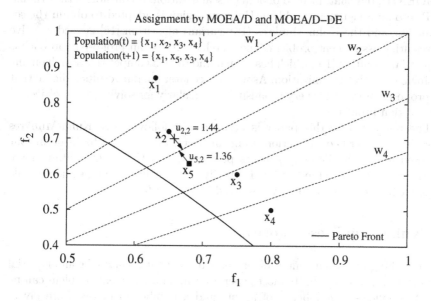

Fig. 2. The new solution x_5 is assigned to the subproblem w_2, therefore, the solution x_2 is replaced by x_5. It is important to note that the solution x_2 is better than solution x_1 for the subproblem w_1. However, MOEA/D and MOEA/D-DE eliminate x_2.

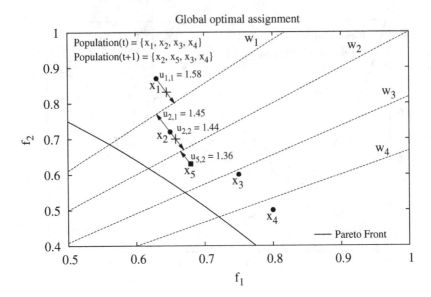

Fig. 3. The new solution x_5 is assigned to the subproblem w_2 and the solution x_2 is assigned to the subproblem w_1. Therefore, the solution x_1 is eliminated.

In this paper, we propose the use of an approach that is conceptually closer to MOEA/D, but that, instead of doing a scalarization, it transforms the original MOP into an assignment problem. Uniform design is adopted to obtain the set of weights, and the Kuhn-Munkres (Hungarian) algorithm [15] is used to solve the resulting assignment problem. The search engine of our proposed approach is differential evolution [19], which has been found to be a competitive search engine for single-objective optimization. As we will see later on, our results indicate that our proposed approach is very promising, particularly for solving many-objective optimization problems.

The remainder of this paper is organized as follows. The Kuhn-Munkres algorithm is described in Section 2 and in Section 3 we describe in detail our proposed approach. The experiments performed and the results obtained are described and discussed in Section 4. Finally, our conclusions and some possible paths for future work are briefly discussed in Section 5.

2   Kuhn-Munkres Algorithm

The matching or assignment problem is a fundamental class of combinatorial optimization problems. In its most general form, an assignment problem can be stated as follows: a number n of agents and a number m of tasks are given, possibly with some restrictions on which agents can perform each particular task. A cost is incurred for each agent performing some task, and the goal is to perform all tasks in such a way that the total cost of the assignment is minimized [15]. The Linear Assignment Problem (LAP) is the simplest of the assignment problems. In the canonical LAP, the number of agents and tasks is the same, and any agent can be assigned to perform any task. Formally, LAP can be formulated as follows.

Definition 1. *Given a set of agents $A = \{a_1, ..., a_n\}$, a set with the same number of tasks $T = \{t_1, ..., t_n\}$ and the cost function $C : A \times T \to \mathbb{R}$, and let $\Phi : A \to T$ the set of all possible bijections between A and T*

$$\underset{\phi \in \Phi}{\text{minimize}} \sum_{a \in A} C(a, \phi(a)) \tag{1}$$

Usually, the cost function is also viewed as a squared real-valued matrix C with elements $C_{ij} = C(a_i, t_j)$, and the set Φ of all possible bijections between A and T as a set of assignment matrices $\mathcal{X}$. The LAP can be expressed as an integer linear program:

$$\underset{x \in \mathcal{X}}{\text{minimize}} \quad \sum_{i=1}^{n} \sum_{j=1}^{n} C_{ij} x_{ij}$$

$$\text{subject to:} \quad \sum_{i=1}^{n} x_{ij} = 1, \ \forall j \in \{1, .., n\}, \tag{2}$$

$$\sum_{j=1}^{n} x_{ij} \leq 1, \ \forall i \in \{1, ..., n\},$$

$$x_{ij} \in \{0, 1\}, \ \forall i, j \in \{1, ..., n\}$$

In 1955, Harold W. Kuhn [15] proposed an algorithm for constructing a maximum weight perfect matching in a bipartite graph. His pioneering work in this area, is a combinatorial optimization algorithm that solves the assignment problem in polynomial time. Kuhn explained how the works of two Hungarian mathematicians, D. König and E. Egerváry, had contributed to the invention of his algorithm, which is the reason why he called it the *Hungarian Method*. James Munkres [17] reviewed Kuhn's work in 1957 and made several important contributions to the theoretical aspects of the algorithm. Munkres found that the algorithm is (strongly) polynomial and proposed an improved version of $O(n^3)$. The contribution of Munkres to the development of the Hungarian algorithm has led to the algorithm which is being referred to as the Kuhn-Munkres algorithm. An extension of this algorithm for rectangular matrices was introduced by Bourgeois and Lassalle in 1971 [3]. The extension to rectangular matrices allows the algorithm to operate in assignment problems where the numbers of agents and tasks are unequal.

3   Our Proposed Approach

We propose here an alternative selection mechanism for MOEAs which is not based on Pareto dominance or on any performance indicator. The main motivation of this work is to avoid the scalability problems of Pareto-based selection schemes as well as the excessive computational cost of adopting the hypervolume contribution for selecting solutions. The algorithm presented here transforms the selection process into a linear assignment problem, which is solved using Kuhn-Munkres algorithm. As we will see, the solution of this LAP allows convergence towards the true Pareto front and, at the same time, a good distribution of solutions along the Pareto front. The proposed MOEA adopts the recombination operators of differential evolution to create new individuals at each generation and the Hungarian algorithm in its selection scheme. Because of this, our proposed approach is called *Hungarian Differential Evolution* (HDE).

At each gth generation of the HDE algorithm we have a parent population P_g of n individuals and a population P_g^* of n offspring obtained from P_g. Let $Q_g = P_g \cup P_g^*$ be the set of $2n$ solutions in the gth generation. Then, a linear assignment problem is created using the k-dimensional objective vectors from Q_g and n weight vectors uniformly spread in objective function space. In the context of a selection mechanism for MOEAs, a LAP can be understood as follows: we have $2n$ individuals and n vectors well-distributed in the $(k-1)$-dimensional unit simplex of the objective space. A cost is incurred for each individual representing some vector in the Pareto Front approximation. The goal is to describe all regions covered by the n vectors using only n individuals in such a way that the total cost of the assignment is minimized. The main task is how to construct a cost matrix such that it minimizes the total cost involved in retaining the solutions which are a good approximation of the Pareto Front. This procedure is described next.

First, the $2n$ vectors of objective values in Q_g are normalized to reduce the current objective space to a unit hypercube, so that we can deal with non-commensurable objective functions. The maximum z^{max} and minimal z^{min} vectors are calculated for this purpose.

$$
\begin{aligned}
\boldsymbol{z}^{max} &= [z_1^{max}, ..., z_k^{max}]^T, & z_i^{max} &= \max_{j=1,...,2n} f_i(\boldsymbol{x}_j), & i = 1, ..., k, \\
\boldsymbol{z}^{min} &= [z_1^{min}, ..., z_k^{min}]^T, & z_i^{min} &= \min_{j=1,...,2n} f_i(\boldsymbol{x}_j), & i = 1, ..., k,
\end{aligned}
\tag{3}
$$

where $f_i(\boldsymbol{x}_j)$ is the ith objective value of the jth individual in Q_g, and its normalized value $\tilde{f}_i(\boldsymbol{x}_j)$ is calculated as:

$$
\tilde{f}_i(\boldsymbol{x}_j) = \frac{f_i(\boldsymbol{x}_j) - z_i^{min}}{z_i^{max} - z_i^{min}}, \quad j = 1, ..., 2n, \; i = 1, ..., k.
\tag{4}
$$

Let W be a set of n weight vectors uniformly scattered in objective space.

$$
W \subset \mathbb{W} = \{\boldsymbol{w} \mid \boldsymbol{w} \in [0, 1]^k, \sum_{i=1}^{k} w_i = 1\}, \; |W| = n,
\tag{5}
$$

The cost C_{rj} of assigning the individual $\boldsymbol{x}_j$ to the weight vector $\boldsymbol{w}_r$ is given by:

$$
C_{rj} = \max_{i=1,...,k} w_{ri} \times \tilde{f}_i(\boldsymbol{x}_j), \; r = 1, ..., n, \; j = 1, ..., 2n.
\tag{6}
$$

Algorithm 1. Hungarian Differential Evolution (HDE)

Input   : MOP, population size (n), maximum number of generations (g_{max}),
              parameters C_r and F for DE/$rand/1/bin$
Output: $P_{g_{max}}$ (approximation of the $\mathcal{P}^*$ and $\mathcal{PF}^*$)

1 Generate initial population P_1 randomly;
2 Evaluate each individual in P_1;
3 $W \leftarrow$ Generate n weight vectors using Algorithm 2;
4 **for** $g = 1$ **to** g_{max} **do**
5       $P_g^* \leftarrow$ Generate offspring using P_g and DE/$rand/1/bin$;
6   Evaluate each individual in P_g^*;
7       $Q_g \leftarrow P_g \cup P_g^*$;
8   Calculate z^{max} and z^{min} by (3) Normalize objectives of each individual in Q_g by (4);
9   Generate the cost matrix C by (6) using Q_g and W;
10      $\mathcal{I} \leftarrow$ Obtain the best assignment in C using the Hungarian Method;
11      $P_{g+1} \leftarrow \{\boldsymbol{x}_i \mid i \in \mathcal{I},\, \boldsymbol{x}_i \in Q_g\}$;
12 **end**

The matrix C indicates how each individual is suitable to represent each region of the Pareto Front approximation. The solution to our assignment problem is found by identifying the combination of values in C resulting in the smallest sum, subject to certain constraints. These conditions are:

1. Exactly one value must be chosen in each row; this ensures that only one individual is assigned to each position on the Pareto Front.
2. At most one value can be selected in each column; this ensures that no individual is assigned to more than one position.

The matrix C and the two above constraints are formally represented by (2) as a linear programming problem. The solution to this problem is obtained by the extended Kuhn-Munkres algorithm for rectangular matrices, presented in Section 2. The matrix that solves (2) represents the individuals assigned to each weight vector such that it minimizes the total cost of the assignment, allowing to retain the best n individuals to approximate the Pareto Front. The pseudo-code of our proposed approach is depicted in Algorithm 1.

3.1   Generation of Weight Vectors Using Uniform Design

There exist several MOEAs [8,18,23] that require a set of weight vectors uniformly scattered on the $(k-1)$-unit simplex to obtain solutions along the entire Pareto Front in a k-objective optimization problem. A variety of methods to obtain an evenly distributed subset of weights in a simplex are available in the specialized literature [10]. The simplex-lattice design method [20] is the approach that has been the most commonly adopted in MOEAs. However, at least three problems can be identified in this method [10]. First, the weight vectors are not very uniformly distributed. Second, there are too many vectors at the boundary of the domain. Furthermore, the number of vectors generated increases nonlinearly with the number of objectives. That is, if H divisions are considered along each objective, the total number of weight vectors (hence the population size) in a k-objective problem is given by: $\binom{H+k-1}{k-1}$. Due to this, some MOEAs have used other methods to generate an arbitrary number of weight vectors well-distributed over a simplex. In [18] a hypervolume-based weight vector generation is proposed. This method produces well-distributed vectors maximizing the hypervolume covered by them in objective space. A different idea was proposed in [21], where the uniform design (UD) [10] and good lattice point (glp) [14] methods are combined to set the weight vectors. Nevertheless, both the hypervolume and the glp method have a high computational cost when the number of objectives grows.

Uniform design is a space filling design method that seeks experimental points to be uniformly scattered on the domain [10]. In uniform design, a set of points is considered uniformly spread throughout the entire domain if it has a small discrepancy, where discrepancy is a numerical measure of scatter. Fang and Wang [10] presented different methods for generating points that can be applied to the generation of a set of space-filling design points. Among them, we have the good lattice point (glp) method and Hammersley method [12], both of which are efficient quasi Monte-Carlo methods.

We propose to generate weight vectors using uniform design combined with Hammersley method. This algorithm allows a more uniform distribution of the weight vectors over the space than the simplex-lattice method, and the population size neither increases nonlinearly with the number of objectives nor considers a formulaic setting. Additionally, Hammersley method provides a set of design points with low discrepancy similar to the glp method, but at a much lower computational cost [10].

The Hammersley method is based on the p-adic representation of natural numbers: Any positive integer m can be uniquely expressed using a prime base $p \geq 2$ as

$$m = \sum_{i=0}^{r} b_i \times p^i, \quad 0 \leq b_i \leq p-1, \quad i = 0, \ldots, r, \tag{7}$$

where $p^r \leq m < p^{r+1}$. Then, for any integer $m \geq 1$ with representation (7), let

$$y_p(m) = \sum_{i=0}^{r} b_i \times p^{-(i+1)}, \tag{8}$$

where $y_p(m) \in (0,1)$ and is known as the radical inverse of m base p. Let $k \geq 2$ and $p_1, \ldots, p_{k-1}$ be $k-1$ distinct prime numbers, the Hammersley set consisting of n points uniformly scattered on $[0,1]^k$ is given by

$$x_i = \left[\frac{2i-1}{2n}, y_{p_1}(i), \ldots, y_{p_{k-1}}(i) \right]^T, \quad i = 1, \ldots, n. \tag{9}$$

In [22], it was proposed to use uniform design for experiments with mixture (UDEM) that seek points to be uniformly scattered in the domain $\mathbb{W}$ defined

---

Algorithm 2. Generation of weight vectors

    Input : number of objectives (k), number of weights (n)
    Output: W (set of weight vectors with low-discrepancy)
1   $p \leftarrow$ array with the first $k - 2$ prime numbers;
2   $U \leftarrow \emptyset$;
3   **for** $i = 1$ **to** n **do**
4      $u_{i1} \leftarrow (2i - 1)/2n$;
5   **for** $j = 2$ **to** $k - 1$ **do**
6          $u_{ij} \leftarrow 0$;
7          $f \leftarrow 1/p_{j-1}$;
8          $d \leftarrow i$;
9   **while** $d > 0$ **do**
10            $u_{ij} \leftarrow u_{ij} + f \times (d \mod p_{j-1})$;
11            $d \leftarrow \lfloor d/p_{j-1} \rfloor$;
12            $f \leftarrow f/p_{j-1}$;
13   **end**
14   **end**
15      $U \leftarrow U \cup \{u\}$;
16 **end**
17 $W \leftarrow$ Apply the transformation (10) to U;

---

by (5). They employed the transformation method for the construction of such uniform design. This method requires a set of vectors $U = \{u_i = [u_{i1}, ..., u_{i(k-1)}]^T,$ $i = 1, ..., n\} \subset [0, 1]^{k-1}$ with small discrepancy. In our proposal, the Hammersley method is used to obtain U and then to apply the transformation

$$w_{ti} = (1 - u_{ti}^{\frac{1}{k-i}}) \prod_{j=1}^{i-1} u_{tj}^{\frac{1}{k-j}}, \quad i = 1, ..., k-1,$$

$$w_{tk} = \prod_{j=1}^{k-1} u_{tj}^{\frac{1}{k-j}}, \quad t = 1, ..., n. \tag{10}$$

Then $\{w_t = [w_{ti}, ..., w_{tk}]^T, \ t = 1, ..., n\}$ is a uniform design on $\mathbb{W}$. The pseudocode of the algorithm used to generate weight vectors is presented in Algorithm 2.

4   Experimental Results

We validated our proposed HDE comparing its performance with respect to two MOEAs representative of the state-of-the-art in the area: the multi-objective evolutionary algorithm based on decomposition [23] (MOEA/D) and the S metric selection Evolutionary Multi-Objective Algorithm [2] (SMS-EMOA). Since the SMS-EMOA requires a considerably large amount of computational time in problems with more than five objectives, we also include in this comparative study a version of this MOEA (called appSMS-EMOA) that uses the algorithm proposed in [1] to approximate the hypervolume contributions using Monte Carlo sampling.

In our experiments, we adopted 12 test problems, consisting of five bi-objective problems taken from the Zitzler-Deb-Thiele (ZDT) test suite [25] and seven test problems having from two to ten objective functions taken from the Deb-Thiele-Laumanns-Zitzler (DTLZ) test suite [9]. In the problems ZDT1-3 the number of decision variables is 30; ZDT4 and ZDT6 have 10 variables. In the DTLZ test problems, the total number of variables is given by $n = m + k - 1$, where $m = 2, ..., 10$ is the number of objectives and k was set to 10 for DTLZ1-6 and 20 for DTLZ7.

In order to assess the performance of each MOEA, we selected the hypervolume indicator as a performance measure. The hypervolume is the size of the space covered by the Pareto optimal solutions, thus capturing both convergence and diversity in a single value [24]. The hypervolume can differentiate between degrees of complete outperformance of two sets [5]. To calculate the hypervolume indicator, we used the reference points $y_{ref} = [y_1, \cdots, y_m]$ such that: $y_i = 1.1$ for all the ZDT problems, and for DTLZ1, DTLZ2 and DTLZ4; $y_i = 3$ for DTLZ3, DTLZ5 and DTLZ6; and $y_i = 7$ for DTLZ7. We also considered the running time of each algorithm. Running times as a measure of computational cost are particularly relevant when increasing the number of objectives. In order to achieve more confident results, each MOEA was executed 30 times for each problem instance, and we report here their average hypervolume values and their average running times.

Our proposed HDE uses the variation operators of differential evolution and, therefore, it uses its same parameters. The parameters adopted in our experiments were: $F = 1.0$ and $Cr = 0.4$. The recombination operators of MOEA/D, SMS-EMOA and appSMS-EMOA are simulated binary crossover and polynomial-based mutation. Their corresponding parameters were set as follows: crossover probability $p_c = 1.0$, mutation probability $p_m = 1/n$, where n is the number of decision variables; the distribution indexes were set as: $\eta_c = 20$ and $\eta_m = 20$. MOEA/D used the Tchebycheff approach with a neighborhood size of 20. The number of samples for the Monte Carlo estimation in appSMS-EMOA was set to 10^4. The algorithms HDE, SMS-EMOA and appSMS-EMOA can use an arbitrary population size, but in MOEA/D the population size increases nonlinearly with the number of objectives. For this reason, we used different population sizes. In the ZDT bi-objective problems the population size was set to 100. For the DTLZ problems with 2, 3, 4 and 8 objectives, the population size was set to 120. For problems having 5 and 6 objectives, the population size was set to 126. Finally, for problems having 7, 9 and 10 objectives, the population size was set to 210, 165 and 220, respectively. The maximum number of generations adopted in the ZDT test problems was 200, and we used 300 for the DTLZ test problems. It is important to note that SMS-EMOA was not applied to problems with more than 5 objectives due to its high computational cost.

4.1   Discussion of Results

First, we will review our results in the ZDT test problems. Table 1 provides the average hypervolume of each compared MOEA for each test problem. The best results are presented in **boldface**. From these results, we can see that our proposed

HDE outperformed all the other MOEAs in all the test problems, except for ZDT3, where SMS-EMOA achieved a slightly higher hypervolume value.

Table 1. Results obtained in the ZDT test problems. We show the average hypervolume values obtained over 30 independent runs.

Problem	HDE	MOEA/D	appSMS-EMOA	SMS-EMOA
ZDT1	**0.871748**	0.863668	0.868213	0.871514
ZDT2	**0.538383**	0.517640	0.528167	0.537282
ZDT3	1.327721	1.298709	1.296754	**1.328633**
ZDT4	**0.833392**	0.602010	0.804054	0.822489
ZDT6	**0.504490**	0.496180	0.487527	0.493889

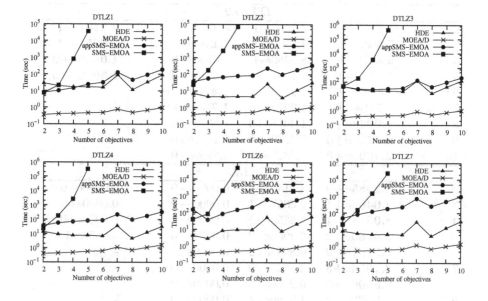

Fig. 4. Average runtime over 30 independent runs of HDE, MOEA/D, SMS-EMOA and appSMS-EMOA in the DTLZ test suite.

In Table 2 we present the average hypervolume for the DTLZ test problems. Figure 4 shows the average runtime for each instance of the DTLZ problems having from two to ten objective functions. In the DTLZ1 and DTLZ3 problems, HDE outperforms the other MOEAs for every number of objective functions. The search space in these two problems contains $(11^k - 1)$ and $(3^k - 1)$ local Pareto fronts, respectively ($k = 10$ in our experiments). This makes difficult to converge to the true Pareto front. SMS-EMOA and appSMS-EMOA are unable to converge to the true Pareto front in any instance of the DTLZ3 problem. Additionally, appSMS-EMOA does not perform well in DTLZ1.

Table 2. Results obtained in the DTLZ test problems. We show the average hypervolume values obtained over 30 independent runs.

N° Obj.	HDE	MOEA/D	appSMS-EMOA	SMS-EMOA
		DTLZ1		
2	**1.0833e+0**	1.0662e+0	1.0286e+0	1.0487e+0
3	**1.3022e+0**	1.2650e+0	8.6233e−1	1.1704e+0
4	**1.4565e+0**	1.2713e+0	2.9529e−2	1.4536e+0
5	**1.6084e+0**	1.3297e+0	0.0000e+0	1.6041e+0
6	**1.7605e+0**	1.5175e+0	1.1103e−4	-
7	**1.9466e+0**	1.9416e+0	0.0000e+0	-
8	**2.1435e+0**	1.9369e+0	0.0000e+0	-
9	**2.3579e+0**	2.2682e+0	0.0000e+0	-
10	**2.5937e+0**	2.5592e+0	0.0000e+0	-
		DTLZ2		
2	4.2060e−1	4.2087e−1	4.2013e−1	**4.2161e−1**
3	7.3603e−1	7.1504e−1	7.4889e−1	**7.6251e−1**
4	9.8341e−1	8.8689e−1	1.0195e+0	**1.0526e+0**
5	1.2229e+0	1.1406e+0	1.2570e+0	**1.3090e+0**
6	1.4462e+0	1.2123e+0	**1.4700e+0**	-
7	1.7219e+0	1.2972e+0	**1.7339e+0**	-
8	**1.9028e+0**	1.2018e+0	1.8572e+0	-
9	**2.1706e+0**	1.3226e+0	2.1051e+0	-
10	**2.4603e+0**	1.4329e+0	2.3859e+0	-
		DTLZ3		
2	**8.2085e+0**	8.1148e+0	0.0000e+0	0.0000e+0
3	**2.6404e+1**	2.6067e+1	0.0000e+0	0.0000e+0
4	**8.0515e+1**	7.6359e+1	0.0000e+0	0.0000e+0
5	**2.4260e+2**	2.2909e+2	0.0000e+0	3.9658e−1
6	**7.2861e+2**	6.8984e+2	0.0000e+0	-
7	**2.1854e+3**	2.1643e+3	0.0000e+0	-
8	**6.5606e+3**	6.2456e+3	0.0000e+0	-
9	**1.9683e+4**	1.9186e+4	0.0000e+0	-
10	**5.9049e+4**	5.7783e+4	0.0000e+0	-
		DTLZ4		
2	4.2050e−1	4.2087e−1	4.2026e−1	**4.2161e−1**
3	7.3349e−1	7.1758e−1	7.4999e−1	**7.6254e−1**
4	9.8347e−1	8.8985e−1	1.0249e+0	**1.0527e+0**
5	1.2290e+0	1.1440e+0	1.2660e+0	**1.3094e+0**
6	1.4518e+0	1.3216e+0	**1.4880e+0**	-
7	1.7290e+0	1.4835e+0	**1.7467e+0**	-
8	**1.9065e+0**	1.3559e+0	1.8973e+0	-
9	**2.1773e+0**	1.4883e+0	2.1206e+0	-
10	**2.4646e+0**	1.5820e+0	2.4016e+0	-

Continued on next page

Table 2. – Continued from previous page

N° Obj.	HDE	MOEA/D	appSMS-EMOA	SMS-EMOA
DTLZ5				
2	8.2106e+0	8.2108e+0	8.2102e+0	**8.2116e+0**
3	2.3979e+1	2.3967e+1	2.3985e+1	**2.3990e+1**
4	7.1549e+1	7.1247e+1	7.1497e+1	**7.1856e+1**
5	2.1419e+2	2.0875e+2	2.1385e+2	**2.1567e+2**
6	**6.4008e+2**	6.1645e+2	6.3956e+2	-
7	**1.9271e+3**	1.8336e+3	1.9188e+3	-
8	5.6978e+3	5.4432e+3	**5.7225e+3**	-
9	**1.7197e+4**	1.6307e+4	1.7171e+4	-
10	**5.1725e+4**	4.8723e+4	5.1545e+4	-
DTLZ6				
2	**8.2108e+0**	8.0197e+0	1.1719e+0	3.1194e+0
3	**2.3982e+1**	2.3487e+1	2.1721e+1	2.3745e+1
4	**7.1345e+1**	6.9232e+1	3.7674e+1	6.7598e+1
5	**2.1324e+2**	1.9631e+2	3.2082e+1	1.9830e+2
6	**6.3859e+2**	5.7393e+2	4.7076e+0	-
7	**1.9204e+3**	1.7051e+3	3.5091e-1	-
8	**5.6590e+3**	4.9692e+3	1.0260e+0	-
9	**1.7070e+4**	1.5059e+4	0.0000e+0	-
10	**5.1444e+4**	4.4826e+4	0.0000e+0	-
DTLZ7				
2	3.1881e+1	3.0554e+1	3.1881e+1	**3.1884e+1**
3	**2.0053e+2**	1.8413e+2	1.9708e+2	1.9931e+2
4	**1.2267e+3**	8.8240e+2	1.1081e+3	1.1598e+3
5	**7.2516e+3**	3.9693e+3	4.5253e+3	6.2818e+3
6	**3.9558e+4**	1.7243e+4	1.4071e+4	-
7	**2.0094e+5**	5.0926e+4	5.2148e+4	-
8	**7.2712e+5**	8.4243e+4	4.6362e+5	-
9	**3.1360e+6**	2.8190e+5	1.6500e+6	-
10	**9.1999e+6**	1.2141e+6	4.5446e+6	-

For DTLZ2 and DTLZ4, SMS-EMOA performs better than the other MOEAs in instances having from two to five objectives, but its runtime is of up to 20 hours in DTLZ2 with five objectives and it reaches up to four days in DTLZ4. appSMS-EMOA obtains the best results in the instances with six and seven objectives, but requires several minutes per run. In DTLZ2 and DTLZ4 with more than seven objectives, HDE outperforms all the other algorithms and requires only seconds per run. A similar observation can be made for the problem DTLZ5, where SMS-EMOA obtains the best results in the instances having from two to five objectives, whereas for more than five objectives HDE performs better than the other MOEAs except for eight objectives.

The main feature of DTLZ5 and DTLZ6 is that the Pareto front is a curve (it loses dimensionality). However, DTLZ6 is considered to be harder to solve than

DTLZ5, because MOEAs tend to have more difficulties to reach the true Pareto front with this problem. In DTLZ6, HDE outperforms the other MOEAs for all the instances having from two to ten objectives, and appSMS-EMOA presented a poor performance. HDE also obtained the best results in DTLZ7, which has a disconnected Pareto front. For all instances with three objectives or more, HDE outperformed the other algorithms; only for two objectives SMS-EMOA achieved a slightly higher hypervolume value.

5   Conclusions and Future Work

We have proposed a novel selection scheme for MOEAs. Our approach transforms the selection mechanism of a MOEA into an assignment problem using a set of well-distributed points on a unit simplex. The obtained assignment problem is solved with the Kuhn-Munkres algorithm. We have also suggested an algorithm based on uniform design to generate a set of weight vectors more uniformly scattered than those obtained by the simplex-lattice method. Our experimental results indicate that our proposed HDE outperforms MOEA/D in several test problems, and is competitive (outperforming it in several instances) with respect to SMS-EMOA, while requiring a significantly lower computational time.

As part of our future work, we intend to study other (computationally inexpensive) uniform design methods to generate a set of points more uniformly distributed. We also plan to analyze other methods for solving assignment problems at a lower computational cost.

References

1. Bader, J., Zitzler, E.: HypE: An Algorithm for Fast Hypervolume-Based Many-Objective Optimization. Evolutionary Computation 19(1), 45–76 (2011)
2. Beume, N., Naujoks, B., Emmerich, M.: SMS-EMOA: Multiobjective selection based on dominated hypervolume. European Journal of Operational Research 181(3), 1653–1669 (2007)
3. Bourgeois, F., Lassalle, J.-C.: An Extension of the Munkres Algorithm for the Assignment Problem to Rectangular Matrices. Commun. ACM 14(12), 802–804 (1971)
4. Bringmann, K., Friedrich, T.: Approximating the least hypervolume contributor: NP-hard in general, but fast in practice. Theoretical Computer Science 425, 104–116 (2012)
5. Brockhoff, D., Friedrich, T., Neumann, F.: Analyzing hypervolume indicator based algorithms. In: Rudolph, G., Jansen, T., Lucas, S., Poloni, C., Beume, N. (eds.) PPSN 2008. LNCS, vol. 5199, pp. 651–660. Springer, Heidelberg (2008)
6. Brockhoff, D., Wagner, T., Trautmann, H.: On the properties of the R2 indicator. In: 2012 Genetic and Evolutionary Computation Conference (GECCO 2012), Philadelphia, USA, pp. 465–472. ACM Press, July 2012
7. Coello Coello, C.A., Lamont, G.B., Van Veldhuizen, D.A.: Evolutionary Algorithms for Solving Multi-Objective Problems, 2nd edn. Springer, New York (2007)

8. Deb, K., Jain, H.: An Evolutionary Many-Objective Optimization Algorithm Using Reference-Point-Based Nondominated Sorting Approach, Part I: Solving Problems With Box Constraints. IEEE Transactions on Evolutionary Computation **18**(4), 577–601 (2014)
9. Deb, K., Thiele, L., Laumanns, M., Zitzler, E.: Scalable test problems for evolutionary multiobjective optimization. In: Abraham, A., Jain, L., Goldberg, R. (eds.) Evolutionary Multiobjective Optimization. Theoretical Advances and Applications, pp. 105–145. Springer, Heidelberg (2005)
10. Fang, K.T., Wang, Y.: Number-Theoretic Methods in Statistics. Chapman & Hall/CRC Monographs on Statistics & Applied Probability. Taylor & Francis (1994)
11. Fleischer, M.: The measure of pareto optima. Applications to multi-objective. In: Fonseca, C.M., Fleming, P.J., Zitzler, E., Deb, K., Thiele, L. (eds.) EMO 2003. LNCS, vol. 2632, pp. 519–533. Springer, Heidelberg (2003)
12. Hammersley, J.M.: Monte-Carlo methods for solving multivariable problems. Annals of the New York Academy of Sciences **86**(3), 844–874 (1960)
13. Ishibuchi, H., Tsukamoto, N., Nojima, Y.: Evolutionary many-objective optimization: A short review. In: 2008 IEEE Congress on Evolutionary Computation CEC 2008 (IEEE World Congress on Computational Intelligence), pp. 2424–2431, Hong Kong, June 2008
14. Korobov, N.M.: The approximate computation of multiple integrals. Doklady Akademii Nauk SSSR **124**, 1207–1210 (1959)
15. Kuhn, H.W.: The Hungarian Method for the Assignment Problem. Naval Research Logistics Quarterly **2**(1–2), 83–97 (1955)
16. Li, H., Zhang, Q.: Multiobjective Optimization Problems With Complicated Pareto Sets, MOEA/D and NSGA-II. IEEE Transactions on Evolutionary Computation **13**(2), 284–302 (2009)
17. Munkres, J.: Algorithms for the Assignment and Transportation Problems. Journal of the Society for Industrial and Applied Mathematics **5**(1), 32–38 (1957)
18. Phan, D. H., Suzuki, J.: R2-IBEA: R2 Indicator based evolutionary algorithm for multiobjective optimization. In: IEEE Congress on Evolutionary Computation (CEC 2013), pp. 1836–1845 (2013)
19. Price, K., Storn, R.M., Lampinen, J.A.: Differential Evolution: A Practical Approach to Global Optimization. Natural Computing Series. Springer, Secaucus (2005)
20. Scheffé, H.: Experiments with mixtures. Journal of the Royal Statistical Society. Series B (Methodological) **20**(2), 344–360 (1958)
21. Tan, Y.-Y., Jiao, Y.-C., Li, H., Wang, X.-K.: MOEA/D + uniform design: A new version of MOEA/D for optimization problems with many objectives. Computers & Operations Research **40**(6), 1648–1660 (2013)
22. Wang, Y., Fang, K.T.: Number-Theoretic Method in Applied statistics (II). Chinese Annals of Mathematics. Serie B **11**, 859–914 (1990)
23. Zhang, Q., Li, H.: MOEA/D: A Multiobjective Evolutionary Algorithm Based on Decomposition. IEEE Transactions on Evolutionary Computation **11**(6), 712–731 (2007)
24. Zitzler, E.: Evolutionary Algorithms for Multiobjective Optimization: Methods and Applications. PhD thesis, Swiss Federal Institute of Technology (ETH), Zurich, Switzerland, November 1999
25. Zitzler, E., Deb, K., Thiele, L.: Comparison of Multiobjective Evolutionary Algorithms: Empirical Results. Evolutionary Computation **8**(2), 173–195 (2000)
26. Zitzler, E., Künzli, S.: Indicator-based selection in multiobjective search. In: Yao, X., et al. (eds.) PPSN 2004. LNCS, vol. 3242, pp. 832–842. Springer, Heidelberg (2004)

An Optimality Theory Based Proximity Measure for Evolutionary Multi-Objective and Many-Objective Optimization

Kalyanmoy Deb[1(✉)], Mohamed Abouhawwash[1], and Joydeep Dutta[2]

[1] Computational Optimization and Innovation (COIN) Laboratory,
Michigan State University, East Lansing, MI 48824, USA
kdeb@egr.msu.edu, mhawwash@msu.edu
http://www.egr.msu.edu/kdeb
[2] Department of Humanities and Social Sciences,
Indian Institute of Technology Kanpur, Kanpur 208016, India
jdutta@iitk.ac.in
http://www.iitk.ac.in/new/index.php/joydeep-dutta

Abstract. Evolutionary multi- and many-objective optimization (EMO) methods attempt to find a set of Pareto-optimal solutions, instead of a single optimal solution. To evaluate these algorithms, performance metrics either require the knowledge of the true Pareto-optimal solutions or, are ad-hoc and heuristic based. In this paper, we suggest a KKT proximity measure (KKTPM) that can provide an estimate of the proximity of a set of trade-off solutions from the true Pareto-optimal solutions. Besides theoretical results, the proposed KKT proximity measure is computed for iteration-wise trade-off solutions obtained from specific EMO algorithms on two, three, five and 10-objective optimization problems. Results amply indicate the usefulness of the proposed KKTPM as a termination criterion for an EMO algorithm.

Keywords: Multi-objective optimization · Evolutionary optimization · Termination criterion · KkT optimality conditions

1 Introduction

Multi-objective optimization problems give rise to a set of trade-off optimal solutions, known as Pareto-optimal solutions [4,11,14]. In solving these problems, one popular approach has been to first find a representative set of Pareto-optimal solutions and then use higher-level information involving one or more decision-makers to choose a preferred solution from the set. Evolutionary multi- (2 or 3-objective) and many-objective (more than 3 objectives) optimization (EMO) methodologies follow this principle of solving multi-objective optimization problems [4] and have received extensive attention for the past two decades. Since multiple solutions are targets for an EMO methodology, it has always been a difficulty to evaluate the performance of an EMO algorithm and therefore to develop a termination criterion for ending a simulation run. In the EMO literature, several performance

© Springer International Publishing Switzerland 2015
A. Gaspar-Cunha et al. (Eds.): EMO 2015, Part II, LNCS 9019, pp. 18–33, 2015.
DOI: 10.1007/978-3-319-15892-1_2

metrics, such as hypervolume measure [2,17] and inverse generational distance (IGD) measure [8], are suggested, but they are not appropriate particularly for the purpose of terminating a simulation run. For the hypervolume measure, there is no pre-defined target that can be set for an arbitrary problem, thereby making it difficult to determine an expected hypervolume value for terminating a run. On the other hand, the IGD measure requires the knowledge of true Pareto-optimal solutions and their corresponding objective values and hence is not applicable to an arbitrary problem.

For terminating a simulation run, it is ideal if some knowledge about the proximity of the current solution(s) from the true optimal solution(s) can be obtained. For single-objective optimization algorithms, recent studies on approximate Karush-Kuhn-Tucker (A-KKT) points have been suggested [1,10,12,16]. The latter studies have also suggested a *KKT proximity measure* that monotonically reduced to zero as the iterates approach a KKT point for a single-objective constrained optimization problem. In this paper, we extend the definition of an A-KKT point for multi-objective optimization using achievement scalarizing function concept proposed in multiple criterion decision making (MCDM) literature and suggest a KKT proximity measure that can suitably used for evaluating an EMO's performance in terms of convergence to the KKT points.

In the remainder of the paper, we present the proposed KKT proximity measure concept in Section 2 by treating every trade-off solution as an equivalent achievement scalarizing problem. The concept is then applied on an illustrative problem to test its working. In Section 3, the KKT proximity measure is computed for the entire objective space on a well-known two-objective ZDT test problems [19] to illustrate that as solutions approach the efficient front, the KKT proximity measure reduces to zero. Thereafter, KKT proximity measure is computed for trade-off sets found using specific EMO algorithms, such as NSGA-II [7] and NSGA-III [8] on two, three, five and 10-objective test problems. Two engineering design problems are also used for the above purpose. Finally, based on this extensive study, conclusions are drawn in Section 4.

2   Proposed KKT Proximity Measure for Multi-objective Optimization

KKT optimality conditions to evaluate the proximity to Pareto-optimal solutions were first used by authors in 2007 [9] and applied in a power dispatch problem in 2008 [5]. The idea was also applied elsewhere [13]. These methods used a straightforward metric measuring the violation of KKT optimality conditions as a KKT error measure, which we revisit in the next subsection.

However, Dutta et al. [10] criticized the above approach due to the *singularity* property of KKT points and relaxed of KKT optimality conditions to define a KKT proximity measure for any iterate $\mathbf{x}^k$ for a single-objective optimization problem of the following type:

$$\text{Minimize}_{(\mathbf{x})}\ f(\mathbf{x}),$$
$$\text{Subject to } g_j(\mathbf{x}) \leq 0, \quad j = 1, 2, \ldots, J. \tag{1}$$

After a lengthy theoretical calculations, they suggested a procedure of computing the KKT proximity measure for an iterate $(\mathbf{x}^k)$ as follows:

$$
\begin{aligned}
\text{Minimize}_{(\epsilon_k, \mathbf{u})} \quad & \epsilon_k \\
\text{Subject to} \quad & \|\nabla f(\mathbf{x}^k) + \sum_{j=1}^{J} u_j \nabla g_j(\mathbf{x}^k)\|^2 \leq \epsilon_k, \\
& \sum_{j=1}^{J} u_j g_j(\mathbf{x}^k) \geq -\epsilon_k, \\
& u_j \geq 0, \quad \forall j,
\end{aligned}
\tag{2}
$$

In this paper, we extend the KKT proximity metric for multi-objective optimization problems.

For an M-objective optimization problem with inequality constraints, the Karush-Kuhn-Tucker (KKT) optimality conditions are given as follows [3, 14]:

$$
\sum_{k=1}^{M} \lambda_k \nabla f_k(\mathbf{x}^k) + \sum_{j=1}^{m} u_j \nabla g_j(\mathbf{x}^k) = \mathbf{0},
\tag{3}
$$

$$
g_j(\mathbf{x}^k) \leq 0, \quad j = 1, 2, \ldots, J,
\tag{4}
$$

$$
u_j g_j(\mathbf{x}^k) = 0, \quad j = 1, 2, \ldots, J,
\tag{5}
$$

$$
u_j \geq 0, \quad j = 1, 2, \ldots, J,
\tag{6}
$$

$$
\lambda_k \geq 0, \quad k = 1, 2, \ldots, M, \text{ and } \boldsymbol{\lambda} \neq \mathbf{0}.
\tag{7}
$$

Multipliers λ_k are non-negative, but at least one of them must be non-zero. The parameter u_j is called the Lagrange multiplier for the j-th constraint and it is also non-negative. Any solution $\mathbf{x}^k$ that satisfies all the above conditions is called a KKT point [15]. Equation (3) is called the *equilibrium equation*. Equation (5) for every constraint is called the *complimentary slackness* equation. Note that (4) ensures feasibility for $\mathbf{x}^k$ while the (6) arises due to the minimization nature of the problem given in (1).

2.1   A Naive Measure from KKT Optimality Conditions

The above KKT conditions can be used to naively define a KKT error measure, which was also used elsewhere [9]. For a given iterate $\mathbf{x}^k$, the parameters λ-vector and $\mathbf{u}$-vector are unknown. A method was proposed to identify suitable λ and $\mathbf{u}$-vectors so that the all inequality constraints (conditions 4, 5, 6 and 7) are satisfied and the equilibrium condition (3) is violated the least:

$$
\begin{aligned}
\text{Minimize}_{(\boldsymbol{\lambda}, \mathbf{u})} \quad & \| \sum_{k=1}^{M} \lambda_k \nabla f_k(\mathbf{x}^k) + \sum_{j=1}^{m} u_j \nabla g_j(\mathbf{x}^k)\|, \\
\text{Subject to} \quad & g_j(\mathbf{x}^k) \leq 0, \quad j = 1, 2, \ldots, J, \\
& u_j g_j(\mathbf{x}^k) = 0, \quad j = 1, 2, \ldots, J, \\
& u_j \geq 0, \quad j = 1, 2, \ldots, J, \\
& \lambda_k \geq 0, \quad k = 1, 2, \ldots, M, \text{ and } \boldsymbol{\lambda} \neq \mathbf{0}.
\end{aligned}
\tag{8}
$$

The operator $\| \cdot \|$ is the l_2-norm. It is clear that if $\mathbf{x}^k$ is a KKT point, the norm would be zero for associated and feasible $(\boldsymbol{\lambda}, \mathbf{u})$-vectors. For a non-KKT but feasible $\mathbf{x}^k$ point, the norm need not be zero, but the above procedure should

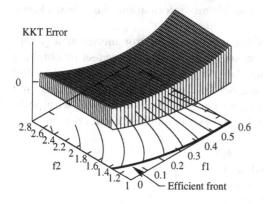

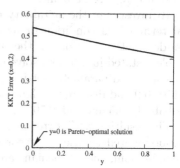

Fig. 1. KKT Error measure for Problem P1

Fig. 2. KKT Error *increases* towards the efficient front along line $x = 0.2$

give us a suitable $(\boldsymbol{\lambda}, \mathbf{u})$-vector that will minimize the norm. This minimum norm value can remain as an indicator to the extent of violation of the KKT optimality conditions. Hence, a KKT Error for a feasible iterate $\mathbf{x}^k$ can be calculated by solving the above problem using an optimization algorithm and computing the following error value:

$$\text{KKT Error}(\mathbf{x}^k) = \| \sum_{k=1}^{M} \lambda_k^* \nabla f_k(\mathbf{x}^k) + \sum_{j=1}^{m} u_j^* \nabla g_j(\mathbf{x}^k) \|, \tag{9}$$

where λ_k^* and u_j^* are optimal values of the problem given in (8).

In order to investigate whether the above KKT error can be used as a metric for evaluating *closeness* of an iterate $\mathbf{x}^k$ to a Pareto-optimal solution, we consider a simple two-variable, bi-objective optimization problem, given as follows:

$$P1 : \begin{cases} \text{Minimize}_{(x,y)} \ \{x, \frac{1+y}{1-(x-0.5)^2}\}, \\ \text{Subject to } 0 \le (x,y) \le 1. \end{cases} \tag{10}$$

Pareto-optimal solutions correspond to $y^* = 0$ and $x^* \in [0, 0.5]$ and the efficient frontier can be represented as $f_2^* = 1/(1 - (f_1^* - 0.5)^2))$. We compute the KKT error for a given point $\mathbf{x}^k = (x^k, y^k)^T$ by solving (8) using Matlab's fmincon() procedure and the resulting error plot is shown in Figure 1. It is clear that the KKT error is zero for all Pareto-optimal solutions ($y = 0$), but it increases (contradictory to a desired decreasing trend) as the points get closer to the Pareto-optimal solution, as shown in Figure 2. This figure is plotted for a fixed value of $x = 0.2$. The KKT error at the point Pareto-optimal solution $(0.2, 0)^T$ is zero, but it jumps to a high value in its vicinity. Ironically, the figure depicts that as points move away from this Pareto-optimal point, the KKT error gets smaller. The above phenomenon suggests that KKT optimality condition is a *singular* set of conditions valid only at KKT points. A minimal violation of equilibrium KKT condition cannot provide any information about how close a solution is to a KKT point.

To investigate the suitability of the above KKT Error metric as a potential termination condition or as a measure for judging closeness of obtained non-dominated solutions to the true Pareto-optimal solutions, we consider the problem stated in (10) and solve using NSGA-II algorithm [7]. NSGA-II is used with standard parameter settings: population size 20, SBX operator with probability 0.9 and distribution index of 20 and polynomial mutation operator with probability 0.5 and index 50.

The resulting KKT Error (ϵ^*) for each non-dominated solution is recorded and the minimum, mean, median, and maximum KKT Error value is plotted in Figure 3 at every generation. It is interesting to observe that although the smallest KKT Error value reduces in the first few generations, the variation of the KKT Error value is too large even up to generation 100 for it to be considered as any stable performance metric. This behavior of the KKT Error measure does not allow us to use it to either as a performance metric or as a termination condition. However,

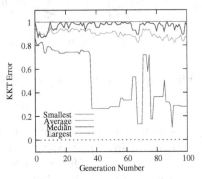

Fig. 3. KKT Error variation for NSGA-II solutions for Problem P1

the modifications suggested in an earlier study [10] allowed a stable closeness information for single-objective optimization problems. In the following section, we extend the idea and develop a KKT proximity measure for multi-objective optimization problems.

2.2   Proposed KKT Proximity Measure

To this effect, we consider a popular multi-criterion decision-making (MCDM) concept often used to find a Pareto-optimal solution using a scalarized method. In this approach, a parameterized *achievement scalarization function (ASF)* optimization problem (similar to L_∞ Tchebyshev scalarization problem [14]) is formulated and solved repeatedly for different parameter values [18]. For a specified reference point $\mathbf{z}$ and a weight vector $\mathbf{w}$ (parameters of the ASF problem), the ASF problem is given as follows:

$$\text{Minimize}_{(\mathbf{x})} \ \text{ASF}(\mathbf{x}, \mathbf{z}, \mathbf{w}) = \max_{i=1}^{M} \left(\frac{f_i(\mathbf{X}) - z_i}{w_i} \right),$$
$$\text{Subject to } g_j(\mathbf{x}) \leq 0, \quad j = 1, 2, \ldots, J. \tag{11}$$

The reference point $\mathbf{z} \in R^M$ is any point in the M-dimensional objective space and the weight vector $\mathbf{w} \in R^M$ is an M-dimensional unit vector for which every $w_i \geq 0$ and $\|\mathbf{w}\| = 1$. Figure 4 illustrates the ASF procedure of arriving at a weak or a strict Pareto-optimal solution. Reference vector $\mathbf{z}$ (marked in the figure) and weight vector $\mathbf{w}$ (marked as $\mathbf{w}$ in the figure) are chosen. For any point $\mathbf{x}$, the objective vector $\mathbf{f}$ is computed (shown as F). Larger of two quantities $((f_1 - z_1)/w_1$ and $(f_2 - z_2)/w_2)$ is then chosen as the ASF value of the point $\mathbf{x}$. Thus, all points along line HG and GK will have the same ASF value as that at point G. A little thought will reveal that a minimization of ASF function will produce point O as the final solution. By keeping the reference point $\mathbf{z}$ fixed and by changing the weight vector $\mathbf{w}$ (treating it like a parameter of the scalarization process), different points on the efficient front can be found by the above ASF minimization process.

For our purpose of estimating a KKT proximity measure for multi-objective optimization, we fix the reference point $\mathbf{z}$ to an utopian point: $z_i = z_i^{\text{ideal}} - \epsilon_i$ for all i, where z_i^{ideal} is the ideal point. We are now left with a systematic procedure for setting the weight vector $\mathbf{w}$. For this purpose, we compute the direction vector from $\mathbf{z}$ to the objective vector $\mathbf{f} = (f_1(\mathbf{x}), f_2(\mathbf{x}), \ldots, f_M(\mathbf{x}))^T$ computed at the current point $\mathbf{x}$ for which the KKT proximity measure needs to be computed. The weight value for the i-th objective function is computed as follows:

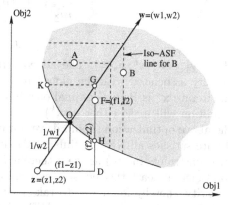

Fig. 4. ASF procedure of finding a Pareto-optimal solution

$$w_i = \frac{f_i(\mathbf{x}) - z_i}{\sqrt{\sum_{j=1}^{M}(f_j(\mathbf{x}) - z_j)^2}}. \tag{12}$$

In the subsequent analysis, we treat the above $\mathbf{w}$-vector fixed for a given solution $\mathbf{x}^k$ and set $\mathbf{w}^k = \mathbf{w}$. To formulate the KKT proximity measure, first, we use a reformulation of optimization problem stated in (11) into a *smooth* problem [14]:

$$\begin{aligned}
&\text{Minimize}_{(\mathbf{x},x_{n+1})} \; F(\mathbf{x}, x_{n+1}) = x_{n+1} \\
&\text{Subject to} \; \left(\frac{f_i(\mathbf{X}) - z_i}{w_i^k}\right) - x_{n+1} \leq 0, \quad i = 1, 2, \ldots, M, \\
&\qquad\qquad g_j(\mathbf{x}) \leq 0, \quad j = 1, 2, \ldots, J.
\end{aligned} \tag{13}$$

A new variable x_{n+1} is added along with M additional inequality constraints to make the problem smooth [14]. Thus, to find the corresponding Pareto-optimal solution, the above optimization problem has $(n+1)$ variables: $\mathbf{y} = (\mathbf{x}; x_{n+1})$.

Since the above optimization problem is a single-objective problem, we can use the KKT proximity metric discussed in the previous section to estimate a proximity measure for any point $\mathbf{x}$. Note that above problem has $M+J$ inequality constraints:

$$G_j(\mathbf{y}) = \left(\frac{f_j(\mathbf{x}) - z_j}{w_j^k}\right) - x_{n+1} \leq 0, \quad j = 1, 2, \ldots, M, \tag{14}$$

$$G_{M+j}(\mathbf{y}) = g_j(\mathbf{x}) \leq 0, \quad j = 1, 2, \ldots, J. \tag{15}$$

The KKT proximity measure can now be computed for a given *feasible* $\mathbf{x}^k$ as follows. It is recognized that the variable x_{n+1} is unknown. We construct the $(n+1)$-dimensional vector $\mathbf{y} = (\mathbf{x}; x_{n+1})$ and formulate the following optimization problem using (13) to compute the KKT proximity measure:

$$
\begin{aligned}
\text{Minimize}_{(\epsilon_k, x_{n+1}, \mathbf{u})} \ & \epsilon_k + \sum_{j=1}^{J} \left(u_{M+j} g_j(\mathbf{x}^k) \right)^2, \\
\text{Subject to } & \|\nabla F(\mathbf{y}|_{\mathbf{x}^k}) + \sum_{j=1}^{M+J} u_j \nabla G_j(\mathbf{y}|_{\mathbf{x}^k})\|^2 \le \epsilon_k, \\
& \sum_{j=1}^{M+J} u_j G_j(\mathbf{y}|_{\mathbf{x}^k}) \ge -\epsilon_k, \\
& \frac{f_j(\mathbf{x}^k) - z_j}{w_j^k} - x_{n+1} \le 0, \quad \forall \ j = 1, 2 \ldots, M, \\
& u_j \ge 0, \quad j = 1, 2, \ldots, (M+J).
\end{aligned}
\tag{16}
$$

Here, $\mathbf{y}|_{\mathbf{x}^k} = (\mathbf{x}^k; x_{n+1})$. The added term in the objective function allows a penalty associated with the violation of complementary slackness condition. If the iterate $\mathbf{x}^k$ is Pareto-optimal or a KKT point, the complementary slackness condition will make either $g_j(\mathbf{x}^k) = 0$ or $u_{M+j} = 0$ (for $j \in [1, J]$) and hence the above optimization should drive towards smallest value of ϵ_k. Since such an iterate satisfies all KKT optimality conditions, the left side expressions of first two constraints in (16) are zero thereby forcing $\epsilon_k = 0$. For non-Pareto-optimal solutions, at least the left side expression of the first constraint is a positive quantity, thereby forcing ϵ_k to take positive value. The third constraint set is added to make sure a positive ASF value (x_{n+1}) is assigned to every feasible point for an ideal or utopian reference point $\mathbf{z}$. The optimal objective value (ϵ_k^*) to the above problem will correspond to the proposed KKT proximity measure. The variables for this problem are ϵ_k, x_{n+1}, and the Lagrange multiplier vector $u_j \in R^{M+J}$. The first constraint requires gradient of F and G functions:

$$
\left\| \begin{pmatrix} 0 \\ 0 \\ \vdots \\ 0 \\ 1 \end{pmatrix} + \sum_{i=1}^{M} u_i \begin{pmatrix} \frac{1}{w_i^k}\frac{\partial f_i(\mathbf{x}^k)}{\partial x_1} \\ \frac{1}{w_i^k}\frac{\partial f_i(\mathbf{x}^k)}{\partial x_2} \\ \vdots \\ \frac{1}{w_i^k}\frac{\partial f_i(\mathbf{x}^k)}{\partial x_n} \\ -1 \end{pmatrix} + \sum_{j=1}^{J} u_{M+j} \begin{pmatrix} \frac{\partial g_j(\mathbf{x}^k)}{\partial x_1} \\ \frac{\partial g_j(\mathbf{x}^k)}{\partial x_2} \\ \vdots \\ \frac{\partial g_j(\mathbf{x}^k)}{\partial x_n} \\ 0 \end{pmatrix} \right\| \le \sqrt{\epsilon_k}.
\tag{17}
$$

Here, the quantity $\frac{\partial f_i(\mathbf{x}^k)}{\partial x_j}$ is the partial derivative of objective function $f_i(\mathbf{x})$ with respect to variable x_j evaluated at the given point $\mathbf{x}^k$. A similar meaning is associated with the partial derivative of the constraint g_j above.

For *infeasible* iterates, we simply compute the constraint violation as the KKT proximity measure. It is observed that for feasible $\mathbf{x}^k$, ϵ_k^* is bounded in [0,1]; hence, the KKT proximity measure is calculated as follows:

$$
\text{KKT Proximity Measure}(\mathbf{x}^k) = \begin{cases} \epsilon_k^*, & \text{if } \mathbf{x}^k \text{ is feasible,} \\ 1 + \sum_{j=1}^{J} \langle g_j(\mathbf{x}^k) \rangle^2, & \text{otherwise.} \end{cases}
\tag{18}
$$

where $\langle \alpha \rangle = \alpha$ if $\alpha > 0$; zero, otherwise.

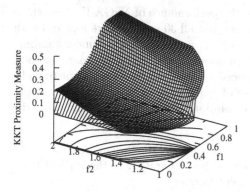

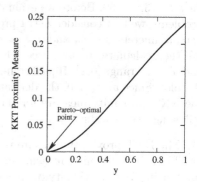

Fig. 5. KKTPM for Problem P1

Fig. 6. KKTPM *reduces* to zero at the Pareto-optimal solution along line $x = 0.2$

We now illustrate the working of the above KKT proximity measure on Problem P1. The ideal point for this problem is $z^{ideal} = (0, 1)^T$. The variable bounds are converted into four inequality constraints and the optimization problem stated in (16) is solved for 10,000 grid points in the variable space. Each optimization problem is solved using Matlab's fmincon() optimization routine. Figure 5 shows the KKT proximity surface and its contour plot on the objective space. The efficient frontier can be clearly seen from the contour plot. Moreover, it is interesting to note how the contour lines become almost parallel to the efficient frontier, thereby indicating that when a set of non-dominated solutions close to the efficient frontier is found, their KKT proximity measure values will be more or less equal to each other. Also, unlike in the naive approach (shown in Figure 1), a monotonic reduction of the KKT proximity measure towards the efficient frontier is evident from this figure and also from Figure 6 plotted for a fixed value of $x = 0.2$ and spanning y in its range [0,1]. The KKT proximity measure monotonically reduces to zero at the corresponding Pareto-optimal point. A comparison of this figure with 2 clearly indicates that the proposed KKT proximity measure is a better and stable metric to assess closeness of points to Pareto-optimal points and also to use it to terminate an EMO simulation run reliably.

3   Results

In this section, we consider two and three-objective test problems to demonstrate the working of the proposed KKT proximity measure with an EMO algorithm.

3.1   Two-Objective ZDT Problems

First, we consider three commonly used two-objective ZDT problems. Problems ZDT1, ZDT2 and ZDT4 each has 30 variables. Efficient solutions occur for $x_i = 0$

for $i = 2, 3, \ldots, 30$. Before we evaluate the performance of NSGA-II [7] on these problems, we first consider ZDT1 problem, but all 30 variables are replaced with two parameters: $x = x_1$ and $y = x_2 = x_3 = \ldots = x_{30}$ for an easy understanding. We then calculate the proposed KKT proximity measure for two parameters x and y in the range $[0, 1]$. 10,000 equi-spaced points are chosen i x-y plane to make the plot. Since at $x_1 = 0$, the derivative of f_2 does not exist, we do not compute the KKT proximity measure for these solutions. Figure 7 clearly indicates the following two aspects:

1. The KKT proximity measure reduces to zero at the efficient frontier.
2. The KKT proximity measure increases almost parallely to the efficient frontier in the local vicinity of the efficient frontier.

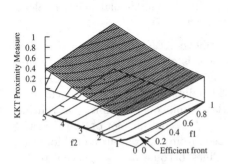

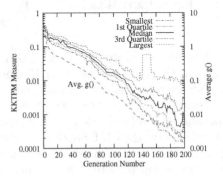

Fig. 7. KKTPM for Problem ZDT1

Fig. 8. Generation-wise KKTPM for NSGA-II populations on ZDT1

We now apply the KKT proximity measure procedure with ASF formulation to non-dominated solutions obtained by NSGA-II at every generation, having standard parameter values [7] and with 40 population members running for 200 generations. The variation of average $g()$ function value plotted in Figure 8 shows that non-dominated points gets closer to the true Pareto-optimal front with increasing generations. The smallest, first quartile, median, third quartile, and largest KKT proximity measure values for all non-dominated solutions at each generation are also plotted. The figure shows that KKTPM value reduces exponentially with generations. The first population member to reach a KKTPM value of 0.01 takes 98 generations and the median KKTPM value of 0.01 is achieved at generation 117. The third quartile of population members takes another nine generations to get to the same KKTPM value, however the last member to get close to the Pareto-optimal front takes 135 generations. Thus, instead of running an EMO algorithm for an arbitrary 200 generations, a track of KKTPM value indicates a theoretical way of terminating an EMO run. Interestingly, EMO algorithms work without any derivative information or without using any KKT optimality conditions in its operations.

Next, we consider 30-variable ZDT2 problem. Considering $x = x_1$ and $y = x_2 = x_3 = \ldots = x_{30}$, we compute KKTPM for 10,000 points in the x-y

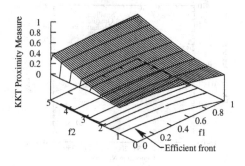

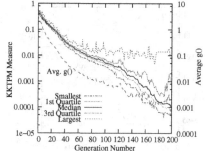

Fig. 9. KKTPM for Problem ZDT2

Fig. 10. Generation-wise KKTPM for NSGA-II populations on ZDT2

space and resulting measure values are plotted on the ZDT2 objective space in Figure 9. ZDT2 has a non-convex efficient front, as shown in the figure. Notice, how the KKTPM with AASF formulation monotonically reduces to zero at the efficient frontier. The contour plots indicate the almost parallel contour lines to the efficient frontier, as the points move away from the efficient frontier. Next, we demonstrate how the non-dominated sets of points obtained from NSGA-II (with identical parameter values) are evaluated by the KKT proximity measure in Figure 10. Once again, a monotonic decrease in the KKTPM indicates its suitability as a termination condition for an EMO algorithm.

The ZDT4 problem is multi-modal and has a number of local efficient fronts and is more difficult to solve than ZDT1 and ZDT2. Using identical NSGA-II parameter values, Figure 11 shows how generation-wise KKTPM with AASF formulation varies for the non-dominated population members of a typical NSGA-II run. Due its complex nature, NSGA-II takes about 225 generations to have all non-dominated close to the true Pareto-optimal front. Once again, a generic decrease in the KKTPM indicates its suitability as a termination condition for an EMO algorithm.

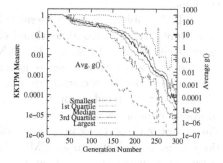

Fig. 11. Generation-wise KKTPM for NSGA-II populations in Problem ZDT4

3.2 Three-Objective DTLZ Problems

DTLZ1 problem has a number of locally efficient fronts on which some points can get stuck, hence it is relatively difficult problem to solve to global optimality. Recently proposed NSGA-III procedure [8] is applied to DTLZ1 problems with 92 population members. Figure 12 shows the variation of KKT proximity measure values with AASF formulation versus the generation counter. Although

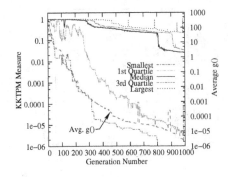

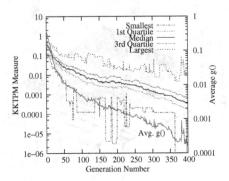

Fig. 12. Generation-wise KKTPM for NSGA-III populations in 3-objective DTLZ1

Fig. 13. Generation-wise KKTPM for NSGA-III populations in 3-objective DTLZ2

the first Pareto-optimal solution is found early (at generation 110), half of the non-dominated set did not find a KKTPM value of 0.01 in 1,000 generations. A variation of $g()$ value indicates a steady progress of points towards the true Pareto-optimal front.

Next, we apply NSGA-III (with 92 population members) to three-objective DTLZ2 problem, which has a concave efficient front. Figure 13 shows the KKTPM values. Interestingly, the convergence is faster than DTLZ1 and the median point took 103 generations to have a KKTPM value of 0.01 and all points took 363 generations.

DTLZ5 problem has a degenerate efficient front having a two-dimensional efficient front. Figure 14 shows the KKT proximity measure values versus generation counter. Despite a degenerate nature of the front, a number of 'redundant' solutions remain as non-dominated with Pareto-optimal solutions and these solutions cause a challenge to an EMO algorithm to converge to the true Pareto-optimal front. The figure shows that although the worst performance fluctuates, the third quartile of points took only 61 generations to have a KKTPM value of 0.01 and 225 generations to have KKTPM value of 0.001.

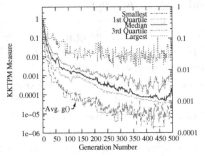

Fig. 14. Generation-wise KKTPM for NSGA-III populations in 3-objective DTLZ5

3.3 Many-Objective Optimization Problems

As the number of objectives increase, DTLZ1 and DTLZ2 problems get more difficult to optimize as evident from Figures 15 and 16. NSGA-III with 212 and 276 population members [8] are run and KKTPM values are plotted. The continual reduction of average $g()$ value indicates the progressive convergence

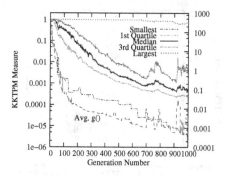 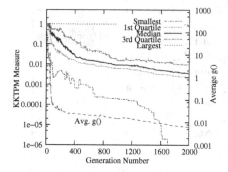

Fig. 15. Generation-wise KKTPM for NSGA-III populations in 5-objective DTLZ1

Fig. 16. Generation-wise KKTPM for NSGA-III populations in 10-objective DTLZ1

of non-dominated points towards the true Pareto-optimal front. While the five-objective DTLZ1 problem takes about 337 generations to have half of all non-dominated solutions to have a KKTPM value of 0.01, the 10-objective problems takes 925 generations. Large value of these generation numbers indicate that DTLZ1 problem gets harder to solve with increasing number of objectives.

Since DTLZ2 problem is relatively easier to solve to Pareto-optimality, KKTPM values for five and 10-objective DTLZ2 problems are found to be have converged faster (302 generations for 5-obj), as evident from Figures 17 and 18, respectively. The above results amply demonstrate the scalability of the proposed KKT proximity metric to many-objective optimization problems as well.

3.4 Constrained Test Problems

We now consider two constrained test problems from the literature. First, we consider the problem TNK, which has two constraints, two variables and two objectives [4]. NSGA-II is run with 12 population members. For the specific run, the initial population has only 2 feasible solutions, but since they were away

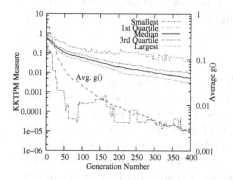 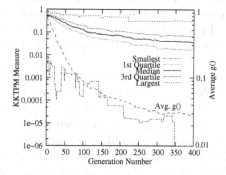

Fig. 17. Generation-wise KKTPM for NSGA-III populations in 5-obj DTLZ2

Fig. 18. Generation-wise KKTPM for NSGA-III populations in 10-obj DTLZ2

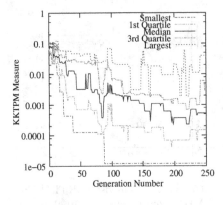

Fig. 19. Generation-wise KKTPM for NSGA-II populations in TNK

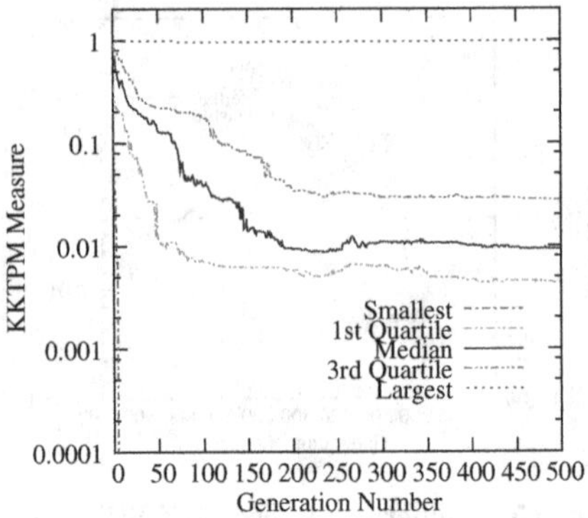

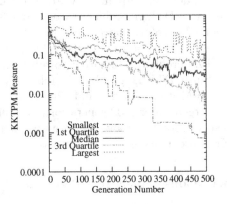

Fig. 20. Generation-wise KKTPM for NSGA-III populations in SRN

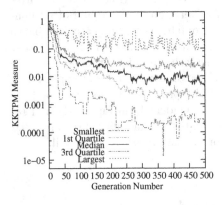

Fig. 21. Generation-wise KKTPM for NSGA-II populations in welded-beam problem

Fig. 22. Generation-wise KKTPM for NSGA-III populations in car side impact problem

from the Pareto-optimal region, their KKTPM value was large. Thereafter, with iterations more and more solutions became feasible and importantly, as shown in Figure 19, the non-dominated solutions start to approach the Pareto-optimal front and all 12 points converge within 0.01 KKTPM value at 160-th generation.

Next, we consider problem SRN which has two variables and two constraints [4]. NSGA-III is run with 200 population members. Figure 20 shows the variation of KKTPM with generation for problem SRN. Although one solution comes close to the efficient front (within 0.01 KKTPM) at the third generation, 50% of the non-dominated solutions take around 189 generations to come close. The NSGA-II points are unable to improve the KKTPM value with generations. This indicates the use of a more focused local search method at around 200 generations to cause a faster convergence of points.

3.5 Engineering Design Problems

Next, we consider two multi-objective optimization problems from practice.

The welded beam design problem has two objectives and a few non-linear constraints [4]. This problem is solved using NSGA-II having 60 population members, initially created at random. Figure 21 shows the variation of KKT proximity measure values of the population with generation counter. 50% of the population comes close (within 0.01 KKTPM value) to the Pareto-optimal front by 265 generations. More results are available elsewhere [6].

Finally, we solve a three-objective car side impact design problem [8] using NSGA-III with 92 population members. Figure 22 shows the variation of KKT proximity measure with generations. The convergence of 25% population within 0.01 KKTPM took 480 generations, while 50% members did not come close even after 500 generations. Compared to the test problems, the rate of convergence of NSGA-II non-dominated points is slow in these two real-world problems. These results indicate that a use of a better constrained handling procedure or a more focused local search may be necessary for making an overall faster approach.

4 Conclusions and Extensions

This paper has extended the concept of approximate KKT point definition proposed in an earlier study for single-objective optimization problem to multi-objective optimization. For each point, an equivalent achievement scalarizing function is developed and a KKT proximity metric is defined, for the first time, to estimate the proximity of a solution from a respective Pareto-optimal solution, without using any knowledge of the Pareto-optimal solution directly. The idea is novel and has been demonstrated to have a reducing proximity measure value as solutions get closer to the true Pareto-optimal front on many test problems and a couple of engineering design problems, providing an idea of dynamics of an EMO algorithm.

The study is also interesting from the following reasons. First, the continual reduction of KKTPM values gives us confidence in its use as a termination condition for an EMO algorithm. Second, solutions having relatively higher KKTPM value at any generation can be improved by using a local search procedure for a faster overall algorithm. Third, the relative convergence pattern of KKT proximity measure values can provide a clear picture about the difficulty in converging to different parts of the search space, an information which could be used in subsequent decision-making tasks. Fourth, the KKTPM can be applied to classical multi-objective optimization algorithms equally well.

The computation of the metric requires gradients of the objective and constraint functions at each point, hence the metric is computable only for differentiable problems. The method also requires to solve an optimization problem to find KKTPM. To reduce the overall computational complexity, we recommend to compute the KKT proximity metric after every five or 10 generations. However, despite these limitations, the convergence of EMO-obtained solutions to theoretical KKT points in multi-objective problems remains as a hallmark achievement of this paper. Although gradients are not used in an EMO algorithm for updating one population, the fact that the overall search process directs the population towards theoretical Pareto-optimal points is interesting and intriguing. Certainly, such studies help close the gap between theoretical and computational optimization studies and should bring EMO's due recognition from classical and mathematical multi-objective optimization fields.

Acknowledgments. Authors acknowledge the efforts of Mr. Haitham Saeda in providing us with results of NSGA-III. The second author's support from the Ministry of Higher Education in Egypt is appreciated.

References

1. Andreani, R., Haeser, G., Martinez, J.M.: On sequential optimality conditions for smooth constrained optimization. Optimization **60**(5), 627–641 (2011)
2. Bader, J., Deb, K., Zitzler, E.: Faster hypervolume-based search using Monte Carlo sampling. In: Proceedings of Multiple Criteria Decision Making (MCDM 2008). LNEMS, vol. 634, pp. 313–326. Springer, Heidelberg (2010)
3. Bector, C.R., Chandra, S., Dutta, J.: Principles of Optimization Theory. Narosa, New Delhi (2005)
4. Deb, K.: Multi-objective optimization using evolutionary algorithms. Wiley, Chichester (2001)
5. Deb, K.: Scope of stationary multi-objective evolutionary optimization: A case study on a hydro-thermal power dispatch problem. Journal of Global Optimization **41**(4), 479–515 (2008)
6. Deb, K., Abouhawwash, M.: An optimality theory based proximity measure for set based multi-objective optimization. COIN Report Number 2014015. Electrical and Computer Engineering, Michigan State University, East Lansing, USA (2014)
7. Deb, K., Agrawal, S., Pratap, A., Meyarivan, T.: A fast and elitist multi-objective genetic algorithm: NSGA-II. IEEE Transactions on Evolutionary Computation **6**(2), 182–197 (2002)
8. Deb, K., Jain, H.: An evolutionary many-objective optimization algorithm using reference-point based non-dominated sorting approach, Part I: Solving problems with box constraints. IEEE Transactions on Evolutionary Computation **18**(4), 577–601 (2014)
9. Deb, K., Tiwari, R., Dixit, M., Dutta, J.: Finding trade-off solutions close to KKT points using evolutionary multi-objective optimization. In: Proceedings of the Congress on Evolutionary Computation (CEC 2007), pp. 2109–2116. IEEE Press, Piscatway (2007)
10. Dutta, J., Deb, K., Tulshyan, R., Arora, R.: Approximate KKT points and a proximity measure for termination. Journal of Global Optimization **56**(4), 1463–1499 (2013)
11. Ehrgott, M.: Multicriteria Optimization. Springer, Berlin (2005)
12. Haeser, G., Schuverdt, M.L.: Approximate KKT conditions for variational inequality problems. Optimization Online (2009)

13. Manoharan, P.S., Kannan, P.S., Baskar, S., Iruthayarajan, M.W.: Evolutionary algorithm solution and kkt based optimality verification to multi-area economic dispatch. Electrical Power and Energy Systems **31**, 365–373 (2009)
14. Miettinen, K.: Nonlinear Multiobjective Optimization. Kluwer, Boston (1999)
15. Rockafellar, R.T.: Convex Analysis. Princeton University Press (1996)
16. Tulshyan, R., Arora, R., Deb, K., Dutta, J.: Investigating ea solutions for approximate KKT conditions for smooth problems. In: Proc. of Genetic and Evolutionary Algorithms Conference (GECCO 2010), pp. 689–696. ACM Press (2010)
17. While, L., Hingston, P., Barone, L., Huband, S.: A faster algorithm for calculating hypervolume. IEEE Trans. on Evolutionary Computation **10**(1), 29–38 (2006)
18. Wierzbicki, A.P.: The use of reference objectives in multiobjective optimization. In: Fandel, G., Gal, T. (eds.) Multiple Criteria Decision Making Theory and Applications, pp. 468–486. Springer, Berlin (1980)
19. Zitzler, E., Deb, K., Thiele, L.: Comparison of multiobjective evolutionary algorithms: Empirical results. Evol. Comput. Journal **8**(2), 125–148 (2000)

U-NSGA-III: A Unified Evolutionary Optimization Procedure for Single, Multiple, and Many Objectives: Proof-of-Principle Results

Haitham Seada and Kalyanmoy Deb[✉]

Department of Computer Science and Engineering, Michigan State University,
428 S. Shaw Lane, 2120 EB, East Lansing, MI 48824, USA
{seadahai,kdeb}@msu.edu
http://www.egr.msu.edu/~kdeb

Abstract. Evolutionary algorithms (EAs) have been systematically developed to solve mono-objective, multi-objective and many-objective problems, respectively, over the past few decades. Despite some efforts in unifying different types of mono-objective evolutionary and non-evolutionary algorithms, there does not exist too many studies to unify all three types of optimization problems together. In this study, we propose an unified evolutionary optimization algorithm U-NSGA-III, based on recently-proposed NSGA-III procedure for solving all three classes of problems. The U-NSGA-III algorithm degenerates to an equivalent and efficient population-based optimization procedure for each class, just from the description of the number of specified objectives of a problem. The algorithm works with usual EA parameters and no additional tunable parameters are needed. The performance of U-NSGA-III is compared with a real-coded genetic algorithm for mono-objective problems, with NSGA-II for two-objective problems, and with NSGA-III for three or more objective problems. Results amply demonstrate the merit of our proposed unified approach, encourage its further application, and motivate similar studies for a richer understanding of the development of optimization algorithms.

Keywords: Mono-objective optimization · Multi-objective optimization · Many-objective optimization · NSGA-II · NSGA-III · Unified algorithms

1 Introduction

During the past two decades, evolutionary multi-objective optimization (EMO) researchers have demonstrated their usefulness in solving optimization problems having two and more objectives. Initial studies concentrated in solving two and three-objective problems and efficient multi-objective optimization algorithms were developed to adaptively distribute its population members along the entire efficient front [4,8,14,16]. Recent many-objective optimization studies have concentrated in solving four or more objectives mostly using an external guidance

© Springer International Publishing Switzerland 2015
A. Gaspar-Cunha et al. (Eds.): EMO 2015, Part II, LNCS 9019, pp. 34–49, 2015.
DOI: 10.1007/978-3-319-15892-1_3

mechanism to help algorithms distribute its population along higher-dimensional efficient front [3,10,12,15]. The multi-objective optimization algorithms do not extend to handle many objectives, mainly due to exponential growth of non-dominated solutions in a population with an increase in number of objectives [6]. Thus, in principle, evolutionary multi-objective and many-objective optimization algorithms are different from each other.

Although certain evolutionary multi-objective optimization methodologies such as NSGA-II [8] does not scale up to solve many-objective optimization problems efficiently, they are found to work well in solving mono-objective optimization problems. Based on NSGA-II framework, an omni-optimizer algorithm [11] was suggested to mono- and multi-objective optimization problems. This is because the domination operator used in NSGA-II's selection mechanism becomes an ordinal comparison operator, which is an essential operation for progressing towards the optimum solution for a mono-objective optimization problem. Thus, these multi-objective optimization methods can be considered as unified methods for solving mono- and multi-objective optimization problems.

However, existing many-objective optimization methods are tested for three and more objective problems and have not been adequately evaluated for their performance in solving mono- and bi-objective optimization problems. One apparent difficulty of their scaling down to solve mono-objective problems is that the objective space becomes one-dimensional and the inherent guidance mechanism which ensures diversity of population members in the objective space becomes defunct. Thus, although these algorithms are efficient in handling many objectives, their working in mono- and bi-objective problems becomes questionable.

In this paper, we make an effort to develop a single unified evolutionary optimization procedure that will solve mono-, multi- and many-objective optimization problems efficiently. Such an algorithm not only will require an user to solve different types of problems, but also an understanding of algorithmic features needed in such an efficient unified approach would be beneficial for EMO researchers. The successful development of an unified approach for handling one to many objectives will also provide a triumph of generic computing concept in the optimization problem solving. The philosophy of computing through a computerized software is to implement an algorithm that is most generic capable of working with multiple and arbitrary number of input data. But when the software is applied to a lower-dimensional data or even to a single data, the software is expected work as a specialized lower-dimensional or single-dimensional algorithm would perform. Unfortunately, optimization literature has traditionally followed an opposite philosophy. A lot of stress has been put in developing mono-objective optimization algorithms and often multi- or many-objective optimization problems are suitably converted to a mono-objective optimization problem so as to use mono-objective optimization algorithms. Our motivation in this paper is to explore the possibility of developing an unified optimization approach that naturally solves many-objective problems having four or more objectives and degenerates to solve one, two or three-objective problems, as efficiently as to other competing lower-objective optimization algorithms.

Since such an unified approach would be generic to solving many-objective optimization problems, we base our algorithm using one of the recently proposed decomposition-based many-objective optimization algorithm – NSGA-III [10]. In the remainder of this paper, we provide a brief description of NSGA-III in Section 2. Thereafter, we present our proposed unified approach U-NSGA-III in Section 3 and explain how the method degenerates to efficient mono- and multi-objective optimization algorithms. Simulation results on a variety of mono, multi- and many-objective test problems are presented using U-NSGA-III and compared with a real-parameter genetic algorithm, NSGA-II and NSGA-III in Section 4. Finally, conclusions are drawn in Section 5.

2   A Brief Introduction to NSGA-III

The proposed U-NSGA-III algorithm is based on the structure of NSGA-III, hence we first give a description of NSGA-III here. NSGA-III starts with a random population of size N and a set of widely-distributed pre-specified reference points H on a unit hyper-plane having a normal vector of ones. The hyper-plane is placed in a manner so that it intersects each objective axis at one. Das and Dennis's technique [5] is used to place $H = \binom{M+p-1}{p}$ reference points on the hyper-plane having $(p + 1)$ points along each boundary. The population size N is chosen to be the smallest multiple of four greater than H, with the idea that for every reference point, one population member is expected to be found.

At a generation t, the following operations are performed. First, the whole population P_t is classified into different non-domination levels, as it is done in NSGA-II as well, following the principle of non-dominated sorting. An offspring population Q_t is created from P_t using recombination and mutation operators. Since only one population member is expected to be found for each reference point, there is no need for any selection operation in NSGA-III. A combined population $R_t = P_t \cup Q_t$ is then formed. Thereafter, points starting from the first non-dominated front is selected for P_{t+1} one at a time until all solutions from a complete front cannot be included. This procedure is also identical to that in NSGA-II. Let us denote the final front that could not be completely selected as F_L. In general, only a few solutions from F_l needs to be selected for P_{t+1}. We describe the niche-preserving operation next. First, each population member of P_{t+1} and F_L is normalized by using the current population spread so that all objective values have commensurate values. Thereafter, each member of P_{t+1} and F_L is associated with a supplied reference point by using the shortest perpendicular distance of each population member with a reference line created by joining the origin with a supplied reference point. Then, a careful niching strategy is employed to choose those F_L members that are associated with the least represented reference points in P_{t+1}. The niching strategy puts an emphasis to select a population member for as many supplied reference points as possible. A population member associated with a under-represented or un-represented reference point is immediately preferred. With a continuous stress for emphasizing non-dominated individuals, the whole process is then expected to find one population member corresponding to each supplied reference point close to the Pareto-optimal front, provided the genetic variation operators (recombination and mutation) are capable of producing respective solutions.

The original NSGA-III study [10] have demonstrated to work well from three to 15-objective DTLZ and other problems. A nice aspect of NSGA-III is that it does not require any additional parameter. The method was also extended to handle constraints without introducing any new parameter. This study has also introduced a computationally fast approach by which the reference point set is adaptively updated on the fly based on the association status of each reference point over a number of generations.

2.1   NSGA-III for Mono- and Multi-objective Problems

NSGA-III was proposed to solve many-objective optimization problems having more than three objectives exclusively, although NSGA-III was demonstrated to work well on three-objective optimization problems. Authors of NSGA-III did not consider any bi-objective or mono-objective problems in the original study. Here, we discuss how NSGA-III would perform in two-objective problems and then highlight its possible working on mono-objective problems.

Here are the differences in working principles of NSGA-II and NSGA-III for their working on two-objective problems.

1. NSGA-III does not use any explicit selection operator in P_t in the process of creating Q_t. On the other hand NSGA-II's selection operator uses non-dominated rank and a crowding distance value to choose a winner between two feasible individuals from P_t.
2. NSGA-III uses a set of reference directions to maintain diversity, while NSGA-II uses the crowding distance value for the same purpose, as illustrated in Figure 1.

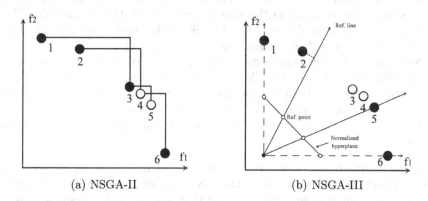

(a) NSGA-II   (b) NSGA-III

Fig. 1. NSGA-II and NSGA-III working principles

If NSGA-III having a population size almost identical to number of chosen reference directions is compared with NSGA-II having an identical population size as in NSGA-III, the former will introduce a milder selection pressure. This is because

on an average each population member in NSGA-III becomes associated with a
different reference direction and is an important individual to be compared with
another individual. The only selection pressure comes from their domination levels. However, the second difference mentioned above may produce a significant difference in their performances. NSGA-III uses a pre-defined guidance mechanism to
choose diverse solutions in the population, whereas NSGA-II uses no pre-defined
guidance and emphasizes relatively diverse solutions on the fly. Thus, if the first
aspect is taken care of somehow and more selection pressure can be introduced,
NSGA-III may become an equivalent or even a better algorithm than NSGA-II for
solving two-objective optimization problems.

Let us now discuss how NSGA-III works if applied to a mono-objective
optimization problem. In mono-objective optimization, the domination concept
degenerates to fitness superiority – a domination check between two solutions
chooses the one having better objective value. At every generation, usually there
is one solution in each non-dominated front in a mono-objective problem and
it is expected to have N fronts in a population of size N. This characteristics of mono-objective problems affect the working of NSGA-III in the following
manner:

1. First, in NSGA-III, there will be only one reference direction (the real line)
   to which all the individuals will be associated. Since the recommended population size is the smallest multiple of 4 greater than the number of reference
   directions, for all mono-objective optimization problems, NSGA-III will use
   a population of size four, which from all practical purposes too small for
   NSGA-III's recombination operator to find useful offspring solutions. This
   is a major issue in developing an unified algorithm that will seamlessly work
   for many to mono-objective problems.
2. Moreover, since no explicit selection operator is used, the algorithm will pick
   a random solution for its recombination and mutation operators. The only
   selection effect comes from the elite-preserving operation for choosing P_{t+1}
   from a combination of P_t and Q_t. This is another major issue, which needs
   to be addressed while developing an unified approach.
3. Note also that the niching operation of NSGA-III is defunct for mono-
   objective problems, as there is no concept of perpendicular distance of a
   function value from the reference direction. Every function value falls on the
   real line, providing an identical perpendicular distance of zero.
4. NSGA-III's normalization also becomes a defunct operation for the same
   above reason.

A unified approach may still have niching and normalization operators for it
to be efficient for multi and many-objective optimization problems, but they
should be implemented in such a manner that automatically becomes defunct
for the mono-objective case without producing any unwanted effects. However
the first two points mentioned above present a challenge. It is now clear that a
straight application of original NSGA-III to mono-objective optimization problems will result in an extremely small population size and a random selection

process, neither of which is recommended for a successful evolutionary optimization algorithm.

3   Proposed Unified Approach: U-NSGA-III

The above discussions suggest that the proposed U-NSGA-III method can retain the features of the original NSGA-III algorithm, as NSGA-III was shown to work well on three or more objectives well. However, the difficulties in scaling down to two and mono-objective problems mentioned above require certain changes in NSGA-III algorithm, but we should be modifying NSGA-III in such a manner that the changes do not affect its working on three and more objective problems.

The difficulty for solving two-objective optimization problems seems to lie in the mild selection pressure that NSGA-III introduces to non-dominated solutions of a population and the difficulty for solving mono-objective problems are small population size and a random selection process. One way to alleviate these difficulties is to a population size N which is larger than the number of reference points (H) and introduce a selection operator. Thus, unlike in NSGA-III, N and H will now be different parameters with a condition that $N \geq H$ and N is a multiple of four. Although this seems to introduce an additional parameter to our proposed U-NSGA-III, but H is the desired number of optimal solutions expected at the end of a simulation run, and hence is not a parameter that needs to be tuned for U-NSGA-III to work well on different problems. We shall soon investigate the effect of this change for different problem sizes, but for mono-objective problems H is always one and N becomes simply the population size which is a generic principle in all mono-objective evolutionary algorithms. For two-objective problems, H can be a handful of solutions (such as 10 or 20) for the decision-makers to consider, while the population size N can be much larger, such as 100 or 200. The population size consideration mainly comes from the complexity of the problem and an adequate sample size needed for the genetic operators to work well. In this case, although N different Pareto-optimal solutions could be present in the final population, the H specific Pareto-optimal solutions, each closest to a different reference direction will be the outcome of the U-NSGA-III algorithm and will be presented to the decision-maker for choosing a single preferred solution. For three or more objective problems, since the number of specified reference directions (H) can already be quite high (due to the increase in H with M according to Das and Dennis's approach [5]), N can be made almost equal to H with the divisibility by four restriction.

Let us now discuss the algorithmic implication of introducing more population members than H for solving mono- and two-objective optimization problems. It is now expected that for each reference direction, there will be more than one population member associated. This then allows to introduce a selection operator to have an adequate selection pressure for good population members. We add a niching-based tournament selection operator as follows. If the two solutions being compared come from two different associated reference directions, one of them chosen at random, thereby introducing preservation of multiple niches in the population. Otherwise, the solution coming from a better non-dominated rank is chosen. In this case, if both solutions belong to the same niche and same non-dominated front, the one closer to the associated reference direction is chosen. Algorithm in Figure 2 presents the niched tournament selection procedure in a pseudo code. As discussed elsewhere [13], the complexity of U-NSGA-III is $O(MN^2)$ which is similar to NSGA-II and NSGA-III procedures.

Input: Two parents: p_1 and p_2
Output: Selected individual, p_s
1: **if** $\pi(p_1) = \pi(p_2)$ **then**
2:   **if** $p_1.rank < p_2.rank$ **then**
3:     $p_s = p_1$
4:   **else**
5:     **if** $p_2.rank < p_1.rank$ **then**
6:       $p_s = p_2$
7:     **else**
8:       **if** $d(p_1) < d(p_2)$ **then**
9:         $p_s = p_1$
10:       **else**
11:         $p_s = p_2$
12:       **end if**
13:     **end if**
14:   **end if**
15: **else**
16:   $p_s = randomPick(p_1, p_2)$
17: **end if**

Fig. 2. Niched selection operator in U-NSGA-III

Input: *Mono-objective* function
Output: Best solution found, p_{best}
1: $P = initialize()$
2: **while** *termination condition* **do**
3:   $Q = \phi$
4:   **while** $|Q| < |P|$ **do**
5:     $p_1 = tournamentSelect(P)$
6:     $p_2 = tournamentSelect(P)$
7:     $(c_1, c_2) = recombination(p_1, p_2)$
8:     $c_1 = mutate(c_1)$
9:     $c_2 = mutate(c_2)$
10:     $Q \cup \{c_1, c_2\}$
11:   **end while**
12:   $P = best(P \cup Q)$
13: **end while**
14: $p_{best} = best(P)$

Fig. 3. Degenerated U-NSGA-III for mono-objective problems

We now present the respective degenerative U-NSGA-III algorithm for solving mono-, multi- and many-objective optimization problems.

3.1 U-NSGA-III for Mono-objective Problems

For mono-objective problems, the flexibility of choosing any N alleviates one of the discussed difficulties. The niched selection operator degenerates to an usual binary tournament selection operator for which the solution having a better objective value becomes the winner. Algorithm in Figure 3 presents a pseudo-code for the resulting U-NSGA-III algorithm when $M = 1$ is specified. The resulting U-NSGA-III is a generational evolutionary algorithm (EA) that uses (i) a binary tournament selection, (ii) recombination and mutation operators, and (iii) an elite-preserving operator. This our proposed U-NSGA-III is similar to other generational EAs, such as elite-preserving real-coded genetic algorithm [9] or the $(\mu/\rho + \lambda)$ evolution strategy [1], where $\mu = \rho = \lambda = N$.

3.2   U-NSGA-III for Multi-objective Problems

For two or three-objective problems, if N is chosen to be greater than H, U-NSGA-III is expected to have multiple population members for each reference direction. For multi-objective problems having two or three objectives, the non-dominated sorting will, in general, divide the population into multiple non-dominated fronts. The proposed niched tournament selection operator of U-NSGA-III then emphasizes (i) non-dominated solutions over dominated solutions and (ii) solutions closer to reference directions over other non-dominated but distant solutions from the reference directions. The rest of the U-NSGA-III algorithm works exactly the same way as NSGA-II would work on multi-objective problems. However, although like in NSGA-II, all final population members of U-NSGA-III are also expected to be non-dominated to each other, the distribution of additional $(N - H)$ population members need not have a good diversity among them. Only H population members closest to each H reference directions are expected to be well distributed.

When N/H is one or close to one, U-NSGA-III algorithm may not have a solution for each reference direction in the early generations, but the selection pressure introduced by the niched tournament selection will emphasize finding and maintaining a single population member for each reference direction until they are all found. In either case of N/H being one or more than one, U-NSGA-III provides a right selection pressure for it to be a good algorithm for handling bi-objective problems. For three-objective problems, the uniform distribution of supplied reference points should find a better distributed efficient points than those by NSGA-II.

3.3   U-NSGA-III for Many-objective Problems

For many-objective optimization problems, most population members are expected to be non-dominated to each other. Hence, the niched tournament selection operator degenerates in choosing the closer of two parent solutions with respect to their associated reference direction, when both parent solutions lie on the same niche. When N/H is much greater than one, this allows an additional filtering of choosing parent solutions closer to reference directions for their subsequent mating operation. This is, in general, a good operation to have particularly when there are multiple population members available around a specific reference direction. However, if N/H is one or close to one, the niched tournament selection, in most cases, becomes a defunct operator resulting into a very similar algorithm as NSGA-III.

4   Results

We now present simulation results of U-NSGA-III applied to mono-, multi- and many-objective problems. For more detailed results, please refer to the original study [13].

4.1   Mono-objective Problems

We choose three mono-objective test problems and compare results of U-NSGA-III with a generational real-parameter genetic algorithm which was used to solve various problems in the past [7]. We employ an elite-preserving operator between parent and offspring populations in the proposed EliteRGA algorithm here. The problems are given below:

$$f_{\text{elp}}(\mathbf{x}) = \sum_{i=1}^{n} i x_i^2, \quad -10 \le x_i \le 10, \quad i = 1, \dots, n, \tag{1}$$

$$f_{\text{ros}}(\mathbf{x}) = \sum_{i=1}^{n-1} [100(x_i^2 - x_{i+1})^2 + (x_i - 1)^2] \ -10 \le x_i \le 10, \ i = 1, \dots, n, \tag{2}$$

$$f_{\text{sch}}(\mathbf{x}) = 418.9829n - \sum_{i=1}^{n} (x_i \sin \sqrt{|x_i|}) \ -500 \le x_i \le 500, \ i = 1, \dots, n. \tag{3}$$

For each problem, $n = 20$ is used and 11 simulations with the same set of parameters but from different initial populations are performed. Although first two problems are unimodal, f_{sch} is multi-modal. We use $N = 48$, 100 and 300 for Ellipsoidal, Rosenbrock and Schwefel problems, respectively, due to handle the increasing complexity in these problems. For all problems, we use $p_c = 0.7$, $p_m = 0.02$, $\eta_c = 0$ and $\eta_m = 20$.

Figures 4(a), 4(b) and 4(c) show the median function value for Ellipsoidal, Rosenbrock, and Schwefel problems over 11 runs for different function evaluations, shown in the x-axis. EliteRGA and U-NSGA-III perform in a similar manner for the unimodal problems, while for the multi-modal problem U-NSGA-III performs the best. Table 1 summarizes the best, median and worst function value over 11 runs for each problem after different function evaluations. The performances of U-NSGA-III and EliteRGA are equivalent. As expected, NSGA-III with $N = 4$ performs well on unimodal problems, but fails miserably on Schwefel problem.

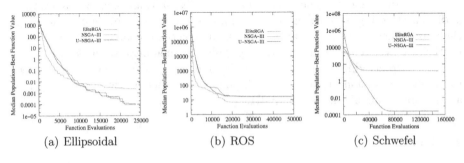

(a) Ellipsoidal   (b) ROS   (c) Schwefel

Fig. 4. Reduction of population-based function value with the number of function evaluations

Table 1. Best, median and worst objective values of three problems using EliteRGA, NSGA-III, and U-NSGA-III

Func.	EliteRGA			NSGA-III			U-NSGA-III		
Eval.	Best	Median	Worst	Best	Median	Worst	Best	Median	Worst
Ellipsoidal Problem									
4,800	0.204	0.849	2.248	0.008	0.058	0.296	**0.169**	**0.436**	**0.803**
14,400	0.000	0.001	0.006	0.001	0.005	0.028	**0.000**	**0.001**	**0.005**
24,000	0.000	0.000	**0.000**	0.000	0.002	0.003	**0.000**	**0.000**	0.002
Rosenbrock Problem									
100,000	21.119	72.412	771.324	**3.974**	**35.241**	**73.086**	20.005	56.201	135.439
300,000	**0.256**	17.005	760.077	0.624	**6.637**	**71.777**	8.569	16.801	78.111
500,000	**0.253**	16.226	760.069	0.176	**6.407**	**71.558**	8.144	16.346	74.871
Schwefel Problem									
300,000	17.875	20.061	**79.408**	710.707	1184.421	1539.744	**0.263**	**1.039**	119.652
900,000	15.285	16.662	**73.784**	710.642	1184.385	1539.702	**0.000**	**0.000**	118.439
1,500,000	1.553	15.735	**71.611**	710.633	1184.384	1539.700	**0.000**	**0.000**	118.439

4.2 Bi-objective Problems

As mentioned before, the performance of NSGA-III was not tested on bi-objective optimization problems in the original study [10]. This section compares the performance of U-NSGA-III, NSGA-III, and NSGA-II on several bi-objective test problems. The criterion of comparison used here is the number of function evaluations required to reach a pre-defined hyper-volume (HV)

Table 2. Fixed HV values used in bi-objective problems (Removal and No Removal)

Problem	No Removal	Removal
$ZDT1$	0.64	0.64
$ZDT2$	0.316	0.316
$ZDT3$	0.516	0.512
$ZDT4$	0.53	0.53

value. Table 2 presents the HV values used in our bi-objective simulations. For all runs of this and subsequent sections, we use SBX $p_c = 0.9$, $\eta_c = 30$ and polynomial mutation $p_m = 1/n$ and $\eta_m = 20$. We use ZDT1, ZDT2, ZDT3 and ZDT4 as our test bed. In all four problems, we use 30 variables and 10 different simulation runs are performed. Figures 5(a), 5(b), 5(c) and 5(d) compare the performance of U-NSGA-III with NSGA-II for different population sizes N. In each case, $H = N$ is used for U-NSGA-III. Each figure is plotted with average number of solution evaluations needed to achieve the respective HV value presented in Table 2. It can be seen from the figures that NSGA-II requires certain minimum population sizes to catch up with the U-NSGA-III's performance. For small number of reference points (H or N), since U-NSGA-III's reference points are uniformly distributed over the entire efficient frontier, the corresponding HV values are better than NSGA-II's adaptive spreading mechanism. The difference in performance is more for the ZDT2 problem. Interestingly, the performance of both methods deteriorate after a certain population size for ZDT2 and ZDT4, due to over-sizing to solve the respective problems. Clearly, there exists an optimal population size for each of these two cases in terms of achieving a certain

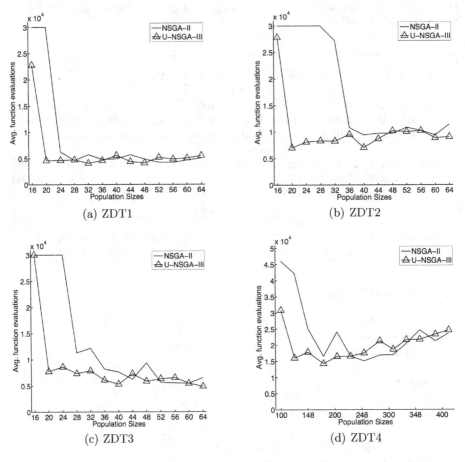

(a) ZDT1

(b) ZDT2

(c) ZDT3

(d) ZDT4

Fig. 5. Comparison of U-NSGA-III and NSGA-II for bi-objective problems for different population sizes (N) with $H = N$

HV value from overall solution evaluation point of view. Nevertheless, the performance similarity of U-NSGA-III and NSGA-II with an adequate population size is clear from these figures.

Next, we compare the performance of U-NSGA-III, NSGA-III, and NSGA-II algorithms in Figures 6(a), 6(b), 6(c) and 6(d) for the four ZDT problems for which H is kept fixed following values: 16 for ZDT1, ZDT2 and ZDT3, and 100 for ZDT4. For U-NSGA-III, we have used different population sizes $N \geq H$. Average number of solution evaluations over 10 runs needed to achieve the pre-specified HV value (Table 2) are plotted in the figures. For runs having a population size larger than number of reference points, the points that are closest to each H reference line in the objective space are used for the HV calculation. For ZDT1 and ZDT3, the use of a larger population size is not found to be beneficial, whereas for ZDT2 and ZDT4 (more difficult problems), a larger population brings in the necessary diversity needed to solve the respective

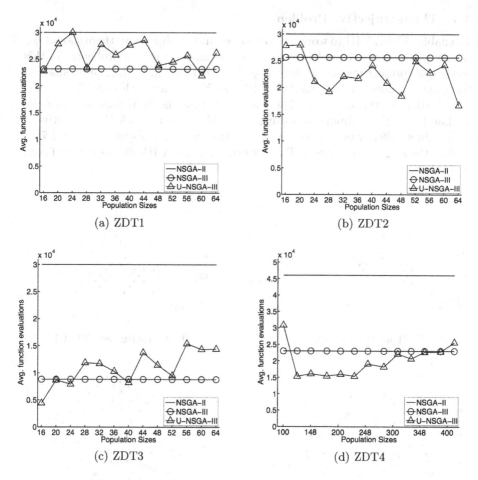

Fig. 6. Comparison of U-NSGA-III, NSGA-III, and NSGA-II for bi-objective problems for different population sizes (N) of U-NSGA-III

problem adequately. NSGA-III and NSGA-II are also applied with $N = H$ for all four problems and the average number of solution evaluations in 10 runs needed to achieve an identical HV value is marked in the respective figure. It is clear that in all problems, there exists certain population sizes, in general, higher than H that make U-NSGA-III to perform better than NSGA-III and NSGA-II. For relatively difficult problems, the difference is quite obvious. In all cases, however, the performance of NSGA-III is better than NSGA-II, due to the use of an external guidance for diversity through a uniformly distributed set of reference points. These results are interesting and demonstrate the usefulness of a larger population size than the number of reference points for the proposed U-NSGA-III algorithm.

4.3 Three-Objective Problems

To enable U-NSGA-III to work well on mono- and bi-objectives, there should not be any performance degradation to three and many-objective problems. In this section, we present results on three-objective DTLZ1 and DTLZ2 problems. For these experiments, we have used $H = 91$ and $N = 92$ for both U-NSGA-III and NSGA-III algorithms. Figures 7(a) and 7(b) show the final population of the median (HV) of ten simulations using U-NSGA-III and NSGA-III, respectively. Only minor differences can be seen across the two plots. Figures 7(c) and 7(d) lead to the same conclusion on DTLZ2 problem. Exact HV values can be found in Table 3.

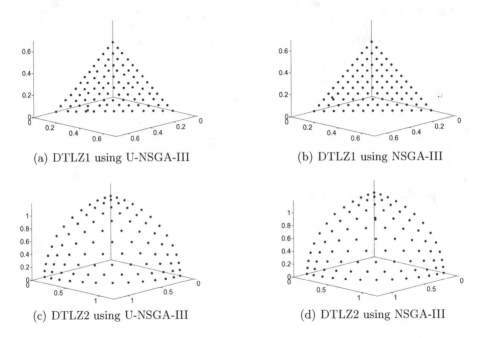

(a) DTLZ1 using U-NSGA-III

(b) DTLZ1 using NSGA-III

(c) DTLZ2 using U-NSGA-III

(d) DTLZ2 using NSGA-III

Fig. 7. U-NSGA-III and NSGA-III points for three-objective DTLZ1 and DTLZ2 problems

4.4 Many-objective Problems

Next, we consider five, eight, and 10-objective DTLZ1 and DTLZ2 problems for our study here. We compare our proposed U-NSGA-III with NSGA-III in terms of hyper-volume values which are computed using the sampling based strategy proposed elsewhere [2] due to the computational complexities involved in the HV calculation in higher dimensions. As discussed before, the difference between U-NSGA-III and NSGA-III are negligible for many-objective optimization problems from a algorithmic point of view. Here, we are interested in investigating how both these methods perform empirically on a series of test problems. In these problems, we use identical values of N and H as those used in NSGA-III study.

Table 3 shows that in most cases an identical HV value is obtained. Even when there is a difference, it is small. The almost-identical PCP plots can be observed for 10-objective DTLZ1 problem in Figures 8(a) and 8(b) for U-NSGA-III and NSGA-III, respectively. Similar results are also obtained for three-objective DTLZ2 in Figures 8(c) and 8(d) by U-NSGA-III and NSGA-III, respectively. All these results demonstrate that the introduction of the niched tournament selection in the original NSGA-III algorithm and the flexibility of using a different population size from the number of reference points do not change its performance in U-NSGA-III on many-objective optimization problems.

Table 3. Comparison of performance of U-NSGA-III and NSGA-III on 3, 5, 8 and 10-objective DTLZ1 and DTLZ2 problems

Problem	#Obj.	Max. Gen.	U-NSGA-III	NSGA-III
DTLZ1	3	400	0.79261	0.79425
	5	600	0.96853	0.96867
	8	750	0.00421	0.00421
	10	1000	0.00108	0.00108
DTLZ2	3	250	0.42956	0.43016
	5	350	0.70806	0.70866
	8	500	0.90406	0.90532
	10	750	1.01819	1.01697

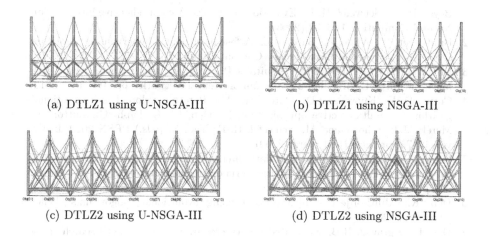

(a) DTLZ1 using U-NSGA-III

(b) DTLZ1 using NSGA-III

(c) DTLZ2 using U-NSGA-III

(d) DTLZ2 using NSGA-III

Fig. 8. U-NSGA-III and NSGA-III points for 10-objective DTLZ1 and DTLZ2 problems

5 Conclusions

In this study, we have developed an unified evolutionary optimization algorithm U-NSGA-III which is a modification to a recently proposed evolutionary many-objective optimization method. U-NSGA-III has been carefully designed so as to solve mono-, multi-, and many-objective optimization problems. Simulation results on a number of mono-objective, two-objective ZDT, and scalable DTLZ problems have demonstrated the efficacy of the proposed unified approach. In each problem type on multiple problem instances, it has been found that the proposed U-NSGA-III performs in a similar manner to a respective specific EA – an elite-preserving rGA for mono-objective problems, NSGA-II for bi-objective problems, and NSGA-III for three and many-objective problems. The ability of one optimization algorithm to solve different types of problems equally efficiently and sometimes better with the added flexibility brought in through a population size control remains a hallmark achievement of this study.

We plan to extend U-NSGA-III for handling constrained problems by modifying its selection operator. MOEA/D and other efficient EMO methods can also be tried to develop an equivalent unified approach. As a further extension, the unified approach can be modified to handle multi-modal problems for finding multiple optimal or Pareto-optimal solutions in a single simulation. The population approach and flexibility of EAs makes such a unified approach possible and further such studies will demonstrate overall usefulness of EAs in solving various optimization problems in an unified manner.

References

1. Beyer, H.-G., Schwefel, H.-P.: Evolution strategies: A comprehensive introduction. Natural Computing **1**, 3–52 (2003)
2. Bradstreet, L., While, L., Barone, L.: A fast incremental hypervolume algorithm. IEEE Transactions on Evolutionary Computation **12**(6), 714–723 (2008)
3. Britto, A., Pozo, A.: I-MOPSO: A suitable PSO algorithm for many-objective optimization. In: 2012 Brazilian Symposium on Neural Networks, pp. 166–171 (2012)
4. Corne, D.W., Knowles, J.D., Oates, M.: The Pareto envelope-based selectional-gorithm for multiobjective optimization. In: Deb, K., Rudolph, G., Lutton, E., Merelo, J.J., Schoenauer, M., Schwefel, H.-P., Yao, X. (eds.) PPSN 2000. LNCS, vol. 1917. Springer, Heidelberg (2000)
5. Das, I., Dennis, J.E.: Normal-boundary intersection: A new method for generating the Pareto surface in nonlinear multicriteria optimization problems. SIAM Journal of Optimization **8**(3), 631–657 (1998)
6. Deb, K.: Multi-objective optimization using evolutionary algorithms. Wiley, Chichester (2001)
7. Deb, K., Agrawal, R.B.: Simulated binary crossover for continuous search space. Complex Systems **9**(2), 115–148 (1995)
8. Deb, K., Agrawal, S., Pratap, A., Meyarivan, T.: A fast and elitist multi-objective genetic algorithm: NSGA-II. IEEE Transactions on Evolutionary Computation **6**(2), 182–197 (2002)
9. Deb, K., Goyal, M.: A robust optimization procedure for mechanical component design based on genetic adaptive search. Transactions of the ASME: Journal of Mechanical Design **120**(2), 162–164 (1998)
10. Deb, K., Jain, H.: An evolutionary many-objective optimization algorithm using reference-point based non-dominated sorting approach, Part I: Solving problems with box constraints. IEEE Transactions on Evolutionary Computation **18**(4), 577–601 (2014)

11. Deb, K., Tiwari, S.: Omni-optimizer: A generic evolutionary algorithm for global optimization. European Journal of Operations Research (EJOR) **185**(3), 1062–1087 (2008)
12. Jain, H., Deb, K.: An evolutionary many-objective optimization algorithm using reference-point based non-dominated sorting approach, Part II: Handling constraints and extending to an adaptive approach. IEEE Transactions on Evolutionary Computation **18**(4), 602–622 (2014)
13. Seada, H., Deb, K.: U-NSGA-III: A unified evolutionary algorithm for single, multiple, and many-objective optimization. COIN Report Number 2014022, Computational Optimization and Innovation Laboratory (COIN), Electrical and Computer Engineering, Michigan State University, East Lansing, USA (2014)
14. Srinivas, N., Deb, K.: Multi-objective function optimization using non-dominated sorting genetic algorithms. Evolutionary Computation Journal **2**(3), 221–248 (1994)
15. Zhang, Q., Li, H.: MOEA/D: A multiobjective evolutionary algorithm based on decomposition. IEEE Transactions on Evolutionary Computation **11**(6), 712–731 (2007)
16. Zitzler, E., Laumanns, M., Thiele., L.: SPEA2: Improving the strength Pareto evolutionary algorithm for multiobjective optimization. In: Giannakoglou, K.C., et al (eds.) Evolutionary Methods for Design Optimization and Control with Applications to Industrial Problems, pp. 95–100. International Center for Numerical Methods in Engineering (CIMNE) (2001)

Clustering Based Parallel Many-Objective Evolutionary Algorithms Using the Shape of the Objective Vectors

Christian von Lücken[1]([⊠]), Carlos Brizuela[2], and Benjamin Barán[1]

[1] Facultad Politécnica, Universidad Nacional de Asunción, San Lorenzo, Paraguay
clucken@pol.una.py
[2] CICESE Research Center, Ensenada, México

Abstract. Multi-objective Evolutionary Algorithms (MOEA) are used to solve complex multi-objective problems. As the number of objectives increases, Pareto-based MOEAs are unable to maintain the same effectiveness showed for two or three objectives. Therefore, as a way to ameliorate this performance degradation several authors proposed preference-based methods as an alternative to Pareto based approaches. On the other hand, parallelization has shown to be useful in evolutionary optimizations. A central aspect for the parallelization of evolutionary algorithms is the population partitioning approach. Thus, this paper presents a new parallelization approach based on clustering by the shape of objective vectors to deal with many-objective problems. The proposed method was compared with random and k-means clustering approaches using a multi-threading framework in parallelization of the NSGA-II and six variants using preference-based relations for fitness assignment. Executions were carried-out for the DTLZ problem suite, and the obtained solutions were compared using the generational distance metric. Experimental results show that the proposed shape-based partition achieves competitive results when comparing to the sequential and to other partitioning approaches.

Keywords: Multi-Objective Evolutionary Algorithms · Many-objective optimization · Parallel evolutionary algorithms

1 Introduction

Multi-objective evolutionary algorithms (MOEAs) are well-suited for solving several problems requiring simultaneous optimization of two or three conflicting objectives [3,4]. In general, MOEAs differ in the fitness assignment method, but some of the most successful of them, such as the NSGA-II [5], use the Pareto dominance concept as the foundation to guide the search towards the optimal solution set.

In the last few years, several researchers have pointed out convergence difficulties that Pareto-based MOEAs face when solving many-objective problems,

A. Gaspar-Cunha et al. (Eds.): EMO 2015, Part II, LNCS 9019, pp. 50–64, 2015.
DOI: 10.1007/978-3-319-15892-1_4

i.e. problems having four or more conflicting objectives [17]. The main source of these difficulties comes from that the proportion of non-dominated individuals in an evolutionary population tends to one as the number of objectives increases. Therefore, for a growing number of objectives, it becomes increasingly difficult to discriminate among solutions to assign fitness values using only the dominance relation [4,12]. In order to distinguish among solutions, some authors propose to replace the Pareto dominance relation by preference relations that use additional information such as the number of objectives for which one solution is better than another [9,12], the size of improvement [12,20], or the number of subspaces in which a given solution remains non-dominated [8].

Even though there are other approaches to deal with many-objective problems using MOEAs [17], the use of alternative relations have shown to be able to improve the quality of the obtained Pareto set approximations without requiring to combine or reduce the number of objectives which, in several cases, may not be adequate or even possible. Also, alternative relations are relatively easy to incorporate into existing (Pareto-based) MOEAs with a minimal computational overhead.

On the other hand, parallelization of existing MOEAs have proven to be an effective mechanism to improve the quality of the obtained approximations in multi-objective problems of increasing difficulty [3,21]. However, to the best of our knowledge, parallelization of methods based on preference relations were not applied and tested in many-objective problems. Moreover, parallel MOEAs (pMOEAs) were mainly developed and tested in distributed memory parallel systems [15]; however, nowadays availability of lower cost shared memory multi-core platforms requires a review of the parallelization methods in the new existing environment in order to leverage the computational power that these platforms may offer.

The island model is the most popular parallelization paradigm for MOEAs, it consists of a number of subpopulations or islands evolving independently provided with a mechanism to interchange individuals exploring for global optima. Using multiple subpopulations has been identified as a simple way to increase the chance of finding better solutions for a problem [2,3]. In general, in the island model, the interchange of individuals is carried out by a migration operator that selects solutions to migrate from one subpopulation to another. The migration strategy definition includes to specify the elapsed time between migrations, as well as the selection criterion and the number of elements to migrate.

Alternatively, instead of conceiving a migration strategy, other island-based methods split the evolutionary population into multiple subpopulations that, after evolving independently, are combined again into a single population, which also can evolve in a single process for a number of iterations, then the division process is repeated [1,13,18,19]. In this work, since division is usually used to create partitions with similar individuals, these methods are generically called clustering based parallel MOEA. Population partitions with similar individuals induce a mating restriction, i.e. recombination of similar individuals, that has shown to be useful to improve the performance of evolutionary algorithms [4]. As it is noted in [2], the main drawback of these pMOEAs based on repeated partitioning of a global population is that the division procedure introduces a strong dependency among the computing units. However, the expected improvement provided by working with multiple subpopulations running in different processors may justify the use of these methods. Most of clustering based pMOEAs were implemented into distributed parallel systems, thus, requiring additional data communication to send subpopulations to the process in charge of join them in a whole population and redistribute it again. However, in nowadays multicore systems with shared memory systems to gather subpopulations requires a reduced communication and synchronization time.

This paper studies clustering based parallelization of NSGA-II variants using preference relations for many-objective problems. In this case, the clustering methods serves to search into different regions simultaneously using preference based MOEAs, while, in each subpopulation, the preference relations are used to improve the rank over similar individuals, which may be useful to improve the overall search. Moreover, a new clustering method based on the shape of the objective vectors is proposed (see Section 3).

In order to validate the proposal in the field of many objective optimization, an empirical comparison of the proposed method against other alternatives for population division for pMOEAs was produced considering the same parallel framework. Using this framework, the NSGA-II and six of its variants based on preference relations were parallelized with three options to divide the population: at random, using the k-means clustering algorithm, and using the proposed shape-based clustering algorithm. Sequential and parallel implementations were used to solve the suite of DTLZ problems with 10 objectives and the obtained results compared regarding the Generational Distance metric [4].

2   Clustering Based Parallelization of MOEAs

2.1   Parallel MOEAs Using Clustering

In general, the island model is used to develop pMOEAs as parallel extensions of existing mono-objective or multi-objective evolutionary algorithms. Besides the algorithm considered for parallelization, island based pMOEAs differ in the mechanism used to interchange information among subpopulations searching for the global Pareto front. As previously explained, an alternative to exchange individuals among islands is to iteratively divide a global population into subpopulations and gather them to repeat the cycle again. In this case, methods may also alternate between the execution of sequential iterations, considering the whole population, with parallel iterations. Some methods to divide the global population into subpopulations considered here as relevant related works are:

- population partitioning based on sorting of a objective function [13,18],
- population partitioning based on the cone separation approach [1],
- population partitioning based on the k-means clustering algorithm [19].

Parallelization approaches based on objective function sorting use the value of one selected criterion to divide and distribute the population in different islands. Thus, each island works on a subpopulation with individuals that are similar regarding the chosen sorting objective. After some evaluations, subpopulations are gathered again and the process is repeated until a stop condition is met. Examples of these approaches the Divided Range Multi-objective Genetic Algorithm [13] and the Parallel Single Front Genetic Algorithm [18].

The Cone Separation Approach [1] parallelizes MOEAs by dividing the search space and mapping different search areas to different processors. In this case, objective values are normalized, then, considering a given reference point, the search space is divided in regular partitions or regions called cones that can be assigned to different processors. As the searching process progresses, the search space partitioning is adapted at regular intervals by normalizations and readjustment of the region assigned to each processor. Region restrictions are applied to each subpopulation and individuals not meeting constraints in a subpopulation migrate to the one where they do not violate the constraints. Migrated individuals are added to the receiving population without deletion. This way, there is not a centralized process that divides the population iteratively but a decentralized one; however, the method is conceptually equal to the methods based on repeated division of a global population.

Streichert et al. [19] use the k-Means clustering algorithm to divide the search space of a given optimization problem in suitable partitions without a priori knowledge about the search space topology. The k-Means procedure is applied over the current Pareto Front of the whole population to produce partitions to be distributed to the available processors or islands. In case the size of Pareto Front being smaller than the number of processors, next levels of Pareto fronts are also used for clustering. Each processor runs NSGA-II until the number of generations reaches a number of iterations, then, solutions are gathered in a master procedure where search space is partitioned and distributed again.

2.2 Preference Relations in MOEAs for Many-Objective Problems

The NSGA-II is a well known MOEA that showed an excellent performance in several multi-objective problems, thus, several researchers considered the NSGA-II [5] as the base algorithm to implement and validate their proposed relations and algorithmic approaches both in multi-objective as well as many-objective problem domains. The NSGA-II assigns the fitness of solutions based on two values: its non-dominance ranking and its crowding distance. Variants of the NSGA-II can be produced by considering alternative methods to calculate fitness. In this paper, the following operators and ranking methods are used to modify the fitness assignment procedure of the NSGA-II in a minimization context, for space limitation reasons the details are not explained here but we refer the interested reader to the corresponding references:

1. **Favour relation [9]:** this relation counts the number of objectives in which a given solution outperforms another. Given a multi-objective problem minimizing a function $\mathbf{F}(\mathbf{x}) = (f_1(\mathbf{x}), \ldots, f_m(\mathbf{x}))$ with m objectives, and let $\mathbf{x}$

and $\mathbf{x}'$ be two vectors in the set of feassible solutions $\mathcal{X}_f$, it is said that $\mathbf{x}$ is favoured than $\mathbf{x}'$, denoted as $\mathbf{x} \prec_{favour} \mathbf{x}'$, if and only if

$$n_b(\mathbf{F}(\mathbf{x}), \mathbf{F}(\mathbf{x}')) > n_b(\mathbf{F}(\mathbf{x}'), \mathbf{F}(\mathbf{x})) \tag{1}$$

where:

$$n_b(\mathbf{F}(\mathbf{x}), \mathbf{F}(\mathbf{x}')) = |\{f_i(\mathbf{x}) \text{ s.t. } f_i(\mathbf{x}) < f_i(\mathbf{x}')\}| \tag{2}$$

In [9], the favour relation is proposed to be used with the Satisfiability Class Ordering classification (SCO) procedure [9] to sort solutions.

2. $\epsilon-$**Preferred Relation [20]**: the $\epsilon-$Preferred relation compares solutions by counting the number of times a solution exceeds user defined limits for each dimension (ϵ_i) and, in case of a tie, it uses the favour relation. Given two solutions $\mathbf{x}$ and $\mathbf{x}' \in \mathcal{X}_f$, $\mathbf{F}(\mathbf{x}) = \mathbf{y} = (y_1, \ldots, y_m)$, $\mathbf{F}(\mathbf{x}') = \mathbf{y}' = (y_1', \ldots, y_m')$, it is said that $\mathbf{x}$ is $\epsilon-$preferred than $\mathbf{x}'$, denoted as $\mathbf{x} \prec_{\epsilon-preferred} \mathbf{x}'$, iff $\mathbf{x} \prec_{\epsilon-exceed} \mathbf{x}' \vee (\mathbf{x}' \not\prec_{\epsilon-exceed} \mathbf{x} \wedge \mathbf{x} \prec_{favour} \mathbf{x}')$, where $\mathbf{x} \prec_{\epsilon-exceed} \mathbf{x}'$ implies that:

$$|\{i : y_i < y_i' \wedge |y_i - y_i'| > \epsilon_i\}| > |\{i : y_i' < y_i \wedge |y_i' - y_i| > \epsilon_i\}|$$

As in [9], in [20] the SCO algorithm is used to rank solutions.

3. **Preference Ordering based on order of efficiency** (PO_k) **[8]**: a solution $\mathbf{x}$ is considered to be efficient of order k if it is Pareto optimal in the $\binom{m}{k}$ subspaces of the objective space taking into account only k out of m objectives at a time. The order of efficiency of a solution $\mathbf{x}$, denoted by $K(\mathbf{x})$ is the minimum k value for which $\mathbf{x}$ is efficient.

4. $-\epsilon$-**DOM [16]**: the $-\epsilon$-DOM distance replaces the NSGA-II crowding distance. The $-\epsilon$-DOM distance of a solution $\mathbf{x}$ is the smallest value such that if subtracted from all objectives of $\mathbf{F}(\mathbf{x}')$, makes $\mathbf{x}$ dominated.

5. $(1 - k)-$**dominance relation [12]**: this relation counts the objectives in which a solution is better or equal than another one. Let $\mathbf{x}$ and $\mathbf{x}' \in \mathcal{X}_f$, $\mathbf{F}(\mathbf{x}) = \mathbf{y}$, $\mathbf{F}(\mathbf{x}') = \mathbf{y}'$, it is said that $\mathbf{x}$ $(1 - k)-$dominates $\mathbf{x}'$ iff : $n_e(\mathbf{y}, \mathbf{y}') < m$ and $n_b(\mathbf{y}, \mathbf{y}') \geq \frac{m - n_e}{k+1}$, where m is the number of objectives, $0 \leq k \leq 1$, n_b is as in Eq. (2), and $n_e(\mathbf{F}(\mathbf{x}), \mathbf{F}(\mathbf{x}'))$ is $|\{f_i(\mathbf{x}) \text{ s.t. } f_i(\mathbf{x}) = f_i(\mathbf{x}')\}|$.

6. **Fuzzy** $(1 - k_F)-$**dominance relation [12]**: the fuzzy extension of $(1 - k)-$dominance is defined by determining membership functions μ_b^i, μ_e^i and μ_w^i for each objective function i.

2.3   A Framework for Clustering Based Parallelization of MOEAs Based on Preference Relations

To study clustering based pMOEAs, Algorithm 1 presents a framework for pMOEAs that use iterative partitioning of a global population for multi-threading systems. Using this framework, by setting the corresponding parameters it is possible to study different options for pMOEAs based on clustering.

Algorithm 1 starts reading its parameters; the MOEA $\mathcal{M}$ to be considered for evolutions and its parameters, the partition method (PM), the number of

Algorithm 1. A framework for clustering based pMOEAs

Read parameters
Set $t = 0$
Create an initial random global evolutionary population P_t
while $t < it_g$ **do**   ▷ it_g is the total number of evolutionary steps performed
   for it_s iterations **do**   ▷ it_s: iterations considering a single population
     Evolve P_t in P_{t+1} using $\mathcal{M}$ ▷ $\mathcal{M}$ is the MOEA to be used
     $t = t + 1$
   end for
   In τ **parallel threads**
   $t' = 0$
   for it_p iterations **do**   ▷ it_p: iterations considering subpopulations in parallel
     if t' mod $it_c = 0$ **then**
       Split P_t in $P_t^1, \ldots, P_t^\tau$ using partition method PM
     end if
     Evolve $P_{t+t'}^{Id}$ in $P_{t+t'+1}^{Id}$ using $\mathcal{M}$
     $t' = t' + 1$
   end for
   End parallel
   $t = t + it_p$
end while
Save non-dominated solutions from P_t

---

islands (τ), the total number of evolutionary steps (it_g), the number of single thread iterations (it_s), the number of parallel iterations (it_p) and the number of iterations before clustering (it_c). Next, the global number of iterations t is set to 0, and the global population P_t is created at random. After initialization, while $t < it_g$, it_s evolutionary iterations are executed in a single thread considering the evolutionary population as a whole. Then, τ threads are created, one for each island, and parallel execution starts. The next step is to split $P(t)$ using the procedure PM in τ subpopulations ($P_t^1 \ldots, P_t^\tau$), this procedure, repeated each it_c iterations, may be implemented in parallel or as a single thread. Each thread has an identifier Id, thus at each thread Id, evolution of P_t^{Id} occurs during it_p iterations. When iterations in all threads end, the global count of iterations t is updated to $t + it_p$ and the cycle continues until the stop condition is met. Finally, the final set of solutions is saved.

3   Partition Approach Based on the Shape of Solutions

In [10] it is presented a similarity measure over Euclidean spaces for high dimensional vectors. To calculate this measure, a real vector $\mathbf{y} = (y_1, \ldots, y_M) \in \mathcal{R}^m$ is divided in a pair $(s(\mathbf{y}), \pi(\mathbf{y}))$, where $s(\mathbf{y})$ is the ordered version (weak) of $\mathbf{y}$ elements, and $\pi(\mathbf{y})$ is the permutation of indexes $\{1, \ldots, m\}$ that produce the sorting which is called as the shape part. The distance between two vectors $\mathbf{y}$ and $\mathbf{y}'$ is defined in [10] by a combination of the distance between $s(\mathbf{y})$ and $s(\mathbf{y}')$, and $\pi(\mathbf{y})$ and $\pi(\mathbf{y}')$. The shape of a vector can be formally defined as follows:

Definition 1. *Shape of a vector in* $\mathcal{R}^m$: *Given a vector* $\mathbf{y} \in \mathcal{R}^m$, *a permutation* $\pi(\mathbf{y}) = \{\pi_1, \ldots, \pi_m\}$, $\pi_i \in \{1, \ldots, m\}$, *is the shape of* $\mathbf{y}$ *iff:*

$$y_{\pi_i} \leq y_{\pi_j}, \forall i < j$$

As an example, let $\mathbf{y} = [0.1, 0.5, 0.3]$ and $\mathbf{y}' = [0.3, 0.1, 0.5]$, then, their shapes are $\pi(\mathbf{y}) = [1, 3, 2]$ and $\pi(\mathbf{y}') = [2, 1, 3]$.

This work proposes the shape of a vector to divide the population in groups of solutions having a similar shape. The number of different shapes grows factorially with the number of objectives, i.e. for three objectives there are 6 different shapes, but for 5 objectives there are 120 possible shapes. Therefore, it is not practical for a large number of objectives to split the solutions according to all possible shapes, but by means of a clustering method considering the similarity between the shapes of objective vectors. There are several measures that can be used to measure the distance between permutations. In this work, the Spearman's rho distance, defined as follows, is considered [11].

Definition 2. *Spearman's rho distance: Given permutations π and π', and interpreting π_i as the position of element i in π, the Spearman's rho distance is defined as:*

$$\rho(\pi, \pi') = \left(\sum_{i=1}^{M} |\pi_i - \pi_i'|^2 \right)^{(1/2)} \tag{3}$$

Algorithm 2 presents a clustering procedure using the shape of objective functions. First, a set $\{C_1, C_2, \ldots, C_\tau\}$ of τ clusters are initialized at empty. Then, for each element $\mathbf{x}$ in the population to be classified P, a normalized objective value $\hat{\mathbf{F}}(\mathbf{x})$ is calculated with its corresponding shape $\pi_{\mathbf{x}} = \pi(\hat{\mathbf{F}}(\mathbf{x}))$. Thereafter, τ different shapes are randomly selected from the set of shapes of the normalized objective values into a set $\mathcal{S}$. If eventually there are less than τ different shapes, $\mathcal{S}$ is completed by repeated elements. An index Id corresponding to each cluster is assigned to each shape in $\mathcal{S}$. Next, for each $\mathbf{x}$ in P, a set $\mathcal{S}'_{\mathbf{x}}$ containing the indexes of the shapes in $\mathcal{S}$ having the minimal Spearman's rho distance to $\pi_{\mathbf{x}}$ is created. The set $\mathcal{S}'_{\mathbf{x}}$ is used to select the index Id of the cluster in which $\mathbf{x}$ will be included. If the cardinality of $\mathcal{S}'_{\mathbf{x}}$ is one, the index is the value of the unique element in $\mathcal{S}'_{\mathbf{x}}$; otherwise, one of the indexes in $\mathcal{S}'_{\mathbf{x}}$ is selected at random. Finally, $\mathbf{x}$ is included in the cluster C_{Id}. The procedure finishes when all individuals are assigned to a cluster.

Algorithm 2. Clustering algorithm using the shape of objective functions

---

Initialize $\{C_1, C_2, \ldots, C_\tau\}$ clusters, s.t. $C_{Id} = \emptyset$ for $Id \in [1, \ldots, \tau]$
For each $\mathbf{x} \in P$ obtain its normalized objective value $\hat{\mathbf{F}}(\mathbf{x})$ and shape $\pi_{\mathbf{x}} = \pi(\hat{\mathbf{F}}(\mathbf{x}))$
Randomly select τ different shapes $\mathcal{S} = \{sh_1, sh_2, \ldots, sh_\tau\}$ from shapes of $\mathbf{x} \in P$
for each $\mathbf{x} \in P$ do
    $\mathcal{S}' = \{ Id \mid sh_{Id} \in \min_{sh \in Sh} \{\rho(\pi_{\mathbf{x}}, sh)\} \}$
    if($|\mathcal{S}'| == 1$) **then** Id is the element of $\mathcal{S}'$
    else select Id at random from $\mathcal{S}'$
    end if
    Set $C_{Id} = C_{Id} \cup \mathbf{x}$
end for

---

4   Experimental Comparison

4.1   Experimental Setup and Metrics

The main focus of this work is to compare the shape-based clustering method to parallelize MOEAs based on preference relations with their sequential counterparts and other population division methods over a set of many-objective optimization problems. Thus, using the framework presented in Subsection 2.3, three partition methods were implemented to parallelize the original NSGA-II and six variants of it considering the relations explained in Subsection 2.2. The considered partition methods are: the proposed clustering method based on the shape of objective vectors (SH), a random (RN) partition and a k-means based partitioning (KM). The test problems used in this work are the DTLZ1 to DTLZ7 problems with 10 objectives [6]. The programs were implemented in C language and the OpenMP library.

For each problem 10 runs were executed for sequential and parallel programs using 2, 4, and 8 threads. The choice of the number of runs was made taking into account other works such as [8,14], and the available time to execute the programs and analyse the results (note that there is a total of 4900 executions). The experimental computational platform was a machine provided with two Intel Xeon quad-core Processors E5640 (12M Cache, 2.66 GHz, 5.86 GT/s) and 16 GB of main memory running the GNU/Linux operating system. The programs consider the following common parameters: population size 400, it_g is 400, binary coding of 32 bits per variable, one point crossover probability of 0.8, mutation probability of 0.002. For the ϵ-Preferred relation, the ϵ value is 0.0001; for the $(1 - k)$−dominance relation, k is 0.5; and, for the $(1 - k_F)$−dominance relation, k_F is also 0.5 and a fuzzy trapezoidal rule is used ($a = -0.001$, $b = 0$, $c = 0$, $d = 0.001$) [12]. The k and k_F values were selected taking into account the test cases in [12], while the values used for ϵ and the fuzzy trapezoidal rule were selected on experimental basis, however, no fine tuning of the above parameters was considered. For parallel methods $it_s = 0$, it_p is set to 400, it_c is 1. The sequential version is also implemented using the framework, but, in this case, $\tau = 1$ and $it_p = 0$.

The obtained results were evaluated using the Generational Distance (GD) and Spread (Δ) metrics. GD measures the average distance between obtained solutions in objective space and the true Pareto Front of the problem, while Spread evaluates the extent of the Pareto Front covered by the obtained set of solutions. Since GD requires a reference $\mathcal{PF}^\star$ to be computed, and equations to produce $\mathcal{PF}^\star$ are known, a set of 2000 optimal solutions was determined analytically. Both metrics are expected to be minimized.

4.2   Experimental Results

Table 1 and Table 2 show the average GD and Spread values performing 10 runs of implemented combinations, for each DTLZ problem. In these tables, for each problem, there are 10 rows corresponding to each execution type: one for the sequential execution (Seq) and 3 for each partition method used for parallelization considering 2, 4, and 8 threads (τ value), while each column is for different MOEAs tested in this work. The values in parenthesis show the ranking obtained by each execution type of the given MOEA for each problem. Note that, until the value is represented by using a reduced number of digits, rankings are calculated using the machine numeric representation. Also, the best value for each column (execution method) is boldfaced. In order to summarize the results, the two last columns indicate the average of the rankings of each row, and regarding the data in this column, the last column labelled "Final" shows the overall rank obtained by each execution type for each problem.

The results in Table 1 and Table 2 may be used to determine the combination of partition method, MOEA and number of threads, performing the best for each problem and metric. However, in this work, the detailed results are not analyzed but the general and average behaviour of the studied partition methods in order to show how clustering may serve as a basis to develop pMOEAs for many-objective problems. At first glance, Table 1 shows that in almost all cases, for each MOEA considered, at least one parallel implementation evaluates better than its corresponding sequential counterparts for the GD metric. Considering the seven DTLZ problems and seven implemented MOEAs, only in 4 out of 49 results obtained by sequential implementations are not improved by parallel executions for GD. The result is remarkable since parallel and sequential implementations were executed using the same number of iterations, and, therefore, the same number of objective function evaluations. Therefore, the source of the benefit is provided by the interactions among individuals in subpopulations and not by executing more evaluations in the same execution time.

Figure 1 and Figure 2 show the final ranking of each combination of partition method and number of threads by problem using data in Table 1 and Table 2, respectivelly. The labels indicate the τ value followed by the partition method, i.e. 2-KM is for implementations using 2 threads and k-means. As it can be noted, in Figure 1 the smaller bars are for the SH implementations. In fact, in 5 out of 7 problems, an SH implementation obtains the best rank value for GD, in one problem a method based on k-means and in another case a method based on random partitioning. Also, the figure indicates that the best performance is for 8-SH obtaining the best ranking positions for almost all problems. In fact, the worst value obtained by the 8-SH is 5, in problem DTLZ6, whereas the other partitioning options receive in at least one problem an overall ranking greater than 5. From a visual inspection of Figure 2, it is not possible to determine which implementation alternative may be considered as the best, however it appears that sequential implementation obtain the best ranking in three problems, and that the 8-KM obtains, in general, the worse values.

Figure 3 presents the ranking distribution for GD metric considering the 49 implementations of each sequential and parallel MOEAs with different number of threads (7 MOEAs times 7 DTLZ problems). According to this figure the results of shape-based implementations concentrates in the first ranks while they have the fewer number of implementations in position 10. Thus, using the Figure 3 as

Table 1. Average values for GD metric with 10 objectives

Test	τ	Part	ε−DOM	ε−Pref	$(1-k)$	$(1-k_F)$	Favour	NSGAII	PO_k	Ranking Avg.	Ranking Final
DTLZ1	1	0	1.05E-1(6)	1.28E+0(10)	2.11E-1(4)	1.99E-1(10)	2.50E-1(2)	3.40E+0(6)	**8.77E-2(1)**	5.57	5
		KM	1.53E-1(10)	2.14E-1(3)	2.31E-1(7)	1.95E-1(9)	3.88E-1(6)	2.85E+0(5)	1.27E-1(6)	6.57	10
	2	RN	9.24E-2(3)	3.87E-1(6)	2.33E-1(8)	1.47E-1(5)	**2.39E-1(1)**	5.31E+0(8)	1.21E-1(4)	5	4
		SH	8.24E-2(2)	6.13E-1(9)	2.00E-1(3)	1.54E-1(6)	2.65E-1(3)	3.75E+0(7)	1.69E-1(10)	5.71	6
		KM	9.26E-2(4)	1.75E-1(2)	3.21E-1(10)	1.43E-1(4)	4.12E-1(7)	1.14E+0(2)	1.10E-1(3)	4.57	2
	4	RN	9.34E-2(5)	3.40E-1(5)	2.34E-1(9)	1.60E-1(7)	3.07E-1(5)	5.47E+0(9)	1.25E-1(5)	6.43	9
		SH	1.14E-1(7)	4.62E-1(8)	2.15E-1(5)	1.19E-1(3)	4.88E-1(8)	2.58E+0(4)	1.62E-1(9)	6.29	8
		KM	1.27E-1(9)	4.22E-1(7)	1.56E-1(2)	1.02E-1(2)	5.11E-1(9)	**1.09E+0(1)**	1.08E-1(2)	4.57	3
	8	RN	**5.35E-2(1)**	2.36E-1(4)	2.28E-1(6)	1.67E-1(8)	2.75E-1(4)	1.11E+1(10)	1.61E-1(8)	5.86	7
		SH	1.22E-1(8)	**1.35E-1(1)**	**1.46E-1(1)**	**9.89E-2(2)**	6.98E-1(10)	1.15E+0(3)	1.47E-1(7)	**4.43**	1
DTLZ2	1	0	6.63E-3(9)	6.18E-5(7)	2.02E-3(10)	1.38E-3(8)	2.90E-5(7)	4.92E-2(8)	5.83E-3(10)	8.43	10
		KM	5.90E-3(7)	3.21E-4(8)	6.42E-4(8)	1.33E-3(4)	4.89E-5(8)	4.59E-2(6)	2.28E-3(8)	7	8
	2	RN	6.84E-3(10)	3.62E-5(6)	1.47E-5(5)	1.96E-3(10)	1.97E-5(5)	4.88E-2(7)	4.86E-3(9)	7.43	9
		SH	5.12E-3(4)	2.55E-5(5)	1.83E-5(6)	1.38E-3(7)	1.66E-5(4)	4.58E-2(5)	2.02E-3(6)	5.29	4
		KM	5.07E-3(3)	1.08E-3(9)	1.15E-3(9)	1.33E-3(3)	6.76E-4(9)	4.16E-2(3)	5.47E-4(3)	5.57	7
	4	RN	6.00E-3(8)	1.92E-5(4)	**1.32E-5(1)**	1.36E-3(6)	1.54E-5(2)	4.97E-2(9)	2.22E-3(7)	5.29	5
		SH	**4.60E-3(1)**	1.86E-5(3)	1.38E-5(3)	1.73E-3(9)	1.98E-5(6)	4.33E-2(4)	8.21E-4(5)	4.43	2
		KM	5.68E-3(6)	1.54E-3(10)	4.42E-4(7)	**1.25E-3(1)**	1.51E-3(10)	**3.47E-2(1)**	3.55E-4(2)	5.29	6
	8	RN	5.37E-3(5)	1.65E-5(2)	1.41E-5(4)	1.36E-3(5)	1.65E-5(3)	5.12E-2(10)	7.92E-4(4)	4.71	3
		SH	4.73E-3(2)	**1.54E-5(1)**	1.38E-5(2)	1.32E-3(2)	**1.42E-5(1)**	3.85E-2(2)	3.32E-4(1)	**1.57**	1
DTLZ3	1	0	6.82E-1(2)	6.01E+0(9)	5.57E+0(10)	1.14E+0(10)	2.70E+0(5)	2.84E+0(6)	6.41E-1(3)	6.43	7
		KM	7.14E-1(4)	6.90E+0(10)	3.22E+0(9)	1.09E+0(7)	3.53E+0(7)	3.06E+0(7)	6.64E-1(4)	6.86	9
	2	RN	7.70E-1(8)	3.97E+0(4)	2.42E+0(6)	1.08E+0(6)	3.75E+0(9)	4.02E+0(8)	7.02E-1(5)	6.57	8
		SH	7.95E-1(9)	4.60E+0(7)	2.79E+0(8)	9.53E-1(5)	3.70E+0(8)	2.50E+0(5)	7.84E-1(7)	7	10
		KM	7.35E-1(5)	4.75E+0(8)	2.02E+0(5)	9.26E-1(3)	3.89E+0(10)	1.30E+0(2)	8.51E-1(9)	6	5
	4	RN	6.95E-1(3)	4.39E+0(6)	1.82E+0(3)	1.13E+0(8)	2.63E+0(4)	5.34E+0(9)	7.41E-1(6)	5.57	4
		SH	7.54E-1(6)	4.16E+0(5)	2.72E+0(7)	9.44E-1(4)	**2.01E+0(1)**	1.92E+0(4)	8.90E-1(10)	5.29	3
		KM	**6.16E-1(1)**	**2.51E+0(1)**	1.84E+0(4)	**7.12E-1(1)**	3.52E+0(6)	**1.14E+0(1)**	**5.50E-1(1)**	2.14	1
	8	RN	9.96E-1(10)	3.00E+0(3)	**1.40E+0(1)**	1.14E+0(9)	2.12E+0(2)	1.87E+1(10)	8.07E-1(8)	6.14	6
		SH	7.68E-1(7)	2.90E+0(2)	1.77E+0(2)	7.77E-1(2)	2.59E+0(3)	1.85E+0(3)	6.27E-1(2)	3	2
DTLZ4	1	0	6.74E-3(9)	2.55E-2(10)	4.62E-2(10)	1.83E-3(3)	4.16E-3(6)	6.79E-2(7)	1.13E-2(4)	7	10
		KM	**3.31E-3(1)**	3.58E-3(3)	9.01E-3(8)	3.14E-3(6)	7.02E-3(10)	6.77E-2(6)	1.81E-2(9)	6.14	7
	2	RN	6.99E-3(10)	6.32E-3(8)	7.61E-3(7)	**1.22E-3(1)**	4.27E-3(7)	7.01E-2(8)	1.39E-2(7)	6.86	9
		SH	5.13E-3(5)	3.70E-3(4)	2.94E-3(4)	2.36E-3(4)	2.64E-3(2)	6.73E-2(5)	1.27E-2(6)	4.29	2
		KM	3.78E-3(2)	3.04E-3(2)	1.38E-2(9)	3.56E-3(7)	3.39E-3(5)	6.36E-2(3)	2.11E-2(10)	5.43	5
	4	RN	5.85E-3(7)	4.78E-3(7)	1.51E-3(1)	2.81E-3(5)	5.28E-3(9)	7.21E-2(9)	1.24E-2(5)	6.14	8
		SH	4.40E-3(4)	**2.64E-3(1)**	3.68E-3(5)	4.53E-3(9)	**2.28E-3(1)**	6.57E-2(4)	1.08E-2(3)	**3.86**	1
		KM	5.61E-3(6)	3.93E-3(5)	4.29E-3(6)	7.66E-3(10)	2.88E-3(3)	5.87E-2(1)	1.49E-2(8)	5.57	6
	8	RN	6.55E-3(8)	9.10E-3(9)	1.51E-3(2)	1.76E-3(2)	3.12E-3(4)	7.35E-2(10)	9.02E-3(2)	5.29	4
		SH	4.17E-3(3)	4.52E-3(6)	2.29E-3(3)	3.65E-3(8)	4.54E-3(8)	6.00E-2(2)	**8.36E-3(1)**	4.43	3
DTLZ5	1	0	3.85E-2(2)	1.34E-5(4)	1.25E-1(7)	**1.30E-5(1)**	1.38E-1(10)	1.20E-1(9)	**1.14E-1(1)**	4.86	2
		KM	4.02E-2(9)	1.95E-5(9)	1.27E-1(8)	1.38E-1(5)	1.29E-1(9)	1.17E-1(5)	1.20E-1(3)	6.86	10
	2	RN	3.95E-2(7)	**1.26E-5(1)**	1.25E-1(5)	1.32E-5(3)	1.28E-1(4)	1.18E-1(7)	1.25E-1(4)	4.43	1
		SH	3.85E-2(3)	1.28E-5(3)	1.25E-1(6)	1.56E-5(7)	1.28E-1(7)	1.18E-1(6)	1.25E-1(5)	5.29	4
		KM	3.86E-2(4)	1.46E-5(7)	1.28E-1(9)	1.58E-5(9)	1.28E-1(6)	1.14E-1(3)	1.17E-1(2)	5.71	7
	4	RN	4.04E-2(10)	1.27E-5(2)	1.25E-1(4)	1.32E-5(2)	1.28E-1(5)	1.18E-1(8)	1.25E-1(6)	5.29	5
		SH	3.91E-2(6)	1.42E-5(6)	1.25E-1(2)	1.56E-5(8)	1.28E-1(8)	1.14E-1(4)	1.25E-1(7)	5.86	8
		KM	**3.81E-2(1)**	2.04E-5(10)	1.32E-1(10)	3.24E-3(10)	1.27E-1(2)	**1.02E-1(1)**	1.29E-1(10)	6.29	9
	8	RN	4.01E-2(8)	1.40E-5(5)	**1.25E-1(1)**	1.36E-5(4)	**1.25E-1(1)**	1.28E-1(10)	1.25E-1(8)	5.29	6
		SH	3.90E-2(5)	1.81E-5(8)	1.25E-1(3)	1.52E-5(6)	1.27E-1(3)	1.11E-1(2)	1.25E-1(9)	5.14	3
DTLZ6	1	0	1.75E-1(7)	2.98E+0(7)	**6.36E-4(1)**	2.69E+0(5)	2.73E-3(7)	4.57E-1(5)	4.19E-3(7)	5.57	7
		KM	1.75E-1(8)	2.63E+0(3)	1.93E-2(10)	3.13E+0(10)	1.23E-2(10)	4.53E-1(2)	5.95E-3(8)	7.29	10
	2	RN	1.64E-1(3)	2.76E+0(5)	7.55E-4(5)	2.77E+0(7)	3.48E-4(6)	4.65E-1(6)	4.14E-3(6)	5.43	6
		SH	1.66E-1(4)	3.00E+0(8)	6.51E-4(2)	2.93E+0(8)	3.35E-4(5)	**4.53E-1(1)**	2.94E-3(3)	**4.43**	1
		KM	1.87E-1(9)	2.40E+0(2)	1.32E-2(9)	2.75E+0(6)	9.25E-3(8)	4.56E-1(3)	1.03E-2(9)	6.57	9
	4	RN	1.58E-1(2)	3.08E+0(9)	8.53E-4(6)	2.20E+0(2)	**2.39E-4(1)**	4.69E-1(8)	3.79E-3(5)	4.71	4
		SH	1.70E-1(5)	2.72E+0(4)	6.79E-4(4)	2.98E+0(9)	2.48E-4(4)	4.57E-1(4)	2.55E-3(2)	4.57	3
		KM	1.88E-1(10)	**1.83E+0(1)**	1.27E-2(8)	2.24E+0(3)	9.49E-3(9)	4.67E-1(7)	1.24E-2(10)	6.86	9
	8	RN	**1.52E-1(1)**	2.90E+0(6)	1.51E-3(7)	**2.11E+0(1)**	2.44E-4(3)	4.75E-1(9)	3.40E-3(4)	4.43	2
		SH	1.71E-1(6)	3.14E+0(10)	6.69E-4(3)	2.35E+0(4)	2.40E-4(2)	4.75E-1(10)	**1.55E-3(1)**	5.14	5
DTLZ7	1	0	8.44E-3(10)	9.11E-5(8)	6.28E-3(10)	1.34E-3(7)	2.93E-4(7)	4.67E-2(8)	7.10E-3(10)	8.57	10
		KM	6.32E-3(6)	5.35E-5(7)	2.14E-3(9)	1.34E-3(5)	3.08E-4(8)	4.32E-2(6)	3.15E-3(6)	6.71	9
	2	RN	7.93E-3(9)	4.53E-5(6)	2.70E-5(6)	1.31E-3(4)	2.37E-5(3)	4.67E-2(7)	5.43E-3(9)	6.29	7
		SH	4.40E-3(2)	4.39E-5(5)	2.67E-5(5)	1.34E-3(6)	2.51E-5(4)	4.30E-2(5)	3.86E-3(7)	4.86	3
		KM	4.75E-3(4)	1.43E-3(9)	1.07E-3(7)	1.25E-3(2)	8.48E-4(9)	3.76E-2(3)	8.80E-4(3)	5.29	5
	4	RN	7.68E-3(8)	3.27E-5(3)	2.46E-5(2)	1.62E-3(9)	2.56E-5(5)	4.97E-2(10)	4.12E-3(8)	6.43	8
		SH	4.65E-3(3)	3.64E-5(4)	**2.44E-5(1)**	1.35E-3(8)	2.79E-5(6)	3.97E-2(4)	1.44E-3(4)	4.29	2
		KM	4.80E-3(5)	2.03E-3(10)	1.66E-3(8)	**1.23E-3(1)**	1.28E-3(10)	**3.15E-2(1)**	**6.09E-4(1)**	5.14	4
	8	RN	6.92E-3(7)	2.93E-5(2)	2.56E-5(3)	1.88E-3(10)	**1.98E-5(1)**	4.94E-2(9)	1.60E-3(5)	5.29	6
		SH	**4.22E-3(1)**	**2.67E-5(1)**	2.57E-5(4)	1.30E-3(3)	2.12E-5(2)	3.60E-2(2)	6.60E-2(2)	**2.14**	1

Table 2. Average values for Spread metric with 10 objectives

Test	τ	Part	ε−DOM	ε−Pref	(1−k)	(1−k_F)	Favour	NSGAII	PO_k	Ranking Avg.	Final
DTLZ1	1	0	5.95E-1(3)	1.07E+0(6)	**1.02E+0(1)**	9.08E-1(8)	1.06E+0(5)	5.47E-1(7)	**5.81E-1(1)**	4.43	4
		KM	6.07E-1(5)	1.06E+0(5)	1.09E+0(6)	8.35E-1(2)	1.04E+0(2)	5.31E-1(6)	7.28E-1(2)	4.00	2
	2	RN	5.75E-1(2)	1.09E+0(8)	1.04E+0(2)	**8.33E-1(1)**	1.05E+0(4)	4.96E-1(4)	8.30E-1(4)	3.57	1
		SH	**5.59E-1(1)**	**1.00E+0(1)**	1.11E+0(9)	9.11E-1(9)	**1.03E+0(1)**	4.66E-1(2)	1.06E+0(7)	4.29	3
		KM	6.42E-1(7)	1.08E+0(7)	1.06E+0(4)	9.75E-1(10)	1.05E+0(3)	6.53E-1(9)	7.47E-1(3)	6.14	8
	4	RND	6.78E-1(8)	1.02E+0(2)	1.06E+0(3)	8.78E-1(6)	1.09E+0(8)	4.79E-1(3)	9.28E-1(5)	5.00	5
		SH	6.03E-1(4)	1.06E+0(4)	1.06E+0(5)	8.40E-1(3)	1.11E+0(9)	5.04E-1(5)	1.11E+0(8)	5.43	6
		KM	7.18E-1(9)	1.16E+0(10)	1.09E+0(7)	8.87E-1(7)	1.08E+0(7)	6.70E-1(10)	9.88E-1(6)	8.00	9
	8	RND	7.32E-1(10)	1.05E+0(3)	1.11E+0(8)	8.72E-1(5)	1.07E+0(6)	**4.12E-1(1)**	1.23E+0(9)	6.00	7
		SH	6.12E-1(6)	1.14E+0(9)	1.18E+0(10)	8.48E-1(4)	1.21E+0(10)	5.85E-1(8)	1.52E+0(10)	8.14	10
DTLZ2	1	0	4.82E-1(3)	**1.00E+0(1)**	1.04E+0(2)	7.34E-1(3)	**1.00E+0(1)**	2.77E-1(10)	**6.92E-1(1)**	3.00	1
		KM	4.86E-1(4)	1.03E+0(3)	1.22E+0(4)	7.31E-1(2)	1.03E+0(2)	2.44E-1(4)	8.58E-1(3)	3.14	2
	2	RN	5.45E-1(7)	1.00E+0(2)	**1.00E+0(1)**	7.51E-1(7)	**1.00E+0(1)**	2.73E-1(9)	7.58E-1(2)	4.14	6
		SH	4.81E-1(2)	1.00E+0(2)	**1.00E+0(1)**	7.61E-1(9)	**1.00E+0(1)**	2.51E-1(7)	1.09E+0(4)	3.71	3
		KM	5.38E-1(6)	1.10E+0(4)	1.34E+0(5)	7.51E-1(6)	1.14E+0(3)	2.35E-1(2)	1.20E+0(6)	4.57	8
	4	RND	5.77E-1(9)	1.00E+0(2)	**1.00E+0(1)**	7.52E-1(8)	**1.00E+0(1)**	2.53E-1(8)	1.17E+0(5)	4.86	10
		SH	**4.78E-1(1)**	1.00E+0(2)	**1.00E+0(1)**	7.82E-1(10)	**1.00E+0(1)**	2.47E-1(5)	1.45E+0(8)	4.00	5
		KM	5.72E-1(8)	1.31E+0(5)	1.17E+0(3)	7.49E-1(5)	1.23E+0(4)	**2.33E-1(1)**	1.32E+0(7)	4.71	9
	8	RND	6.54E-1(10)	1.00E+0(2)	**1.00E+0(1)**	**7.30E-1(1)**	**1.00E+0(1)**	2.49E-1(6)	1.56E+0(9)	4.29	7
		SH	4.92E-1(5)	1.00E+0(2)	**1.00E+0(1)**	7.38E-1(4)	**1.00E+0(1)**	2.39E-1(3)	1.63E+0(10)	3.71	4
DTLZ3	1	0	5.84E-1(5)	1.00E+0(2)	1.00E+0(4)	7.37E-1(5)	**1.00E+0(1)**	4.83E-1(8)	**5.76E-1(1)**	3.29	1
		KM	5.92E-1(6)	1.00E+0(4)	1.01E+0(3)	7.48E-1(7)	1.01E+0(5)	5.11E-1(9)	7.41E-1(3)	5.29	8
	2	RN	6.40E-1(7)	1.00E+0(5)	1.01E+0(4)	7.71E-1(9)	1.03E+0(6)	4.61E-1(6)	6.20E-1(2)	5.57	9
		SH	**5.47E-1(1)**	1.00E+0(3)	1.01E+0(6)	7.12E-1(2)	1.00E+0(4)	4.69E-1(7)	9.17E-1(5)	4.00	3
		KM	5.74E-1(3)	1.03E+0(6)	1.00E+0(2)	7.16E-1(3)	1.05E+0(7)	4.24E-1(4)	1.01E+0(6)	5.14	6
	4	RND	6.95E-1(8)	**1.00E+0(1)**	1.00E+0(2)	7.44E-1(6)	1.00E+0(2)	4.28E-1(5)	8.98E-1(4)	4.00	4
		SH	5.83E-1(4)	**1.00E+0(1)**	1.00E+0(2)	7.25E-1(4)	1.00E+0(2)	3.87E-1(3)	1.28E+0(8)	3.43	2
		KM	7.91E-1(9)	1.13E+0(7)	1.16E+0(8)	**6.93E-1(1)**	1.22E+0(8)	5.66E-1(10)	1.13E+0(7)	7.14	10
	8	RND	9.72E-1(10)	**1.00E+0(1)**	1.00E+0(2)	8.09E-1(10)	1.00E+0(2)	3.67E-1(2)	1.28E+0(9)	5.14	7
		SH	5.63E-1(2)	**1.00E+0(1)**	1.01E+0(5)	7.70E-1(8)	1.00E+0(3)	**3.66E-1(1)**	1.56E+0(10)	4.29	5
DTLZ4	1	0	1.65E+0(4)	1.21E+0(6)	1.29E+0(6)	1.29E+0(4)	1.00E+0(2)	2.49E-1(10)	1.17E+0(4)	5.71	6
		KM	1.67E+0(10)	1.40E+0(8)	1.46E+0(8)	1.54E+0(7)	1.19E+0(7)	2.24E-1(2)	9.47E-1(2)	6.29	7
	2	RN	1.59E+0(3)	1.05E+0(2)	1.22E+0(4)	**1.23E+0(1)**	1.09E+0(3)	2.40E-1(9)	1.22E+0(6)	4.00	2
		SH	1.64E+0(5)	1.16E+0(5)	**1.02E+0(1)**	1.26E+0(3)	**1.00E+0(1)**	2.28E-1(5)	1.20E+0(5)	3.57	1
		KM	1.65E+0(7)	1.44E+0(9)	1.53E+0(10)	1.67E+0(10)	1.29E+0(9)	**2.22E-1(1)**	**9.46E-1(1)**	6.71	8
	4	RND	**1.50E+0(1)**	1.09E+0(3)	1.25E+0(5)	1.36E+0(5)	1.15E+0(5)	2.33E-1(8)	1.24E+0(7)	4.86	4
		SH	1.61E+0(4)	**1.03E+0(1)**	1.13E+0(3)	1.50E+0(6)	1.16E+0(6)	2.27E-1(4)	1.32E+0(8)	4.57	3
		KM	1.66E+0(9)	1.61E+0(10)	1.52E+0(9)	1.66E+0(9)	1.57E+0(10)	2.24E-1(3)	1.14E+0(3)	7.57	10
	8	RND	1.56E+0(2)	1.09E+0(4)	1.30E+0(7)	1.25E+0(2)	1.14E+0(4)	2.30E-1(6)	1.36E+0(9)	4.86	5
		SH	1.64E+0(6)	1.24E+0(7)	1.04E+0(2)	1.57E+0(8)	1.23E+0(8)	2.30E-1(7)	1.44E+0(10)	6.86	9
DTLZ5	1	0	4.60E-1(8)	1.00E+0(2)	1.02E+0(5)	**1.00E+0(1)**	1.01E+0(7)	3.94E-1(9)	1.06E+0(7)	5.57	8
		KM	4.91E-1(10)	1.00E+0(5)	1.14E+0(8)	1.00E+0(2)	1.19E+0(8)	3.12E-1(4)	1.10E+0(8)	6.43	9
	2	RN	4.41E-1(5)	1.00E+0(3)	1.02E+0(3)	**1.00E+0(1)**	1.00E+0(6)	3.56E-1(7)	1.03E+0(4)	4.14	6
		SH	4.50E-1(7)	1.00E+0(4)	**1.00E+0(1)**	1.00E+0(3)	1.00E+0(5)	3.12E-1(5)	1.01E+0(3)	3.86	5
		KM	4.38E-1(3)	**1.00E+0(1)**	1.14E+0(9)	**1.00E+0(1)**	1.21E+0(10)	3.07E-1(3)	1.21E+0(10)	5.29	7
	4	RND	4.34E-1(2)	1.00E+0(3)	1.02E+0(2)	**1.00E+0(1)**	1.00E+0(2)	3.43E-1(6)	1.03E+0(5)	3.00	1
		SH	4.47E-1(6)	1.00E+0(6)	1.05E+0(7)	**1.00E+0(1)**	**1.00E+0(1)**	2.85E-1(2)	1.01E+0(2)	3.57	3
		KM	4.71E-1(9)	1.00E+0(4)	1.16E+0(10)	1.04E+0(3)	1.20E+0(9)	4.40E-1(10)	1.14E+0(9)	7.71	10
	8	RND	4.38E-1(4)	1.00E+0(4)	**1.01E+0(1)**	**1.00E+0(1)**	1.00E+0(3)	3.68E-1(8)	**1.00E+0(1)**	3.14	2
		SH	**4.31E-1(1)**	1.00E+0(7)	1.04E+0(6)	**1.00E+0(1)**	1.00E+0(4)	**2.80E-1(1)**	1.03E+0(6)	3.71	4
DTLZ6	1	0	6.57E-1(5)	**9.95E-1(1)**	1.02E+0(7)	1.01E+0(8)	1.01E+0(7)	4.00E-1(10)	1.02E+0(7)	6.43	7
		KM	6.50E-1(4)	1.01E+0(8)	1.37E+0(10)	9.97E-1(2)	1.29E+0(10)	3.13E-1(5)	1.05E+0(8)	6.71	8
	2	RN	7.05E-1(8)	9.99E-1(4)	1.01E+0(2)	1.00E+0(5)	1.00E+0(6)	3.69E-1(9)	1.01E+0(4)	5.43	6
		SH	6.72E-1(6)	9.96E-1(2)	1.01E+0(6)	1.00E+0(6)	1.00E+0(5)	3.20E-1(7)	**1.00E+0(2)**	3.57	3
		KM	6.46E-1(2)	1.03E+0(9)	1.25E+0(9)	1.04E+0(10)	1.27E+0(9)	2.75E-1(2)	1.10E+0(9)	7.14	10
	4	RND	7.66E-1(9)	1.00E+0(7)	**1.00E+0(1)**	9.97E-1(1)	1.00E+0(3)	3.21E-1(8)	1.01E+0(6)	5.00	5
		SH	6.49E-1(3)	1.00E+0(5)	1.01E+0(6)	9.98E-1(3)	1.00E+0(4)	2.82E-1(3)	**9.94E-1(1)**	3.57	1
		KM	**6.34E-1(1)**	1.08E+0(10)	1.25E+0(8)	1.04E+0(9)	1.27E+0(8)	**2.57E-1(1)**	1.19E+0(10)	6.71	9
	8	RND	8.10E-1(10)	9.99E-1(3)	1.01E+0(3)	1.00E+0(4)	1.00E+0(2)	3.15E-1(6)	1.01E+0(5)	4.71	4
		SH	6.76E-1(7)	1.00E+0(6)	1.01E+0(4)	1.01E+0(7)	**1.00E+0(1)**	2.85E-1(4)	1.00E+0(3)	4.57	2
DTLZ7	1	0	5.02E-1(3)	**1.00E+0(1)**	1.05E+0(4)	7.22E-1(5)	**1.00E+0(1)**	2.79E-1(10)	**5.99E-1(1)**	3.57	1
		KM	**4.86E-1(1)**	1.00E+0(3)	1.27E+0(5)	7.14E-1(4)	1.02E+0(4)	2.53E-1(7)	7.87E-1(3)	3.86	3
	2	RN	5.78E-1(8)	1.00E+0(4)	**1.00E+0(1)**	7.08E-1(3)	1.00E+0(2)	2.57E-1(8)	6.87E-1(2)	4.00	4
		SH	5.11E-1(5)	1.00E+0(2)	**1.00E+0(1)**	7.28E-1(7)	1.00E+0(2)	2.46E-1(5)	8.24E-1(4)	3.71	2
		KM	5.36E-1(6)	1.07E+0(8)	1.29E+0(6)	**6.95E-1(1)**	1.23E+0(6)	2.35E-1(2)	8.58E-1(5)	4.86	5
	4	RND	5.66E-1(7)	1.00E+0(5)	1.00E+0(3)	7.25E-1(6)	1.00E+0(2)	2.59E-1(9)	8.61E-1(6)	5.43	8
		SH	4.89E-1(2)	1.00E+0(7)	1.00E+0(4)	7.32E-1(9)	1.00E+0(3)	2.46E-1(4)	1.13E+0(8)	4.86	6
		KM	6.03E-1(10)	1.26E+0(9)	1.35E+0(7)	6.98E-1(2)	1.22E+0(5)	**2.35E-1(1)**	1.04E+0(7)	5.86	9
	8	RND	6.01E-1(9)	1.00E+0(6)	**1.00E+0(1)**	7.30E-1(8)	1.00E+0(2)	2.49E-1(6)	1.36E+0(9)	5.86	10
		SH	5.06E-1(4)	1.00E+0(6)	1.00E+0(2)	7.32E-1(10)	1.00E+0(2)	2.39E-1(3)	1.53E+0(10)	5.29	7

a complement to the average ranking values in the last two columns of Table 1, we can state that parallelization using shape-based clustering can be considered in general as the best parallelization option for GD improvement for the set of problems and MOEAs considered.

To statistically support the claims about the convenience of parallelization using shape-based clustering, Table 3 presents the number of wins among the implementations indicated in each column regarding implementations in each row for the Sign test pairwise comparisons [7] the obtained ranking using the GD and Spread metrics. In case that detected differences exists these are indicated in parenthesis, and the data is boldfaced. The values 32, 34, 36, and 37 are considered as the critical number of wins needed to achieve levels of significance of 0.1, 0.05, 0.02, and 0.01, respectively. In case of ties, the count is split evenly between the pair of compared algorithms. This table shows that the 8-SH has a significant improvement over the parallelization options using less than 8 threads. Also, as it is indicated, parallel approaches based on shape clustering clearly improve the results of the sequential implementations. For the Δ metric, there is a similar number of wins among sequential and paralelization approaches based on random and shape based partitioning, thus it can not be stated that one algorithm performs better than another considering this metric. However, for the Δ metric, all alternatives are better than the k-means parallelization using 8 threads.

Finally, Figure 3 also shows that considering the number of threads, the execution using 8 threads (clusters) clearly obtains a large number of GD rankings one and two than the other alternatives. The results suggest that using a large number of clusters may aid to improve the GD results of parallel executions using clustering based pMOEAs.

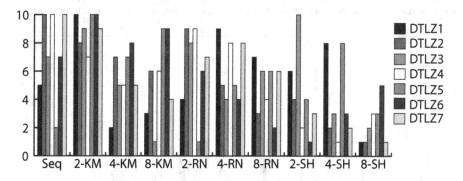

Fig. 1. Final GD ranking of each implementation option by problem

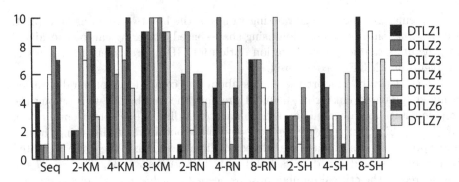

Fig. 2. Final Δ ranking of each implementation option by problem

Table 3. Number of wins and detected differences for Sing test for pairwise distribution of each implementation approach considering ranking values in Table 1 and Table 2

	Seq	2-KM	2-RN	2-SH	4-KM	4-RN	4-SH	8-KM	8-RN	8-SH
				Generational distance(GD)						
Seq	24.5	24	30	**35(0.05)**	27	29	**34(0.05)**	30	31	**36(0.02)**
2-KM	25	24.5	28	**34(0.05)**	30	28	**36(0.02)**	**33(0.1)**	30	**39(0.01)**
2-RN	19	21	24.5	29	26	24	**32(0.1)**	29	27	**38(0.01)**
2-SH	14	15	20	24.5	23	22	29	24	25	**37(0.01)**
4-KM	22	19	23	26	24.5	25	23	28	26	**33(0.1)**
4-RN	20	21	25	27	24	24.5	28	26	26	**33(0.1)**
4-SH	15	13	17	20	26	21	24.5	23	23	**35(0.05)**
8-KM	19	16	20	25	21	23	26	24.5	25	27
8-RN	18	19	22	24	23	23	26	24	24.5	31
8-SH	13	10	11	12	16	16	14	22	18	24.5
				Spread (Δ)						
	Seq	2-KM	2-RN	2-SH	4-KM	4-RN	4-SH	8-KM	8-RN	8-SH
Seq	24.5	15	26	29.5	15.5	24	28	9	25	21
2-KM	**34(0.05)**	24.5	29	**32(0.1)**	18	29	**34(0.05)**	12	28	25
2-RN	23	20	24.5	29	19.5	22	24.5	9	22	21.5
2-SH	19.5	17	20	24.5	14.5	20.5	24.5	12	23	17.5
4-KM	**33.5(0.1)**	31	29.5	**34.5(0.05)**	24.5	29.5	30.5	16	27.5	**32.5(0.1)**
4-RN	25	20	27	28.5	19.5	24.5	23.5	11	21	22
4-SH	21	15	24.5	24.5	18.5	25.5	24.5	15	22	15.5
8-KM	**40(0.01)**	**37(0.01)**	**40(0.01)**	**37(0.01)**	**33(0.1)**	**38(0.01)**	**34(0.05)**	24.5	**32.5(0.1)**	**34(0.05)**
8-RN	24	21	27	26	21.5	28	27	16.5	24.5	19.5
8-SH	28	24	27.5	31.5	16.5	27	**33.5(0.1)**	15	29.5	24.5

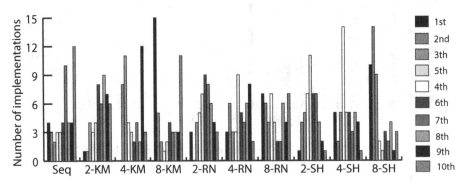

Fig. 3. Ranking distribution of GD by partition method and number of threads

5 Conclusions and Future Work

In evolutionary algorithms dealing with the simultaneous optimization of more than 4 objectives, the use of parallelization may be useful to improve their performances. One key aspect of this parallelization is the correct partitioning of the population into subpopulations that are to be distributed among processors. In this paper, we have proposed a method to do such partitioning based on what is known as the shape of the objective vector. We present an experimental comparison of the performance of the resulting parallel MOEAs and their sequential counterpart as well as a comparison of different partition methods.

The obtained results have shown that, in most of the studied cases, at least one parallel MOEAs outperform their sequential counterparts (45 out of 49 cases). Comparison results have also shown that, for the considered experimental setting and metrics, the parallel implementations based on the clustering of solutions using the shape of objective vectors obtains the best average rank in almost all considered problems (5 out of 7 cases).

Future work is aimed at extending the evaluation considering the hypervolume metric and increasing the number of objectives.

References

1. Branke, J., Schmeck, H., Deb, K., Reddy, M.: Parallelizing multi-objective evolutionary algorithms: cone separation. In: 2004 Congress on Evol. Comput., vol. 2, pp. 1952–1957. IEEE, Portland (2004)
2. Cheshmehgaz, H.R., Haron, H., Sharifi, A.: The review of multiple evolutionary searches and multi-objective evolutionary algorithms. Artificial Intelligence Review, 1–33 (2013)
3. Coello Coello, C.A., Lamont, G., Van Veldhuizen, D.: Evolutionary Algorithms for Solving Multi-Objective Problems, 2nd edn. Springer, New York (2007). ISBN: 978-0-387-33254-3
4. Deb, K.: Multi-objective optimization using evolutionary algorithms. Wiley (2001)
5. Deb, K., Pratap, A., Agarwal, S., Meyarivan, T.: A fast and elitist multiobjective genetic algorithm: NSGA-II. IEEE Trans. on Evol. Comput. (2002)
6. Deb, K., Thiele, L., Laumanns, M., Zitzler, E.: Scalable multi-objective optimization test problems. In: Proc. of the 2002 Congr. on Evol. Comput. (2002)
7. Derrac, J., García, S., Molina, D., Herrera, F.: A practical tutorial on the use of nonparametric statistical tests as a methodology for comparing evolutionary and swarm intelligence algorithms. Swarm and Evolutionary Computation 1(1), 3–18 (2011)
8. di Pierro, F., Khu, S., Savić, D.: An investigation on Preference Order ranking scheme for multiobjective evolutionary optimization. IEEE Trans. on Evol. Comput. (2007)
9. Drechsler, N., Drechsler, R., Becker, B.: Multi-objective optimisation based on relation favour. In: First Int. Conf. on Evol. Multi-Criterion Optim. Springer (2001)
10. Egecioglu, Ö.: Parametric approximation algorithms for high-dimensional euclidean similarity. In: Siebes, A., De Raedt, L. (eds.) PKDD 2001. LNCS (LNAI), vol. 2168, pp. 79–90. Springer, Heidelberg (2001)

11. Fagin, R., Kumar, R., Sivakumar, D.: Comparing top k lists. In: Proc. of the 14th Annual ACM-SIAM Symp. on Discrete Algorithms. p. 36. SIAM (2003)

12. Farina, M., Amato, P.: On the Optimal Solution Definition for Many-criteria Optimization Problems. In: Proc. of the NAFIPS-FLINT Int. Conf. 2002, pp. 233–238. IEEE Service Center (2002)

13. Hiroyasu, T., Miki, M., Watanabe, S.: The new model of parallel genetic algorithm in multi-objective optimization problems: divided range multi-objective genetic algorithm. In: 2000 Congress on Evol. Comput., vol. 1, pp. 333–340. IEEE, New Jersey (2000)

14. Ishibuchi, H., Tsukamoto, N., Nojima, Y.: Evolutionary many-objective optimization: A short review. In: 2008 IEEE Congr. on Evol. Comput. (2008)

15. Jaimes, A.L., Coello Coello, C.A.: Applications of parallel platforms and models in evolutionary multi-objective optimization. In: Lewis, A., Mostaghim, S., Randall, M. (eds.) Biologically-Inspired Optimisation Methods. SCI, vol. 210, pp. 23–49. Springer, Heidelberg (2009)

16. Köppen, M., Yoshida, K.: Substitute distance assignments in NSGA-II for handling many-objective optimization problems. In: Obayashi, S., Deb, K., Poloni, C., Hiroyasu, T., Murata, T. (eds.) EMO 2007. LNCS, vol. 4403, pp. 727–741. Springer, Heidelberg (2007)

17. von Lücken, C., Barán, B., Brizuela, C.: A survey on multi-objective evolutionary algorithms for many-objective problems. Comput. Optim. and Appl. 1(1), 1–50 (2014)

18. Negro, F.D.T., Ortega, J., Ros, E., Mota, S., Paechter, B., Martín, J.M.: PSFGA: parallel processing and evolutionary comput. for multiobjective optim. Parallel Comput. (2004)

19. Streichert, F., Ulmer, H., Zell, A.: Parallelization of multi-objective evolutionary algorithms using clustering algorithms. In: Coello Coello, C.A., Hernández Aguirre, A., Zitzler, E. (eds.) EMO 2005. LNCS, vol. 3410, pp. 92–107. Springer, Heidelberg (2005)

20. Sülflow, A., Drechsler, N., Drechsler, R.: Robust multi-objective optimization in high dimensional spaces. In: Obayashi, S., Deb, K., Poloni, C., Hiroyasu, T., Murata, T. (eds.) EMO 2007. LNCS, vol. 4403, pp. 715–726. Springer, Heidelberg (2007)

21. Talbi, E.-G., Mostaghim, S., Okabe, T., Ishibuchi, H., Rudolph, G., Coello Coello, C.A.: Parallel approaches for multiobjective optimization. In: Branke, J., Deb, K., Miettinen, K., Słowiński, R. (eds.) Multiobjective Optimization. LNCS, vol. 5252, pp. 349–372. Springer, Heidelberg (2008)

Faster Exact Algorithms for Computing Expected Hypervolume Improvement

Iris Hupkens, André Deutz, Kaifeng Yang, and Michael Emmerich$^{(\boxtimes)}$

LIACS, Leiden University, Niels Bohrweg 1, Leiden, The Netherlands
m.t.m.emmerich@liacs.leidenuniv.nl
http://moda.liacs.nl

Abstract. This paper is about computing the expected improvement of the hypervolume indicator given a Pareto front approximation and a predictive multivariate Gaussian distribution of a new candidate point. It is frequently used as an infill or prescreening criterion in multiobjective optimization with expensive function evaluations where predictions are provided by Kriging or Gaussian process surrogate models. The expected hypervolume improvement has good properties as an infill criterion, but exact algorithms for its computation have so far been very time consuming even for the two and three objective case. This paper introduces faster exact algorithms for computing the expected hypervolume improvement for independent Gaussian distributions. A new general computation scheme is introduced and a lower bound for the time complexity. By providing new algorithms, upper bounds for the time complexity for problems with two as well as three objectives are improved. For the 2-D case the time complexity bound is reduced from previously $O(n^3 \log n)$ to $O(n^2)$. For the 3-D case the new upper bound of $O(n^3)$ is established; previously $O(n^4 \log n)$. It is also shown how an efficient implementation of these new algorithms can lead to a further reduction of running time. Moreover it is shown how to process batches of multiple predictive distributions efficiently. The theoretical analysis is complemented by empirical speed comparisons of C++ implementations of the new algorithms to existing implementations of other exact algorithms.

Keywords: Expected improvement · Time complexity · Global multiobjective optimization · Hypervolume indicator · Kriging surrogate models

1 Introduction

In simulator-based or experimental optimization a recurring problem is that objective function evaluations can be costly. A common remedy to this problem is the partial replacement of exact objective function evaluations by predictions from Gaussian processes or Kriging models. These models provide a predictive distribution, consisting of a mean value and variance for each objective function value for each candidate point. The expected improvement [10,11] takes this

© Springer International Publishing Switzerland 2015
A. Gaspar-Cunha et al. (Eds.): EMO 2015, Part II, LNCS 9019, pp. 65–79, 2015.
DOI: 10.1007/978-3-319-15892-1_5

information into account when computing a figure of merit for each given candidate solution. It thereby provides a robust pre-screening criterion and algorithms with attractive theoretical properties [12].

Recently, the expected improvement was generalized for multiobjective optimization using different approaches [7]. Among these the *expected hypervolume improvement* (EHVI) [14] had interesting monotonicity properties [7] and a good benchmark performance [6]. It is also widely used in applications, such as design engineering [2,4,5], algorithm tuning [17] and for optimizing controllers [3]. The EHVI represents the expected improvement in the hypervolume measure relative to the current approximation of the Pareto front [7] given the probability distribution of possible function values.

Besides Monte Carlo integration [14,18], exact computation of the numerical integral was proposed [1,8]. The time complexity for these original schemes was $O(n^{m+1} \log n)$ for $m = 2, 3$ objectives and n points. The high computational effort was so-far widely regarded as a disadvantage of the EHVI [3,7,17]. A significant acceleration of exact algorithms by an efficient decomposition scheme was recently achieved by Couckuyt et al. [6], however, without providing new complexity bounds.

This paper introduces new exact algorithms for computing the expected hypervolume improvement. After providing the preliminaries, new upper bounds for the computational complexity of $O(n^2)$ for the 2-D case (Section 3) and of $O(n^3)$ for the 3-D computation (Section 4) will be established by providing algorithms. Moreover, lower complexity bounds and a new formulation of the exact computation for the general case are introduced. Theoretical analysis is followed by empirical studies on benchmarks (Section 5) comparing the approach to the algorithm proposed in [6]. It is also studied to what extent the recycling of data structures and reuse of nonlinear function evaluations can further increase speed.

2   Expected Hypervolume Improvement

The maximization of $m \geq 1$ objective functions $f_1 : X \to \mathbb{R}, \ldots, f_m : X \to \mathbb{R}$ is considered. The attention will be on points and probability distributions of points in the objective space $\mathbb{R}^m$ (equipped with the Pareto dominance order). With $P = \{p_1, \ldots, p_n\} \subset \mathbb{R}^m$ we will denote a set of n mutually non-dominated points. The family of such sets we denote with A (*approximation sets*).

Definition 1 (Hypervolume Indicator). *The hypervolume indicator (HI) of $P \in A$ is defined as the Lebesgue measure of the subspace dominated by P and cut from below by a reference point r: $\mathrm{HI}(P) := \lambda\{p \in \mathbb{R}^m \mid p \text{ dominates } r \land \exists p_0 \in P : p_0 \text{ dominates } p\} = \lambda(\cup_{p \in P}[r, p])$, with λ being the Lebesgue measure in $\mathbb{R}^m$.*

The set containing the part of the objective space that is dominated by the points in P will be referred to as DomSet(P). Note that in the entire article we consider a fixed reference point[1] that is dominated by all points in the Pareto front approximation.

---

[1] The choice of the reference point influences the placing of the hypervolume optimal population. The methods in this paper work for every valid reference point.

Definition 2 (Expected hypervolume improvement (EHVI)). *The hypervolume improvement* $I_H(p, P)$ *of a point* $p \in \mathbb{R}^m$ *is defined as the increment of the hypervolume indicator after* p *is added to* P, *i.e.* $I_H(p) = HI(P \cup \{p\}) - HI(P)$. *Given the probability density function (PDF) of a predictive distribution of objective function vectors the expected hypervolume improvement (EHVI) is defined as:*

$$\int_{y \in \mathbb{R}^m} I_H(y, P) \cdot PDF(y) \mathrm{d}y.$$

We will only consider independent multivariate Gaussian distributions with independent components and mean values $\mu_1, \ldots, \mu_m$ and standard deviations $\sigma_1, \ldots, \sigma_m$, as they are commonly obtained from multivariate Kriging or Gaussian process surrogate models.

Example 1. An illustration of the EHVI is displayed in Figure 1. The gray area is the dominated subspace of $P = \{p_1, p_2, p_3\}$ cut by the reference point $r \in \mathbb{R}^m$. The PDF of the predictive Gaussian distribution is indicated as a 3-D plot. For some sample from the multivariate Gaussian distribution, indicated as p, the dark shaded area is the hypervolume improvement $I_H(p)$. The variable x stands for the f_1 value and y for the f_2 value.

In order to calculate the EHVI, we need to calculate many integrals that have the form of a partial one-dimensional improvement. In [8] a function was derived that could be used for this purpose. In the following we introduce a useful shorthand named ψ.

Definition 3. *The function* $\phi(s) = 1/\sqrt{2\pi} e^{-\frac{1}{2}s^2}, s \in \mathbb{R}$ *is the PDF of the standard normal distribution and* $\Phi(s) = \frac{1}{2}\left(1 + erf\left(\frac{s}{\sqrt{2}}\right)\right)$ *is its cumulative probability distribution function (CDF). The general normal distribution with mean* μ *and variance* σ *has the PDF* $\phi_{\mu,\sigma}(s) = \frac{1}{\sigma}\phi(\frac{s-\mu}{\sigma})$ *and the CDF* $\Phi_{\mu,\sigma}(s) = \Phi(\frac{s-\mu}{\sigma})$. *We define*

$$\psi(a, b, \mu, \sigma) := \sigma \cdot \phi(\frac{b-\mu}{\sigma}) + (a - \mu)\Phi(\frac{b-\mu}{\sigma}) \tag{1}$$

Integrals of the form $\int_{z=b}^{\infty}(z - a)\frac{1}{\sigma}\phi(\frac{z-\mu}{\sigma})$ are equal to $\sigma\phi(\frac{b-\mu}{\sigma} + (\mu - a)[1 - \Phi(\frac{b-\mu}{\sigma})]$. Integrals whose upper limit is less than ∞ and lower limit greater than $-\infty$ can be written as the difference of two such integrals, allowing partial expected improvements over an interval $[l, u] \subset \mathbb{R}$, $l \geq f'$ to be calculated. This difference can be neatly expressed in terms of ψ: $\int_{z=l}^{u}(z - f')\frac{1}{\sigma}\phi(\frac{z-\mu}{\sigma})\mathrm{d}z = \int_{z=l}^{\infty}(z - f')\frac{1}{\sigma}\phi(\frac{z-\mu}{\sigma})\mathrm{d}z - \int_{z=u}^{\infty}(z - f')\frac{1}{\sigma}\phi(\frac{z-\mu}{\sigma})\mathrm{d}z = \psi(f', l, \mu, \sigma) - \psi(f', u, \mu, \sigma)$ The value f' in this case is the current best function value recorded by an optimization process. We will use the abbreviations $\phi_x(s) := \phi_{\mu_x,\sigma_x}(s)(= \frac{1}{\sigma_x}\phi(\frac{s-\mu_x}{\sigma_x}))$ and $\Phi_x(s) := \Phi_{\mu_x,\sigma_x}(s)(= \Phi(\frac{s-\mu_x}{\sigma_x}))$, where μ_x and σ_x are the mean and variance of the x-component of the normal distribution associated to a point in the search space. Analogously, we use abbreviations ϕ_y, Φ_y and ϕ_z, Φ_z for the y and the z coordinate. In case the dimension of the objective space is more than 3 we revert to natural number subscripts or to variables which vary over $1, \cdots, m$, where m is the dimension of the objective space.

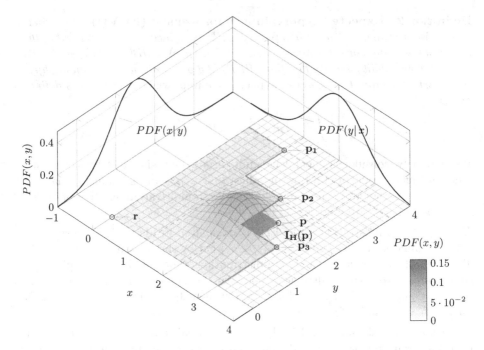

Fig. 1. 2-D expected hypervolume improvement: $\sigma = (0.7, 0.6), \mu = (2, 1.5)$

3   Algorithm for the Bi-objective Case

The aim is to calculate the expected hypervolume improvement for a point for which we have the mean (μ_x, μ_y) and standard deviation (σ_x, σ_y) of a predictive distribution. Its EHVI can be computed by piecewise integration over a set of half-open rectangular interval boxes (cells) formed by the horizontal and vertical lines going through the points in approximation set P and through r. The final EHVI is the sum of the contributions calculated for all grid cells. In Figure 1 a grid is indicated by the dashed lines. Formally, individual grid cells will be denoted by $\mathcal{C}(a, b)$, where $0 \leq a \leq n$ and $0 \leq b \leq n$, n represents the number of segments in each dimension. Let $Q = P \cup \{(\infty, r_y)\} \cup \{(r_x, \infty)\}$, with Q^x denoting Q sorted in order of ascending x coordinate, and Q^y denoting Q sorted in order of ascending y coordinate. Let $\mathcal{C}$ be the set of grid cells representing the interval boxes. The numbers a and b represent positions in the sorting order of Q, starting with 0. Then, a is the position of elements of Q^x and b is the position of elements of Q^y. The lower left corner of a cell will have the coordinates $(Q_a^x.x, Q_b^y.y)$. The upper right corner of the grid cell will have the coordinates $(Q_{a+1}^x.x, Q_{b+1}^y.y)$. The integration contributions for cells that have an upper right corner that is dominated by or equal to some point in P are zero. The set of all other cells (i.e., the non-dominated ones) we will call $\mathcal{C}_{nd}$. It holds that $\forall (\mathcal{C}(a, b) \in \mathcal{C}_{nd}, p \in P) : p.x > Q_a^x.x \Rightarrow Q_b^y.y \geq p.y$ and, analogously, $p.y > Q_b^y.y \Rightarrow Q_a^x.x \geq p.x$. Due to the definition of Q, we know that for $p \in P$ it holds that $p = Q_k^x = Q_{n+1-k}^y$ for some $0 < k \leq n$. Furthermore $k > a$ and $n+1-k > b$, if and only if p dominates $C(a,b)$. From this we get the following equivalence: $a < n - b \Leftrightarrow C(a,b)$ is dominated by some point $p \in P$. Thus C_{nd} consists of all cells satisfying $a \geq n - b$. There are $\frac{(n+1)(n+2)}{2}$ of such cells. If we call the lower corner of the cell l and the upper corner u, the contribution of a grid cell to the integral is: $\int_{p_y=l_y}^{u_y} \int_{p_x=l_x}^{u_x} I_H(p, P) \, \phi_x(p_x) \, \phi_y(p_y) \, dp_x \, dp_y$. Dominated cells have a contribution of 0 to the integral, and for cells which are non-dominated, $I_H(p)$ can be calculated as a rectangular volume from which a correction term is subtracted (see [8]):

$$\int_{p_y=l_y}^{u_y} \int_{p_x=l_x}^{u_x} ((p_x - v_x)(p_y - v_y) - S_{minus}) \, \phi_x(p_x) \, \phi_y(p_y) \, dp_x \, dp_y$$
$$= \int_{p_y=l_y}^{u_y} \int_{p_x=l_x}^{u_x} (p_x - v_x)(p_y - v_y) \, \phi_x(p_x) \, \phi_y(p_y) \, dp_x \, dp_y -$$
$$\int_{p_y=l_y}^{u_y} \int_{p_x=l_x}^{u_x} S_{minus} \, \phi_x(p_x) \, \phi_y(p_y) \, dp_x \, dp_y$$
$$= (\psi(v_x, l_x, \mu_x, \sigma_x) - \psi(v_x, u_x, \mu_x, \sigma_x)) \cdot (\psi(v_y, l_y, \mu_y, \sigma_y) - \psi(v_y, u_y, \mu_y, \sigma_y))$$
$$- S_{minus} \cdot (\Phi_x(u_x) - \Phi_x(l_x)) \cdot (\Phi_y(u_y) - \Phi_y(l_y)).$$

The last step is by construction of ψ (Section 2) and Fubini's Theorem [13]. It can be seen that the formula is of the form $c_1 - S_{minus} \cdot c_2$, where c_1 and c_2 are calculations which are performed in constant time with respect to n for a single cell. The correction term S_{minus} is equal to the hypervolume contribution of $S \subseteq P$, where S consists of those points dominated by or equal to the lower corner of the cell (Figure 2, (left)). Calculating the dominated hypervolume of a set in the two-dimensional plane has a time complexity of $O(n \log n)$. This complexity results from needing to find the neighbors of each point in order to calculate the rectangular area it adds to the 2-D hypervolume. Sorting the set has a time complexity of $O(n \log n)$, after which the dominated hypervolume calculation itself is done in $O(n)$ by iterating over each point and performing an $O(1)$ calculation using the points that come before and after it in the sorting order. When calculating S_{minus}, the points for which the dominated hypervolume is to be calculated come from P, which was already sorted. This brings the complexity of this step down to $O(n)$, but it can be brought down to $O(1)$ when the order of calculations is chosen carefully (Figure 2, (right)), giving the algorithm a total complexity of $O(n^2)$. The ordering of the cells makes sure that for any step at most just one point is added or deleted at the boundary of S as compared to the set S for a neighboring cell for which S_{minus} was previously computed (or, initially, the empty set). This neighbor can be found in constant time. Adding or deleting the additional area at the boundary also takes only constant time. See [19] for additional examples and further details. A lower bound for the time complexity of computing the EHVI is $\Omega(n \log n)$. This can be shown by a linear time reduction of 2-D hypervolume computation (with time complexity $\Omega(n \log n)$ [9]) to EHVI: Set the values of σ_d to zero, and μ_d to the maximal coordinates of each dimension $d = 1, \ldots, m$ found in P. Let E_{max} denote the resulting $EHVI$. Then $HI(P) = \prod_{i=1}^{m} (\mu_i - r_i) - E_{max}$.

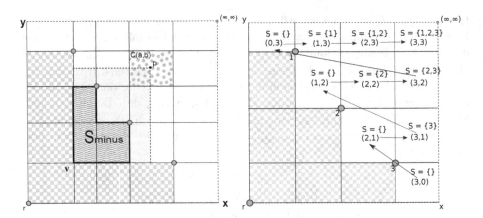

Fig. 2. Definition of S_{minus} (left). Example showing the order of iterations which allows the hypervolume contribution of S to be updated in constant time (right).

4   Algorithm for Three and More Objective Functions

The algorithm given in [8] for exactly calculating the expected hypervolume improvement is not correct[2] when the number of dimensions is higher than two. This is because the shape of the hypervolume improvement becomes more complex when the number of dimensions increases. We will derive a new formula by first decomposing the calculation into parts with less complex shapes, and then simplifying the resulting formula for the sake of more convenient calculation.

In higher dimensions, the search space can be divided into cells the same way it is done in two dimensions, except instead of the boundaries being given by lines going through the points in P and the reference point r, now the cells are separated from each other by $(m-1)$-dimensional hyperplanes (where m is the number of objective functions). Each cell is denoted by $\mathcal{C}(a_1, a_2, \ldots, a_m)$ where a_1 through a_m are integers from 0 to $|P|$ denoting the labeling of the cell. Then the left lower corner $l \in \mathbb{R}^m$ and right upper corner $u \in \mathbb{R}^m$ of the cell with label $a_1, \ldots, a_m$ are defined as follows: Let $P' = \{r\} \cup P \cup (\infty, \ldots, \infty)^T$ and let $s_d[0], \ldots, s_d[|P| + 1]$ denote the d-th components of the vectors in P' sorted in ascending order. Note that a_d refers to the a_d-th point of P' in case P' is sorted according to the d-th coordinate and thus $l_d = s_d[a_d]$ and $u_d = s_d[a_d + 1]$ for $d = 1, \ldots, m$. In other words, corners of this cell complex are given as the intersection points of all axis-parallel $m-1$ dimensional hyperplanes through points in P'.

The hypervolume improvement of a new point p with respect to the current Pareto front approximation P is given by the function $\lambda\,(A \setminus \mathrm{DomSet}(P))$, where A is the dominated subspace covered by p and cut from below by the reference point. This is the same as calculating $\lambda(A) - \lambda\,(\mathrm{DomSet}(P) \cap A)$. We will denote the set of dimension indexes by $D = \{1, 2, \ldots, m\}$. As exhibited in Figure 3,

---

[2] The formula presented in [8] wrongly omitted a correction term for $m \geq 3$.

one can decompose the calculation of the hypervolume improvement of a point $p \in \mathcal{C}(a_1, a_2, \ldots, a_m)$ as follows:

$$I_H(p) = \sum_{C \subseteq D} I_C, \text{ where } I_C := \lambda(A_C) - \lambda(\text{DomSet}(P) \cap A_C)$$

and the A_Cs are given by:

$$A_C := \left[\begin{pmatrix} v_1 \\ v_2 \\ \vdots \\ v_m \end{pmatrix}, \begin{pmatrix} w_1 \\ w_2 \\ \vdots \\ w_m \end{pmatrix} \right], v_d = \begin{cases} l_d & \text{if } d \in C \\ r_d & \text{if } d \notin C \end{cases}, w_d = \begin{cases} p_d & \text{if } d \in C \\ l_d & \text{if } d \notin C \end{cases}$$

λ denotes the Lebesgue measure in $\mathbb{R}^m$. Note that it can happen that the dimension of A_C is strictly less than m. In this case $\lambda(A_C) = 0$. We can make a similar remark about $\lambda(\text{DomSet}(P) \cap A_C)$. For the Lebesgue measure for $\mathbb{R}^s$ with $s < m$ we will use the symbol with subindex s, i.e. λ_s.

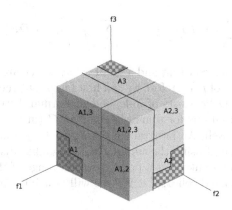

Fig. 3. An example showing how the quantities A_C for $C \subseteq \{1, 2, 3\}$ are defined in a three-dimensional objective space. $A_\emptyset$ is hidden within the rectangular volume. The checkered volumes are dominated by the points in the Pareto front approximation.

The values of r_d and l_d are constant for all points that fall within a given interval box (cell): r is the reference point and is, of course, always constant, while l represents the position of the lower corner of the cell. From this, it follows that I_C represents the portion of the hypervolume improvement which is constant with regards to the values of $p_d, d \notin C$, and which is variable with regards to the values of $p_d, d \in C$. In fact, it is *linearly related* to these values. This is a direct consequence of the way the cell boundaries are defined.

Let Sec_C denote the cross-section of $\text{DomSet}(P) \cap A_C$ which goes through p. This cross-section is defined by a projection to the dimensions not in C. If C consists of k dimensions, the slice will be $(m - k)$-dimensional. The projection of DomSet(P) uses only those points in P for which the function values in the dimensions given by C are larger than the corresponding function values of p. We shall call this selection P_{sel}. No points in P can fall between cell boundaries in any dimension, so the composition of P_{sel} must be the same for all points within a cell. The projection of A_C to the dimensions not in C is constant for all points within a cell as well, because the coordinates defining A_C are independent of p in all dimensions not in C. $\lambda_{m-k}(\text{Sec}_C)$ is constant as a result. Because A_C does not span across cell boundaries in the dimensions in C, $\lambda(\text{DomSet}(P) \cap A_C)$ is equal to the hypervolume of Sec_C multiplied by the length of A_C in each dimension in C, and those lengths are given by $(p_d - l_d)$ with $d \in C$. There is one quantity I_C for which $C = D$. This quantity I_D is special because it is linearly related to all values of p. I_D falls entirely within the cell, and as such, instead of projecting P onto a zero-dimensional space, it can simply be said that $\lambda(\text{DomSet}(P) \cap A_D) = \lambda(A_D)$ if the cell is not dominated, and $\lambda(A_D \cap \text{DomSet}(P)) = 0$ if it is. Therefore, $I_D = \lambda(A_D)$ for non-dominated cells.

By decomposing the calculation of the hypervolume improvement, we can use the sum rule to decompose the calculation of a cell's contribution to the EHVI as well.

$$\int_{p=l}^{u} \sum_{C \subseteq D} I_C \cdot PDF(p)\mathrm{d}p = \sum_{C \subseteq D} \int_{p=l}^{u} I_C \cdot PDF(p)\mathrm{d}p.$$

I_C is the product of a constant and a set of values which are linearly related to exactly one coordinate of p, therefore we can first factor out the calculation of the constant part. The PDF consists of independent normal distributions, allowing the probability distributions for dimensions not in C (in which I_C is constant) to be factured out as well. An integral consisting solely of a normal distribution can be exactly calculated using the cumulative probability distribution function Φ to calculate the probability that a point is within range of the cell. For a fixed subset C of D and a fixed cell $[l, u] \subset \mathbb{R}^m$ we will now integrate:

$$\int_{p=l}^{u} I_C \cdot PDF(p)\mathrm{d}p.$$

As stated earlier I_C (which depends on p) is,

$$I_C = \lambda(A_C) - \lambda(\text{DomSet}(P) \cap A_C) =$$

$$\prod_{d \in C}(p_d - l_d) \cdot ((\prod_{d \in D \setminus C}(l_d - r_d)) - \lambda_{|D|-|C|}(\text{Sec}_C)).$$

So the needed integral is equal to:

$$\int_{p=l}^{u}((\prod_{d \in D \setminus C}(l_d - r_d)) - \lambda_{|D|-|C|}(\text{Sec}_C)) \cdot \prod_{d \in C}(p_d - l_d) \cdot PDF(p)\mathrm{d}p,$$

where PDF is the m-dimensional density. Denoting $(\prod_{d \in D \setminus C}(l_d - r_d)) - \lambda_{|D|-|C|}$ (Sec$_C$) by I_C^{const}, we can write for the integral:

$$\int_{p=l}^{u} I_C^{const} \cdot \prod_{d \in C}(p_d - l_d) \cdot \text{PDF}(p)\mathrm{d}p.$$

Now we can unravel this as follows.

$$I_C^{const} \int_{l_1}^{u_1} \cdots \int_{l_m}^{u_m} \prod_{d \in C}(p_d - l_d)\phi_{\mu_1,\sigma_1}(p_1) \cdots \phi_{\mu_m,\sigma_m}(p_m)\mathrm{d}p_1 \cdots \mathrm{d}p_m.$$

This in turn is equal to (using Fubini's theorem):

$$I_C^{const} \cdot \prod_{d \in C} \int_{l_d}^{u_d}(p_d - l_d)\phi_{\mu_d,\sigma_d}(p_d)\mathrm{d}p_d \cdot \prod_{d \in D \setminus C} \int_{l_d}^{u_d} \phi_{\mu_d,\sigma_d}(p_d)\mathrm{d}p_d,$$

According to Definition 3, we obtain:

$$I_C^{const} \prod_{d \in C}(\psi(l_d, l_d, \mu_d, \sigma_d) - \psi(l_d, u_d, \mu_d, \sigma_d)) \prod_{d \in D \setminus C}(\Phi_{\mu_d,\sigma_d}(u_d) - \Phi_{\mu_d,\sigma_d}(l_d)).$$

Fubini's theorem [13] states that iterated integration, performed in any order, can be used to calculate a multiple integral under the condition that the multiple integral is absolutely convergent. The partial integrals making up the cell's contribution to the EHVI all converge to finite numbers, so we can safely use it.

4.1 Calculation of 3-D EHVI

In Section 3 we showed that calculating the 2-D expected hypervolume improvement is possible with time complexity $O(n^2)$. Although the algorithm described in that subsection made use of characteristics of a 2-D Pareto approximation set which are not present in higher dimensions, this subsection will show that there is also a way to calculate the 3-D EHVI with time complexity $O(n^3)$. In other words: the calculations necessary for computing the partial expected hypervolume improvement of each grid cell will be performed in constant time. The trade-off is that we will need $O(n^2)$ extra memory.

The only calculations which have a complexity higher than constant time are the Lebesgue measure computations for the dominated parts of cross sections. Three sets of correction terms are needed to calculate the partial expected hypervolume improvement of a cell:

- $S_\emptyset^-$, a constant correction term which requires a three-dimensional hypervolume calculation.
- S_x^-, S_y^- and S_z^-, which each require a two-dimensional hypervolume calculation. We will call the 2-D areas used in the calculation of these correction terms $xslice$, $yslice$ and $zslice$, respectively.

– S_{xy}^-, S_{xz}^- and S_{yz}^-, which requires a 'one-dimensional' hypervolume calculation.

Instead of calculating these correction terms afresh for each cell, it is possible to perform all necessary calculations in only $O(n^3)$ time total. For this we create a data structure which allows us to check whether or not a cell is dominated in $O(1)$ time. This can simply be a two-dimensional array holding the highest value of z for which the cell is dominated, which we shall call H_z. A simple way to fill this array is to iterate over all points $q \in P$ in order of ascending z value, setting the array value $H_z(a_1, a_2)$ to z for each q which dominates the lower corner of $\mathcal{C}(a_1, a_2, 0)$ (overwriting previous ones). The complexity of this operation is in $O(n^2 n + n \log n) = O(n^3)$. This only needs to be done once, so the $O(n^3)$ time complexity does not increase the total asymptotic time complexity of computing the EHVI in 3-D. Figure 4 shows an example. Besides containing information that allows constant-time evaluation of whether a cell is dominated, the value of S_{xy}^- for a cell $\mathcal{C}(a_1, a_2, a_3)$ that is not dominated is also given by $H_z(a_1, a_2)$. If we build two more height arrays H_x and H_y where we use the highest value of x and y instead of z, we can determine the results of all three of the one-dimensional hypervolume calculations in constant time during cell calculations.

Now, only the two-dimensional hypervolume calculations represented by $xslice$, $yslice$ and $zslice$, and the three-dimensional hypervolume calculation represented by $S_{\emptyset}^-$, still have a complexity greater than constant time. For notational simplicity, we have omitted their dependence on a particular cell from the notation until now, but in order to show the relations between correction terms of different cells, we will write '$S_{\emptyset}^-$ belonging to $\mathcal{C}(a_1, a_2, a_3)$' as $\mathcal{C}(a_1, a_2, a_3).S_{\emptyset}^-$. For the area of two-dimensional slices we proceed likewise.

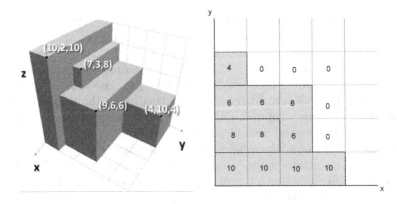

Fig. 4. Example height array H_z for a population consisting of 4 points, which is visualized on the left. Cells on the outermost edge of the integration area (which stretch out to ∞ in some dimension) are always non-dominated.

The value of $S_\emptyset^-$ is related to the values of $xslice$, $yslice$ and $zslice$ in the following way:

$$\mathcal{C}(a_1, a_2, a_3).xslice = \frac{\mathcal{C}(a_1+1, a_2, a_3).S_\emptyset^- - \mathcal{C}(a_1, a_2, a_3).S_\emptyset^-}{u_x - l_x}$$

$$\mathcal{C}(a_1, a_2, a_3).yslice = \frac{\mathcal{C}(a_1, a_2+1, a_3).S_\emptyset^- - \mathcal{C}(a_1, a_2, a_3).S_\emptyset^-}{u_y - l_y}$$

$$\mathcal{C}(a_1, a_2, a_3).zslice = \frac{\mathcal{C}(a_1, a_2, a_3+1).S_\emptyset^- - \mathcal{C}(a_1, a_2, a_3).S_\emptyset^-}{u_z - l_z}$$

With our height array H_z, we can calculate all values of $zslice$ for a given value of a_3 in $O(n^2)$ time. We can also calculate all values of $S_\emptyset^-$ for a given value of a_3 in $O(n^2)$ time, provided $a_3 = 0$ or we have both $S_\emptyset^-$ and $zslice$ for the cells where a_3 is one lower. The details of these calculations will be given below. If we go through our cells in the right order (with a_3 starting at 0, incrementing it only after we have performed the calculations for all cells with a given value of a_3), we only need to update the values of $zslice$ and $S_\emptyset^-$ n times, resulting in an algorithm for the full computation with complexity in $O(n^3)$. If we know the value of $S_\emptyset^-$ for all cells with a given value of a_3, we can use the formulas given above to calculate $xslice$ and $yslice$ in constant time whenever we need them, so we do not need to calculate these constants in advance. The details of calculating $zslice$ using the height array are as follows. We will iterate through the possible values of a_1 and a_2 in ascending order. We know that $zslice(a_1, a_2)$ is 0 if $a_1 = 0$ or $a_2 = 0$. If our height array shows that $\mathcal{C}(a_1 - 1, a_2 - 1, a_3)$ is dominated, $zslice(a_1, a_2)$ is set equal to the area of the 2-D rectangle from its lower corner to (r_x, r_y). Else, if that cell is not dominated, $zslice(a_1, a_2)$ is set equal to $zslice(a_1 - 1, a_2) + zslice(a_1, a_2 - 1) - zslice(a_1 - 1, a_2 - 1)$. The value of $zslice(a_1 - 1, a_2 - 1)$ is removed as this is the area which is overlapping, causing it to be added twice otherwise. For an example, refer to Figure 5.

5   Empirical Comparison

We compare the new algorithm (IRS) to the publicly available MATLAB implementation of the algorithm by Couckuyt et al. [6] (CDD13) and the MATLAB 2-D implementation of [8] (EDK11). A fast implementation of IRS, called IRS-fast, was made by reusing previous computations of the Gaussian error function (erf) when evaluating it for the same grid coordinate and the same candidate point. Moreover, we stop computations when the result will be zero and compute the data for a single cell for several (BatchSize) predictive distributions at once.

All implementations are made available at http://moda.liacs.nl/index.php?page=code, except [6] which was obtained from http://www.sumo.intec.ugent.be/software. Average runtimes (10 runs) for Pareto front sizes $|P| \in \{10, 100\}$ are computed. Moreover, different number of predictions (candidate points) {1 point, 1000 points} are processed as a batch. The 3-D test problems from Emmerich and Fonseca [20] are used. The data for $|P| = 100$ is visualized in Figure 6.

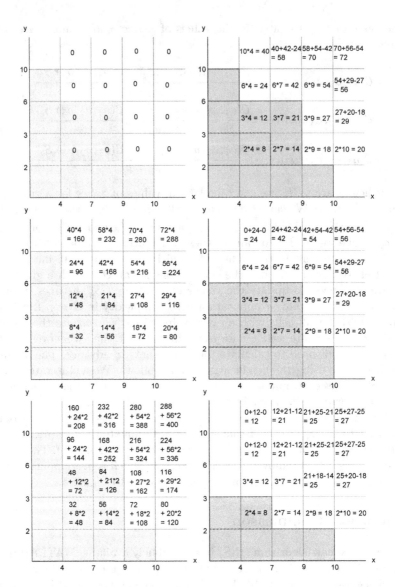

Fig. 5. Some values of *zslice* and S^- for the example shown in Figure 4, with $a_3 = 0$, 1 and 2, respectively. The x and y values of each cell's lower corner are shown on the axes. The grids with the values of S^- are on the left and the grids with the values of *zslice* are on the right.

For 2-D a uniform sample of points on a uniform sphere is created analogously and the CLIFF3D problem is omitted as there is no such problem in 2-D. The parameters in the experiments: $\sigma_d = 2.5, \mu_d = 10, d = 1, \ldots, m$ was used. All codes were validated against each other and produced the same results. For

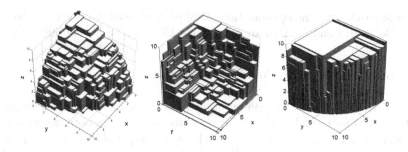

Fig. 6. Randomly generated fronts of type CONVEXSPHERICAL, CONCAVESPHERICAL, and CLIFF3D from [20] with $n = 100$ (above, left to right)

Table 1. Empirical comparison of strategies for 3-D EHVI

Problem	Type	$|P|$	Batch Size	Time_average(s) EDK11	IRS	CDD13	IRSfast
	CONVEX2D	10	1	0.0056	0.0004	0.0097	0.0001
	CONVEX2D	10	1000	5.4336	0.1131	0.0206	0.0071
	CONVEX2D	100	1	0.6438	0.0030	0.0601	0.0012
2D	CONVEX2D	100	1000	901.4221	2.1934	0.1720	0.1052
	CONCAVE2D	10	1	0.0281	0.0003	0.0103	0.0001
	CONCAVE2D	10	1000	7.0726	0.1155	0.0192	0.0073
	CONCAVE2D	100	1	0.8896	0.0020	0.0616	0.0012
	CONCAVE2D	100	1000	920.2566	2.1264	0.1722	0.1053
	CONVEX3D	10	1	N.A.	0.0018	0.0747	0.0010
	CONVEX3D	10	1000	N.A.	1.6187	0.6501	0.1461
	CONVEX3D	100	1	N.A.	0.7813	9.6047	0.3388
	CONVEX3D	100	1000	N.A.	787.1110	432.5856	12.1998
	CONCAVE3D	10	1	N.A.	0.0016	0.2213	0.0011
3D	CONCAVE3D	10	1000	N.A.	1.5791	0.6562	0.1434
	CONCAVE3D	100	1	N.A.	0.7143	7.1798	0.2653
	CONCAVE3D	100	1000	N.A.	726.5219	281.4245	9.8314
	CLIFF3D	10	1	N.A.	0.0016	0.2357	0.0014
	CLIFF3D	10	1000	N.A.	1.5598	0.7121	0.1477
	CLIFF3D	100	1	N.A.	0.7212	7.5086	0.3525
	CLIFF3D	100	1000	N.A.	734.1832	692.7653	9.7818

validation by Monte Carlo integration, see also [19]. The hardware was CPU: Intel(R) Core(TM) i7-4800MQ 2.7GHz RAM: 32GB, 1600MHz. The software used is: Microsoft Windows 7, 64 bit, MATLAB: 8.2.0.701 (R2013b), except IRS code MinGW: MinGW_x86_64 4.9.1, gcc 4.9.1., compiler flag -Ofast. The results are displayed in Table 1.

The results show that the new code IRSfast is consistently faster in 2-D and 3-D for all test problems. The difference is particularly high for small batch size.

For large batch sizes the optimizations IRSfast were needed to achieve the best result. Both, IRS and CDD13 are faster than EDK11, the first implementation.

6   Conclusions and Future Research

In this paper we discussed the computation of the expected hypervolume improvement in $O(n^2)$ for the 2-D case and $O(n^3)$ in the 3-D case. In both cases the time complexity was improved by a factor of $n \log n$ as compared to [8] by accomplishing $O(1)$ computations per grid cell. A significant speed up can be achieved. A significant improvement of the empirical runtime performance was achieved by reusing computations of nonlinear functions on grid coordinates. The memory complexity of the proposed algorithms is $O(n^2)$. Moreover, it has been shown that the memory efficiency is $O(n^2)$ and the computational time complexity is lower bounded by $\Omega(n \log n)$ for $m \geq 2$. Empirical comparisons on randomly generated Pareto fronts of different shape show that the new algorithm is by a factor of 5 to 200 faster than previously existing implementations.

It seems possible that the same ideas that have been used for the construction of 2-D and 3-D algorithms could be used for construction of efficient computation schemes for higher dimensions. So far, however, the algorithms by Couckuyt et al. [6] remain the fastest available algorithms for $m \geq 4$.

Acknowledgments. Kaifeng Yang acknowledges financial support from China Scholarship Council (CSC), CSC No.201306370037. The authors also acknowledge the feedback of Tobias Wagner on an early version of the algorithm. Michael Emmerich gratefully acknowledges support by the Dutch STW foundation for the project "Excellent Buildings via Forefront MDO".

References

1. Hupkens, I.: Complexity Reduction and Validation of Computing the Expected Hypervolume Improvement, Master's Thesis, published as LIACS, Internal Report Nr. 2013–12, August 2013. http://www.liacs.nl/assets/Masterscripties/2013-12IHupkens.pdf
2. Shir, O.M., Emmerich, M., Bäck, T., Vrakking, M.J.: The application of evolutionary multi-criteria optimization to dynamic molecular alignment. In: Proc. of IEEE CEC 2007, pp. 4108–4115 (2007)
3. Zaefferer, M., Bartz-Beielstein, T., Naujoks, B., Wagner, T., Emmerich, M.: A case study on multi-criteria optimization of an event detection software under limited budgets. In: Purshouse, R.C., Fleming, P.J., Fonseca, C.M., Greco, S., Shaw, J. (eds.) EMO 2013. LNCS, vol. 7811, pp. 756–770. Springer, Heidelberg (2013)
4. Shimoyama, K., Sato, K., Jeong, S., Obayashi, S.: Comparison of the criteria for updating Kriging response surface models in multi-objective optimization. In: Proc. of IEEE CEC 2012, pp. 1–8 (2012)
5. Shimoyama, K., Jeong, S., Obayashi, S.: Kriging-surrogate-based optimization considering expected hypervolume improvement in non-constrained many-objective test problems. In: Proc. of IEEE CEC 2013, pp. 658–665 (2013)

6. Couckuyt, I., Dirk, D., Tom, D.: Fast calculation of multiobjective probability of improvement and expected improvement criteria for Pareto optimization. Journal of Global Optimization **60**, 575–594 (2013)
7. Wagner, T., Emmerich, M., Deutz, A., Ponweiser, W.: On expected-improvement criteria for model-based multi-objective optimization. In: Schaefer, R., Cotta, C., Kołodziej, J., Rudolph, G. (eds.) PPSN XI. LNCS, vol. 6238, pp. 718–727. Springer, Heidelberg (2010)
8. Emmerich, M., Deutz, A.H., Klinkenberg, J.W: Hypervolume-based expected improvement: Monotonicity properties and exact computation. In: Proc. of IEEE CEC 2011, pp. 2147–2154 (2011)
9. Beume, N., Fonseca, C.M., Lo'pez-Iba'ez, M., Paquete, L.: On the complexity of computing the hypervolume indicator. IEEE Transactions on Evolutionary Computation **13**(5), 1075–1082 (2009)
10. Mockus, J., Tiesis, V., Zilinskas, A.: The application of Bayesian methods for seeking the extremum. In: Dixon, L., Szego, G. (eds.) Towards Global Optimization (1978). vol. 2, pp. 117–129 (1978)
11. Donald, R.J., Matthias, S., William, J.W.: Efficient Global Optimization of Expensive Black-Box Functions. Journal of Global Optimization **13**(4), 455–492 (1998)
12. Emmanuel, V., Julien, B.: Convergence properties of the expected improvement algorithm with fixed mean and covariance functions. Journal of Statistical Planning and Inference **140**, 3088–3095 (2010)
13. Fubini, G.: Sugli integrali multipli. Opere scelte, vol. 2. Cremonese, pp. 243–249
14. Emmerich, M. (2005). Single-and multi-objective evolutionary design optimization assisted by gaussian random field metamodels. Dissertation, TU Dortmund, Informatik, Eldorado. http://hdl.handle.net/2003/21807
15. Zitzler, E., Thiele, L.: Multiobjective optimization using evolutionary algorithms - A comparative case study. In: Eiben, A.E., Bäck, T., Schoenauer, M., Schwefel, H.-P. (eds.) PPSN 1998. LNCS, vol. 1498, pp. 292–301. Springer, Heidelberg (1998)
16. Laniewski-Wollk, P., Obayashi S., Jeong, S.: Development of expected improvement for multi-objective problems, In: Proceedings of 42nd Fluid Dynamics Conference/Aerospace Numerical, Simulation Symposium (CD ROM), June 2010
17. Koch, P., Wagner T., Emmerich M., Bäck Th., Konen W.: Efficient Multi-criteria Optimization on Noisy Machine Learning Problems, Applied Soft Computing, (in print) (2014)
18. Luo, C., Shimoyama, K., Obayashi, S.: Kriging model based many-objective optimization with efficient calculation of expected hypervolume improvement. In: Proc. of IEEE CEC 2014, pp. 1187–1194 (2014)
19. Hupkens, I., Emmerich, M. and Deutz, A.: Faster Computation of Expected Hypervolume Improvement. arXiv preprint arXiv: 1408.7114 (2014)
20. Emmerich, M.T.M., Fonseca, C.M.: Computing hypervolume contributions in low dimensions: Asymptotically optimal algorithm and complexity results. In: Takahashi, R.H.C., Deb, K., Wanner, E.F., Greco, S. (eds.) EMO 2011. LNCS, vol. 6576, pp. 121–135. Springer, Heidelberg (2011)

A GPU-Based Algorithm for a Faster Hypervolume Contribution Computation

Edgar Manoatl Lopez, Luis Miguel Antonio, and Carlos A. Coello Coello[✉]

Departamento de Computación, CINVESTAV-IPN
(Evolutionary Computation Group), 07300 Mexico D.F., Mexico
{emanoatl,lmiguel}@computacion.cs.cinvestav.mx,
ccoello@cs.cinvestav.mx

Abstract. The hypervolume has become very popular in current multi-objective optimization research. Because of its highly desirable features, it has been used not only as a quality measure for comparing final results of multi-objective evolutionary algorithms (MOEAs), but also as a selection operator (it is, for example, very suitable for *many-objective optimization problems*). However, it has one serious drawback: computing the exact hypervolume is highly costly. The best known algorithms to compute the hypervolume are polynomial in the number of points, but their cost grows exponentially with the number of objectives. This paper proposes a novel approach which, through the use of Graphics Processing Units (GPUs), computes in a faster way the hypervolume contribution of a point. We develop a highly parallel implementation of our approach and demonstrate its performance when using it within the *S-Metric Selection Evolutionary Multi-Objective Algorithm* (SMS-EMOA). Our results indicate that our proposed approach is able to achieve a significant speed up (of up to 883x) with respect to its sequential counterpart, which allows us to use SMS-EMOA with exact hypervolume calculations, in problems having up to 9 objective functions.

1 Introduction

Several recent studies have shown that Pareto-based multi-objective evolutionary algorithms (MOEAs) do not perform properly when dealing with problems having more than three objectives (the so-called *many-objective optimization problems*) [12]. This has motivated the development of new selection schemes from which the use of quality assessment indicators is one of the most promising choices. The idea when using this sort of scheme is to maximize a quality assessment indicator that provides a good ordering among sets that represent Pareto approximations. From the many indicators currently available, the hypervolume has been the most popular choice, mainly because it is the only unary quality indicator that is known to be Pareto compliant [20]. The nice mathematical properties of the hypervolume has motivated the development of several

The third author gratefully acknowledges support from a Cátedra Marcos Moshinsky 2014 and from CONACyT project no. 221551.

A. Gaspar-Cunha et al. (Eds.): EMO 2015, Part II, LNCS 9019, pp. 80–94, 2015.
DOI: 10.1007/978-3-319-15892-1_6

hypervolume-based MOEAs (see for example [2, 19]). However, these approaches have a very high computational cost which normally becomes unaffordable for problems having five or more objectives. Although there exist proposals to estimate the hypervolume contribution using sampling (see for example [7]), these approaches are known to have a poor performance with respect to those that use exact hypervolume calculations [13]. Here, we propose a parallel approach, which is implemented in graphics processing units (GPUs), and is coupled to a hypervolume-based MOEA: the *S-Metric Selection Evolutionary Multi-Objective Algorithm* (SMS-EMOA).

The remainder of this paper is organized as follows. Section 2 provides an introduction to the hypervolume, including a short review of the main algorithms that have been proposed to compute it. Our proposed approach is described in Section 3. The experimental results are presented in Section 4, including the methodology and a short discussion of our main findings. Finally, conclusions and some possible paths for future research are provided in Section 5.

2   About the Hypervolume

The hypervolume indicator has become widely used in recent years [18]. This indicator encapsulates in a single unary value a measure of the spread of the solutions along the Pareto front, as well as the distance of the approximation set from the true Pareto optimal front. Whenever one approximation completely dominates another approximation, the hypervolume of the former will be greater than the hypervolume of the latter. Also, the hypervolume is maximized if, and only if, the set of solutions contains all Pareto optimal points. The hypervolume is defined as the n-dimensional space that is contained by an n-dimensional set of points. When applied to multi-objective optimization, the n-dimensional objective values for solutions are treated as points for the computation of such space. That is, the hypervolume is obtained by computing the volume (in objective function space) of the non-dominated set of solutions Q that minimize a MOP. For every solution $i \in Q$, a hypercube v_i is generated with a reference point W and the solution i as its diagonal corner of the hypercube.

$$\mathcal{S} = Vol \left(\bigcup_{i=1}^{|Q|} v_i \right) \tag{1}$$

The hypervolume has important advantages over other set measures [18]:

- It is sensitive to any type of improvements, i.e., whenever an approximation set A dominates another approximation set B, then the hypervolume has a strictly better quality value for the former than for the latter set.
- As a result from the first property, the hypervolume measure guarantees that any approximation set A that achieves the maximally possible quality value for a particular problem contains all Pareto-optimal objective vectors.

– The ranking of the solutions that it generates is invariant to the linear scaling of the objective functions.

In spite of its nice features, the use of the hypervolume is limited by its high computational cost. Hypervolume computation has been proven to be #P-hard (analogous to NP-hard for counting problems) in the number of objectives [3]. As a result, hypervolume algorithms have been used primarily for performance assessment.

Many algorithms have been created to compute hypervolume, each of which has a different worst-case complexity. Next, we introduce the main algorithms that have been proposed for this sake, and we briefly discuss their time complexities.

2.1   Inclusion-Exclusion Algorithm

The Inclusion-Exclusion hypervolume algorithm [17] is perhaps the easiest method for calculating the hypervolume. It works by the inclusion-exclusion principle in the following way: the algorithm adds volumes of rectangular polytopes (n-dimensional rectangular volumes) dominated by each point individually, then subtracts the volumes dominated by intersections of pairs of points; after that, it adds back in volumes dominated by intersections of three points, and so on. Unfortunately, while simple, this method has a time complexity of $\mathcal{O}(n2^m)$ that makes it infeasible on all but the smallest sets.

2.2   LebMeasure Hypervolume Algorithm

The LebMeasure algorithm was proposed by Fleischer [8]. He realized that for any space covered by a set of non-dominated points, one can always identify a rectangular polytope that does not intersect with any other region, so that this region can be lopped off. Then, the hypervolume contributions of these lopped off regions can be easily computed. The hypervolume of the space dominated by these polytopes is then added to a hypervolume accumulator and new points are spawned to reflect the removal of such region. This process can then be repeated until the remaining polytopes no longer dominate any region of space. LebMeasure was initially thought to have polynomial time performance, however it was later demonstrated empirically to exhibit exponential time complexity in the number of objectives and later was proved that the lower bound for LebMeasure's worst case complexity is $\mathcal{O}(2^{n-1})$. Thus, it is also exponential in the number of objectives [15].

2.3   HSO Hypervolume Algorithm

Another hypervolume calculation algorithm is HSO (Hypervolume by Slicing Objectives) [16]. It manages a front by processing one objective at a time and slicing it along the chosen objective. This is known as a *dimension-sweep* algorithm. HSO is given with a front that is pre-sorted with respect to the first objective.

Point values in this objective are used to create cross-sectional slices along this objective. When sweeping along an objective, each point in the list is visited in turn. A list of points is maintained which is sorted in the $(n-1)^{th}$-objective, containing points that have been processed so far, i.e., the points contributing to the current slice. At each slice, there is an $n-1$-objective hypervolume, its hypervolume is calculated recursively and multiplied by the depth of the slice, i.e., the difference between the current point value and the next point value. The point is then added to the $n-1$-objective slice, after removing any points that it dominates. This process is repeated until every point in the list has been visited. While et al. [16] proved that HSO's computational cost is exponential in the number of objectives with a lower bound of $\mathcal{O}(m^{n-1})$, so it still performs poorly for data sets with high dimensionality.

2.4   The FPL Hypervolume Algorithm

The FPL hypervolume algorithm [10] is another dimension-sweep algorithm which improves upon HSO. It adds a new linked data structure which reduces the work required to maintain the fronts built iteratively by HSO. Dominated points must be retained, as points must be reinserted in the reverse order of their deletion. Therefore, dominated points are marked instead of deleted and are skipped over in lower objectives. This data structure improves performance by minimizing the number of comparisons necessary to maintain the sorting within the $n-1$-dimensional slices. It reuses previous calculations when a smaller dimensional slice has already been calculated. Also, hypervolumes are stored along with the current coordinate in the current objective. As these values become staled, bound values which keep track of reusable hypervolumes are updated whenever points are deleted or reinserted. The worst-case complexity of FPL is $\mathcal{O}(m^{n-2}\log m)$. Although all previously described exact hypervolume algorithms and recent ones have led to improved feasibility or better worst-case time complexities, hypervolume calculation remains $\#P$-hard and exponential in the number of objectives [3].

2.5   Hypervolume within MOEAs

The most common way to use the hypervolume as a selection method in MOEAs is through the measure of how much an individual contributes to the hypervolume value of the whole set it belongs to. Then, the solutions that contribute the least to the hypervolume of a front are discarded. The contributing hypervolume of an individual a which belongs to a population P can then be stated in the following way:

$$C_a = \mathcal{S}(\mathcal{P}, y_{ref}) - \mathcal{S}(\mathcal{P} \backslash \{a\}, y_{ref}) \tag{2}$$

Nowadays, there exist several MOEAs that incorporate the hypervolume in their selection mechanism [2,19]. However, these approaches have a high computational overload and this creates the necessity to develop alternative strategies to deal with this problem. Because of this, approximation approaches have also been proposed (e.g., [7] which uses Monte Carlo sampling to approximate expensive hypervolume calculations). Another example is an approach by Bringmann and Friedrich [3] that has a polynomially bounded error and shows promise. These types of methods are faster when the samples are small. However these approaches do not guarantee a bound on the error and most of them deteriorate their behavior as the number of objectives of the problem increases. In fact, in some cases the number of samples needed to produce a good approximation is too large, turning these approaches impractical for many-objective optimization.

3   Proposed Approach

The main idea of our proposed approach is to use all the available hardware resources to calculate the exact contributing hypervolume in a more efficient way, in order to alleviate the high computational overload that current hypervolume-based MOEAs present. Since the computation of the exact hypervolume involves a high computational complexity, in this work we try to circumvent this problem by developing a faster way of computing the exact contributing hypervolume, instead of the hypervolume itself. We propose a way of saving unnecessary hypervolume points computations and a model that is efficient and highly parallelizable. For the descriptions provided next, we assume that we are dealing with sets of non-dominated solutions. Next, we describe our proposed approach, which is implemented in CUDA-C.[1] For the bi-objective case, we take the points of the non-dominated front and sort them in ascending order according to the values of the first objective function f_1. We get then, at the same time, a sequence that is additionally sorted in descending order concerning the f_2 values, because the points are mutually non-dominated. Given a sorted front $SF = \{p_1, \ldots, p_{|SF|}\}$ the contributing hypervolume of a point $C_{\mathcal{P}_i}$ is given by:

$$C_{\mathcal{P}_i} = \Delta f_1 * \Delta f_2 = (p_{i+1,1} - p_{i,1}) * (p_{i,2} - p_{i-1,2}) \tag{3}$$

The graphical representation of this computation can be seen in Figure 2. The parallelization of this two-dimensional approach is done in the following way: using the SIMD[2] model, we work in each thread of the GPU with the computation of the contribution of a point. First, a sorting procedure is performed. For this purpose, we used the so called *bitonic sort* [1], which is a parallel sorting algorithm, originally created for sorting networks. Once the non-dominated front is sorted in ascending order, according to the values of the first objective,

---

[1] The GPU platform and API developed by Nvidia called CUDA [14] (Computer Unified Device Architecture), which is the one adopted in this work, is based on the CUDA-C language, which is an extension to C that allows development of GPU routines called *kernels*. Each kernel defines instructions that are executed on the GPU by many threads at the same time.

[2] SIMD (Single Instruction Multiple Data) is a computer architecture which can handle only one instruction but applies it to many data streams simultaneously [9].

the ith thread computes the C_{p_i}, so that a set of threads $\mathcal{TH}$ can obtain $C_\mathcal{P}$. We consider the case where there might exist more non-dominated points in a set than threads in the GPU and the assignment of the number of threads created in the GPU is done with that in mind. This procedure is presented in Algorithm 1 which shows the architecture of the kernel used for the GPU implementation. The communication between host (CPU) and device (GPU) is done in a synchronous way, since we first need to send the whole set of non-dominated solutions to the device's global memory in order to compute the whole set contributions $C_\mathcal{P}$; once the $C_\mathcal{P}$ set is ready, it is sent back to the host.

Input: A non-dominated set $\mathcal{P}$ with $\|\mathcal{P}\| = k$, where $\mathcal{P}_i = (p_{i,1}, p_{i,2})$ and a reference point $\mathcal{R} = (r_1, r_2)$
Output: A hypervolumen contribution set $C_\mathcal{P}$

Assign an Id for each thread ;
Assign the number of threads created in the GPU to $Dimblock$;
$C_\mathcal{P} \leftarrow 0$;
if $Id = 0$ **then**
  | Add the reference point $\mathcal{R}$ to $\mathcal{P}$;
end
$k \leftarrow k + 1$;
Sort in ascending order the set $\mathcal{P}$ in the first objective. $i \leftarrow Id + 1$;
while $i < k$ **do**
  | $C_{\mathcal{P}_i} \leftarrow (p_{i+1,1} - p_{i,1}) * (p_{i,2} - p_{i-1,2})$;
  | $i \leftarrow i + Dimblock$;
end
return $C_\mathcal{P}$;

Algorithm 1. Computation of the hypervolumen contribution set $\mathcal{P}$ for two dimensions in a GPU.

For the case of problems with three or more objectives we propose a model in which we try to save unnecessary hypervolume points computations by finding which points in the non-dominated front are not needed for the computation of the contributing hypervolume of an individual. Having a set of non-dominated solutions of μ individuals, the contributing hypervolume of each of the individuals in the whole set can be expressed in the following way:

$$\forall \mathcal{P}_i \in \mathcal{P}, C_{\mathcal{P}_i} = \mathcal{S}(\mathcal{P}, y_{ref}) - \mathcal{S}(\mathcal{P} \backslash \{\mathcal{P}_i\}, y_{ref}) \tag{4}$$

This means that we will need to compute the hypervolume $\mu + 1$ times. Here, we try to discard the points that are unnecessary for computing the hypervolume contribution of a point, in order to compute volumes of subsets with less dimensions, thus reducing the cost. So, instead of computing each hypervolume contribution $C_{\mathcal{P}_i}$ with the aforementioned formulation 4, we are able to calculate this contribution $C_{\mathcal{P}_i}$ in the following way:

$$C_{\mathcal{P}_i} = \prod_{k=0}^{n} (|\mathcal{P}_i[k] - y_{ref}[k]|) - Vol \left(\bigcup_{y \in \mathcal{P}_i'} \{y' | y \prec y' \prec y_{ref}\} \right) \tag{5}$$

where $\mathcal{P}'_i$ is a set of points shifted to be delimited by the non-dominated point $\mathcal{P}_i$. So, we can reformulate the contributing hypervolume as the volume delimited by $\mathcal{P}_i$ and a reference point y_{ref}, which we will call $Vol(\mathcal{P}_i)$, less the volume of the set $\mathcal{P}'_i$, which is a subset of $Vol(\mathcal{P}_i)$, that we will call $Vol(\mathcal{P}'_i)$, also with the same reference point. This idea is presented in Figure 1, where we show a graphical representation of the way in which the hypervolume contribution of a point is computed. The $\mathcal{P}'_i$ set contains the points that we want to find for each non-dominated solution $\mathcal{P}_i$ of a set $\mathcal{P}$, so that we can compute the hypervolume contribution $\mathcal{C}_{\mathcal{P}_i}$ of a point P_i as: $\mathcal{C}_{\mathcal{P}_i} = Vol(\mathcal{P}_i) - Vol(\mathcal{P}'_i)$.

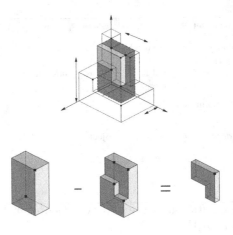

Fig. 1. Computation of the hypervolume contribution of a point b which belongs to a set $\mathcal{P} = \{a, b, c, d, e\}$ of non-dominated solutions with the use of a set of points $\mathcal{P}' = \{a', c', d', e'\}$ delimited by $\mathcal{P}_i$

This is possible since it is always the case that $Vol(\mathcal{P}'_i) \leq Vol(\mathcal{P}_i)$, because $\mathcal{P}'_i \subset \mathcal{P}$. So, what we want is to cut the volume given by the set of non-dominated solutions $\mathcal{P}$ and the reference point y_{ref}, in order to compute the volume $Vol(\mathcal{P}'_i)$ of reduced sets of points $\mathcal{P}'_i, i = 1, 2, \ldots, |P|$. This is less costly than the way the original aforementioned approach works, for obtaining each individual's contribution. This new $\mathcal{P}'$ set of points, created from a specific non-dominated solution $\mathcal{P}_i$ of a set, might have dominated solutions, so it is necessary to find only the non-dominated solutions of this new set before performing the hypervolume computation of this. The kernel procedure for the computation of the shifted set of points $\mathcal{P}_i'$, delimited by a point $\mathcal{P}_i$ of the non-dominated set, is shown in Algorithm 2.

Input: A non-dominated set $\mathcal{P} \subseteq \mathbb{R}^d$ with $\|\mathcal{P}\| = k$, where $\mathcal{P}_i = (p_{i,1}, \cdots, p_{i,k})$ and the current index is i for the point $\mathcal{P}_i$
Output: A set $\mathcal{P}'$

Assign an Id for each thread ;
Assign the number of threads created in the GPU to $Dimblock$;
$Size \leftarrow d * k$;
$i \leftarrow I$;
// Restrict all points which are in the box delimited by $\mathcal{P}_i$
while $i < Size$ **do**
     $l \leftarrow i/d$;
     $m \leftarrow i\%d$;
     if $\mathcal{P}[I][m] > \mathcal{P}[l][m]$ **then**
       | $\mathcal{P}'[l][m] \leftarrow \mathcal{P}[l][m]$;
     end
     $i \leftarrow i + Dimblock$;
end
// Filter out all points which are covered by other points
$i \leftarrow Id$;
for $i < k$ **do**
     if $\exists \mathcal{P}'_j \in \mathcal{P}' | \mathcal{P}'_j \not\preceq \mathcal{P}'_i$ **then**
       | Remove $\mathcal{P}'_i$ of $\mathcal{P}'$;
     end
     $i \leftarrow i + Dimblock$;
end
Sort in ascending order the set $\mathcal{P}'$ in the first objective.
return $\mathcal{P}'$;

Algorithm 2. Computation of a set $\mathcal{P}'_i$ for a point $\mathcal{P}_i$.

So, with this idea, we can develop a parallelization of this approach. The model is implemented in two kinds of parallelism: using data-level parallelism[3] and transaction-level parallelism[4] through the use of *streams*[5]. Each *stream* in the whole procedure is responsible for the kernel execution, and for sending a set $\mathcal{P}'_i$ of reduced points solutions from the GPU to the CPU, as well as of the computation of the hypervolume contribution $\mathcal{C}_{\mathcal{P}_i}$ of a point P_i, all of which is done in a parallel way. The number of *streams* adopted depends of each GPU's architecture and on the number of available *streams*. We perform a dynamic task assignation to the *streams* so that the overhead generated is minimal. Figure 3 shows the way the assignation policy is done. The use of *streams* allows us to have an asynchronous communication between the CPU and the GPU, since while a $stream_i$ performs the $\mathcal{C}_{\mathcal{P}_2}$ computation, another $stream_j$ executes the *kernel* with its set of parameters and, at the same time, another $stream_l$ downloads data from the GPU.

The overall procedure works in the following way: First, a set of *streams* is created in the host to perform concurrent operations with the GPU. Then, the set of non-dominated solutions $\mathcal{P}$ and the reference point y_{ref} are sent to the

---

[3] In data-level parallelism, instructions from a single stream operate concurrently on several data.
[4] In a transaction-level parallelism, multiple threads/processes from different transactions can be executed concurrently.
[5] A stream is a sequence of commands, possibly issued by different host threads, that are executed in a certain order.

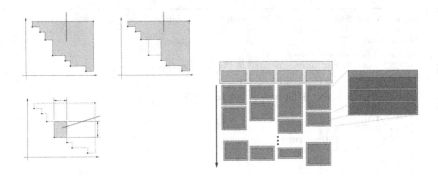

Fig. 2. Contributing hypervolume computation of a point   **Fig. 3.** Task assignation to *Streams*

GPU. Next, each kernel, executed by an specific *stream*, finds the set $\mathcal{P}'_i$ for a point $\mathcal{P}_i$. Once the set is computed, $\mathcal{P}'_i$ is sent to the host in order to compute the hypervolume contribution $C_{\mathcal{P}_i}$ of such point. Each hypervolume is computed making use of the FPL Hypervolume Algorithm, which is the fastest existent algorithm for the hypervolume computation. The procedure goes on until all the contributions of each point in the non-dominated set had been computed. The pseudocode of the whole procedure is shown in Algorithm 3.

In order to apply our approach, we adopted SMS-EMOA [2], since it fits perfectly with our parallel approach. SMS-EMOA is a steady state evolutionary algorithm in which each newly created solution is ranked and a solution is removed from the worst ranked front in order to maintain the population size. The solution that contributes the least to the hypervolume of the worst ranked front is then discarded (see [2] for details).

4 Experimental Results

For purposes of this study, we adopted MOPs from two benchmarks: the Deb-Thiele-Laumanns-Zitzler (DTLZ) test suite [6] (DTLZ2, DTLZ3 and DTLZ4) and the Walking-Fish-Group (WFG) test suite [11] (WFG1, WFG2 and WFG3). We compare our proposed approach with respect to a sequential version of SMS-EMOA which uses the FPL algorithm. In our experiments, we used from 2 to 9 objectives. Our proposed approach is called *S-Metric GPU Selection Evolutionary Multi-Objective Algorithm* (SMGPUS-EMOA).

4.1 Methodology

For our comparative study, we decided to adopt several performance measures as described next.

One of the most important actions in parallel computing is to actually measure how much faster a parallel algorithm runs with respect to the best sequential

---

Input: A non-dominated set $\mathcal{P} \subseteq \mathbb{R}^d$ with $\|\mathcal{P}\| = k$, where $\mathcal{P}_i = (p_{i,1}, \cdots, p_{i,k})$
and a reference point $\mathcal{R} = (r_1, \cdots, r_k)$
Output: The set of hypervolumen contributions $\mathcal{C}_\mathcal{P}$ of the non-dominated set $\mathcal{P}$

Create s for asynchronous managing of the data ;
Assign an Id for each stream of the GPU;
$\mathcal{C}_\mathcal{P} \leftarrow 0$;
if $IdStream = 0$ **then**

> Send the non-dominated set $\mathcal{P}$ to the GPU;
> Send the reference point $\mathcal{R}$ to the GPU;

end
Wait until all the data are sent to the GPU by the CPU;
$IdSig \leftarrow 0$;   // Where IdSig is a shared memory for streams
// Start of the transaction parallelism
for *each stream of the GPU in parallel manner* **do**

> **while** $IdSig < k$ **do**
>
>> $i \leftarrow IdSig$;
>> **if** $i < NumStreams$ **then**
>>
>>> $i \leftarrow IdStream$;
>>
>> **end**
>> // Start of the data parallelism
>> Launch the kernel $\mathcal{P}'_i\_computation <<<$
>> $NumSMs/NumStreams, SizeBlock >>> (\mathcal{P}, i, k, d)$;
>> Copy the set $\mathcal{P}'_i$ to the CPU ;
>> Calculate the $Vol(\mathcal{P}_i)$;
>> Calculate the $Vol(\mathcal{P}')$;
>> $\mathcal{C}_{\mathcal{P}_i} \leftarrow Vol(\mathcal{P}_i) - Vol(\mathcal{P}'_i)$;
>
> **end**
> // Wait until the shared resource is available
> **while** *IdStream is not increased* **do**
>
>> **if** *IdSig is available* **then**
>>
>>> $IdSig \leftarrow IdSig + 1$
>>
>> **end**
>
> **end**

end
return $\mathcal{C}_\mathcal{P}$;

Algorithm 3. Computation of the hypervolumen contribution set $\mathcal{P}$ for three or more dimensions in a GPU.

one. This measure is known as **speedup**. For a problem of size n, the expression for speedup is:

$$S_p = \frac{T_s(n, 1)}{T(n, p)} \tag{6}$$

where $T_s(n, 1)$ is the time of the best sequential algorithm and $T(n, p)$ is the time of the parallel algorithm with p processors, both solving the same problem.

In order to measure the uniformity of the solutions produced by a MOEA, we adopted **spacing** [4]. This indicator is computed using:

$$S = \sqrt{\frac{1}{|Q|} \sum_{i=1}^{|Q|} (d_i - \bar{d})^2} \tag{7}$$

where $d_i = \min_{k \in Q \wedge k \neq i} \sum_{j=1}^{m} |f_j^i - f_j^k|$ and $\bar{d}$ is the mean value of the above distance measure $\bar{d} = \sum_{i=1}^{|Q|} \frac{d_i}{|Q|}$.

Low values of this indicator reflect a good (uniform) distribution of solutions.

We also analyze its convergence rate with respect to that of the sequential version of SMS-EMOA. For this purpose, we adopted the hypervolume. The reference points used for each of the problems are shown in Table 1. The aim of this study is to identify which of the MOEAs being compared is able to reach the results in a faster way. So, we decided to run each of the MOEAs being analyzed, until reaching a maximum number of function evaluations. At that point, we applied the performance measures previously indicated.

It is worth noting that the sequential version of SMS-EMOA may be unable to achieve the desired number of function evaluations in a reasonable computational time. For that reason, we used an additional stopping criterion: if an algorithm hasn't reached the desired number of objective function evaluations after 8000 minutes, then we stop it. We performed 25 independent runs for each algorithm, problem instance and (given) number of objective functions. The number of objectives used for each problem is shown in Table 2.

Table 1. Reference points used for the hypervolume indicator

Problem	Reference points
DTLZ1	$(1, 1, 1, \ldots, 1)$
DTLZ2-4	$(2, 2, 2, \ldots, 2)$
WFG1-3	$(3, 5, 7, \ldots, 2m + 1)$

4.2 Parameterization

The parameters of each MOEA used in our study were chosen in such a way that we could do a fair comparison among them. Thus, for both approaches we used the same parameter values since they are similar in everything but the way the hypervolume contribution is computed. The distribution indexes for the SBX and polynomial-based mutation operators [5], used by SMS-EMOA were set as: $\eta_c = 20$ and $\eta_m = 20$, respectively. The crossover probability is $p_c = 0.9$ and the mutation probability is $p_m = 1/L$, where L is the number of decision variables. The internal population size and the maximum number of function evaluations for each problem is defined as indicated in Table 2. For the case of our SMGPUS-EMOA, the number of threads used within the GPU was set in the next way: for problems with two objective functions, 1024 threads were used, with just one stream in the whole process. For the case of three up to 9 objective functions, 1024 threads and seven streams were used. The main characteristics of the hardware used for the experiments are the following: An Intel Core i7-3930k CPU running at 3.20 GHz, with 8GB of RAM 1600 MHz DDR3. Our GPU was a Geforce GTX 680, and we ran our experiments in Fedora 18 (64-bit version).

Table 2. Parameterization values

Problem	Objectives	Population size	Generations	Function evaluations
DTLZ1	2 to 4 and 8	100	250	25000
	5 to 7	95	263	24985
	9	90	277	24930
DTLZ2	2 to 8	100	150	15000
	9	90	166	14940
DTLZ3, WFG1 and WFG3.	2 to 6	100	750	75000
	7 and 8	95	789	74955
	9	90	830	74700
DTLZ4 and WFG2	2 to 7	100	500	50000
	8	110	454	49940
	9	90	555	49950

4.3 Results

Table 3 shows the mean of the spacing and hypervolume values obtained for each final result obtained by the two versions of SMS-EMOA used in our study. Additionally, we show the average time, in minutes, needed to perform the maximum number of function evaluations in each case. When no value is shown in the table for any of the algorithms, this means that it was not able to perform the maximum number of function evaluations after 8000 minutes. The results show that our proposed approach SMGPUS-EMOA is considerably faster and that it achieves a speedup of up to 883x with respect to the sequential algorithm (this speed up is achieved in WFG2). We are also able to obtain the same results as the sequential version, which verifies that our parallel implementation is working as expected.[6]

5   Conclusions and Future Work

We have proposed a new approach for computing the hypervolume contribution of a point. The core idea of our proposed algorithm is to exploit the parallelization provided by the use of GPUs, combined with a novel implementation that allows us to save unnecesary computations.[7] The proposed algorithm was incorporated within SMS-EMOA, and was tested in several well-known test problems

---

[6] There are a few differences in the spacing indicator, which are due to stochastic variations produced by a few isolated runs, which affected the computation of the mean values.

[7] The source code of our proposed approach is available from the first author, upon request.

Table 3. Experimental results

Objectives	SMS-EMOA			SMGPUS-EMOA			Speed-up
	Hypervolume	Spacing	Time (mins)	Hypervolume	Spacing	Time (mins)	
DTLZ1							
▮	**0.873238**	**0.001283**	0.1784	0.873162	0.001457	**0.1354**	1.3186
3	**0.973716**	**0.007665**	0.6921	0.97325	0.009791	**0.5578**	1.2409
4	0.992412	0.017341	9.927	**0.994162**	**0.016266**	**0.9542**	10.404
5	**0.998515**	**0.026013**	146.5268	0.99851	0.02788	**1.9638**	74.6154
6	0.999553	0.038846	1954.4977	**0.99958**	**0.034769**	**9.4954**	205.8373
7	0.999841	0.050285	25632.8153	**0.999876**	**0.041381**	**52.7094**	486.3049
8	–	–	–	**0.99993**	**0.059126**	**437.9792**	–
9	–	–	–	**0.99997**	**0.056958**	**1673.3501**	–
DTLZ2							
▮	**3.210879**	0.007702	0.2784	3.210868	**0.007602**	**0.1339**	2.0806
3	**7.426076**	0.041653	1.6382	7.426069	**0.040691**	**0.2926**	5.5995
4	15.5753	0.067581	11.1038	**15.575445**	**0.06661**	**0.673**	16.5008
5	31.677782	**0.084248**	168.761	**31.678164**	0.086913	**1.3908**	121.3476
6	63.75333	**0.133544**	2420.9844	**63.753512**	0.135595	**10.4016**	232.7513
7	–	–	–	**127.79839**	**0.181875**	**49.7891**	–
8	–	–	–	**255.837758**	**0.192845**	**525.0347**	–
9	–	–	–	**511.846819**	**0.168345**	**1713.133**	–
DTLZ3							
▮	3.208826	**0.007444**	0.6189	**3.20903**	0.0076	**0.4559**	1.3577
3	7.420476	**0.041679**	1.8316	**7.421543**	0.041999	**1.4572**	1.257
4	**15.574479**	**0.06765**	27.1425	15.57325	0.067938	**2.1452**	12.6527
5	31.67833	0.088963	434.3908	**31.678284**	**0.085466**	**4.2416**	102.4135
6	63.740603	**0.146409**	6040.9012	**63.74412**	0.161621	**23.9932**	251.7766
7	–	–	–	**127.788183**	**0.223037**	**133.1065**	–
8	–	–	–	**255.815976**	**0.192676**	**1021.2014**	–
9	–	–	–	**511.840859**	**0.180436**	**4437.1247**	–
DTLZ4							
▮	2.87192	**0.005499**	0.5385	**3.065692**	0.006722	**0.3511**	1.5339
3	6.887928	0.026107	3.1228	**7.145613**	0.032301	**2.5011**	1.2486
4	14.99058	**0.049317**	35.218	**15.406114**	0.057918	**3.4136**	10.3172
5	30.189868	**0.050782**	520.502	**30.787056**	0.062325	**5.6853**	91.5528
6	62.447761	**0.07819**	7446.6796	**63.368461**	0.098488	**26.3014**	283.129
7	–	–	–	**127.726866**	**0.177763**	**125.5394**	–
8	–	–	–	**255.731938**	**0.194006**	**1629.4346**	–
9	–	–	–	**511.694486**	**0.184917**	**4199.4646**	–
WFG1							
▮	**6.735278**	**0.006432**	1.6154	6.611425	0.009437	**0.6304**	2.5626
3	**66.595412**	**0.047712**	7.3843	64.496605	0.049191	**4.2319**	1.745
4	550.509191	**0.077903**	108.9946	**635.368636**	0.070628	**5.0127**	21.7439
5	5811.563198	0.077009	2047.7594	**5981.306146**	**0.075655**	**6.6857**	306.2917
6	–	–	–	**69980.43159**	**0.072144**	**28.7109**	–
7	–	–	–	**975295.6509**	**0.074721**	**154.8892**	–
8	–	–	–	**16252223**	**0.071669**	**992.7482**	–
9	–	–	–	**260018586.1**	**0.082601**	**5850.1223**	–
WFG2							
▮	**10.655908**	0.008643	0.6955	10.569597	**0.008441**	**0.3571**	1.9476
3	86.215292	**0.04886**	3.7612	**86.827259**	0.052103	**2.4849**	1.5137
4	786.873748	**0.096958**	74.153	**793.125358**	0.101117	**3.3485**	22.1456
5	**8860.244206**	**0.082606**	1458.6645	8566.72997	0.07059	**4.5121**	323.2813
6	110240.8108	**0.100501**	17867.2959	**112357.7782**	0.108096	**20.2233**	883.5017
7	–	–	–	**1646787.124**	**0.178076**	**134.0804**	–
8	–	–	–	**27973328.67**	**0.240691**	**1288.5644**	–
WFG3							
▮	**10.954985**	0.004739	1.648	10.954594	**0.00423**	**0.6439**	2.5597
3	**76.42553**	0.132885	6.3563	76.426786	**0.132637**	**4.4548**	1.4269
4	683.216698	0.442542	38.2058	**683.589267**	**0.439745**	**7.8809**	4.848
5	7474.344326	**0.704298**	188.9923	**7484.96526**	0.706353	**35.5841**	5.3112
6	96821.66284	0.939095	880.7624	**96950.48865**	**0.905224**	**212.272**	4.1493
7	–	–	–	**1365800.643**	**1.030017**	**779.205**	–
8	–	–	–	**22781648.93**	**1.332752**	**5323.5093**	–

having up to 9 objectives. Our results indicate that our proposed approach is able to achieve a speed up of up to 883x with respect to the sequential version of SMS-EMOA, using the FPL algorithm.

As part of our future work, we would like to incorporate our proposed approach into other hypervolume-based MOEAs (e.g., IBEA [19]). We would also like to develop indicator-based MOEAs that combine the use of the hypervolume with another (less computational intensive) indicator.

References

1. Batcher, K.E.: Sorting networks and their applications. In: Proceedings of the April 30-May 2, 1968, Spring Joint Computer Conference, AFIPS 1968 (Spring), pp. 307–314. ACM, New York (1968)
2. Beume, N., Naujoks, B., Emmerich, M.: SMS-EMOA: Multiobjective selection based on dominated hypervolume. European Journal of Operational Research **181**(3), 1653–1669 (2007)
3. Bringmann, K., Friedrich, T.: Approximating the volume of unions and intersections of high-dimensional geometric objects. Computational Geometry-Theory and Applications **43**(6–7), 601–610 (2010)
4. Coello Coello, C.A., Lamont, G.B., Van Veldhuizen, D.A.: Evolutionary Algorithms for Solving Multi-Objective Problems, 2nd edn. Springer, New York (2007). ISBN: 978-0-387-33254-3
5. Deb, K., Pratap, A., Agarwal, S., Meyarivan, T.: A Fast and Elitist Multiobjective Genetic Algorithm: NSGA-II. IEEE Transactions on Evolutionary Computation **6**(2), 182–197 (2002)
6. Deb, K., Thiele, L., Laumanns, M., Zitzler, E.: Scalable test problems for evolutionary multiobjective optimization. In: Abraham, A., Jain, L., Goldberg, R. (eds.) Evolutionary Multiobjective Optimization. Theoretical Advances and Applications, pp. 105–145. Springer, USA (2005)
7. Everson, R.M., Fieldsend, J.E., Singh, S.: Full elite sets for multi-objective optimisation. In: Parmee, I. (ed.) Proceedings of the Fifth International Conference on Adaptive Computing Design and Manufacture (ACDM 2002), University of Exeter, Devon, UK, April 2002, vol. 5, pp. 343–354. Springer-Verlag (2002)
8. Fleischer, M.: The measure of pareto optima. In: Fonseca, C.M., Fleming, P.J., Zitzler, E., Deb, K., Thiele, L. (eds.) EMO 2003. LNCS, vol. 2632, pp. 519–533. Springer, Heidelberg (2003)
9. Flynn, M.J.: Some computer organizations and their effectiveness. IEEE Trans. Comput. **21**(9), 948–960 (1972)
10. Fonseca, C.M., Paquete, L., López-Ibáñez, M.: An improved dimension-sweep algorithm for the hypervolume indicator. In: 2006 IEEE Congress on Evolutionary Computation (CEC 2006), Vancouver, BC, Canada, July 2006, pp. 3973–3979. IEEE (2006)
11. Huband, S., Hingston, P., Barone, L., While, L.: A Review of Multiobjective Test Problems and a Scalable Test Problem Toolkit. IEEE Transactions on Evolutionary Computation **10**(5), 477–506 (2006)
12. Ishibuchi, H., Tsukamoto, N., Nojima, Y.: Evolutionary many-objective optimization: A short review. In: 2008 Congress on Evolutionary Computation (CEC 2008), Hong Kong, pp. 2424–2431. IEEE Service Center, June 2008

13. Menchaca-Mendez, A., Coello Coello, C.A.: A new selection mechanism based on hypervolume and its locality property. In: 2013 IEEE Congress on Evolutionary Computation (CEC 2013), Cancún, México, pp. 924–931. IEEE Press, June 20–23, 2013. ISBN: 978-1-4799-0454-9
14. NVIDIA Corporation. Cuda zone (2014)
15. While, L.: A new analysis of the lebmeasure algorithm for calculating hypervolume. In: Coello Coello, C.A., Hernández Aguirre, A., Zitzler, E. (eds.) EMO 2005. LNCS, vol. 3410, pp. 326–340. Springer, Heidelberg (2005)
16. While, L., Hingston, P., Barone, L., Huband, S.: A Faster Algorithm for Calculating Hypervolume. IEEE Transactions on Evolutionary Computation 10(1), 29–38 (2006)
17. Wu, J., Azarm, S.: Metrics for Quality Assessment of a Multiobjective Design Optimization Solution Set. Transactions of the ASME, Journal of Mechanical Design 123, 18–25 (2001)
18. Zitzler, E., Brockhoff, D., Thiele, L.: The hypervolume indicator revisited: on the design of pareto-compliant indicators via weighted integration. In: Obayashi, S., Deb, K., Poloni, C., Hiroyasu, T., Murata, T. (eds.) EMO 2007. LNCS, vol. 4403, pp. 862–876. Springer, Heidelberg (2007)
19. Zitzler, E., Künzli, S.: Indicator-based selection in multiobjective search. In: Yao, X., et al. (eds.) PPSN 2004. LNCS, vol. 3242, pp. 832–842. Springer, Heidelberg (2004)
20. Zitzler, E., Thiele, L., Laumanns, M., Fonseca, C.M., da Fonseca, V.G.: Performance Assessment of Multiobjective Optimizers: An Analysis and Review. IEEE Transactions on Evolutionary Computation 7(2), 117–132 (2003)

A Feature-Based Performance Analysis in Evolutionary Multiobjective Optimization

Arnaud Liefooghe[1](✉), Sébastien Verel[2], Fabio Daolio[3], Hernán Aguirre[3], and Kiyoshi Tanaka[3]

[1] CRIStAL (UMR CNRS 9189) — Inria Lille-Nord Europe,
Université Lille 1, Villeneuve-d'Ascq, France
arnaud.liefooghe@univ-lille1.fr
[2] LISIC, Université du Littoral Côte d'Opale, Calais, France
verel@lisic.univ-littoral.fr
[3] Faculty of Engineering, Shinshu University, Nagano, Japan
{fdaolio,ahernan,ktanaka}@shinshu-u.ac.jp

Abstract. This paper fundamentally investigates the performance of evolutionary multiobjective optimization (EMO) algorithms for computationally hard 0–1 combinatorial optimization, where a strict theoretical analysis is generally out of reach due to the high complexity of the underlying problem. Based on the examination of problem features from a multiobjective perspective, we improve the understanding of the efficiency of a simple dominance-based EMO algorithm with unbounded archive for multiobjective NK-landscapes with correlated objective values. More particularly, we adopt a statistical approach, based on simple and multiple linear regression analysis, to enquire the expected running time of global SEMO with restart for identifying a $(1 + \varepsilon)-$approximation of the Pareto set for small-size enumerable instances. Our analysis provides further insights on the EMO search behavior and on the most important features that characterize the difficulty of an instance for this class of problems and algorithms.

1 Introduction

Black-box multiobjective combinatorial optimization problems are characterized by a discrete solution space and by multiple objective functions, such as cost, profit, or risk, that are ill-defined, computationally expensive, or for which an analytical form is not available. Due to the black-box nature of the objective functions, problem-specific algorithms are usually excluded to identify or approximate the Pareto set, so that an increasing number of general-purpose evolutionary multiobjective optimization (EMO) algorithms and other randomized search heuristics have been proposed in recent years [3]. However, the overall amount of understanding about the pros and cons of different EMO algorithm designs and configurations with respect to a given problem structure is rather scarce. Due to the increasing number and complexity of black-box multiobjective optimization problems and algorithms, one of the most difficult challenges is to devise and exhibit a number

© Springer International Publishing Switzerland 2015
A. Gaspar-Cunha et al. (Eds.): EMO 2015, Part II, LNCS 9019, pp. 95–109, 2015.
DOI: 10.1007/978-3-319-15892-1_7

of general-purpose problem characteristics and statistical methodologies allowing to explain the dynamics and the performance of EMO algorithms.

Recently, a few attempts to explain the performance of randomized search heuristics based on relevant fitness landscape features have been proposed for single-objective optimization problems of continuous and combinatorial nature; see *e.g.* [4,6,12]. In this paper, we address the issue of feature-based performance analysis for EMO algorithms according to the main characteristics of 0–1 multiobjective optimization problems. We first extend our previous works by summarizing a number of problem properties and fitness landscape features for black-box 0–1 multiobjective optimization [11]. They include features extracted from the problem input data, like variable correlation, objective correlation, and objective space dimension [17], as well as features from the Pareto set [1,8], the Pareto graph [14] and the ruggedness and multimodality of the fitness landscape [17]. Then, we analyze the correlation between those features and the performance of an EMO algorithm. More particularly, we investigate the expected running time of the global SEMO algorithm [9] with restart to identify a $(1+\varepsilon)-$approximation of the Pareto set on a large number of small-size enumerable multiobjective NK-landscapes with objective correlation, *i.e.* ρMNK-landscapes [17]. Our analysis shows the relative influence of each *individual* problem feature on the algorithm performance. In particular, the running time of global SEMO appears to be predominantly impacted by the ruggedness of the fitness landscape, more than other features like the number of Pareto optimal solutions. Additionally, we investigate different formulations of a multiple linear regression model. This allows us to discuss the *joint* effect of different subsets of features in capturing the dynamics of the algorithm. The ruggedness and the multimodality, but also the number of objectives, the correlation between them, and the hypervolume of the Pareto front turn out to be the most impactful characteristics that allow to explain the performance of global SEMO for ρMNK-landscapes.

The remainder of the paper is organized as follows. Section 2 details the problem and algorithm settings of our analysis. Section 3 summarizes the problem features under consideration in the paper. Section 4 introduces different regression models to explain the performance of global SEMO for enumerable ρMNK-landscapes. Section 5 concludes the paper and suggests further research.

2   Problem and Algorithm Settings

In this paper, we are interested in the ability of evolutionary multiobjective optimization (EMO) algorithms to identify a Pareto set approximation for black-box multiobjective combinatorial optimization problems. In particular, we investigate the (estimated) running time of global SEMO [9] with restart to identify a $(1+\varepsilon)-$approximation of the Pareto set on a large bench of enumerable ρMNK-landscapes with different structural properties. We consider the maximization of an objective function vector $f = (f_1, \ldots, f_m)$ over the discrete set of solutions $X = \{0,1\}^n$, where m is the number of objectives, and n is the problem size. X is the *solution space*, and $Z = f(X) \subseteq \mathbb{R}^m$ is the *objective space*. A solution

$x \in X$ is dominated by a solution $x' \in X$ if $\forall i \in \{1, \ldots, m\}$, $f_i(x) \leqslant f_i(x')$ and $\exists i \in \{1, \ldots, m\}$ such that $f_i(x) < f_i(x')$. The set of solutions that are not dominated by any other is the *Pareto set*, and its image in the objective space is the *Pareto front*.

2.1 ρMNK-Landscapes

The family of ρMNK-landscapes constitutes a problem-independent model used for constructing multiobjective multimodal landscapes with objective correlation [17]. They extend single-objective NK-landscapes [7] and multiobjective NK-landscapes with independent objective functions [1]. Feasible solutions are binary strings of size n, *i.e.* the solution space is $X = \{0, 1\}^n$. The parameter k refers to the number of variables that influence a particular position from the bit-string. The objective function vector $f = (f_1, \ldots, f_i, \ldots, f_m)$ is defined as $f : \{0, 1\}^n \rightarrow [0, 1]^m$ such that each objective function f_i is to be maximized. The problem can be formalized as follows.

$$
\begin{aligned}
\max \ & f_i(x) = \frac{1}{n} \sum_{j=1}^{n} f_{ij}(x_j, x_{j_1}, \ldots, x_{j_k}) , \quad i \in \{1, \ldots, m\} \\
\text{s.t. } & x_j \in \{0, 1\} \quad\quad\quad\quad\quad\quad\quad\quad\quad\quad , \quad j \in \{1, \ldots, n\}
\end{aligned}
\tag{1}
$$

As in the single-objective case, each separate objective function value $f_i(x)$ of a solution $x = (x_1, \ldots, x_j, \ldots, x_n)$ is an average value of the individual contributions associated with each variable x_j. Indeed, for each objective f_i, $i \in \{1, \ldots, m\}$, and each variable x_j, $j \in \{1, \ldots, n\}$, a component function $f_{ij} : \{0, 1\}^{k+1} \rightarrow [0, 1]$ assigns a real-valued contribution for every combination of x_j and its k *epistatic interactions* $\{x_{j1}, \ldots, x_{jk}\}$. These f_{ij}-values are uniformly distributed in the range $[0, 1]$. As a consequence, the individual contribution of a variable x_j depends on the value of x_j, as well as on the values of $k < n$ other variables $\{x_{j_1}, \ldots, x_{j_k}\}$. In this work, the epistatic interactions, *i.e.* the k variables that influence the contribution of x_j, are set uniformly at random among the $(n - 1)$ variables other than x_j, following the random neighborhood model from [7]. By increasing the number of epistatic interactions k from 0 to $(n - 1)$, problem instances can be gradually tuned from smooth to rugged. In ρMNK-landscapes, f_{ij}-values additionally follow a multivariate uniform distribution of dimension m, defined by an $m \times m$ positive-definite symmetric covariance matrix (c_{pq}) such that $c_{pp} = 1$ and $c_{pq} = \rho$ for all $p, q \in \{1, \ldots, m\}$ with $p \neq q$, were $\rho > \frac{-1}{m-1}$ defines the objective correlation degree; see [17] for details. The positive (respectively negative) data correlation ρ allows to decrease (respectively increases) the degree of conflict between the objective function values. The same correlation coefficient ρ is then defined between all pairs of objectives, and the same epistatic degree k and epistatic interactions are set for all the objectives.

2.2 Global SEMO

Global SEMO [9], or G-SEMO for short, is a simple elitist steady-state EMO algorithm for black-box 0–1 optimization problems dealing with an arbitrary objective function vector defined as $f : \{0,1\}^n \rightarrow Z$ such that $Z \subseteq \mathbb{R}^m$, like in ρMNK-landscapes. It maintains an *unbounded* archive A of non-dominated solutions found so far. The archive is initialized with one random solution from the solution space. At each iteration, one solution is chosen at random from the archive $x \in A$. Each binary variable from x is independently flipped with a rate of $\frac{1}{n}$ in order to produce an offspring solution x'. This mutation operator is ergodic, meaning that there is a non-zero probability of jumping from any point to any other point in the solution space. The archive is then updated by keeping the non-dominated solutions from $A \cup \{x'\}$. In its general form, the G-SEMO algorithm does not have any explicit stopping rule [9]. In this paper, we are interested in its running time, in terms of a number of function evaluations, until an $(1 + \varepsilon)-$approximation of the Pareto set has been identified and is contained in the internal memory A of the algorithm, subject to a maximum budget of function evaluations.

2.3   Performance Measure

Let ε be a constant value such that $\varepsilon \geqslant 0$. The (multiplicative) ε-dominance relation ($\preceq_\varepsilon$) can be defined as follows. For all $x, x' \in X$, $x \preceq_\varepsilon x'$ if $f_i(x) \leqslant (1 + \varepsilon) \cdot f_i(x')$, $\forall i \in \{1, \ldots, m\}$. A set $X^\varepsilon \subseteq X$ is an $(1 + \varepsilon)-$approximation of the Pareto set if for any solution $x \in X$, there is one solution $x' \in X^\varepsilon$ such that $x \preceq_\varepsilon x'$. This is equivalent to finding an approximation set whose multiplicative epsilon quality indicator value with respect to the (exact) Pareto set is lower than $(1 + \varepsilon)$, see *e.g.* [18]. Interestingly, under some general assumptions, there always exists an $(1 + \varepsilon)$-approximation, for any given $\varepsilon \geqslant 0$, whose cardinality is both polynomial in the problem size and in $\frac{1}{\varepsilon}$ [13].

Following a conventional methodology from single-objective continuous black-box optimization benchmarking [5], the expected number of function evaluations to identify an $(1 + \varepsilon)-$approximation is here chosen as a performance measure. However, as any EMO algorithm, G-SEMO can either succeed or fail to reach an accuracy of ε in a single simulation run. In case of a success, the running time is the number of function evaluations until an $(1 + \varepsilon)-$approximation was found. In case of a failure, we simply *restart* the algorithm at random. We then obtain a *"simulated running time"* [5] from a set of given trials of G-SEMO on a given instance. Such a performance measure allows to take into account both the success rate $p_s \in (0, 1]$ and the convergence speed of the G-SEMO algorithm with restarts. Indeed, after $(t - 1)$ failures, each one requiring T_f evaluations, and the final successful run with T_s evaluations, the total running time is $T = \sum_{i=1}^{t-1} T_f + T_s$. By taking the expectation value and by considering that the probability of success after $(t - 1)$ failures follows a Bernoulli distribution of parameter p_s, we have:

$$\mathbb{E}[T] = \left(\frac{1 - p_s}{p_s}\right) \mathbb{E}[T_f] + \mathbb{E}[T_s] \tag{2}$$

In our case, the success rate p_s is estimated with the ratio of successful runs over the total number of executions ($\hat{p}_s$), the expected running time for unsuccessful runs $\mathbb{E}[T_f]$ is set to a constant limit on the number of function evaluation calls T_{max}, and the expected running time for successful runs $\mathbb{E}[T_s]$ is estimated with the average number of function evaluations performed by successful runs.

$$\text{ert} = \left(\frac{1 - \hat{p}_s}{\hat{p}_s}\right) T_{max} + \frac{1}{t_s} \sum_{i=1}^{t_s} T_i \tag{3}$$

where t_s is the number of successful runs, and T_i is the number of evaluations required for successful run i. For more details, we refer to [5].

2.4   Parameter Setting

In the following, we investigate ρMNK-landscapes with an epistatic degree $k \in \{2, 4, 6, 8, 10\}$, an objective space dimension $m \in \{2, 3, 5\}$, and an objective correlation $\rho \in \{-0.9, -0.7, -0.4, -0.2, 0.0, 0.2, 0.4, 0.7, 0.9\}$, such that $\rho > \frac{-1}{m-1}$. The problem size is set to $n = 18$ in order to enumerate the solution space exhaustively. The solution space size is then 2^{18}. A set of 30 different landscapes, independently generated at random, are considered for each parameter combination: ρ, m, and k. They are made available at the following URL: http://mocobench.sf.net.

We set a target $\varepsilon = 0.1$. The time limit is set to $T_{max} = 2^n \cdot 10^{-1} < 26\,215$ function evaluations without identifying an $(1 + \varepsilon)$−approximation. The G-SEMO algorithm is executed 100 times *per* instance. For a given instance, the success rate and the expected number of evaluations for successful runs are estimated from those 100 executions. However, let us note that G-SEMO was not able to identify a $(1 + \varepsilon)$−approximation set for any of the runs on one instance with $m = 3$, $\rho = 0.2$ and $k = 10$, one instance with $m = 3$, $\rho = 0.4$ and $k = 10$, ten instances with $m = 5$, $\rho = 0.2$ and $k = 10$, six instances with $m = 5$, $\rho = 0.4$ and $k = 10$, as well as two instances with $m = 5$, $\rho = 0.7$ and $k = 10$. Moreover, G-SEMO was not able to solve the following instances due to an overload of CPU resources available: $m = 5$ and $\rho \in \{-0.2, 0.0\}$. Those experiments have then been discarded due to missingness. Overall, this represents a total amount of $2\,980$ instances times 100 executions, that is $298\,000$ simulation runs.

3   Features to Characterize Problem Difficulty

In Table 1, we give a number of general-purpose problem features, either directly extracted from the problem instance (*low-level features*), or computed from the enumerated Pareto set and solution space (*high-level features*). Obviously, since the features require the solution space to be completely enumerated, they are not practical for performance prediction purposes. However, we still include them in order to examine their impact on the algorithm performance. For the case of ρMNK-landscapes, the neighborhood is induced by the *bit-flip* operator, which is directly related to the Hamming distance between solutions. For the computation of the hypervolume, the reference point is set to the origin $z^\star = (0.0, \ldots, 0.0)$. For

Table 1. Summary of low-level and high-level features investigated in the paper

low-level features

(k) Number of variable (epistatic) interactions
(m) Number of objective functions
(ρ) Correlation between the objective function values

high-level features

(npo) Number of Pareto optimal solutions	[1,8]
(hv) Hypervolume value [18] of a the Pareto set	[1]
(avgd) Average distance between Pareto optimal solutions	[11]
(maxd) Maximum distance between Pareto optimal solutions	[11]
(nconnec) Number of connected components in the Pareto set	[14]
(lconnec) Proportion of the largest connected component of the Pareto set	[10]
(kconnec) Minimal Hamming distance to connect the Pareto set	[14]
(nplo) Number of Pareto local optimal solutions	[15]

a more comprehensive explanation of those features and a correlation analysis between them, we refer to [11]. In the next section, we relate the value of those features for enumerable ρMNK-landscapes to the performance of G-SEMO.

4   Problem Features *vs.* Algorithm Performance

In this section, we conduct a linear regression analysis on the correlation between the problem features presented in the previous section and the performance of G-SEMO. The algorithm performance is defined as the expected running time ert, in terms of the number of evaluation function calls, required by the algorithm to identify a $(1 + \varepsilon)-$approximation of the Pareto set. We first detail our methodological setup. Then, we analyze the *individual* as well as the *joint* impact of problem features on the algorithm performance. At last, we compare the accuracy of regression models for different objective space dimensions.

4.1   Methodological Setup

Linear Regression. In order to provide an explanatory model for the algorithm performance, we perform a linear regression, whose general model can be formalized as follows:

$$y = \beta_0 + \beta_1 \cdot v_1 + \beta_2 \cdot v_2 + \ldots + \beta_p \cdot v_p + e \tag{4}$$

where y is the response variable, $(v_1, v_2, \ldots, v_p)$ are the explanatory variables, and e is the usual error term. In our case, the response variable to be explained is the expected running time of G-SEMO: log(ert), and the p explanatory variables correspond to selected problem features as detailed in Section 3. The response of the linear model is here log-transformed in order to better approach linearity; see Fig. 1 (top-left). Also, an order of magnitude is in general sufficiently relevant for

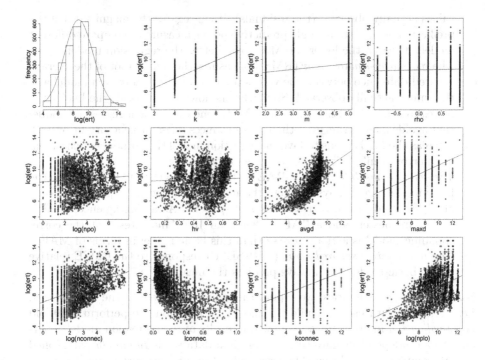

Fig. 1. Histogram of the distribution of log(**ert**)-values over all the instances (top-left), as well as scatter plots and regression lines for each feature *vs.* log(**ert**).

the running time of an EMO algorithm. Such linear regression models are usually fitted using an ordinary least-square minimization. A least-square estimator $\hat{\beta}_i$ is then produced for each regression coefficient β_i, $i \in \{1, \ldots, p\}$, by minimizing the sum of squared residuals between the fitted and the observed values. The case of a single explanatory variable ($p = 1$) is known as *simple* linear regression, whereas the extension to more than one explanatory variables ($p > 1$) is known as *multiple* linear regression. Notice that, although a linear regression model might not be able to reliably catch the existing correlation between explanatory variables as well as non-linear dependencies with the response variable, it has the advantage of being simple and easily interpretable.

Regression Accuracy. In the following, when measuring the accuracy of a linear regression model, we will be interested in the following statistics:

- The absolute *correlation coefficient* (r) measures the linear association between the predicted and the actually observed values (the conventional Pearson correlation coefficient is here used). Its absolute value ranges from 0 to 1. The closer r to 1.0, the better the fitting. Actually, an r-value of 1.0 indicates that the linear regression line perfectly fits the data.
- The *mean absolute error* (MAE) measures the average value of the absolute difference between the values predicted by the regression model and the

values actually observed (the residuals). It aggregates the magnitudes of the prediction errors into a single predictive power measure to compare different models. Clearly, the lower the MAE, the better the regression model.

- The *root mean-square error* (RMSE) measures the square root of the average squared difference between the values predicted by the regression model and the values actually observed. Similarly, the lower the RMSE, the better the regression model. Notice that the RMSE tends to favor a regression model that avoids large errors even though it produces a less satisfactory fit overall, whereas the MAE tends to favor a regression model that produces occasional large errors while being reasonably good on average.
- The *relative absolute error* (%RAE) corresponds to the MAE relative to the basic model that always predicts the mean, *i.e.* when no explanatory variables are used in the regression model ($p = 0$). As a consequence, smaller values are better, and a RAE higher than 100% indicates that the corresponding model is actually *worse* than this basic model in terms of MAE.
- The *root relative squared error* (%RRSE) corresponds to the RMSE relative to the basic model that always predicts the mean.

For each of those statistics, we report the values observed on the *training set*, *i.e.* the set of data used to build the model. In addition, we perform a 10-fold *cross-validation* in order to assess how the results of the regression model generalize to an independent data set. The original data set is *randomly* partitioned into 10 samples of equal size. The cross-validation process is repeated 10 times, with each of the samples being used exactly once as the validation data. For each sample, the above statistics are computed, and then averaged in order to produce a cross-validated r, MAE, RMSE, %RAE, and %RRSE value.

Data Preparation. Although features might usually have to be normalized appropriately in order to get rid of scaling issues and for a fair comparison between them, we here chose *not* to normalize them in order to ease the interpretation of the different models. Notice that normalizing the features would only result in a change on the value of the estimated regression coefficient $\hat{\beta}_i$, $i \in \{1, \ldots, p\}$. Moreover, this is not an issue within our experiments since we explicitly generate instances of the same size ($n = 18$) that takes their objective values in the same hyper-box $[0, 1]^m$.

4.2   Individual Impact of Problem Features

The scatter plots between each feature and the log-transformed estimated running time of G-SEMO log(ert) is reported in Fig. 1. Notice that some features (npo, nconnec, nplo) have been log-transformed in order to better approach linearity. Additionally, the statistics of all possible simple linear regression models, one for each feature, are reported in Table 2, from the lowest to the highest absolute correlation value. The individual impact of each feature is analyzed below.

First of all, four features are not directly linearly correlated to the expected running time of G-SEMO: the number of objective functions m, the objective correlation ρ, the cardinality of the Pareto set log(npo), and the hypervolume **hv**.

Table 2. Summary statistics of simple linear regression models, each one being based on a single problem feature. Values are rounded to 10^{-2}.

feature	\multicolumn{5}{c}{training set}	\multicolumn{5}{c}{10-fold cross validation}
	r	MAE	RMSE	%RAE	%RRSE	r	MAE	RMSE	%RAE	%RRSE
ρ	0.03	1.52	1.89	99.66	99.94	0.01	1.53	1.88	99.75	100.00
hv	0.04	1.52	1.89	99.66	99.94	0.02	1.53	1.89	99.73	99.99
log(npo)	0.08	1.54	1.88	100.61	99.70	0.07	1.54	1.88	100.66	99.74
m	0.21	1.51	1.85	98.63	97.86	0.20	1.51	1.85	98.69	97.91
kconnec	0.37	1.38	1.76	89.99	93.03	0.37	1.38	1.76	90.01	93.06
log(nconnec)	0.40	1.44	1.73	94.21	91.47	0.40	1.44	1.73	94.27	91.52
log(nplo)	0.46	1.41	1.67	92.53	88.56	0.46	1.42	1.67	92.57	88.59
maxd	0.47	1.30	1.67	85.07	85.07	0.47	1.30	1.67	85.09	88.36
lconnec	0.49	1.31	1.65	85.67	87.35	0.49	1.31	1.65	85.69	87.37
avgd	0.60	1.15	1.51	75.31	79.95	0.60	1.15	1.51	75.33	80.00
k	0.85	0.77	1.00	50.67	52.95	0.85	0.78	1.00	50.67	52.96
none	0.00	1.53	1.89	100.00	100.00	0.04	1.53	1.89	100.00	100.00

For each of these features, the absolute correlation coefficient value is under 0.25, and the prediction error is around the one of the most basic model that always predicts the mean. Surprisingly, there is *no* direct connection with the two main low-level features from problem input data m and ρ. At least, the link between the running time and those features is not a direct linear correlation, but a more complex model will be analyzed in the next section. As well, the cardinality and the hypervolume of the Pareto set, features closely related to the final goal of the search process, do not explain the variance of log(ert) by themselves.

The features related to the connectedness of the Pareto set are all weakly correlated to log(ert). The absolute correlation coefficients of the number of connected components log(nconnec), the proportional size of the largest connected component lconnec, and the minimum distance to be connected kconnec are between 0.37 and 0.49. The more connected the Pareto set, the smaller the running time of G-SEMO. The algorithm performance is also moderately correlated with the average and maximal distance between Pareto optimal solutions, avgd and maxd (the absolute correlation coefficient values are 0.60 and 0.47, respectively). The larger the distance between Pareto optimal solutions in the solution space, the larger the running time of G-SEMO. Interestingly, the cardinality of the Pareto set has a smaller impact on the performance of G-SEMO than the distance between solutions in the Pareto set. Moreover, the multimodality of the landscape, in terms of the number of Pareto local optimal solutions log(nplo), is moderately correlated to the running time of G-SEMO: the more Pareto local optima, the longer the running time (the correlation coefficient is 0.46).

At last, the only strong correlation appears with the feature related to the ruggedness of the landscape. Indeed, the number of epistatic interactions k is highly correlated to the efficiency of G-SEMO (the correlation coefficient is 0.85). The more rugged the landscape, the longer it takes to identify a $(1+\varepsilon)-$approximation of the Pareto set. In other words, by taking the features individually, the model based on k is the one that gives the highest accuracy. On average, it allows to predict the logarithm of the runtime of G-SEMO within ± 0.77 of the observed value. Since the RMSE is much larger (1.00), this suggests that the deviation to this average value might be large, as we can see also on Fig.1 (first line, second column). Such a regression accuracy is around twice better than the most basic model that always predicts the mean (± 1.53), and largely better than the second more accurate simple linear regression model based on avgd (± 1.15).

Having the ruggedness of the landscape k as a more important (individual) feature than the number of objectives m and the objective correlation ρ might be surprising at first sight. Indeed, the number of Pareto optimal solutions is known to increase exponentially with the number of objectives and the degree of conflict between them; see *e.g.* [17]. However, let us remind that the algorithm under consideration in the paper (G-SEMO) actually handles an *unlimited* approximation set size. It is then only slightly affected by the minimum number of solutions required to obtain a $(1 + \varepsilon)$−approximation of the Pareto set [13]. Actually, depending on ρ, the estimated expected running time of G-SEMO is 23 to 118 times larger for rugged two-objective instances than smoother five-objective instances. For instance, when $\rho = 0.2$, the average ert−value is equal to 61 836 for $k = 10$ and $m = 2$ while it is equal to 2 691 for $k = 2$ and $m = 5$. Similarly, for $\rho = 0.9$, the average ert−value is 33 922 for $k = 10$ and $m = 2$, and only 287 for $k = 2$ and $m = 5$. In accordance with known results from single-objective optimization [2], the ruggedness of the landscape k seems to largely impact the running time of EMO algorithms.

Overall, analyzing the individual impact of problem features supports the hypothesis that the structural properties identified in the previous section can help to understand the performance of a simple dominance-based EMO algorithm like G-SEMO for ρMNK-landscapes. In the next section, different multiple linear regression models are examined in order to better explain the running time of G-SEMO, based on the *joint effect* of these problem-related characteristics.

4.3   Joint Impact of Problem Features

We start by fitting the response variable log(ert) against all the low-level and high-level features presented in Section 3. The statistics related to the model accuracy are provided in Table 3 (line 1). For this complete multiple linear regression model, the correlation coefficient is over 0.9. Overall, it allows to explain the performance of G-SEMO with a much higher accuracy compared to the simple linear regression model based on k only, and outperforms the basic model that always predicts the mean by around 60%. Moreover, the model has a high degree of generalization. Indeed, the RMSE and the cross-validated RMSE are very close to each other. The same happens with the MAE (the difference between both is always under 10^{-2}). The scatter plot of the actual *vs.* the predicted performance values is given in Fig. 2 (left). This allows us to visualize how the model accuracy varies depending on the hardness of the problem instance: The model seems to slightly underestimate the runtime for easier and harder instances whereas it rather overestimates it for an intermediate instance difficulty.

Although the regression coefficients related to each feature are not interpretable due to the different scaling of the metric values, not all regression coefficients are

Table 3. Summary statistics of the multiple linear regression model with backward-elimination feature selection. Values are rounded to 10^{-2}.

			training set				10-fold cross validation			
	r	MAE	RMSE	%RAE	%RRSE	r	MAE	RMSE	%RAE	%RRSE
all features	0.91	0.58	0.76	37.74	40.34	0.91	0.58	0.76	37.86	40.48
\ maxd	0.91	0.58	0.76	37.73	40.38	0.91	0.58	0.76	37.84	40.50
\ log(nconnec)	0.91	0.58	0.76	37.73	40.49	0.91	0.58	0.77	37.83	40.60
\ log(npo)	0.91	0.58	0.77	37.72	40.53	0.91	0.58	0.77	37.80	40.62
\ lconnec	0.91	0.58	0.77	37.74	40.54	0.91	0.58	0.77	37.81	40.61
\ avgd	0.91	0.58	0.77	37.83	40.58	0.91	0.58	0.77	37.90	40.64
\ kconnec	0.91	0.58	0.77	37.95	40.66	0.91	0.58	0.77	38.00	40.71
\ ρ	0.91	0.61	0.80	39.98	42.20	0.91	0.61	0.80	40.04	42.26
\ hv	0.89	0.67	0.87	43.59	46.29	0.89	0.67	0.87	43.63	46.33
\ log(nplo)	0.88	0.70	0.91	45.83	48.19	0.88	0.70	0.91	45.86	48.21
\ m	0.85	0.77	1.00	50.67	52.95	0.85	0.78	1.00	50.67	52.96
\ k	0.00	1.53	1.89	100.00	100.00	0.04	1.53	1.89	100.00	100.00

statistically significant in a general multiple linear regression model. In order to eliminate the influence of the less significant regression coefficients, we proceed by *backward elimination*. Starting from the inclusion of all features, we iteratively remove the feature that has the lowest impact on the increase of the MAE until no feature remains. This allows us to produce a ranked list of features by traversing the feature space from one side to the other and recording the order that attributes are deleted. Hence, the attributes that are deleted in the last steps have a more meaningful impact on the model. The steps of the backward elimination are sketched in Table 3, where one feature is removed at every line. Notice that a forward selection, that does the opposite procedure of iteratively adding attributes, ends up with a similar ranking on the importance of features, except that maxd is in the latter case more important than log(npo) and log(nconnec) (detailed results are not reported due to space limitation).

This feature selection analysis allows us to gain further insights about which subset of features obtains the highest accuracy. Indeed, the error increase is almost insignificant until the deletion of ρ in the model (line 8 in Table 3), where the correlation coefficient drops from 0.91 to 0.89, the MAE rises from 0.58 to 0.61 and the RMSE rises from 0.77 to 0.80. We can then conclude that ρ and subsequent attributes constitute the most significant subset for explaining the algorithm performance. Actually, a more compact model, with only the five most significant features, constitute an acceptable alternative, and has almost the same accuracy than the full model; see also Fig. 2 (right). Once again, the most important feature seems to be k, now followed by m, log(nplo), hv and ρ, in the order of importance. As a consequence, although they are not able to catch all the variations of log(ert) individually (see Section 4.2), the joint effect of all three low-level features from the problem input data (k, m and ρ) is relevant for explaining the running time of G-SEMO. Moreover, there is one high-level feature related to the hypervolume of the Pareto set hv and to the multimodality log(nplo). As well, the number of Pareto optimal solutions is *not* a significant addition to the regression model. We attribute this to the fact that the hypervolume incorporates a more relevant information related to the Pareto front for the algorithm behavior.

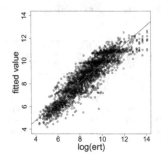
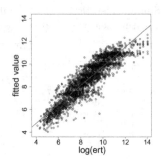

Fig. 2. log(ert) *vs.* fitted values for (left) the model with all features; and (right) the model with a selected subset of features (*i.e.* $k, m, \log(\mathtt{nplo}), \mathtt{hv}, \rho$).

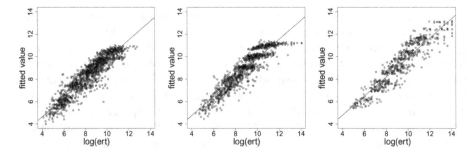

Fig. 3. log(ert) *vs.* fitted values for (left) $m = 2$, (middle) $m = 3$, (right) $m = 5$. All features are part of the models.

4.4 Explanatory Models *vs.* Objective Space Dimension

In this section, we build a separate regression model for each objective space dimension $m \in \{2, 3, 5\}$. The statistics related to the model accuracy are provided in Table 4. The information for the complete model mixing all m-values is also given in the tables in order to facilitate the comparison between the different models. Additionally, the scatter plot of the actual *vs.* the predicted performance values is given in Fig. 3.

First of all, the regression model for a particular number of objectives is always slightly more accurate than the global model for all m-values (whatever

Table 4. Summary statistics of the multiple linear regression models for all objective space dimensions ($\forall m$) and for each individual objective space dimension $m \in \{2, 3, 5\}$. All features are part of the models. Values are rounded to 10^{-2}.

		training set				10-fold cross validation				
	r	MAE	RMSE	%RAE	%RRSE	r	MAE	RMSE	%RAE	%RRSE	#inst
$\forall m$	0.91	0.58	0.76	37.74	40.34	0.91	0.58	0.76	37.86	40.48	2980
$m = 2$	0.92	0.52	0.68	35.86	38.62	0.92	0.53	0.68	36.15	38.95	1350
$m = 3$	0.92	0.52	0.69	36.12	39.02	0.92	0.53	0.70	36.60	39.51	1048
$m = 5$	0.93	0.64	0.81	36.86	37.72	0.92	0.65	0.83	37.62	38.53	582

the m-value, there is an improvement in terms of correlation, %RAE or %RRSE over the complete model). Actually, the worsening in terms of MAE or RMSE for larger m-values is only an artefact that the expected running time of G-SEMO increases with the number of objectives; see Fig. 1 (first line, third column). This means that constructing a regression model *per* objective space dimension only allows to reduce the prediction error to a very small extent.

As reported in Table 5, by focusing on the five most relevant features identified in the previous section ($k, m, \log(\texttt{nplo}), \texttt{hv}$ and ρ, following the order of importance), we are also able to construct a more compact regression model for each m-value with a satisfactory response compared to the model using all features. The difference in the correlation coefficient and the average error between the models with all features and their compact counterpart is always less than $2 \cdot 10^{-2}$.

In fact, applying a feature selection procedure by backward elimination for each of the models until the MAE increases by more than 10^{-2} ends up with following subset feature selection, following the order of importance (detailed results are omitted due to space restriction):

- $k, \texttt{hv}$ and $\log(\texttt{nplo})$ for $m = 2$;
- $k, \log(\texttt{nplo})$ and hv for $m = 3$;
- k and ρ for $m = 5$.

Interestingly, this means that the objective correlation ρ is *not* a relevant feature for the models with $m \in \{2, 3\}$. We attribute this to the fact that the number of Pareto local optimal solutions $\log(\texttt{nplo})$ increases with ρ in these cases, whereas the correlation between both is much lower for $m = 5$. For the same reason, ρ is selected over hv in the latter case. In addition, given that the proportion of Pareto local optimal solutions in the solution space increases exponentially with m (we know, for instance, that more than 95% of the solution space correspond to Pareto local optimal solutions for $\rho = -0.2$ and $m = 5$ in average [17]), the multimodality of the landscape, corresponding to $\log(\texttt{nplo})$, is not relevant anymore for larger objective space dimensions. However, in all cases, the ruggedness of the landscape k is again the most relevant feature.

Table 5. Summary statistics of the multiple linear regression models for all objective space dimensions ($\forall m$) and for each individual objective space dimension $m \in \{2, 3, 5\}$. Only the subset of significant features ($k, m, \rho, \texttt{hv}, \log(\texttt{nplo})$) are part of the models. Values are rounded to 10^{-2}.

| | training set | | | | | 10-fold cross validation | | | | | |
	r	MAE	RMSE	%RAE	%RRSE	r	MAE	RMSE	%RAE	%RRSE	#inst
$\forall m$	0.91	0.58	0.77	37.95	40.66	0.91	0.58	0.77	38.00	40.71	2980
$m = 2$	0.92	0.53	0.69	36.14	39.17	0.92	0.53	0.69	36.25	39.29	1350
$m = 3$	0.92	0.54	0.71	37.35	39.89	0.92	0.55	0.71	37.62	40.14	1048
$m = 5$	0.92	0.64	0.82	37.30	38.23	0.92	0.65	0.83	37.54	38.52	582

5   Conclusions

In this paper, we investigated the impact of problem features on the running time of a simple dominance-based EMO algorithm with restart, that maintains an unbounded archive of non-dominated solutions found so far. The topology of an arbitrary problem instance, in terms of ruggedness, multimodality, objective space dimension, objective correlation, cardinality and hypervolume of the Pareto set, as well as distance and connectedness between non-dominated solutions, has been examined for a large set of enumerable multiobjective NK-landscapes with objective correlation. First, a simple linear regression analysis revealed that the ruggedness of the landscape had the more critical effect on the algorithm performance. Second, a more-advanced multiple linear regression analysis allowed us to highlight the more significant subset of problem features. As in the single-objective case [2,7], the ruggedness and the multimodality of the landscape affect the algorithm running time to a large extent. Additionally, the number of objectives, the correlation between them, and the hypervolume of the Pareto front to be covered are all jointly impactful in the multiobjective case. At last, although problem features have a different impact depending on the objective space dimension, the degree of explanation they are able to provide together is always as meaningful for the algorithm performance. Overall, our feature-based analysis was able to highlight the main relationships between the structural properties of the landscape and the performance of the algorithm. This allowed us to better understand the behavior and the performance of this EMO algorithm class.

The problem characteristics under analysis in the paper validate the relevance of our methodology for explaining the performance of EMO approaches. However, it remains an open question if there exist supplementary features that could better capture the problem difficulty for different problem and algorithm classes, and if more general regression models would allow to better apprehend the correlations among the features as well as their (non-linear) dependencies with the algorithm performance. Furthermore, the goal of the paper was on *understanding* the algorithm behavior and performance rather than blindly recommending the best-performing approach, but a natural extension for future research is to investigate the *prediction* power of the regression models proposed in the paper, based on existing works from single-objective optimization [6]. Following the *algorithm selection problem* formulated by Rice in the 1970s [16], this would allow us to design a portfolio approach for selecting the most appropriate algorithm configuration, based on a relevant structural characterization of the multiobjective problem instance to be solved. For that purpose, extending our paradigm with more-advanced regression models based on problem features that can be estimated inexpensively for large-size instances is currently under investigation.

Acknowledgments. This work was partially supported by the Japanese-French research project "Global Research on the Framework of Evolutionary Solution Search to Accelerate Innovation"

References

1. Aguirre, H.E., Tanaka, K.: Working principles, behavior, and performance of MOEAs on MNK-landscapes. Eur. J. Oper. Res. **181**(3), 1670–1690 (2007)
2. Barnett, L.: Ruggedness and neutrality - the NKp family of fitness landscapes. In: Sixth International Conference on Artificial Life (ALIFE VI), pp. 18–27 (1998)
3. Coello Coello, C.A., Lamont, G.B., Van Veldhuizen, D.A.: Evolutionary Algorithms for Solving Multi-Objective Problems, 2nd edn. Springer, (2007)
4. Daolio, F., Verel, S., Ochoa, G., Tomassini, M.: Local optima networks and the performance of iterated local search. In: Genetic and Evolutionary Computation Conference (GECCO 2012), pp. 369–376 (2012)
5. Hansen, N., Auger, A., Ros, R., Finck, S., Pošík, P.: Comparing results of 31 algorithms from the black-box optimization benchmarking BBOB-2009. In: Genetic and Evolutionary Computation Conference (GECCO 2010), pp. 1689–1696 (2010)
6. Hutter, F., Xu, L., Hoos, H.H., Leyton-Brown, K.: Algorithm runtime prediction: Methods & evaluation. Artif. Intell. **206**, 79–111 (2014)
7. Kauffman, S.A.: The Origins of Order. Oxford University Press (1993)
8. Knowles, J.D., Corne, D.W.: Instance generators and test suites for the multiobjective quadratic assignment problem. In: Fonseca, C.M., Fleming, P.J., Zitzler, E., Deb, K., Thiele, L. (eds.) EMO 2003. LNCS, vol. 2632, pp. 295–310. Springer, Heidelberg (2003)
9. Laumanns, M., Thiele, L., Zitzler, E.: Running time analysis of evolutionary algorithms on a simplified multiobjective knapsack problem. Nat. Comput. **3**(1), 37–51 (2004)
10. Liefooghe, A., Paquete, L., Figueira, J.R.: On local search for bi-objective knapsack problems. Evol Comput **21**(1), 179–196 (2013)
11. Liefooghe, A., Verel, S., Aguirre, H., Tanaka, K.: What makes an instance difficult for black-box 0–1 evolutionary multiobjective optimizers? In: Legrand, P., Corsini, M.-M., Hao, J.-K., Monmarché, N., Lutton, E., Schoenauer, M. (eds.) EA 2013. LNCS, vol. 8752, pp. 3–15. Springer, Heidelberg (2013)
12. Mersmann, O., Bischl, B., Trautmann, H., Wagner, M., Bossek, J., Neumann, F.: A novel feature-based approach to characterize algorithm performance for the traveling salesperson problem. Ann. Math. Artif. Intell. **69**(2), 151–182 (2013)
13. Papadimitriou, C.H., Yannakakis, M.: On the approximability of trade-offs and optimal access of web sources. In: Symposium on Foundations of Computer Science (FOCS 2000), pp. 86–92 (2000)
14. Paquete, L., Stützle, T.: Clusters of non-dominated solutions in multiobjective combinatorial optimization: An experimental analysis. In: Multiobjective Programming and Goal Programming, LNMES, vol. 618, pp. 69–77. Springer (2009)
15. Paquete, L., Schiavinotto, T., Stützle, T.: On local optima in multiobjective combinatorial optimization problems. Ann. Oper. Res. **156**(1), 83–97 (2007)
16. Rice, J.R.: The algorithm selection problem. Adv. Comput. **15**, 65–118 (1976)
17. Verel, S., Liefooghe, A., Jourdan, L., Dhaenens, C.: On the structure of multiobjective combinatorial search space: MNK-landscapes with correlated objectives. Eur. J. Oper. Res. **227**(2), 331–342 (2013)
18. Zitzler, E., Thiele, L., Laumanns, M., Fonesca, C.M., Grunert da Fonseca, V.: Performance assessment of multiobjective optimizers: An analysis and review. IEEE Trans. Evol. Comput. **7**(2), 117–132 (2003)

Modified Distance Calculation in Generational Distance and Inverted Generational Distance

Hisao Ishibuchi, Hiroyuki Masuda, Yuki Tanigaki, and Yusuke Nojima[✉]

Department of Computer Science and Intelligent Systems,
Graduate School of Engineering, Osaka Prefecture University,
1-1 Gakuen-cho, Naka-ku, Sakai, Osaka 599-8531, Japan
{hisaoi,nojima}@cs.osakafu-u.ac.jp,
{hiroyuki.masuda,yuki.tanigaki}@ci.cs.osakafu-u.ac.jp

Abstract. In this paper, we propose the use of modified distance calculation in generational distance (GD) and inverted generational distance (IGD). These performance indicators evaluate the quality of an obtained solution set in comparison with a pre-specified reference point set. Both indicators are based on the distance between a solution and a reference point. The Euclidean distance in an objective space is usually used for distance calculation. Our idea is to take into account the dominance relation between a solution and a reference point when we calculate their distance. If a solution is dominated by a reference point, the Euclidean distance is used for their distance calculation with no modification. However, if they are non-dominated with each other, we calculate the minimum distance from the reference point to the dominated region by the solution. This distance can be viewed as an amount of the inferiority of the solution (i.e., the insufficiency of its objective values) in comparison with the reference point. We demonstrate using simple examples that some Pareto non-compliant results of GD and IGD are resolved by the modified distance calculation. We also show that IGD with the modified distance calculation is weakly Pareto compliant whereas the original IGD is Pareto non-compliant.

Keywords: Evolutionary Multiobjective Optimization · Performance indicators · Generational distance · Inverted generational distance · Pareto compliance

1 Introduction

Evolutionary multiobjective optimization (EMO) has been an active research area in the last two decades [3], [6], [20]. One important issue in this area is performance evaluation of EMO algorithms. Since a set of non-dominated solutions is obtained by a single run of an EMO algorithm, performance evaluation in the EMO community usually means the comparison of different non-dominated solution sets. Various performance indicators have been proposed to evaluate the quality of a non-dominated solution set [8], [14], [15], [26]. Among them, the hypervolume indictor [25] has been most frequently used. This is mainly because no other indicators are Pareto compliant [24]. It has been repeatedly pointed out in the literature [14], [15], [18], [24], [26] that

© Springer International Publishing Switzerland 2015
A. Gaspar-Cunha et al. (Eds.): EMO 2015, Part II, LNCS 9019, pp. 110–125, 2015.
DOI: 10.1007/978-3-319-15892-1_8

Pareto non-compliant misleading results can be obtained from some other performance indicators.

For example, Zitzler et al. [26] clearly illustrated that misleading results can be obtained from the generational distance (GD) indicator [21] using a simple example of a two-objective minimization problem in Fig. 1 with a reference point set $Z = \{(1, 0), (0, 10)\}$ and three solution sets $A = \{(2, 5)\}$, $B = \{(3, 9)\}$ and $C = \{(10, 10)\}$. GD is the average distance from each solution to its closest reference point. Thus the solution set $B = \{(3, 9)\}$ is evaluated as being the best since $(3, 9)$ has the minimum distance to its nearest reference point among the three solution sets (i.e., A, B and C). However, it is clear from Fig. 1 that $A = \{(2, 5)\}$ is the best among the three solution sets since $(3, 9)$ in B and $(10, 10)$ in C are dominated by $(2, 5)$ in A. A similar example of a two-objective minimization problem was used in Schütze et al. [18], which is shown in Fig. 2 with a reference point set $Z = \{(0, 1), (10, 0)\}$ and two solutions sets $A = \{(5, 2)\}$ and $B = \{(11, 3)\}$. In this example, the solution set B is evaluated as being better than the solution set A by the GD indicator whereas $(11, 3)$ is dominated by $(5, 2)$.

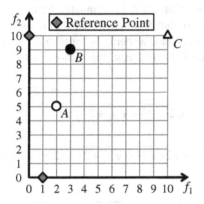
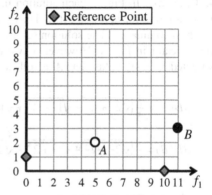

Fig. 1. Example 1 (Zitzler et al. [26])   **Fig. 2.** Example 2 (Schütze et al. [18])

These misleading results are not obtained by the hypervolume indicator since it is Pareto compliant [24]. One difficulty of the hypervolume indicator is its heavy computation load. Recently evolutionary many-objective optimization has attracted increasing attention [12]. Test problems with ten or more objectives are used for performance evaluation in recent studies on evolutionary many-objective optimization [7], [9], [10], [23]. The use of the hypervolume indicator for those test problems is often impractical from a viewpoint of computation time whereas its fast calculation [17], [22] as well as its efficient approximation [1] has been actively studied. Among other indicators, the inverted generational distance (IGD [4], [19]) is most frequently used for performance evaluation of EMO algorithms in evolutionary many-objective optimization studies [7], [9], [23]. IGD is the average distance from each reference point to its nearest solution. When a set of well-distributed reference points over the entire Pareto front is used, a small value of the IGD indicator suggests the good convergence of solutions to the Pareto front and their good distribution over the entire Pareto front.

In the above-mentioned examples in Fig. 1 and Fig. 2, the solution set A is correctly evaluated as being the best by the IGD indicator. Whereas IGD looks a more appropriate indicator than GD in Fig. 1 and Fig. 2, both are Pareto non-compliant. Let us consider another solution set $D = \{(2, 1)\}$ in Fig. 3. It is clear from Fig. 3 that the solution set D is evaluated as being the best among the four solution sets A, B, C and D by the GD indicator. However, the solution set $A = \{(2, 5)\}$ is evaluated as being better than D by the IGD indicator with the Euclidean distance in Fig. 3 as follows:

$$IGD(A) = \frac{1}{2}\left(\sqrt{(2-0)^2 + (5-10)^2} + \sqrt{(2-1)^2 + (5-0)^2} \right) = 5.24, \tag{1}$$

$$IGD(D) = \frac{1}{2}\left(\sqrt{(2-0)^2 + (1-10)^2} + \sqrt{(2-1)^2 + (1-0)^2} \right) = 5.32. \tag{2}$$

In Fig. 4, we show another example of a two-objective minimization problem with $Z = \{(0, 10), (1, 6), (2, 2), (6, 1), (10, 0)\}$, $A = \{(2, 4), (3, 3), (4, 2)\}$ and $B = \{(2, 8), (4, 4), (8, 2)\}$. In Fig. 4, each solution in the solution set B is dominated by at least one solution in the solution set A. Thus we can say that A is better than B in the sense of Pareto dominance. The solution set A is also evaluated as being better than B by the GD indicator in Fig. 4. However, if we use the IGD indicator, the solution set B is evaluated as being better than A as follows:

$$IGD(A) = \frac{1}{5}\left(\sqrt{2^2 + 6^2} + \sqrt{1^2 + 2^2} + \sqrt{1^2 + 1^2} + \sqrt{2^2 + 1^2} + \sqrt{6^2 + 2^2} \right) = 3.71, \tag{3}$$

$$IGD(B) = \frac{1}{5}\left(\sqrt{2^2 + 2^2} + \sqrt{1^2 + 2^2} + \sqrt{2^2 + 2^2} + \sqrt{2^2 + 1^2} + \sqrt{2^2 + 2^2} \right) = 2.59. \tag{4}$$

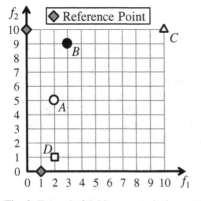 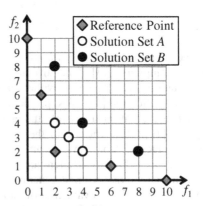

Fig. 3. Example 3 with a new solution set D **Fig. 4.** Example 4 with misleading IGD

In this paper, first we discuss why these misleading results are obtained by the GD and IGD indicators. Then we propose an idea of modifying the distance calculation between a solution and a reference point in the GD and IGD indicators by taking into account the Pareto dominance relation between them. If a solution is dominated by a reference point, we use the Euclidean distance with no modification. However, if they are non-dominated with each other, we calculate the minimum distance from the reference point to the dominated region by the solution. This distance can be viewed as an amount of the inferiority of the solution (i.e., the insufficiency of its objective values) in comparison with the reference point. Only inferior objective values of the solution to the reference point are used in their distance calculation. In our former study [11], we suggested our idea (i.e., modified distance calculation) as a trick to remedy a severe sensitivity of the IGD indicator to the specification of a reference point set. In this paper, we explain our idea in a more general setting and propose its use in both the GD and IGD indicators. We also show a theoretical property of the IGD measure with the modified distance calculation: weak Pareto compliance.

This paper is organized as follows. In Section 2, we briefly explain multiobjective optimization, Pareto dominance relations, and performance indicators. In Section 3, we explain our idea of modifying the distance calculation in the GD and IGD indicators in detail. In Section 4, we demonstrate that the Pareto non-compliant results in Figs. 1-4 are resolved by the use of the modified distance calculation. Then we show that the IGD indicator with the modified distance calculation is weakly Pareto compliant in Section 5. Finally, we conclude this paper in Section 6.

2   Multiobjective Optimization and Performance Indicators

Let us consider the following m-objective minimization problem with a decision vector x and its feasible region X:

$$\text{Minimize } z = f(x) = (f_1(x), f_2(x), ..., f_m(x)) \text{ subject to } x \in X. \tag{5}$$

In this formulation, z is an m-dimensional objective vector: $z = (z_1, z_2, ..., z_m)$. The feasible region Z of the objective vector z is defined as $Z = \{z = f(x) \mid x \in X\}$ using the feasible region X of the decision vector x.

Let us denote two objective vectors as $a = (a_1, a_2, ..., a_m)$ and $b = (b_1, b_2, ..., b_m)$. They are two points in the m-dimensional objective space. The Pareto dominance relation " $\succ$ " and the weak Pareto dominance relation " $\succeq$ " are defined for the minimization problem between the two objective vectors a and b as follows:

Pareto Dominance: $a \succ b \Leftrightarrow \forall i, a_i \leq b_i$ and $\exists j, a_j < b_j$, $\qquad(6)$

Weak Pareto Dominance: $a \succeq b \Leftrightarrow \forall i, a_i \leq b_i$. $\qquad(7)$

The Pareto dominance relation $a \succ b$ means that b is dominated by a (i.e., a is better than b). The second condition " $\exists j, a_j < b_j$ " in (6) can be replaced with $a \neq b$: $a \succ b \Leftrightarrow \forall i, a_i \leq b_i$ and $a \neq b$. The weak Pareto dominance relation $a \succeq b$ means that b is weakly dominated by b (i.e., a is better than or equal to b). The weak Pareto dominance relation $a \succeq b$ includes $a = b$ while $a = b$ is excluded from the Pareto dominance relation $a \succ b$.

If an objective vector $z^* = f(x^*)$ is not dominated by any other feasible objective vectors in Z, x^* is called a Pareto optimal solution. A set of all Pareto optimal solutions is the Pareto optimal solution set. The projection of the Pareto optimal solution set onto the objective space is called the Pareto front. When x^* is a Pareto optimal solution, $z^* = f(x^*)$ is a Pareto optimal objective vector.

Let A be a set of objective vectors. When no objective vector in A is dominated by any other objective vector in A, A is called a non-dominated set. Let us denote two non-dominated sets of objective vectors as $A = \{a_1, a_2, ..., a_{|A|}\}$ and $B = \{b_1, b_2, ..., b_{|B|}\}$ where $|A|$ and $|B|$ are the cardinality of A and B, respectively.

In Zitzler et al. [26], the Pareto dominance relations between objective vectors were extended to the following relations between objective vector sets (also see [8]):

Pareto Dominance for Sets: $A \succ B \iff \forall b_j \in B, \exists a_i \in A: a_i \succ b_j,$ (8)

Weak Pareto Dominance for Sets: $A \succeq B \iff \forall b_j \in B, \exists a_i \in A: a_i \succeq b_j.$ (9)

$A \succ B$ and $A \succeq B$ mean that "B is dominated by A" and "B is weakly dominated by A", respectively. $A \succ B$ does not allow the existence of any shared objective vector in A and B. That is, $A \succ B$ requires $(A \cap B) = \phi$. Whereas $A \succ B$ does not allow any overlap between A and B, $A \succeq B$ allows $A = B$ (i.e., A and B can be the same).

In order to handle partially overlapping sets, Zitzler et al. [26] defined an intermediate relation called "better" denoted by " $\triangleright$ " as follows (also see [8]):

Relation "better" for Sets: $A \triangleright B \iff A \succeq B$ and $A \neq B$. (10)

This relation $A \triangleright B$ means that A is better than B [26]. The concept of the Pareto compliance [24] of an indicator $I(.)$ can be defined using this relation as follows (it is assumed that a smaller value of the indicator $I(.)$ means a better set):

Pareto Compliant Indicator [24]: Whenever $A \triangleright B$ holds between two non-dominated sets A and B, $I(A) < I(B)$ always holds: $A \triangleright B \implies I(A) < I(B)$.

In this definition, the indicator $I(.)$ is a mapping from a set of objective vectors to a real number. Only the hypervolume is known as being Pareto compliant. In this paper, we also use the following weaker version of the Pareto compliance (some indicators such as the D1 [8] and the unary additive-ε [26] are weakly Pareto compliant):

Weak Pareto Compliant Indicator: Whenever $A \succeq B$ holds between two non-dominated sets A and B, $I(A) \leq I(B)$ always holds: $A \succeq B \implies I(A) \leq I(B)$.

As we have already explained in Section 1, the GD and IGD indicators evaluate the quality of an objective vector set using a reference point set. Let $Z = \{z_1, z_2, ..., z_{|Z|}\}$ be a given reference point set where $|Z|$ is the cardinality of Z. The original definition of the GD indicator can be written for a non-dominated objective vector set $A = \{a_1, a_2, ..., a_{|A|}\}$ and the reference point set $Z = \{z_1, z_2, ..., z_{|Z|}\}$ as follows [21]:

$$\textbf{Generational Distance:} \quad GD(A) = \frac{1}{|A|} \left(\sum_{i=1}^{|A|} d_i^{\,p} \right)^{1/p}, \qquad (11)$$

where d_i is the Euclidean distance from a_i to its nearest reference point in Z, and p is an integer parameter. In this paper, we always specify p as $p = 1$ in GD (and IGD).

The IGD indicator is an inverted version of the GD indicator, which is defined as

$$\textbf{Inverted Generational Distance:} \quad IGD(A) = \frac{1}{|Z|} \left(\sum_{j=1}^{|Z|} \hat{d}_j^{\,p} \right)^{1/p}, \qquad (12)$$

where $\hat{d}_j$ is the Euclidean distance from z_j to its nearest objective vector in A.

To the best of our knowledge, the term of "inverted generational distance (IGD)" was first used in 2004 by Coello & Sierra [4] and Sierra & Coello [19]. However, similar indicators had already been used since Czyzak & Jaszkiewicz [5] in 1998. In Czyzak & Jaszkiewicz [5], the weighted achievement scalarizing function was used as the distance between an objective vector and a reference point. Their indicator was denoted as D1 [8], D1$_R$ [14] and I_D [26]. The IGD indicator with the Euclidean distance was used in [2], [13] in 2003 without referring to it as IGD.

Recently, Schütze et al. [18] proposed the following modification of GD and IGD:

$$GD_p(A) = \left(\frac{1}{|A|} \sum_{i=1}^{|A|} d_i^{\,p} \right)^{1/p} \quad \text{and} \quad IGD_p(A) = \left(\frac{1}{|Z|} \sum_{j=1}^{|Z|} \hat{d}_j^{\,p} \right)^{1/p}. \qquad (13)$$

They also proposed a new indicator $\Delta_p(A) = \max\{GD_p(A), IGD_p(A)\}$. This new indicator was used in an indicator-based EMO algorithm in [16].

In this paper, we always specify the value of p as $p = 1$. This is because (i) it makes the meaning of GD and IGD clear, (ii) it has often been used in the literature, and (iii) the modified GD_p and IGD_p [18] in (13) become the same as their original definitions when $p = 1$. The GD with $p = 1$ is the average Euclidean distance from each objective vector to its nearest reference point (and the IGD with $p = 1$ is the average Euclidean distance from each reference point to its nearest objective vector).

3 Modified Distance Calculation

As explained in Section 1, the GD and IGD indicators are Pareto non-compliant. In Fig. 5 (a), we show another example where both GD and IGD are misleading: $GD(A) = 5.10 > GD(B) = 4.33$ and $IGD(A) = 5.24 > IGD(B) = 4.85$. That is, the solution set B is evaluated as being better than the solution set A by the GD and IGD indicators

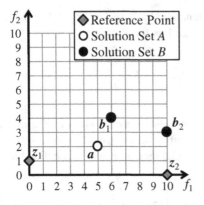
(a) Example 5 with misleading GD and IGD

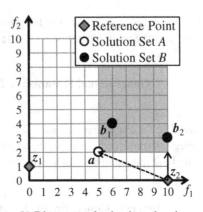
(b) Distance to the dominated region

Fig. 5. Example 5 with misleading GD and IGD, and modified distance calculation

whereas B is dominated by A (i.e., $A \succ B$). These calculations also show that Δ_p is not Pareto compliant since $\Delta_p(A) = 5.24 > \Delta_p(B) = 4.85$ for $p = 1$.

In our former study [11], we suggested an idea to calculate the distance from each reference point to the dominated region by a solution set in the IGD indicator. This idea is illustrated in Fig. 5 (b) where the distance from the reference point z_2 to the dominated region by the solution set A is calculated as shown by the vertical solid arrow. The dotted arrow from z_2 to a in Fig. 5 (b) shows the standard distance calculation from the reference point z_2 to the objective vector a.

In this paper, we formulate this idea in a more general setting so that the modified distance calculation can be used in both the GD and IGD indicators. We also explain the motivation behind the modified distance calculation and its meaning in detail.

In Fig. 5, all objective vectors a, b_1 and b_2 are dominated by the reference point z_1 at $(0, 1)$: $z_1 \succ a$, $z_1 \succ b_1$ and $z_1 \succ b_2$. The two objective vectors b_1 and b_2 in B are also dominated by a in A: $a \succ b_1$ and $a \succ b_2$. In this case, the distance from the reference point z_1 is consistent with the Pareto dominance relations among a, b_1 and b_2 as $d(z_1, a) < d(z_1, b_1)$ and $d(z_1, a) < d(z_1, b_2)$ where $d(a, b)$ is the Euclidean distance between a and b. Actually we can easily prove the following properties:

$$z \succ a \succ b \Rightarrow d(z, a) < d(z, b), \tag{14}$$

$$z \succeq a \succeq b \Rightarrow d(z, a) \leq d(z, b). \tag{15}$$

When $z \succ a \succ b$ holds among the three vectors a, b and z, we have the following relations for their elements a_i, b_i and z_i $(i = 1, 2, ..., m)$ from $z \succ a \succ b$:

$$\forall i, 0 \leq a_i - z_i \leq b_i - z_i \quad \text{and} \quad \exists j, 0 \leq a_j - z_j < b_j - z_j. \tag{16}$$

In this case, $d(z, a) < d(z, b)$ always holds. When $z \succeq a \succeq b$ holds, we have the inequality relations " $\forall i, 0 \leq a_i - z_i \leq b_i - z_i$ " from the definition of $z \succeq a \succeq b$. In this case, $d(z, a) \leq d(z, b)$ always holds.

The two properties in (14) and (15) suggest that the GD and IGD indicators can be Pareto compliant under some special conditions. However, the condition $z \succ a \succ b$ does not hold in general as shown by a and z_2 in Fig. 5.

Let us further discuss the distance calculation from a reference point to an objective vector. In Fig. 6, we show contour lines of the Euclidean distance from the reference point z. The shaded region in Fig. 6 (a) shows that all objective vectors b in this region are dominated by a. From the contour lines in Fig. 6 (a), we can see that $d(z, a) < d(z, b)$ holds for all objective vectors b in the shaded region. This means that the Euclidean distance calculation is consistent with the Pareto dominance relation when $z \succ a \succ b$ holds: $z \succ a \succ b \Rightarrow d(z, a) < d(z, b)$. However, in Fig. 6 (b), b has a shorter Euclidean distance than a whereas $a \succ b$ holds. That is, the dominated objective vector b is evaluated as being better than a by the Euclidean distance from z. The contour lines in Fig. 6 (b) show that every objective vector b in the shaded region has a shorter Euclidean distance than a while b is dominated by a. This inconsistency can be resolved in Fig. 6 (b) by calculating the minimum distance from z to the shaded area instead of the distance between z and a. This modification corresponds to the short vertical arrow in Fig. 5 (b) from z_2 to the dominated region by a (i.e., our modified distance calculation).

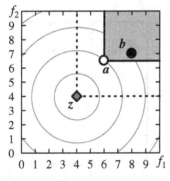

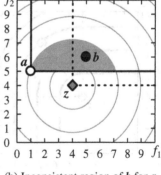

(a) Non-dominated region by a (b) Inconsistent region of b for a

Fig. 6. Contour lines of the Euclidean distance from the reference point z

In the GD and IGD indicators, smaller values mean better solution sets. The best value of each indicator is zero. Thus the distance $d(z, a)$ between the reference point z and the objective vector a used in GD and IGD can be viewed as an error or a penalty to be minimized. The distance can be also interpreted as an amount of the inferiority of a (i.e., the insufficiency of the objective values of a) in comparison with z. This interpretation is consistent with the Pareto dominance relation when a is dominated by z as in Fig. 6 (a). In Fig. 6 (a), the decrease in the distance $d(z, a)$ by moving a towards z always improves the two objectives of a. However, when a and z are non-dominated with each other, the decrease in the distance does not always improve the two objectives of a. Actually the move from a towards z in Fig. 6 (b) degrades the first objective whereas it improves the second objective. Moreover, we do not know which is better between a and z since they are non-dominated with each other.

From these discussions, we can see that the distance $d(z, a)$ cannot be viewed as an amount of the inferiority to be minimized when a and z are non-dominated with each other. As shown in Fig. 7, the distance $d(z, a)$ is the length of the vector $d = a - z$. Each element d_i of d (i.e., $d_i = a_i - z_i$) shows how a_i is inferior to (i.e., larger than) z_i with respect to the ith objective. Thus a positive value of d_i can be viewed as an amount of the inferiority (i.e., insufficiency) of a_i to z_i. However, if d_i is negative, a_i is superior to (i.e., smaller than) z_i. In this case, a negative values of d_i is viewed as having no inferiority (i.e., no insufficiency). As a result, we define an inferiority (i.e., insufficiency) vector $d^+ = (d_1^+, d_2^+, ..., d_m^+)$ as follows:

$$d_i^+ = \max\{a_i - z_i, 0\}, \quad i = 1, 2, ..., m. \tag{17}$$

When $z \succ a$ holds, d^+ is the same as $d = a - z$ (see Fig. 7 (a)). However, when $z \succ a$ does not hold, d^+ is different from $d = a - z$ since only the positive elements of d remain in d^+. In Fig. 7 (b), the vector d^+ is shown by the solid vertical arrow together with the dotted arrow d. It should be noted that the definition in (17) is replaced with $d_i^+ = \max\{z_i - a_i, 0\}$ for multiobjective maximization problems.

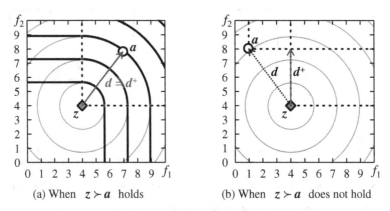

(a) When $z \succ a$ holds   (b) When $z \succ a$ does not hold

Fig. 7. Illustration of the two vectors d and d^+

Using (17), we propose the use of the following modified distance calculation $d^+(z, a)$ in the GD and IGD indicators instead of the Euclidean distance $d(z, a)$:

Modified Distance Calculation for Minimization Problems:

$$d^+(z, a) = \sqrt{d_1^{+^2} + \cdots + d_m^{+^2}} = \sqrt{(\max\{a_1 - z_1, 0\})^2 + \cdots + (\max\{a_m - z_m, 0\})^2}. \tag{18}$$

Modified Distance Calculation for Maximization Problems:

$$d^+(z, a) = \sqrt{d_1^{+2} + \cdots + d_m^{+2}} = \sqrt{(\max\{z_1 - a_1, 0\})^2 + \cdots + (\max\{z_m - a_m, 0\})^2} . \quad (19)$$

Contour lines of the modified distance from z are shown by solid bold lines in Fig. 7 (a). In the right upper region of z where $z \succ a$, the modified distance is the same as the Euclidean distance. However, in the left upper region of z in Fig. 7 (a), the contour lines are horizontal parallel straight lines since only a_2 is used (and vertical parallel straight lines since only a_1 is used in the right lower region of z).

In this paper, we denote the GD and IGD indicators with the modified distance calculation in (18) by GD^+ and IGD^+, respectively. This is because only the positive elements of $d = a - z$ are used in the distance calculation in (18).

The vector d^+ defined by (17) can be viewed as showing the minimum amount of the increase from z so that $z + d^+$ is weakly dominated by the objective vector a. Let us assume that $z + u$ is weakly dominated by a. That is, $a_i \le z_i + u_i$ for all i's. If $z \succ a$ holds, u with the minimum length is obtained from $u_i = a_i - z_i$ for all i's. However, if z and a are non-dominated with each other, there exists at least a pair of a_i and z_i with $a_i < z_i$. Such a z_i does not have to be increased from its current value. Thus $u_i = 0$ if $0 > a_i - z_i$. For the other a_i's with $0 \le a_i - z_i$, u_i is specified as $u_i = a_i - z_i$ so that $a_i \le z_i + u_i$ holds with the minimum increase u_i. This definition of u is the same as the definition of d^+ in (17). In Fig. 7 (a), the move of d is needed to make $z + d$ be weakly dominated by a. However, the move of d is not needed in Fig. 7 (b). This is because $z + d^+$ is dominated by a in Fig. 7 (b).

These discussions can be summarized as the following minimization problem of the Euclidean norm $\|u\|$, which can explain d^+ in (17) and $d^+(z, a)$ in (18):

$$\text{Minimize } \|u\| = \sqrt{u_1^2 + u_2^2 + \cdots + u_m^2} \text{ subject to } a \succeq z + u . \quad (20)$$

The vector d^+ defined in (17) is the optimal solution u^* of this problem. The modified distance $d^+(z, a)$ in (18) is the corresponding optimal value. The standard Euclidean distance corresponds to the optimal value of (20) with the equality constraint $a = z + u$ instead of the weak dominance constraint. For multiobjective maximization problems, the constraint $a \succeq z + u$ in (20) is replaced with $a \succeq z - u$ where u shows the decrease from z so that $z - u$ can be weakly dominated by the objective vector a. The optimal value of the minimization problem of $\|u\|$ with $a \succeq z - u$ corresponds to the modified distance calculation in (19).

4   Effects of Modified Distance Calculation

We examine the effects of the modified distance calculation in (18) using the six examples in Figs. 1-5 (i.e., Examples 1-5) and Fig. 6 (b). The value of p is always specified as $p = 1$. In Tables 1-6, we show the values of GD, IGD, GD^+ and IGD^+ together with the dominance relation among the given solution sets. It should be noted that GD_p and IGD_p [18] with $p = 1$ are the same as GD and IGD, respectively.

In each table, a better indicator value (i.e., smaller value) is highlighted by bold. These tables show that all Pareto non-compliant results are removed by the modified distance calculation. However, the GD^+ indicator is not Pareto compliant as shown in Fig. 8 and Table 7. In Fig. 8, the solution set A dominates the solution set B. However, B is evaluated as being better than A by GD and GD^+ in Table 7.

The IGD^+ indicator is consistent with the Pareto dominance relation (i.e., if $A \succeq B$ holds, the inconsistent result $IGD^+(A) > IGD^+(B)$ is not obtained). This property will be explained in the next section. However, the IGD^+ indicator is not Pareto compliant in the strict sense as shown in Fig. 9 and Table 8. In Fig. 9, the solution set A is better than the solution set B (i.e., $A \triangleright B$). However, in Table 8, $IGD^+(A) < IGD^+(B)$ does not hold. Actually, $I(A) = I(B)$ holds for IGD and IGD^+. This is because the nearest objective vector $(2, 2)$ from the reference point $z = (0, 0)$ is shared by the two solution sets A and B in Fig. 9: $A = \{(1, 8), (2, 2), (8, 1)\}$, $B = \{(2, 2)\}$ and $Z = \{(0, 0)\}$.

Table 1. Example 1 in Fig. 1 ($A \succ B$)

Indicator	$I(A)$	$I(B)$	$I(A) < I(B)$
GD	5.099	**3.162**	Inconsistent
GD^+	**2.000**	3.000	OK
IGD	**5.242**	6.191	OK
IGD^+	**3.550**	6.110	OK

Table 2. Example 2 in Fig. 2 ($A \succ B$)

Indicator	$I(A)$	$I(B)$	$I(A) < I(B)$
GD	5.099	**3.162**	Inconsistent
GD^+	**2.000**	3.162	OK
IGD	**5.242**	7.171	OK
IGD^+	**3.550**	7.171	OK

Table 3. Example 3 in Fig. 3 ($D \succ A$)

Indicator	$I(D)$	$I(A)$	$I(D) < I(A)$
GD	**1.414**	5.099	OK
GD^+	**1.414**	2.000	OK
IGD	5.317	**5.242**	Inconsistent
IGD^+	**1.707**	3.550	OK

Table 4. Example 4 in Fig. 4 ($A \succ B$)

Indicator	$I(A)$	$I(B)$	$I(A) < I(B)$
GD	**1.805**	2.434	OK
GD^+	**1.138**	2.276	OK
IGD	3.707	**2.591**	Inconsistent
IGD^+	**1.483**	2.260	OK

Table 5. Example 5 in Fig. 5 ($A \succ B$)

Indicator	$I(A)$	$I(B)$	$I(A) < I(B)$
GD	5.099	**4.328**	Inconsistent
GD^+	**2.000**	3.500	OK
IGD	5.242	**4.854**	Inconsistent
IGD^+	**3.550**	4.854	OK

Table 6. Example in Fig. 6 (b) ($A \succ B$)

Indicator	$I(A)$	$I(B)$	$I(A) < I(B)$
GD	3.162	**2.236**	Inconsistent
GD^+	**1.000**	2.236	OK
IGD	3.162	**2.236**	Inconsistent
IGD^+	**1.000**	2.236	OK

Table 7. Example in Fig. 8 ($A \succ B$)

Indicator	$I(A)$	$I(B)$	$I(A) < I(B)$
GD	6.318	**5.000**	Inconsistent
GD$^+$	6.318	**5.000**	Inconsistent
IGD	**2.828**	5.000	OK
IGD$^+$	**2.828**	5.000	OK

Table 8. Example in Fig. 9 ($A \triangleright B$)

Indicator	$I(A)$	$I(B)$	$I(A) < I(B)$
GD	6.318	**2.828**	Inconsistent
GD$^+$	6.318	**2.828**	Inconsistent
IGD	2.828	2.828	$I(A) = I(B)$
IGD$^+$	2.828	2.828	$I(A) = I(B)$

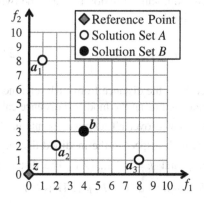

Fig. 8. Example with misleading GD$^+$ **Fig. 9.** Example with overlapping solutions

5   Weak Pareto Compliance of the IGD$^+$ Indicator

In this section, we show that the IGD$^+$ indicator is weakly Pareto compliant (i.e., $A \succeq B \Rightarrow I(A) \le I(B)$). That is, we show that $IGD^+(A) \le IGD^+(B)$ always holds whenever $A \succeq B$ holds between two non-dominated sets A and B. It should be noted that the IGD$^+$ indicator is not Pareto compliant in the strict sense: Even when $A \triangleright B$ holds, $IGD^+(A) < IGD^+(B)$ does not always hold (see Fig. 9 and Table 8 where $IGD^+(A) = IGD^+(B)$ and $A \triangleright B$).

Before showing the weak Pareto compliance property of the IGD$^+$ indicator, we first show that $d^+(z, a) \le d^+(z, b)$ always holds whenever $a \succeq b$ holds. From the definition of the weak Pareto dominance $a \succeq b$, we have $\forall i, a_i \le b_i$. Thus we have $\forall i, a_i - z_i \le b_i - z_i$. Then the following relation is obtained:

$$\forall i, \ 0 \le d_i^+(z, a) = \max\{a_i - z_i, 0\} \le \max\{b_i - z_i, 0\} = d_i^+(z, b) . \quad (21)$$

From (21), we can see that $d^+(z, a) \le d^+(z, b)$ holds. That is, $d^+(z, a) \le d^+(z, b)$ always holds whenever $a \succeq b$ holds:

$$a \succeq b \quad \Rightarrow \quad d^+(z, a) \le d^+(z, b).$$ (22)

It should be noted that this relation holds for an arbitrarily specified reference point z since we do not use any assumption on z. For example, (22) holds when z is non-dominated with a and b. It also holds even when z is dominated by a and b.

Using (22), we show that the IGD$^+$ indicator is weakly Pareto compliant. Let us assume that $A = \{a_1, a_2, ..., a_{|A|}\}$ and $B = \{b_1, b_2, ..., b_{|B|}\}$ are non-dominated sets where $A \succeq B$ holds. We also assume that $Z = \{z_1, z_2, ..., z_{|Z|}\}$ is a non-dominated reference point set. From the definition of the weak Pareto dominance relation between the two sets A and B (i.e., $A \succeq B$), the following relation holds:

$$\forall b_j \in B, \exists a_i \in A : a_i \succeq b_j .$$ (23)

$IGD^+(B)$ is calculated in the following manner. First the distance from each reference point z_k to the nearest objective vector in B is calculated using the modified distance calculation $d^+(z, b)$. Then the average value is calculated over all reference points in Z. Let $b_{j(k)}$ be the nearest objective vector in B to z_k with respect to the modified distance calculation $d^+(z, b)$ where $j(k) \in \{1, 2, ..., |B|\}$ and $k = 1, 2, ..., |Z|$. The distance from each z_k in Z to its nearest objective vector $b_{j(k)}$ in B is $d^+(z_k, b_{j(k)})$. Thus $IGD^+(B)$ is calculated as

$$IGD^+(B) = \frac{1}{|Z|} \sum_{k=1}^{|Z|} d^+(z_k, b_{j(k)}).$$ (24)

From (23), there exists at least one $a_{i(j(k))}$ in A that satisfies $a_{i(j(k))} \succeq b_{j(k)}$ for each $b_{j(k)}$ where $i(j(k)) \in \{1, 2, ..., |A|\}$ and $k = 1, 2, ..., |Z|$. That is, we can choose $a_{i(j(k))}$ for each $b_{j(k)}$ for $k = 1, 2, ..., |Z|$ in (24) such that $a_{i(j(k))} \succeq b_{j(k)}$. From (22), we have

$$a_{i(j(k))} \succeq b_{j(k)} \quad \Rightarrow \quad d^+(z_k, a_{i(j(k))}) \le d^+(z_k, b_{j(k)}).$$ (25)

Since $a_{i(j(k))} \succeq b_{j(k)}$ holds for $k = 1, 2, ..., |Z|$, $d^+(z_k, a_{i(j(k))}) \le d^+(z_k, b_{j(k)})$ also holds for $k = 1, 2, ..., |Z|$. Thus we obtain the following inequality relation:

$$IGD^+(A) \le \frac{1}{|Z|} \sum_{k=1}^{|Z|} d^+(z_k, a_{i(j(k))}) \le \frac{1}{|Z|} \sum_{k=1}^{|Z|} d^+(z_k, b_{j(k)}) = IGD^+(B).$$ (26)

The first inequality in (26) holds since the distance from z_k to its nearest objective vector in A is equal to or smaller than $d^+(z_k, a_{i(j(k))})$. When $a_{i(j(k))}$ is the nearest objective vector in A to z_k for all k's, the equality holds between the first two terms in (26). The second inequality in (26) holds from (25).

6   Conclusions

In this paper, we proposed the use of the modified distance calculation instead of the Euclidean distance in the GD and IGD indicators. The Pareto dominance relation between a reference point and an objective vector is taken into account in our modified distance calculation. Using simple numerical examples of two-objective minimization problems, we demonstrated that some Pareto non-compliant results of the GD and IGD indicators are resolved by the use of our modified distance calculation. We also showed that the IGD indicator with our modified distance calculation, which is called the IGD^+ indicator, is weakly Pareto compliant whereas IGD is Pareto non-compliant. One advantage of IGD^+ over the frequently-used hypervolume indictor is its computational efficiency. No heavy computation is added to IGD in the modified distance calculation. That is, a theoretical property is added to IGD in the IGD^+ indicator with no severe increase in its computation load.

As shown in this paper, the Pareto compliant property between two objective vectors (i.e., $a \succ b \Rightarrow d(z, a) < d(z, b)$) does not always hold when the Euclidean distance is used. That is, if a reference point z and an objective vector a are non-dominated with each other, an inconsistent result $d(z, a) > d(z, b)$ can be obtained for a and b with $a \succ b$. This inconsistency leads to Pareto non-compliant results of the GD and IGD indicators. However, when our modified distance calculation is used, the weak Pareto compliant property always holds: $a \succeq b \Rightarrow d^+(z, a) \le d^+(z, b)$. Good results of GD^+ and IGD^+ were obtained from this property. Especially, it was shown that the IGD^+ indicator is weakly Pareto compliant.

One may feel some similarity between our modified distance calculation and the epsilon indicator [26]. In its additive version, ε is used for all elements of all reference points. That is, the maximum distance over all reference points (and over all objectives of each reference point) is calculated instead of the average distance in IGD^+. One may also feel some similarity between our modified distance calculation and the weighted achievement scalarizing function used in the D1 indicator [8]. In IGD^+, the Euclidean distance is usually used as shown in Fig. 7 (a).

Future research topics include theoretical and experimental studies on the effects of the modified distance calculation on evaluation results by the GD and IGD indicators of EMO algorithms. Only a few experimental results were reported in [11]. It may be an interesting study to re-evaluate recently reported performance evaluation results of EMO algorithms on many-objective problems using the IGD^+ indicator.

References

1. Bader, J., Zitzler, E.: HypE: An algorithm for fast hypervolume-based many-objective optimization. Evolutionary Computation 19, 45–76 (2011)
2. Bosman, P.A.N., Thierens, D.: The balance between proximity and diversity in multiobjective evolutionary algorithms. IEEE Trans. on Evolutionary Computation 7, 174–188 (2003)

3. Coello, C.A.C., Lamont, G.B.: Applications of Multi-Objective Evolutionary Algorithms. World Scientific, Singapore (2004)
4. Coello, C.A.C., Reyes Sierra, M.: A study of the parallelization of a coevolutionary multi-objective evolutionary algorithm. In: Monroy, R., Arroyo-Figueroa, G., Sucar, L.E., Sossa, H. (eds.) MICAI 2004. LNCS (LNAI), vol. 2972, pp. 688–697. Springer, Heidelberg (2004)
5. Czyzak, P., Jaszkiewicz, A.: Pareto simulated annealing - A metaheuristic technique for multiple-objective combinatorial optimization. J. Multi-Criteria Decision Analysis 7, 34–47 (1998)
6. Deb, K.: Multi-objective optimization using evolutionary algorithms. John Wiley & Sons, Chichester (2001)
7. Deb, K., Jain, H.: An evolutionary many-objective optimization algorithm using reference-point based nondominated sorting approach, Part I: Solving problems with box constraints. IEEE Trans. on Evolutionary Computation 18, 577–601 (2014)
8. Hansen, M.P., Jaszkiewicz, A.: Evaluating the quality of approximations of the non-dominated set, Technical Report IMM-REP-1998-7, Technical Univ. of Denmark (1998)
9. He, Z., Yen, G.G., Zhang, J.: Fuzzy-based Pareto optimality for many-objective evolutionary algorithms. IEEE Trans. on Evolutionary Computation 18, 269–285 (2014)
10. Ishibuchi, H., Akedo, N., Nojima, Y.: Behavior of multi-objective evolutionary algorithms on many-objective knapsack problems. IEEE Trans. on Evolutionary Computation (in press)
11. Ishibuchi, H., Masuda, H., Tanigaki, Y., Nojima, Y.: Difficulties in specifying reference points to calculate the inverted generational distance for many-objective optimization problems. In: Proc. of MCDM 2014 (under IEEE SSCI 2014), pp. 170–177 (2014) http://www.cs.osakafu-u.ac.jp/ci/Papers/DownloadablePapers.php
12. Ishibuchi, H., Masuda, H., Tanigaki, Y., Nojima, Y.: Evolutionary many-objective optimization: A short review. In: Proc. of IEEE CEC 2008, pp. 2424–2431 (2008)
13. Ishibuchi, H., Yoshida, T., Murata, T.: Balance between genetic search and local search in memetic algorithms for multiobjective permutation flowshop scheduling. IEEE Trans. on Evolutionary Computation 7, 204–223 (2003)
14. Knowles, J.D., Corne, D.W.: On metrics for comparing non-dominated sets. In: Proc. of CEC 2002, pp. 711–716 (2002)
15. Knowles, J.D., Thiele, L., Zitzler, E.: A tutorial on the performance assessment of stochastic multiobjective optimizers. TIK Report No. 214, ETH Zurich (2006)
16. Martínez, S.Z., Hernández, V.A.S., Aguirre, H., Tanaka, K., Coello, C.A.C.: Using a family of curves to approximate the pareto front of a multi-objective optimization problem. In: Bartz-Beielstein, T., Branke, J., Filipič, B., Smith, J. (eds.) PPSN 2014. LNCS, vol. 8672, pp. 682–691. Springer, Heidelberg (2014)
17. Russo, L.M.S., Francisco, A.P.: Quick hypervolume. IEEE Trans. on Evolutionary Computation 18, 481–502 (2014)
18. Schütze, O., Esquivel, X., Lara, A., Coello, C.A.C.: Using the averaged Hausdorff distance as a performance measure in evolutionary multiobjective optimization. IEEE Trans. on Evolutionary Computation 16, 504–522 (2012)
19. Sierra, M.R., Coello, C.A.C.: A new multi-objective particle swarm optimizer with improved selection and diversity mechanisms, Technical Report, CINVESTAV-IPN (2004)
20. Tan, K.C., Khor, E.F., Lee, T.H.: Multiobjective Evolutionary Algorithms and Applications. Springer, London (2005)

21. Van Veldhuizen D.A.: Multiobjective Evolutionary Algorithms: Classifications, Analyses, and New Innovations. Ph.D. Thesis, Graduate School of Engineering, Air Force Institute of Technology, Wright-Patterson AFB, Ohio, USA (1999)
22. While, L., Bradstreet, L., Barone, L.: A fast way of calculating exact hypervolumes. IEEE Trans. on Evolutionary Computation **16**, 86–95 (2012)
23. Yang, S., Li, M., Liu, X., Zheng, J.: A grid-based evolutionary algorithm for many-objective optimization. IEEE Trans. on Evolutionary Computation **17**, 721–736 (2013)
24. Zitzler, E., Brockhoff, D., Thiele, L.: The hypervolume indicator revisited: on the design of pareto-compliant indicators via weighted integration. In: Obayashi, S., Deb, K., Poloni, C., Hiroyasu, T., Murata, T. (eds.) EMO 2007. LNCS, vol. 4403, pp. 862–876. Springer, Heidelberg (2007)
25. Zitzler, E., Thiele, L.: Multiobjective optimization using evolutionary algorithms - a comparative case study. In: Eiben, A.E., Bäck, T., Schoenauer, M., Schwefel, H.-P. (eds.) PPSN 1998. LNCS, vol. 1498, pp. 292–301. Springer, Heidelberg (1998)
26. Zitzler, E., Thiele, L., Laumanns, M., Fonseca, C.M., Fonseca, V.G.: Performance assessment of multiobjective optimizers: An analysis and review. IEEE Trans. on Evolutionary Computation **7**, 117–132 (2003)

On the Behavior of Stochastic Local Search Within Parameter Dependent MOPs

Víctor Adrián Sosa Hernández[1](✉), Oliver Schütze[1],
Heike Trautmann[2], and Günter Rudolph[3]

[1] Computer Science Department, CINVESTAV-IPN, Av. IPN 2508,
C.P. 07360, Col. San Pedro Zacatenco, Mexico City, Mexico
msosa@computacion.cs.cinvestav.mx, schuetze@cs.cinvestav.mx
[2] Department of Information Systems, University of Münster,
Leonardo-Campus 3, 48149 Münster, Germany
trautmann@uni-muenster.de
[3] Fakultät für Informatik, Technische Universität Dortmund,
44221 Dortmund, Germany
Guenter.Rudolph@tu-dortmund.de

Abstract. In this paper we investigate some aspects of stochastic local search such as pressure toward and along the set of interest within parameter dependent multi-objective optimization problems. The discussions and initial computations indicate that the problem to compute an approximation of the entire solution set of such a problem via stochastic search algorithms is well-conditioned. The new insights may be helpful for the design of novel stochastic search algorithms such as specialized evolutionary approaches. The discussion in particular indicates that it might be beneficial to integrate the set of external parameters directly into the search instead of computing projections of the solution sets separately by fixing the value of the external parameter.

Keywords: Parameter dependent multi-objective optimization · Stochastic local search · Evolutionary algorithms

1 Introduction

In many applications the problem arises that several objectives have to be optimized concurrently leading to a *multi-objective optimization problem* (MOP). Furthermore, it can happen that the MOP contains one or several external parameters $\lambda \in \Lambda$ such as the environmental temperature of a given mechanical system. Such parameters cannot be 'optimized', but on the other hand the decision maker cannot neglect them. Since for every fixed value of λ the problem acts as a 'classical' MOP, the solution set of such a *parameter dependent MOP* (PMOP) is given by an entire family P_Λ of Pareto sets. One question that arises is to compute a finite size representation of P_Λ which has been addressed so far in some works using specialized evolutionary algorithms ([2,3,5,7,14,16]).

© Springer International Publishing Switzerland 2015
A. Gaspar-Cunha et al. (Eds.): EMO 2015, Part II, LNCS 9019, pp. 126–140, 2015.
DOI: 10.1007/978-3-319-15892-1_9

In this work, we address one facet of this problem via investigating the behavior of stochastic local search (SLS) within PMOPs. By utilizing a certain relation of SLS with line search methods as used in mathematical programming we will see that—under certain (mild) assumptions on the model—both pressure toward and along the set of interest (in objective space) is already inherent in SLS. Initial studies on a simple set based method that includes SLS underline that the problem to compute a finite size representation of the entire solution set via stochastic search methods such as evolutionary algorithms (EAs) is a well-conditioned problem. We hope that the obtained insights will be valuable for future designs of specialized EAs. The results in particular suggest that it might make sense to integrate the entire λ-space into the search which will allow to compute the desired approximation in *one* run of the algorithm which is in contrast to the current works which consider 'λ-slices' in each run.

The remainder of this paper is organized as follows: in Section 2 we briefly state the problem at hand and discuss the related work. In Section 3, we consider some aspects of SLS within PMOPs which we underline by some computations, and finally draw our conclusions in Section 4.

2   Background and Related Work

In the following we consider continuous parameter dependent multi-objective optimization problem (PMOPs) of the form

$$\min_{x \in S} F_\lambda(x). \tag{1}$$

Hereby, F_λ is defined as a vector of objective functions

$$
\begin{aligned}
F_\lambda &: S \to \mathbb{R}^k, \\
F_\lambda(x) &= (f_{1,\lambda}(x), \cdots, f_{k,\lambda}(x)),
\end{aligned}
\tag{2}
$$

where $S \subset \mathbb{R}^n$ is the domain (here we will consider unconstrained problems, i.e., $S = \mathbb{R}^n$) and $\lambda \in \Lambda \subset \mathbb{R}^l$ specifies the external parameters to the objective functions.

Note that for every fixed value of λ problem (1) can be seen as a classical multi-objective problem (MOP) (for the discussion on classical MOPs we refer e.g. to [6]). Thus, the solution set of (1) consists of an entire family of Pareto sets which is defined as follows:

$$P_{S,\Lambda} := \{(x, \lambda) \in \mathbb{R}^{n+l}, \text{ s.t } x \text{ is a Pareto point of } F_\lambda, \ \lambda \in \Lambda\}. \tag{3}$$

The according family of Pareto fronts is denoted by $F(P_{S,\Lambda})$. Both sets typically— i.e., under mild assumption on the model— form $(k - 1 + l)$-dimensional objects.

Probably the first study in the field of parameter dependent optimization has been published by Manne in the year of 1953 [1]. In the following we summarize some important works in the evolutionary multi-objective optimization literature.

A classification of dynamic MOPs (which are particular PMOPs where the value of λ changes in time) is presented in [13]. This work focuses on the components that lead to the observed dynamic behavior. The work of Farina, Deb, and Amato [7] also deals with dynamic MOPs. It contains some test case applications as well as many results related to problems which depend of an external parameter. Further, a classification of dynamic MOPs is established. The work in [15] gives a good insight into PMOPs, but only treats problems with uniquely one external parameter by using numerical path following algorithms. Also some geometrical properties of the solution sets are discussed as well as connections to bifurcation theory are provided. In [2] the authors provide a survey over the evolutionary techniques that tackle dynamic optimization problems. They mention four main ways to master such problems: (i) increasing the diversity after the change of the solution set, (ii) maintaining the diversity over the complete run of the evolutionary algorithm to detect the changes in the solution set, (iii) memory based approaches, and finally, (iv) multi-population approaches which are the ones that reduce the main problem into subproblems or 'slices' in order to maintain a small population until the family of solution sets is reached.

The idea to use slices, or multi-population approaches, in the evolution of an evolutionary algorithm is used for example in [3]. There, an algorithm is proposed that solves the problem by dividing the objective landscape into subpopulations in order to reach all the solutions over the external parameter (in this case time). Another work related to PMOPs can be found in [12]. Here the authors use a parallel version of the NSGA-II in order to solve a dynamic optimization problems to reduce the energy consumption when solving this kind of problem. They divide the complete problem into nodes and then the algorithm NSGA-II is executed in each node to compute the solution set. Finally, in [5] the authors present a taxonomy of the ways to treat PMOPs and also mention several similarities and differences between PMOPs and MOPs. Here again the multi-population idea is used and adapted by using migration methods.

In the current literature, the investigation of stochastic local search for continuous PMOPs is neglected so far. This paper makes a first attempt to fill this gap.

3   Behavior of Stochastic Local Search Within PMOPs

For our considerations it is advantageous to treat λ —at least formally— within PMOPs as a 'normal' parameter leading to the following problem:

$$F : \mathbb{R}^{n+l} \rightarrow \mathbb{R}^{k+l}$$

$$F(\boldsymbol{x}, \lambda) = \begin{pmatrix} f_1(\boldsymbol{x}, \lambda) \\ \vdots \\ f_k(\boldsymbol{x}, \lambda) \\ \lambda \end{pmatrix} = \begin{pmatrix} g_1(\boldsymbol{x}, \lambda) \\ \vdots \\ g_{k+l}(\boldsymbol{x}, \lambda) \end{pmatrix}, \tag{4}$$

where $g_i : \mathbb{R}^{n+l} \to \mathbb{R}$, $i = 1, \ldots, k + l$. The Jacobian is given by

$$J(\boldsymbol{x}, \lambda) = \begin{pmatrix} \nabla_{\boldsymbol{x}} f_1(\boldsymbol{x}, \lambda)^T & \nabla_{\lambda} f_1(\boldsymbol{x}, \lambda)^T \\ \vdots & \vdots \\ \nabla_{\boldsymbol{x}} f_k(\boldsymbol{x}, \lambda)^T & \nabla_{\lambda} f_k(\boldsymbol{x}, \lambda)^T \\ 0 & I_l \end{pmatrix} := \begin{pmatrix} J_{\boldsymbol{x}} & J_{\lambda} \\ 0 & I_l \end{pmatrix} \in \mathbb{R}^{(k+l) \times (n+l)}, \quad (5)$$

where

$$J_{\boldsymbol{x}} = \begin{pmatrix} \nabla_{\boldsymbol{x}} f_1(\boldsymbol{x}, \lambda)^T \\ \vdots \\ \nabla_{\boldsymbol{x}} f_k(\boldsymbol{x}, \lambda)^T \end{pmatrix} \in \mathbb{R}^{k \times n}, \quad J_{\lambda} = \begin{pmatrix} \nabla_{\lambda} f_1(\boldsymbol{x}, \lambda)^T \\ \vdots \\ \nabla_{\lambda} f_k(\boldsymbol{x}, \lambda)^T \end{pmatrix} \in \mathbb{R}^{k \times l}, \quad (6)$$

and where I_l denotes the $(l \times l)$-identity matrix.

To understand the behavior of SLS it is advantageous to see its relation to line search as it is used in mathematical programming: if a point $z_1 = (\boldsymbol{x}_1, \lambda_1)$ is chosen (at random) from a small neighborhood of $z_0 = (\boldsymbol{x}_0, \lambda_0)$, then z_1 can be written as

$$z_1 = z_0 + 1(z_1 - z_0) = z_0 + ||z_1 - z_0|| \frac{z_1 - z_0}{||z_1 - z_0||}. \quad (7)$$

Thus, the selection of z_1 can be viewed as a search in direction $v := (z_1 - z_0)/||z_1 - z_0||$. For infinitesimal steps in a direction $\nu \in \mathbb{R}^{n+l}$ (in decision space) the related change in objective space is given by $J(\boldsymbol{x}_0, \lambda_0)\nu$. To see this, consider the i-th component of $J(\boldsymbol{x}_0, \lambda_0)\nu$:

$$(J(\boldsymbol{x}_0, \lambda_0)\nu)_i = \lim_{t \to 0} \frac{g_i((\boldsymbol{x}_0, \lambda_0) + t\nu) - g_i(\boldsymbol{x}_0, \lambda_0)}{t} = \langle \nabla g_i(\boldsymbol{x}_0, \lambda_0), \nu \rangle, \quad (8)$$

$$i = 1, \ldots, k + l.$$

For problem (4) this direction is given by

$$J\nu = \begin{pmatrix} J_{\boldsymbol{x}} & J_{\lambda} \\ 0 & I_l \end{pmatrix} \begin{pmatrix} \nu_{\boldsymbol{x}} \\ \nu_{\lambda} \end{pmatrix} = \begin{pmatrix} J_{\boldsymbol{x}} \nu_{\boldsymbol{x}} + J_{\lambda} \nu_{\lambda} \\ \nu_{\lambda} \end{pmatrix}, \quad (9)$$

where $J = J(\boldsymbol{x}_0, \lambda_0)$ and $\nu = (\nu_{\boldsymbol{x}}, \nu_{\lambda})$ with $\nu_{\boldsymbol{x}} \in \mathbb{R}^n$ and $\nu_{\lambda} \in \mathbb{R}^l$.

Based on these considerations, we now consider different scenarios for SLS that occur in different stages within an evolutionary algorithm.

(a) $(\boldsymbol{x}, \lambda)$ 'far away' from $P_{S,\Lambda}$. Here we use an observation made in [4] for classical MOPs namely that the 'objectives gradients' may point into similar directions when the decision point $(\boldsymbol{x}, \lambda)$ is far from the Pareto set. We assume here the extreme case namely that all gradients point into the same direction. For this, let $g := \nabla_{\boldsymbol{x}} f_i(\boldsymbol{x}, \lambda)$ and assume that

$$\nabla_{\boldsymbol{x}} f_i(\boldsymbol{x}, \lambda) = \mu_i g, \quad i = 1, \ldots, k, \quad (10)$$

where $\mu_i > 0$ for $i = 1, \ldots, k$. Then

$$J_x \nu_x = \begin{pmatrix} \nabla_x f_1(\boldsymbol{x}, \lambda)^T \nu_x \\ \vdots \\ \nabla_x f_k(\boldsymbol{x}, \lambda)^T \nu_x \end{pmatrix} = g^T \nu_x \begin{pmatrix} \mu_1 \\ \vdots \\ \mu_k \end{pmatrix}. \tag{11}$$

That is, the movement is 1-dimensional regardless of ν_x which is n-dimensional. Since $J_x \nu_x = 0$ iff $\nu_x \perp g$, the probability is one that for a randomly chosen ν_x either dominated or dominating solutions are found (and in case a dominated solution is found, the search has simply to be flipped to find dominating solutions).

Thus, for $\nu_\lambda = 0$, which means that the value of λ is not changed in the local search, we obtain for $\mu = (\mu_1, \ldots, \mu_k)^T$

$$J\nu = \begin{pmatrix} g^T \nu_x \mu \\ 0 \end{pmatrix}. \tag{12}$$

For $\nu_\lambda \neq 0$, i.e., in the case that the value of λ is changed within the local search, no such physical meaning exists to the best of our knowledge. Nevertheless, the investigation of this problem will be one topic for future research.

As a general example we consider here the following PMOP ([9]):

$$F_\lambda : \mathbb{R}^2 \to \mathbb{R}^2 \\ F_\lambda(x) := (1 - \lambda) F_1(\boldsymbol{x}) + \lambda F_2(\boldsymbol{x}), \tag{13}$$

where $\lambda \in [0, 1]$ and $F_1, F_2 : \mathbb{R}^2 \to \mathbb{R}^2$,

$$F_1(x_1, x_2) = \begin{pmatrix} (x_1 - 1)^4 + (x_2 - 1)^2 \\ (x_1 + 1)^2 + (x_2 + 1)^2 \end{pmatrix},$$

$$F_2(x_1, x_2) = \begin{pmatrix} (x_1 - 1)^2 + (x_2 - 1)^2 \\ (x_1 + 1)^2 + (x_2 + 1)^2 \end{pmatrix}.$$

This problem is a convex homotopy of the MOPs F_1 and F_2 which have both convex Pareto fronts. Figures 1 and 2 show the behavior of SLS for 100 uniformly randomly chosen points near $(\boldsymbol{x}, \lambda) = (10, 45.2, 0.7)$ for $\nu_\lambda \neq 0$ and $\nu_\lambda = 0$. As neighborhood we have chosen the infinity norm with radius $r_x = 2$ in $\boldsymbol{x}$-space and $r_\lambda = 0.3$ (respectively $r_\lambda = 0$) in λ-space. For the case $\nu_\lambda = 0$ a clear movement toward/against $F(P_{S,\Lambda})$ can be observed while this is not the case for $\nu_\lambda \neq 0$. Thus, it may make sense to exclude the change of the value of λ in early stages of the search process where the individuals of the populations are supposed to be far away from the set of interest.

(b) $(\boldsymbol{x}, \lambda)$ *'near' to* $P_{S,\Lambda}$. Here we consider again the extreme case, namely that $\boldsymbol{x}$ is a Karush-Kuhn-Tucker (KKT) point of F_λ. That is, assume that there exists a convex weight $\alpha \in \mathbb{R}^k$ such that

$$\sum_{i=1}^k \alpha_i \nabla_x f_i(\boldsymbol{x}, \lambda) = J_x^T \alpha = 0. \tag{14}$$

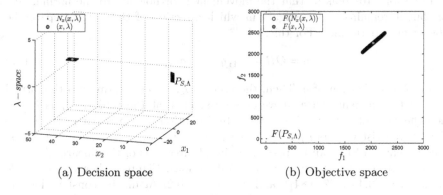

(a) Decision space (b) Objective space

Fig. 1. SLS for a point that is 'far away' from $P_{S,A}$ using $\nu_\lambda = 0$

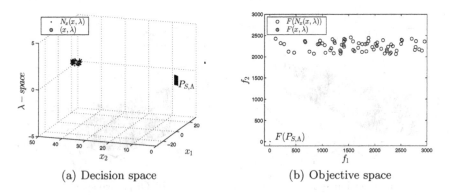

(a) Decision space (b) Objective space

Fig. 2. SLS for a point that is 'far away' from $P_{S,A}$ using $\nu_\lambda \neq 0$

It can be shown ([8]) that the normal vector to the linearized set $F(P_{(S,A)})$ at $(\boldsymbol{x}, \lambda)$ is given by

$$\eta = \begin{pmatrix} \alpha \\ -J_\lambda{}^T \alpha \end{pmatrix}. \tag{15}$$

We obtain

$$\langle J\nu, \eta \rangle = \langle \nu, J^T \eta \rangle = \langle \nu, \begin{pmatrix} J_x^T & 0 \\ J_\lambda^T & I_l \end{pmatrix} \begin{pmatrix} \alpha \\ -J_\lambda{}^T \alpha \end{pmatrix} \rangle = \langle \nu, \begin{pmatrix} J_x^T \alpha \\ J_\lambda^T \alpha - J_\lambda^T \alpha \end{pmatrix} \rangle = 0. \tag{16}$$

That is, it is either (i) $J\nu = 0$ or (ii) $J\nu$ is a movement orthogonal to η and thus along the linearized set at $F(\boldsymbol{x}, \lambda)$. If we assume that the rank of J_x is $k - 1$, then the rank of J is $k - 1 + l$ and the dimension of the kernel of J is $n - k + l$. Hence, for a randomly chosen ν the probability is 1 that event (ii) happens.

Equation (16) tells us that the movement is orthogonal to the normal vector, but it remains to investigate in which direction of the tangent space the movement is performed. For this, let

$$\eta = QR = (q_1, q_2, \ldots, q_{k+l})R \qquad (17)$$

be a QR factorization of η. Then, the vectors $q_2, \ldots, q_{k+l}$ form an orthonormal basis of the tangent space. If we assume again that the rank of J_x is $k - 1$, then the rank of J is $k - 1 + l$. Since by Equation (16) η is not in the image of J, there exist vectors $\nu_q, \ldots, \nu_{k+l}$ such that $J\nu_i = q_i$, $i = 2, \ldots, k + l$. Thus, a movement via SLS can be performed in all directions of the linearized family of Pareto fronts (i.e., both in x- and λ-direction). Figure 3 shows an example for $(x, \lambda) = (0.44, 0.47, 0.84)^T$ and $r_x = r_\lambda = 0.2$. Again, by construction, no structure in decision space can be observed, but a clear movement along the set of interest can be seen in objective space.

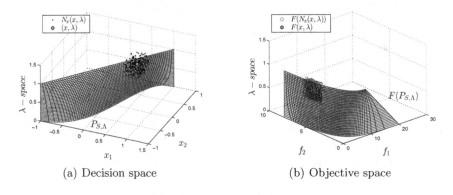

(a) Decision space   (b) Objective space

Fig. 3. SLS for a point that is 'near' to $P_{S,\Lambda}$

(c) (x, λ) 'in between'. Apparently, points (x, λ) do not have to be far away from nor near to the set of interest but can be 'in between'. In this case, no clear preference of the movement in objective space can be detected. However, this 'opening' of the search compared to the 1-dimensional movement in early stages of the search is a very important aspect since it allows in principle to find (in the set based context and given a suitable selection mechanism) and spread the solutions. In this case for finding multiple connected components. Figure 4 depicts such a scenario for $(x, \lambda) = (1, -1, 0.5)^T$ and $r_x = r_\lambda = 0.2$.

(d) Simple Neighborhood Search (SNS) within set based search. As next step we investigate the influence of SNS within set based methods. In order to prevent interferences with other effects we have thus to omit all other operators (as, e.g., crossover). The Simple Neighborhood Search for PMOPs takes this into consideration: initially, a generation $A_0 \subset \mathbb{R}^{n+l}$ is chosen at random, where Λ

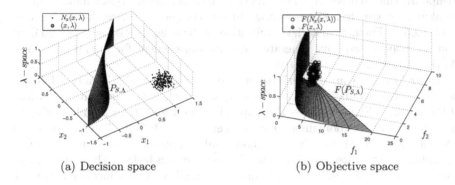

(a) Decision space   (b) Objective space

Fig. 4. SLS for a point that is 'in between' using $\nu_\lambda \neq 0$

is discretized into $\tilde{\Lambda} = \{\lambda_1, \ldots, \lambda_s\}$. In the iteration process, for every element $(a_x, a_\lambda) \in A_i$, a new element (b_x, b_λ) is chosen via SLS, where b_λ has to take one of the values of $\tilde{\Lambda}$. The given archive A_i and the set of newly created solutions B_i are the basis for the sequences of candidate solutions A_i^l, $l = 1, \ldots, s$, and the new archive A_{i+1}: for A_i^l the non-dominated solutions from $A_i \cup B_i$ with λ-value λ_l are taken, and A_{i+1} is the union of these sets (plus the respective λ values). Algorithm 1 shows the pseudo code of SNS. Hereby, $nondom(A)$ denotes the non-dominated elements of a set A, $\pi(A, \lambda_i) := \{a : (a, \lambda_i) \in A\}$ denotes the x-values of the elements of A with λ-value λ_i, and $(A, \lambda) := \{(a, \lambda) : a \in A\}$.

Algorithm 1. SNS for PMOPs

Require: Neighborhood $N_i(x, \lambda)$ of a given point (x, λ) in iteration i.
Ensure: Sequence A_i^l of candidate solutions for F_{λ_l}, $l = 1, \ldots, s$
 1: Generate $A_0 \subset {}^{n+l}$ at random
 2: **for** $i = 0, 1, 2, \ldots$ **do**
 3:    $B_i^l := \emptyset$
 4:    **for all** $(a_x, a_\lambda) \in A_i$ **do**
 5:       choose $(b_x, b_\lambda) \in N_i(a_x, a_\lambda)$
 6:       $B_i := B_i \cup (b_x, b_\lambda)$
 7:    **end for**
 8:    $A_{i+1}^l := nondom(\pi(A_i \cup B_i, \lambda_l))$, $l = 1, \ldots, s$
 9:    $A_{i+1} := \bigcup_{l=1}^s (A_{i+1}^l, \lambda_l)$
10: **end for**

For sake of a small comparison we also investigate here the global counterpart of SNS, the Simple Global Search (GS), where all points are chosen uniformly at random from the entire domain. That is, GS can be viewed as an application of SNS where the neighborhood N_i in Line 5 of Algorithm 1 is chosen as the entire domain. In order to reduce the overall number of candidate solutions we have not stored all non-dominated solutions but have used *ArchiveUpdateTight2* ([11])

to update the archives A_i^l. The archiver *ArchiveUpdateTight2* aims, roughly speaking, for gap free ϵ-Pareto sets. In our computations, we have used $\epsilon = (0.05, 0.05)^T$. Further, in each computation we have used 10 equally spaced divisions in λ-space and have generated one random element for the initial archive A_0 (i.e., $|A_0| = 10$).

Figure 5 shows some numerical results by solving PMOP (13) in two different angles. In this example we have used a budget of 3,000 function evaluations (FEs) for SNS. Figure 6 shows the respective result for GS for a budget of 3,000 and 10,000 FEs. As domain we have chosen $S = [-10, 10]^2$. The superiority of SNS can be detected visually since those final archives are evenly spread around the solution sets. Compared to this, the result of GS lacks both in spread and convergence, though more than 3 times the number of FEs has been spent to get this result. This observation is confirmed by the values in Table 1 where we show the distances (measured in terms of Δ_2 [10]) between the outcome sets and the union of the 10 Pareto fronts (i.e., our discretized set of interest).

Table 1. Comparative results for PMOPs (13) and (18) using Δ_2 value between the final archives and the union of the 10 Pareto fronts for both SNS and GS for a budget of 3,000 and 5,000 FEs. Shown are worst, average, and best value for 20 different runs.

Algorithm	$\lambda = 0.0$	$\lambda = 0.11$	$\lambda = 0.22$	$\lambda = 0.33$	$\lambda = 0.44$	$\lambda = 0.55$	$\lambda = 0.66$	$\lambda = 0.77$	$\lambda = 0.88$	$\lambda = 1.0$
SNS	1.934	2.434	2.453	1.743	1.234	1.673	1.563	0.984	0.546	0.986
	1.494	**1.977**	**2.126**	**1.526**	**1.032**	**1.402**	**1.220**	**0.550**	**0.381**	**0.385**
PMOP (13)	1.101	1.103	1.132	1.341	0.532	0.643	0.992	0.314	0.249	0.148
GS	25.342	14.349	24.342	6.342	11.433	7.322	6.424	3.221	9.534	5.213
	17.220	7.349	22.440	5.826	9.356	5.963	3.226	2.217	6.241	3.320
PMOP (13)	14.425	7.023	20.213	4.342	9.003	4.095	2.657	2.043	5.141	2.134
SNS	0.632	0.567	0.453	0.578	0.375	0.198	0.297	0.246	0.186	0.123
	0.365	**0.221**	**0.136**	**0.104**	**0.049**	**0.047**	**0.069**	**0.056**	**0.077**	**0.045**
PMOP (18)	0.242	0.201	0.103	0.098	0.019	0.043	0.034	0.022	0.062	0.032
GS	0.992	0.832	0.567	0.693	0.700	0.422	0.596	0.834	0.532	0.423
	0.692	0.620	0.454	0.496	0.684	0.329	0.552	0.821	0.446	0.351
PMOP (18)	0.432	0.597	0.353	0.394	0.592	0.239	0.539	0.739	0.422	0.311

Next we consider a second PMOP which is again a convex homotopy of two MOPs. In this case, one of the Pareto fronts is convex while the other one is concave.

$$F_\lambda : \mathbb{R}^2 \to \mathbb{R}^2$$
$$F_\lambda(x) := (1 - \lambda)F_1(x) + \lambda F_2(x), \tag{18}$$

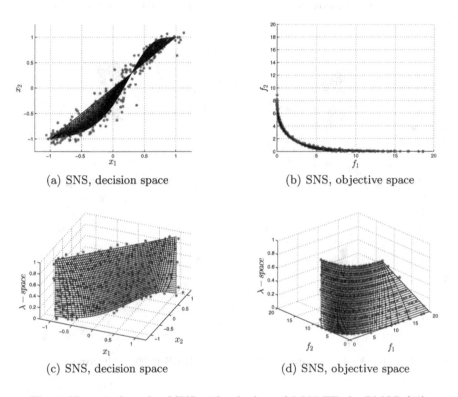

(a) SNS, decision space

(b) SNS, objective space

(c) SNS, decision space

(d) SNS, objective space

Fig. 5. Numerical result of SNS with a budget of 3,000 FEs for PMOP (13)

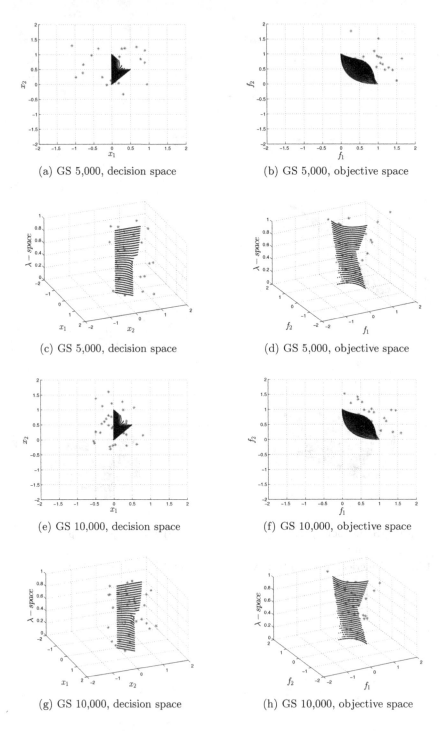

(a) GS 5,000, decision space

(b) GS 5,000, objective space

(c) GS 5,000, decision space

(d) GS 5,000, objective space

(e) GS 10,000, decision space

(f) GS 10,000, objective space

(g) GS 10,000, decision space

(h) GS 10,000, objective space

Fig. 6. Numerical result of GS with a budget of 3,000 FEs and 10,000 FEs for PMOP (13)

where $\lambda \in [0, 1]$, $a_1 = 0$, $a_2 = 1$ and $F_1, F_2 : \mathbb{R}^2 \to \mathbb{R}^2$,

$$F_1(x) = \begin{pmatrix} (x_1^2 + x_2^2)^{0.125} \\ ((x_1 - 0.5)^2 + (x_2 - 0.5)^2)^{0.25} \end{pmatrix},$$

$$F_2(x) = \begin{pmatrix} x_1^2 + x_2^2 \\ (x_1 - a_1)^2 + (x_2 - a_2)^2 \end{pmatrix}.$$

Figures 7 and 8 show some exemplary numerical results of SNS (5,000 FEs) and GS (5,000 and 10,000 FEs) using $S = [-10, 10]^2$. Again, SNS, though less FEs were used, is superior with respect to spread and convergence according the indicator Δ_2 in Table 1.

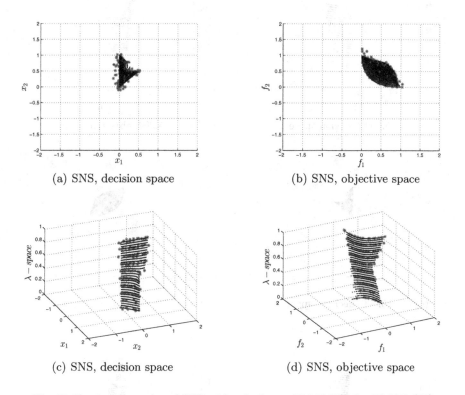

(a) SNS, decision space   (b) SNS, objective space

(c) SNS, decision space   (d) SNS, objective space

Fig. 7. Numerical results of SNS with a budget of 5,000 FEs for PMOP (18)

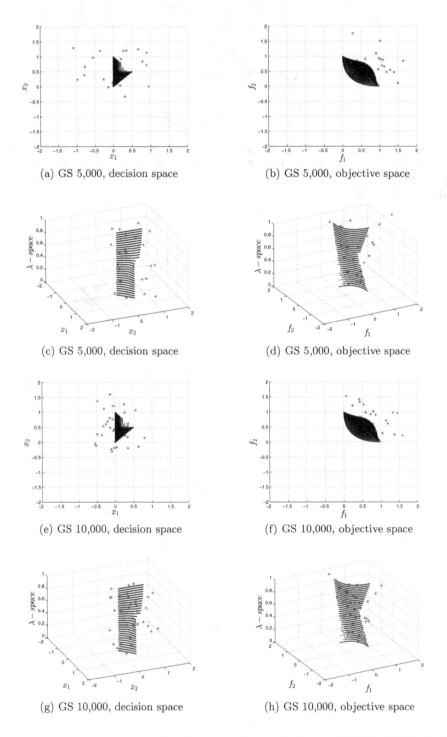

(a) GS 5,000, decision space

(b) GS 5,000, objective space

(c) GS 5,000, decision space

(d) GS 5,000, objective space

(e) GS 10,000, decision space

(f) GS 10,000, objective space

(g) GS 10,000, decision space

(h) GS 10,000, objective space

Fig. 8. Numerical results of SNS (5,000 FEs) and GS (10,000 FEs) for PMOP (18)

4   Conclusions and Future Work

In this paper we have investigated some aspects of the behavior of stochastic local search (SLS) within parameter dependent multi-objective optimization problems (PMOPs) which can to a certain extent be explained by considering line search with infinitesimal step sizes. By this, we have seen that both the movement toward as well as along the set of interest (in objective space) are inherent in SLS which we have done by considering the extreme cases in the stages of a search process. Further, we have conjectured that there is also a kind of 'opening' of the search in objective space which allows to find in principle all regions of the solution set. Some first tests on a simple set based neighborhood search, called SNS, confirmed these statements on two PMOPs. Thus, the discussions indicate that the problem to find an approximation of the entire solution set of a PMOP is a well-conditioned problem for set based probabilistic algorithms such as evolutionary algorithms (EAs).

For future work, there are some issues that are interesting to address. For instance, we intend to extend the above considerations to constrained problems. Further, the design of a specialized EA for the approximation of the family of Pareto fronts of a PMOP might be interesting. Based on the above insight it might be in particular advantageous to develop strategies that allow to compute the set of interest in *one* run of the algorithm instead of the consideration in 'λ-slices' as done so far. Finally, for the efficient comparison of methods a better adaption of performance indicators would be desirable to assess the final approximation of the family of Pareto fronts.

Acknowledgments. Víctor Adrián Sosa Hernández acknowledges support by the Consejo Nacional de Ciencia y Tecnología (CONACYT). Oliver Schütze acknowledges support from CONACYT project no. 128554. Heike Trautmann acknowledges support by the European Center of Information Systems (ERCIS). All authors acknowledge support from CONACYT project no. 174443 and DFG project no. TR 891/5-1.

References

1. Bank, B., Guddat, J., Klatte, D., Kummer, B., Tammer, K.: Non-Linear Parametric Optimization. Akademie-Verlag, Berlin (1982)
2. Branke, J.: Evolutionary approaches to dynamic optimization problems-updated survey. In: GECCO Workshop on Evolutionary Algorithms for Dynamic Optimization Problems, pp. 27–30 (2001)
3. Branke, J., Kaußler, T., Smidt, C., Schmeck, H.: A multi-population approach to dynamic optimization problems. In: Evolutionary Design and Manufacture, pp. 299–307. Springer (2000)
4. Brown, M., Smith, R.E.: Directed multi-objective optimization. International Journal of Computers, Systems, and Signals 6(1), 3–17 (2005)
5. Bu, Z., Zheng, B.: Perspectives in dynamic optimization evolutionary algorithm. In: Cai, Z., Hu, C., Kang, Z., Liu, Y. (eds.) ISICA 2010. LNCS, vol. 6382, pp. 338–348. Springer, Heidelberg (2010)

6. Deb, K.: Multi-objective optimization using evolutionary algorithms, vol. 16. John Wiley & Sons (2001)
7. Farina, M., Deb, K., Amato, P.: Dynamic Multiobjective Optimization Problems: Test Cases, Approximations, and Applications. IEEE Transactions on Evolutionary Computation 8(5), 425–442 (2004)
8. Hernandez, V.A.S.: On the Numerical Treatment of Parametric Multi-Objective Optimization Problems and Memetic Evolutionary Algorithms. Dissertation, CINVESTAV-IPN, December 2013
9. Schütze, O.: Set Oriented Methods for Global Optimization. Dissertation, Universitt Paderborn, December 2004
10. Schütze, O., Esquivel, X., Lara, A., Coello Coello, C.A.: Using the averaged Hausdorff distance as a performance measure in evolutionary multiobjective optimization. IEEE Transactions on Evolutionary Computation 16(4), 504–522 (2012)
11. Schütze, O., Laumanns, M., Tantar, E., Coello, C., Talbi, E.: Computing gap free pareto front approximations with stochastic search algorithms. Evolutionary Computation 18(1), 65–96 (2010)
12. Tantar, A.-A., Danoy, G., Bouvry, P., Khan, S.U.: Energy-efficient computing using agent-based multi-objective dynamic optimization. In: Kim, J.H., Lee, M.J. (eds.) Green IT: Technologies and Applications. Non-series, vol. 77, pp. 267–287. Springer, Heidelberg (2011)
13. Tantar, A.-A., Tantar, E., Bouvry, P.: A classification of dynamic multi-objective optimization problems. In: Proceedings of the 13th Annual Conference Companion on Genetic and Evolutionary Computation, pp. 105–106. ACM (2011)
14. Tinós, R., Yang, S.: A self-organizing random immigrants genetic algorithm for dynamic optimization problems. Genetic Programming and Evolvable Machines 8(3), 255–286 (2007)
15. Witting, K.: Numerical Algorithms for the Treatment of Parametric Multiobjective Optimization Problems and Applications. Dissertation, Universität Paderborn, February 2012
16. Yang, S.: Genetic algorithms with elitism-based immigrants for changing optimization problems. In: Giacobini, M. (ed.) EvoWorkshops 2007. LNCS, vol. 4448, pp. 627–636. Springer, Heidelberg (2007)

An Evolutionary Approach to Active Robust Multiobjective Optimisation

Shaul Salomon[1]([✉]), Robin C. Purshouse[1],
Gideon Avigad[2], and Peter J. Fleming[1]

[1] Department of Automatic Control and Systems Engineering,
University of Sheffield, Mappin Street, Sheffield S1 3JD, UK
{s.salomon,r.purshouse,p.fleming}@sheffield.ac.uk
[2] Department of Mechanical Engineering,
ORT Braude College of Engineering, Karmiel, Israel
gideona@braude.ac.il

Abstract. An Active Robust Optimisation Problem (AROP) aims at finding robust adaptable solutions, i.e. solutions that actively gain robustness to environmental changes through adaptation. Existing AROP studies have considered only a single performance objective. This study extends the Active Robust Optimisation methodology to deal with problems with more than one objective. Once multiple objectives are considered, the optimal performance for every uncertain parameter setting is a set of configurations, offering different trade-offs between the objectives. To evaluate and compare solutions to this type of problems, we suggest a robustness indicator that uses a scalarising function combining the main aims of multi-objective optimisation: proximity, diversity and pertinence. The Active Robust Multi-objective Optimisation Problem is formulated in this study, and an evolutionary algorithm that uses the hypervolume measure as a scalarasing function is suggested in order to solve it. Proof-of-concept results are demonstrated using a simplified gearbox optimisation problem for an uncertain load demand.

Keywords: Robust optimisation · Uncertainties · Multi-objective optimisation · Adaptation · Gearbox · Design

1 Introduction

When solving real-world optimisation problems, the physical system is represented by a model to predict the future performance of candidate solutions. As a result, uncertainties become an inseparable part of the optimisation process, and solutions need to be robust in addition to having good predicted performance. A solution is considered as robust if it is less affected by the negative effects of uncertainties.

The ability of many products to adapt to environment changes provides them with active robustness to uncertain operating conditions. The active robust optimisation (ARO) methodology was suggested in order to evaluate the added value

© Springer International Publishing Switzerland 2015
A. Gaspar-Cunha et al. (Eds.): EMO 2015, Part II, LNCS 9019, pp. 141–155, 2015.
DOI: 10.1007/978-3-319-15892-1_10

of adaptability [1]. Till date, ARO dealt with improvement of a single performance metric through adaptation. Since the majority of real-world optimisation problems involve several, often conflicting, objectives, this study extends the ARO methodology to deal with multi-objective optimisation problems (MOPs).

1.1   Robust Multi-objective Optimisation

A MOP can be formulated as:

$$\min_{\mathbf{x} \in \mathcal{X}} \mathbf{f}(\mathbf{x}, \mathbf{p}), \tag{1}$$

where $\mathbf{f}$ is a vector of performance measures, $\mathbf{x}$ is a vector of design variables, $\mathcal{X}$ is the feasible domain defined by a set of equality and inequality constraints, and $\mathbf{p}$ is a vector of parameters that cannot be determined by the designer.

Since uncertainties exist in all real-world optimisation problems, they should be accommodated within the optimisation procedure. Uncertainties might be *epistemic*, resulting from discrepancies between the model used for optimisation and the real system, or *aleatory*, where the variables within the system inherently change from unit to unit or time to time.

In their review on robust optimisation, Beyer and Sendhoff [2] classified the sources of uncertainties as follows:

Type A uncertainties occur when the environmental parameters $\mathbf{p}$ are unknown (epistemic) or may change within an expected range (aleatory).

Type B uncertainties are present when the actual values of design variables $\mathbf{x}$ differ from their nominal values, identified by the optimisation procedure. The deviation might occur upon production (manufacturing tolerances) or during operation (deterioration).

Type C uncertainties relate to model inaccuracies in predicting the performance $\mathbf{f}$ of the candidate design. This may result from an incorrect or simplified description of the relationship between variables within the model.

If the uncertainties are not addressed during the optimisation, solution identified as 'optimal' may poorly perform when implemented in real life. Over the past two decades, robust optimisation (RO) has gained increasing popularity, with many studies aiming at identifying robust solutions rather than optimal solutions. When formulating a robust optimisation problem, robustness criteria are specified to determine how candidate solutions should be evaluated with respect to the uncertainties involved.

We use upper case letters to distinguish random variates from deterministic values. Whenever uncertainties of either Type A-C are concerned, the objective vector $\mathbf{f}$ becomes a random variate $\mathbf{F}$. In a robust optimisation scheme, the aim is to optimise the robustness criterion $I[\mathbf{F}]$, that holds some information about the distribution of $\mathbf{F}$:

$$\min_{x \in \mathcal{X}} I[\mathbf{F}(\mathbf{X}, \mathbf{P})]. \tag{2}$$

The most common robustness criteria are the worst-case scenario (e.g., [3–5]), and aggregated values such as the mean value or the variance (e.g., [6–9]). Other criteria also exist, for example, the probability for the objective functions to be better than some predefined threshold [10], a minimum confidence level in performance [5], or performing within a predefined neighbourhood of some nominal performance vector [8].

Most of the existing evolutionary algorithms for multi-objective RO consider Type C uncertainties, represented by added noise to the nominal function values. The first evolutionary algorithm for robust MOPs were suggested in 2001 by Teich [6] and Hughes [7]. Teich suggested probabilistic dominance as an alternative to the dominance relation [6]. Hughes suggested a ranking scheme based on the sum of probabilities for each solution to be dominated [7]. Since then, several evolutionary optimisers were designed to account for Type C uncertainties [11–15].

Perturbation in design variables (Type B uncertainty) was addressed by [8,16], where each design was represented by a sampled set of designs within its neighbourhood. An algorithm aiming for reducing the amount of function evaluations for this scheme was introduced in [9].

To our knowledge, apart from previous work by the authors [1], there are no studies that explicitly treat Type A uncertainties with an evolutionary RO scheme. Instead, uncertain and dynamic environments are considered in the scope of *dynamic optimisation*, where the aim is to track a moving optimum, and remain optimal as the environment changes [17]. In dynamic optimisation the problem is deterministic, but it has to be re-solved every time the environment changes.

1.2 Active Robust Optimisation Methodology

The ARO methodology [1], is a special case of robust optimisation, where the product has some adjustable properties that can be modified by the user after the optimised design has been realised. These adjustable variables allow the product to adapt to variations of the uncontrolled parameters, so it can actively suppress their negative effect. The methodology makes a distinction between three types of variables: design variables $\mathbf{x}$, adjustable variables $\mathbf{y}$ and uncertain parameters $\mathbf{P}$, which cannot be controlled. A single realised vector of uncertain parameters from the random variate $\mathbf{P}$ is denoted as $\mathbf{p}$.

In a single-objective robust optimisation problem with Type A uncertainties, each realisation $\mathbf{p}$ is associated with a corresponding objective function value $f(\mathbf{x}, \mathbf{p})$, and a solution $\mathbf{x}$ is associated with a distribution of objective function values that correspond to the variate of the uncertain parameters $\mathbf{P}$. We denote this distribution as $F(\mathbf{x}, \mathbf{P})$. In ARO, for every realisation of the uncertain environment, the performance also depends on the value of the adjustable variables $\mathbf{y}$, i.e., $f \equiv f(\mathbf{x}, \mathbf{y}, \mathbf{p})$. Since the adjustable variables' values can be selected after $\mathbf{p}$ is realised, the solution can improve its performance by adapting its adjustable variables to the new conditions. In order to evaluate the solution's performance according to the robust optimisation methodology, it is conceivable that the $\mathbf{y}$

vector that yields the best performance for each realisation of the uncertainties will be selected. This can be expressed as the optimal configuration $\mathbf{y}^\star$:

$$\mathbf{y}^\star = \underset{\mathbf{y}\in\mathcal{Y}(\mathbf{x})}{\arg\min}\, f(\mathbf{x},\mathbf{y},\mathbf{p}), \tag{3}$$

where $\mathcal{Y}(\mathbf{x})$ is the solution's domain of adjustable variables. it is also termed as the solution's adaptability.

Considering the entire environmental uncertainty, a one-to-one mapping between the scenarios in $\mathbf{P}$ and the optimal configurations in $\mathcal{Y}(\mathbf{x})$ can be defined as:

$$\mathbf{Y}^\star = \underset{\mathbf{y}\in\mathcal{Y}(\mathbf{x})}{\arg\min}\, F(\mathbf{x},\mathbf{y},\mathbf{P}). \tag{4}$$

Assuming a solution will always adapt to its optimal configuration, its performance can be described by the following variate:

$$F(\mathbf{x},\mathbf{P}) \equiv F(\mathbf{x},\mathbf{Y}^\star,\mathbf{P}). \tag{5}$$

Following the above, the *Active Robust Opimisation Problem* is formulated:

$$\underset{\mathbf{x}\in\mathcal{X}}{\min}\, I[F(\mathbf{x},\mathbf{Y}^\star,\mathbf{P})], \tag{6a}$$

$$\text{where:}\quad \mathbf{Y}^\star = \underset{\mathbf{y}\in\mathcal{Y}(\mathbf{x})}{\arg\min}\, F(\mathbf{x},\mathbf{y},\mathbf{P}). \tag{6b}$$

It is a bi-level optimisation problem. In order to compute the objective function F in Eq. (6a), the problem in Eq. (6b) has to be solved for every solution $\mathbf{x}$, with the entire environment universe $\mathbf{P}$. To evaluate F, one may consider one or more robustness criteria $I[F]$.

2   Methodology

This study extends the single objective AROP in Eq. (6) to the following multi-objective formulation:

$$\underset{\mathbf{x}\in\mathcal{X}}{\min}\, I[\underline{\mathbf{F}}(\mathbf{x},\underline{\mathbf{Y}}^\star,\mathbf{P})], \tag{7a}$$

$$\text{where:}\quad \underline{\mathbf{Y}}^\star = \underset{\mathbf{y}\in\mathcal{Y}(\mathbf{x})}{\arg\min}\, \mathbf{F}(\mathbf{x},\mathbf{y},\mathbf{P}), \tag{7b}$$

where $\arg\min\, \mathbf{F}$ is defined in terms of Pareto optimality, and the underscore notation is used to distinguish a set from a single point.

The most notable difference between Eq. (6) and Eq. (7) is that the solution $\underline{\mathbf{Y}}^\star$ in Eq. (7b) is a variate of Pareto optimal sets, rather than the variate of a single optimal configuration in Eq. (6b). Instead of a one-to-one mapping between $\mathbf{P}$ and $\mathbf{Y}^\star$, Eq. (7b) consists of a one-to-many mapping. As a result, Eq. (7a) minimises the variate of Pareto optimal frontiers $\underline{\mathbf{F}}(\mathbf{x},\underline{\mathbf{Y}}^\star,\mathbf{P})$.

The difference between $f(\mathbf{x}, \mathbf{y}^\star, \mathbf{p})$ and $\underline{f}(\mathbf{x}, \underline{\mathbf{y}}^\star, \mathbf{p})$ is illustrated in Fig. 1. A candidate solution $\mathbf{x}$ is evaluated for two scenarios of the uncertain parameter space $\mathbf{P}$ (Fig. 1(a)). The performance of the solution for every scenario $\mathbf{p}$ (star or triangle) depends on its configuration $\mathbf{y}$ (Fig. 1(b)). In Fig. 1(c), f_1 is the only objective. All possible objective values are bounded by the solid and dashed lines for the star and triangle scenarios, respectively. The black stars and triangles in Figures 1(b) and 1(c) mark the optimal configuration and objective value for each scenario ($\mathbf{y}^\star$ and $f(\mathbf{x}, \mathbf{y}^\star, \mathbf{p})$, respectively). In Fig. 1(d) an additional objective is considered. Now all possible objective values are bounded by the solid and dashed contours, and the optimal configuration for each scenario consists of a set rather than a single configuration, denoted by the additional white shapes.

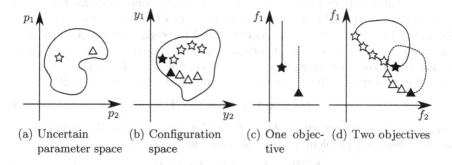

(a) Uncertain   (b) Configuration   (c) One objec-   (d) Two objectives
parameter space   space   tive

Fig. 1. Optimal configurations of a candidate solution $\mathbf{x}$ for two scenarios of the uncertainties, associated with the environmental parameters

The problem in Eq. (7) is termed here as an Active Robust Multi-objective Optimisation Problem (ARMOP). It introduces a very challenging question: *How can adaptable products be evaluated and compared according to their variates of Pareto frontiers $\underline{\mathbf{F}}(\mathbf{x}, \underline{\mathbf{Y}}^\star, \mathbf{P})$?* In Section 2.1 we introduce a first attempt to address this challenge, and suggest a set-based robustness indicator. In Section 4 we demonstrate how this indicator can be integrated into an evolutionary algorithm in order to solve an ARMOP.

2.1   Evaluating a Variate of Sets

In order to evaluate a candidate solution for an ARMOP, we suggest using a robustness criterion that quantifies the variate of Pareto frontiers with a single scalar value. Keeping in mind there is no way to avoid the loss of meaningful information when using a scalarising function, we strive to extract as much information as possible regarding the quality of the trade-off surfaces $\mathbf{F}(\mathbf{x}, \underline{\mathbf{Y}}^\star, \mathbf{P})$. Following the motivation in evolutionary multiobjective optimisation (EMO), an approximated solution to a MOP is evaluated according to three major qualities [18]: **proximity** of the approximated front to the true Pareto front (PF), **diversity** of the solutions, and **pertinence** to the preferred region of interest.

One of the well-known quality indicators for approximation sets is the hyper-volume (HV), defined as the volume of objective space enclosed by the Pareto front and a reference point [19]. The HV measure provides an integrated measure of proximity, diversity and pertinence, although it is sensitive to the choice of a reference point [20]. Despite this drawback, we use it to demonstrate the concept of the robustness indicator suggested in this study.

Hypervolume-Based Robustness Indicator. Without loss of generality, we consider the variate $\mathbf{P}$ as a finite set of sampled scenarios $\mathbf{p}$. The HV of solution $\mathbf{x}$ for scenario $\mathbf{p}$ is denoted as $hv(\mathbf{x}, \mathbf{p})$. It is calculated according to the ideal vector $\mathbf{f}^*$ and the worst objective vector $\mathbf{f}^w$, which are the vectors with minimum and maximum objective values, respectively, amongst all known solutions and scenarios. The robustness indicator I_{hv} is derived as follows:

First, the objectives of $\mathbf{f}$ are normalised in a manner that supports DM's preferences (e.g., setting $\mathbf{f}^*$ to zero and $\mathbf{f}^w$ to a vector of weights between 0-1).

Next, the hypervolume $hv(\mathbf{x}, \mathbf{p})$ is calculated for each scenario $\mathbf{p} \in \mathbf{P}$, using the worst objective vector as a reference point. The variate of the hypervolume measure that corresponds to the variate $\mathbf{P}$ is denoted as $HV(\mathbf{x}, \mathbf{P})$.

Finally, a robustness criterion is used to evaluate the variate $HV(\mathbf{x}, \mathbf{P})$:

$$I_{hv}[\mathbf{F}(\mathbf{x}, \mathbf{P})] = I[HV(\mathbf{x}, \mathbf{P})]. \tag{8}$$

Since the aim is to maximise $HV(\mathbf{x}, \mathbf{P})$ and its value is bounded between 0-1, in a minimisation problem, the complement can be used:

$$I_{hv}[\mathbf{F}(\mathbf{x}, \mathbf{P})] = I[1 - HV(\mathbf{x}, \mathbf{P})]. \tag{9}$$

Fig 2 demonstrates the above procedure for a population of two solutions. Three scenarios of $\mathbf{p}$ are considered, where the Pareto frontiers of the two solutions are depicted in stars and circles. For scenario $\mathbf{p}_3$, dashed contours show the domains in objective space that include the performances of all evaluated configurations. The worst objective vector is calculated according to the objective vectors of all configurations, including non optimal ones. The variate $HV(\mathbf{x}, \mathbf{P})$ is shown as the collection of three HVs for $\mathbf{x}$.

3   Case Study – Gearbox Optimisation Problem

We demonstrate our approach with a gearbox optimisation task for an uncertain load demand. A load with inertia J_L needs to be rotated at speed ω_L with a torque τ_L. All of the load parameters above may vary within known intervals. The torque is transmitted to the load from a geared motor system consisting of a DC motor and a two staged transmission with five gears. The gearbox optimisation problem, formulated as an ARMOP, searches for the number of teeth in each gear to minimise energy consumption and acceleration time. The system is evaluated at both steady state, i.e., operating at the load-speed scenarios (which are uncertain), and during transient conditions when accelerating from rest to each scenario. Three objectives are considered: power consumption in steady state (P), energy required to accelerate to steady state speed (E), and time to accelerate to steady state speed (T).

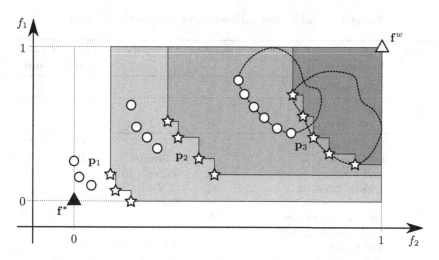

Fig. 2. Pareto frontiers $\underline{\mathbf{F}}(\mathbf{x}, \underline{\mathbf{Y}}^{\star}, \mathbf{P})$ of two solutions ($\mathbf{x}$ and $\mathbf{x}_o$) for three scenarios. The ideal vector is marked with a black triangle and the worst objective vector with a white triangle. The hypervolumes $hv(\mathbf{x}, \mathbf{p}_1)$, $hv(\mathbf{x}, \mathbf{p}_2)$ and $hv(\mathbf{x}, \mathbf{p}_3)$ are shown in the figure.

3.1 Mathematical Model

The variables and parameters of the motor and gear system are described in Table 1. The values are based on the Maxon A-max 32 DC motor specifications.

At steady state, the power consumption of a geared DC motor is [21]:

$$P = V * I, \tag{10a}$$

$$\text{where:} \quad I = \frac{\left(J_L + J_g + n_2^2 J_l + n^2 J_m\right)\dot{\omega} + \left(\nu_g + n^2 \nu_m\right)\omega + \tau}{nk_t}, \tag{10b}$$

$$V = RI + nk_v\omega. \tag{10c}$$

When the load is accelerated from rest, it is possible to calculate the speed trajectory, for given trajectories of input voltage and speed reduction, by solving the following differential equation:

$$\dot{\omega}(t) = \frac{n(t)k_t V(t) - n(t)^2 k_v k_t \omega(t)}{\left(J_L + J_g + n_2(t)^2 J_l + n(t)^2 J_m\right)R} - \frac{\left(\nu_g + n(t)^2 \nu_m\right)\omega(t) + \tau}{J_L + J_g + n_2(t)^2 J_l + n(t)^2 J_m}, \tag{11}$$

where $\omega(0) = 0$ is used as a starting condition.

Table 1. Variables and parameters for the gearbox ARMOP

Type	Variable/ Parameter	Symbol	Units	Lower limit	Upper limit
x	no. of teeth	z_g		19	61
y	gear no.	i		1	5
	input voltage	V	V	0	12
p	load speed	ω	s^{-1}	16.5	295
	load torque	τ	Nm$\cdot 10^{-3}$	10	260
	load inertia	J_L	Kg$\cdot$m$^2 \cdot 10^{-3}$	5	10
	velocity constant	k_v	V$\cdot$s$\cdot 10^{-3}$		24.3
	torque constant	k_t	Nm$\cdot$A$^{-1} \cdot 10^{-3}$		24.3
	armature resistance	R	Ω		2.23
	motor damping coefficient	ν_m	Nm$\cdot$s$\cdot 10^{-6}$		3.16
	motor inertia	J_m	Kg$\cdot$m$^2 \cdot 10^{-6}$		4.17
	max nominal current	I_{nom}	A		1.8
	gear damping coefficient	ν_g	Nm$\cdot$s$\cdot 10^{-6}$		30
	first reduction ratio	n_1			3.21
	transmission no. of teeth	N_t			80
	maximum acceleration time	t_{max}	s		20
derived	armature current	I	A	0	5.39
	second reduction ratio	n_2		0.311	3.21
	total reduction ratio	n		1	10.3
	layshaft inertia	J_l	Kg$\cdot$m$^2 \cdot 10^{-6}$	15.9	64.5
	load shaft inertia	J_g	Kg$\cdot$m$^2 \cdot 10^{-6}$	5.21	53.7

The total energy required for acceleration E can be derived from Eq. (10):

$$E = \int_0^T \frac{V(t)\Big(V(t) - n(t)k_v\omega(t)\Big)}{R} dt, \tag{12}$$

where T is the time ω reaches the required speed.

3.2   Problem Formulation

According to the ARO methodology, introduced in Section 1.2, the problem variables are sorted in Table 1 to three types: **x**, **y** and **p**. Most of the parameters in this problem are considered as having deterministic values, but some (ω, τ and J_L) possess uncertain values. The random variates of ω, τ and J_L are denoted as Ω, $\mathcal{T}$ and $\mathcal{J}_L$, respectively. The resulting variate of **p** is denoted as **P**.

A gearbox is required to perform well both in steady state and during acceleration. These two requirements can be considered as different operation modes, with different configuration spaces. The configuration space in steady state includes the choice of the gear i and the input voltage V. During acceleration, it consists of trajectories in time of $i(t)$ and $V(t)$. Therefore, the search for the optimal configuration can be separated to $\mathbf{y}_{ss}^{\star}$ that minimises P, and to $\underline{\mathbf{y}_t^{\star}}$ that minimises E and T. Since the latter is a solution to a MOP, it is expected to be a set. The variates of $\mathbf{y}_{ss}^{\star}$ and $\underline{\mathbf{y}_t^{\star}}$ that correspond to the variate $\mathbf{P}$, are denoted as $\mathbf{Y}_{ss}^{\star}$ and $\underline{\mathbf{Y}_t^{\star}}$, respectively.

Following the above, the AROP is formulated:

$$\min_{\mathbf{x}\in\mathcal{X}}\left[P(\mathbf{x},\mathbf{Y}_{ss}^{\star},\mathbf{P}),\underline{E}\left(\mathbf{x},\underline{\mathbf{Y}_t^{\star}},\mathbf{P}\right),\underline{T}\left(\mathbf{x},\underline{\mathbf{Y}_t^{\star}},\mathbf{P}\right)\right],\tag{13a}$$

$$\text{where}:\quad \mathbf{Y}_{ss}^{\star}=\operatorname*{argmin}_{\mathbf{y}\in\mathcal{Y}(\mathbf{x})}P(\mathbf{y},\mathbf{P}),\tag{13b}$$

$$\underline{\mathbf{Y}_t^{\star}}=\operatorname*{argmin}_{\mathbf{y}\in\mathcal{Y}(\mathbf{x})}\left[E(\mathbf{y},\mathbf{P}),T(\mathbf{y},\mathbf{P})\right],\tag{13c}$$

$$\mathbf{x}=[z_i],\quad i=1,\ldots,5,\tag{13d}$$

$$\mathbf{y}=[i,V],\tag{13e}$$

$$\mathbf{P}=[\Omega,\mathcal{T},\mathcal{J}_L,k_v,k_t,R,\nu_m,I_{\text{nom}},\nu_g,n_1,N_t,J_m,J_l,J_G,t_{\max}],\tag{13f}$$

$$\text{s.t.}:\quad z_{g,i}+z_{l,i}=N_t,\quad i=1,\ldots,5,\tag{13g}$$

$$I_{ss}\leq I_{\text{nom}},\tag{13h}$$

$$T\leq t_{\max}.\tag{13i}$$

The steady state current constraint is evaluated according to Eq. (10b), and the objectives according to Equations (10a), (11) and (12).

Since the ARMOP consists of separable configuration spaces, it can be decoupled into two subproblems, one that searches for $\mathbf{Y}_{ss}^{\star}$ and $P(\mathbf{x},\mathbf{Y}_{ss}^{\star},\mathbf{P})$, and another that searches for $\underline{\mathbf{Y}_t^{\star}}$ and $\left[\underline{E}\left(\mathbf{x},\underline{\mathbf{Y}_t^{\star}},\mathbf{P}\right),\underline{T}\left(\mathbf{x},\underline{\mathbf{Y}_t^{\star}},\mathbf{P}\right)\right]$. The former problem is a single-objective AROP, and the latter is an ARMOP. Using robustness indicators, Eq. (13a) can be converted to the following bi-objective problem that simultaneously minimises the steady-state AROP and the transient ARMOP:

$$\min_{\mathbf{x}\in\mathcal{X}}\left[I\left[P(\mathbf{x},\mathbf{Y}_{ss}^{\star},\mathbf{P})\right],I_{hv}\left[\underline{E}\left(\mathbf{x},\underline{\mathbf{Y}_t^{\star}},\mathbf{P}\right),\underline{T}\left(\mathbf{x},\underline{\mathbf{Y}_t^{\star}},\mathbf{P}\right)\right]\right].\tag{14}$$

4   Optimiser Design

The problem was solved by a bi-level EMOA whose structure is described in Algorithm 1.

First, the uncertain domain is sampled N_p times. These samples serve as the same representation of uncertainties to evaluate all solutions.

Next, Eq. (13b) is solved for the entire design space, and $\mathbf{Y}_{ss}^{\star}$ and $P(\mathbf{x},\mathbf{Y}_{ss}^{\star},\mathbf{P})$ are stored in an archive for every feasible solution. It is possible to find the optimal steady-state configuration of every solution for all sampled load scenarios because the design space is discrete and the objective and constraints are simple expressions. The search space consists of 962,598 different combinations of gears (choice of 5 gears from 43 possibilities). The constraints and objective functions depend on the number of teeth z, so they only have to be evaluated 43 times for each of the sampled scenarios. A feasible solution is a gearbox that has at least one gear that does not violate the constraints for each of the scenarios (i.e., $I \leq I_{nom}$ and $V \leq V_{max}$).

Next, a multi-objective search is conducted amongst the feasible solutions to solve Eq. (14). The solutions to Eq. (13c) for every sampled scenario are obtained by the evolutionary algorithm described in Section 4.1. The solutions to Eq. (13b) are already stored in an archive.

Algorithm 1.. Pseudo algorithm for solving the ARMOP

---

sample the uncertain domain
evaluate all possible solutions for steady state (s.s)
initialise nadir and ideal points for transient objectives (limits)
generate an initial population
while stopping criterion not satisfied **do**
    for every scenario **do**
        for every new solution **do**
            optimise for time–energy and store PF
        end for
    end for
    if limits have changed **then**
        update limits
        calculate HV of entire population
    else
        calculate HV of new feasible solutions
    end if
    assign scalar indicator values for s.s and transient
    evolve new population (selection, cross-over and mutation)
    re-mutate solutions that were already evaluated / infeasible for s.s
end while

4.1 EMOA for Identifying Optimal Gearing Sequences

For every load scenario, a multi-objective optimisation is conducted for each candidate solution to identify the optimal shift sequence that minimises energy and acceleration time. Early experiments revealed that maximum voltage results in better values for both objectives, regardless of the candidate solution or the load scenario. Therefore, the input voltage was considered as constant V_{max}, and the only search variable is $i(t)$, the selected gear at time t. A certain trajectory $i(t)$ results in a gearing ratio trajectory $n(t)$ that depends on the gearbox $\mathbf{x}$ that is being evaluated.

The trajectory $i(t)$ is coded as a vector of time intervals $\mathbf{dt} = [dt_1, \ldots, dt_N]$ defining the duration of each gear in the sequence from first gear to the Nth, with N being the optimal gear at steady state for the load scenario under consideration. The sum of all time intervals is equal to $t_{\max}$, and this relation is enforced whenever a new solution is created by setting:

$$\mathbf{dt} \leftarrow \frac{\mathbf{dt}}{\|\mathbf{dt}\|_1} t_{\max}. \tag{15}$$

Plugging $n(t)$ into Eq. (11) results in a trajectory $\omega(t)$, which can be used to calculate E, T or whether the gearbox failed to reach the desired speed before $t_{\max}$. A multi-objective evolutionary algorithm was used to estimate:

$$\mathbf{y}_t^\star = \underset{n(t)}{\arg\min} \left[E(\mathbf{x}, n(t), \mathbf{p}), T(\mathbf{x}, n(t), \mathbf{p}) \right], \tag{16}$$

where both $\mathbf{x}$ and $\mathbf{p}$ are fixed during the entire optimisation run.

Solving the differential equation (11) repeatedly to obtain $\mathbf{y}_t^\star$ is the most expensive part of the algorithm in terms of computational resources. Therefore, all of the solutions to (16) are stored in an archive to avoid repeated computations.

4.2   Calculating the Set-Based Robustness Indicator

The ARMOP's indicator I_{hv} uses a dynamic reference point. At every generation, after the approximated Pareto frontiers $\mathbf{\underline{F}}(\mathbf{x}, \mathbf{\underline{Y}^\star}, \mathbf{P})$ are identified for all evaluated solutions, the ideal and worst objective vectors are re-evaluated to include the objective vectors of the new solutions. If neither the ideal nor the worst objective vectors have changed, I_{hv} is calculated only for the recently evaluated solutions according to the procedure described in Section 2.1. Otherwise, the indicator values of the entire current population are recalculated as well, in order to allow for fair comparisons between new and old candidate solutions. No preferences were considered in this case study, hence, the objectives were normalised by setting $\mathbf{f}^w$ to one.

5   Simulation Results

5.1   Parameter Setting

The ARMOP described in Section 3 was solved with the proposed evolutionary algorithm. Two robustness criteria were considered: I^w considers the worst case scenario, meaning the upper limits of the uncertain load parameters, as given in Table 1. I^m considers the mean value over a set of sampled load scenarios. For both cases the same criterion was used for the steady state and transient indicators of Eq. 14, i.e., either I^w and I_{hv}^w or I^m and I_{hv}^m.

A standard elitist MOEA [22] with a fixed number of generations was used for both stages of the problem (referred to as outer and inner).

Parameter setting of the outer algorithm: population size $N = 100$, 50 genera-tions, integer coded, One-point crossover with crossover rate $p_c = 1$, polynomial mutation with mutation rate $p_m = 1/n_x = 0.2$ and distribution index $\eta_m = 20$. *Parameter setting of the inner algorithm:* population size $N = 50$, 30 gener-ations, real coded, SBX crossover with crossover rate $p_c = 1$ and distribution index $\eta_c = 15$, polynomial mutation with mutation rate $p_m = 1/n_y = 0.2$ and distribution index $\eta_m = 20$.

Both stages used sequential tournament selection, considering constraint vio-lation, non-dominance rank and niche count, and had an elite population size of $N_E = 0.4N$. The uncertain load domain was sampled 25 times using Latin hypercube sampling.

5.2   Results

The approximated Pareto frontiers for both worst-case and mean-value criteria are depicted in Fig. 3. For the worst-case criterion, the PS consists of only two, almost identical, solutions. In a close-up view on the approximated PF for mean performance, the extreme solutions are marked as A and B. Mean performances of the approximated Pareto set for the worst-case problem are also shown.

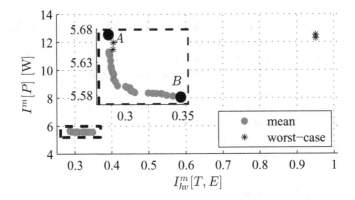

Fig. 3. Approximated Pareto frontiers for the worst-case and mean-value criteria. A close-up of the robust mean Pareto front is shown with the extreme solutions marked as A and B, and the mean performance of the approximated set according to I^w.

Details on the solutions for both robustness criteria are summarised in Table 2. Note the similarity in both design and objective spaces between the two solu-tions of the worst-case problem, and the difference between Solutions A and B. Also note that the best solutions found for a certain robustness criterion, are dominated for another. Solution B performs well in most steady state scenarios, since it has a large variety of high gears (small reduction ratio), but its ability to efficiently accelerate the load is limited from the same reason. Solution B becomes infeasible when the worst-case is considered. This was not detected

Table 2. Optimisation Results

Goal	Solution	Reduction Ratios					I^m [P]	I^m_{hv} [T, E]	I^w [P]	I^w_{hv} [T, E]
		1st	2nd	3rd	4th	5th				
I^m	A	9.02	4.34	2.62	1.93	1.30	5.672	0.2857	13.10	0.9631
	B	2.76	2.25	1.92	1.73	1.64	5.577	0.3481	infeasible	
I^w		7.06	3.38	2.14	1.55	1.14	5.649	0.2896	12.30	0.9511
		7.49	3.38	2.03	1.46	1.14	5.660	0.2899	12.52	0.9510

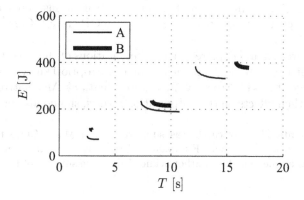

Fig. 4. Approximated Pareto frontiers $\underline{\mathbf{F}}(\mathbf{x}, \underline{\mathbf{Y}}^\star, \mathbf{P})$ of two solutions (A and B) for three scenarios (of 25). Solution A dominates Solution B in all evaluated scenarios.

while optimising for the mean value since the worst-case scenario was not sampled. This result highlights the impact of the choice of robustness criterion, and the challenge in optimising for the worst-case (see [3]).

The dynamic performances of Solutions A and B for three load scenarios are depicted in Fig 4. Solution A's superiority for both dynamic objectives is well captured by the I_{hv} indicator values.

6   Discussion and Future Work

This study introduced a new optimisation problem, the Active Robust Multiobjective Optimisation Problem. It enables a designer to examine the effectiveness of design adaptability to improve performance in an uncertain environment. The ARMOP introduces several challenges, some of which were addressed in this study, and others which need to be further explored.

The approach taken in this study to solve an ARMOP is to use a scalarising function to represent the variate of Pareto frontiers of every candidate solution. This approach was found useful for the gearbox case study – solutions with better Pareto frontiers were assigned with a better indicator value. However, whenever a set is represented by a scalar value, some of its information must be lost. As a result, setting a robustness criterion for the utility indicator value does not automatically imply that the individual objectives will also be robust.

Being a bi-level optimisation problem, an AROP requires many function evaluations. An ARMOP is even harder to solve, because the inner problem is a MOP. The strategy for obtaining robust solutions taken in this study was based on Monte Carlo simulations to represent the uncertain variables. This representation requires a large set of samples to adequately capture the nature of the uncertainties involve, and to gain confidence in the robustness of the obtained solutions. Due to limited computational resources, the approach was demonstrated in this study with a small set of sampled scenarios, only to provide a proof of concept. Even for these minimal optimiser settings, almost 70 million function evaluations were conducted. It took approximately three days to compute on a 3.40GHz Intel® Core™ i7-4930K CPU, running Matlab® on 12 cores.

Future research should explore other representations of the uncertainties that involve more efficient sampling approaches and use of a-priori knowledge; as well as optimisation algorithms for expensive function evaluations. Alternative scalarising functions, and their effects on the optimisation results, should also be explored.

Acknowledgments. This research was supported by a Marie Curie International Research Staff Exchange Scheme Fellowship within the 7th European Community Framework Programme. The first author acknowledges the support of the Anglo-Israel Association.

References

1. Salomon, S., Avigad, G., Fleming, P.J., Purshouse, R.C.: Active Robust Optimization - Enhancing Robustness to Uncertain Environments. IEEE Transactions on Cybernetics **44**(11), 2221–2231 (2014)
2. Beyer, H.G., Sendhoff, B.: Robust Optimization - A Comprehensive Survey. Computer Methods in Applied Mechanics and Engineering **196**(33–34), 3190–3218 (2007)
3. Branke, J., Rosenbusch, J.: New approaches to coevolutionary worst-case optimization. In: Rudolph, G., Jansen, T., Lucas, S., Poloni, C., Beume, N. (eds.) PPSN X. LNCS, vol. 5199, pp. 144–153. Springer, Heidelberg (2008)
4. Avigad, G., Coello, C.A.: Highly Reliable Optimal Solutions to Multi-Objective Problems and Their Evolution by Means of Worst-Case Analysis. Engineering Optimization **42**(12), 1095–1117 (2010)
5. Alicino, S., Vasile, M.: An evolutionary approach to the solution of multi-objective min-max problems in evidence-based robust optimization. In: 2014 IEEE Congress on Evolutionary Computation (CEC), pp. 1179–1186 (2014)
6. Teich, J.: Pareto-front exploration with uncertain objectives. In: Zitzler, E., Deb, K., Thiele, L., Coello Coello, C.A., Corne, D. (eds.) EMO 2001. LNCS, vol. 1993, pp. 314–328. Springer, Heidelberg (2001)
7. Hughes, E.J.: Evolutionary multi-objective ranking with uncertainty and noise. In: Zitzler, E., Thiele, L., Deb, K., Coello Coello, C.A., Corne, D. (eds.) EMO 2001. LNCS, vol. 1993, pp. 329–343. Springer, Heidelberg (2001)
8. Deb, K., Gupta, H.: Introducing Robustness in Multi-Objective Optimization. Evolutionary Computation **14**(4), 463–494 (2006)

9. Saha, A., Ray, T.: Practical Robust Design Optimization Using Evolutionary Algorithms. Journal of Mechanical Design **133**(10), 101012 (2011)
10. Beyer, H.G., Sendhoff, B.: Functions with noise-induced multimodality: a test for evolutionary robust Optimization-properties and performance analysis. IEEE Transactions on Evolutionary Computation **10**(5), 507–526 (2006)
11. Fieldsend, J.E., Everson, R.M.: Multi-objective optimisation in the presence of uncertainty. In: The 2005 IEEE Congress on Evolutionary Computation, vol. 1, pp. 243–250 (2005)
12. Bui, L.T., Abbass, H.A., Essam, D.: Fitness inheritance for noisy evolutionary multi-objective optimization. In: Proceedings of the 7th Annual Conference on Genetic and Evolutionary Computation, GECCO 2005, pp. 779–785, New York. ACM (2005)
13. Goh, C.K., Tan, K.C.: An Investigation on Noisy Environments in Evolutionary Multiobjective Optimization. IEEE Transactions on Evolutionary Computation **11**(3), 354–381 (2007)
14. Knowles, J., Corne, D., Reynolds, A.: Noisy multiobjective optimization on a budget of 250 evaluations. In: Ehrgott, M., Fonseca, C.M., Gandibleux, X., Hao, J.-K., Sevaux, M. (eds.) EMO 2009. LNCS, vol. 5467, pp. 36–50. Springer, Heidelberg (2009)
15. Fieldsend, J.E., Everson, R.M.: The Rolling Tide Evolutionary Algorithm: A Multi-Objective Optimiser for Noisy Optimisation Problems. IEEE Transactions on Evolutionary Computation **PP**(99), 1 (2014)
16. Paenke, I., Branke, J., Jin, Y.: Efficient Search for Robust Solutions by Means of Evolutionary Algorithms and Fitness Approximation. IEEE Transactions on Evolutionary Computation **10**(4), 405–420 (2006)
17. Cruz, C., González, J.R., Pelta, D.A.: Optimization in Dynamic Environments: A Survey on Problems, Methods and Measures. Soft Computing **15**(7), 1427–1448 (2011)
18. Fleming, P.J., Purshouse, R.C., Lygoe, R.J.: Many-objective optimization: an engineering design perspective. In: Coello Coello, C.A., Hernández Aguirre, A., Zitzler, E. (eds.) EMO 2005. LNCS, vol. 3410, pp. 14–32. Springer, Heidelberg (2005)
19. Zitzler, E.: Evolutionary Algorithms for Multiobjective Optimization: Methods and Applications. Phd dissertation, Swiss Federal Institute of Technology Zurich (1999)
20. Knowles, J., Corne, D.: On metrics for comparing nondominated sets. In: Proceedings of the 2002 Congress on Evolutionary Computation, CEC 2002, pp. 711–716. IEEE (2002)
21. Krishnan, R.: Electric Motor Drives - Modeling, Analysis, And Control. Prentice Hall (2001)
22. Salomon, S., Avigad, G., Goldvard, A., Schütze, O.: PSA – a new scalable space partition based selection algorithm for MOEAs. In: Schütze, O., Coello Coello, C.A., Tantar, A.-A., Tantar, E., Bouvry, P., Del Moral, P., Legrand, P. (eds.) EVOLVE - A Bridge between Probability. AISC, vol. 175, pp. 137–151. Springer, Heidelberg (2012)

Linear Scalarization for Pareto Front Identification in Stochastic Environments

Madalina M. Drugan[(⊠)]

Artificial Intelligence Lab, Vrije Universiteit Brussels,
Pleinlaan 2, 1050 Brussels, Belgium
Madalina.Drugan@vub.ac.be

Abstract. Multi-objective multi-armed bandits (MOMAB) is a multi-arm bandit variant that uses stochastic reward vectors. In this paper, we propose three MOMAB algorithms. The first algorithm uses a fixed set of linear scalarization functions to identify the Pareto front. Two topological approaches identify the Pareto front using linear weighted combinations of reward vectors. The *weight hyper-rectangle decomposition* algorithm explores a convex shape Pareto front by grouping scalarization functions that optimise the same arm into weight hyperrectangles. It is generally acknowledged that linear scalarization is not able to identify all the Pareto front for non-convex shapes. The *hierarchical PAC* algorithm iteratively decomposes the Pareto front into a set of convex shapes to identify the entire Pareto front. We compare the performance of these algorithms on a bi-objective stochastic environment inspired from a real life control application.

Keywords: Multi-objective multi-armed bandits · Scalarization functions · Pareto front identification · Topological decomposition

1 Introduction

Multi-armed bandits [1] (MAB) is a machine learning paradigm used to study and analyse resource allocation in stochastic and uncertain environments. The multi-armed bandit problem that considers reward vectors and imports techniques from multi-objective optimisation into the multi-armed bandits algorithms is referred to as *multi-objective multi-armed bandits* (MOMAB) [3].

A reward vector can be optimal in one objective and sub-optimal in the other objectives. The *Pareto front* contains several arms considered to be the best according to their reward vectors. Scalarization functions transform the reward vectors into scalar reward values in order to use MAB algorithms. In the case of linear scalarization functions, each scalarization instance corresponds to a weight vector. Due to its simplicity, *linear scalarization* is the most popular scalarization function used in designing both multi-objective optimisation and reinforcement learning algorithms [9,11,13]. Section 2 gives a short introduction to the MOMAB problem.

A common approach in multi-objective optimisation selects a number of weight vectors that are uniform randomly spread in the weighted space [3,4] with the goal

© Springer International Publishing Switzerland 2015
A. Gaspar-Cunha et al. (Eds.): EMO 2015, Part II, LNCS 9019, pp. 156–171, 2015.
DOI: 10.1007/978-3-319-15892-1_11

to minimize the regret of choosing a sub-optimal arm. In Section 3 we propose a baseline algorithm that is basically *probably approximatively correct* (PAC) algorithm [6] meaning that it identifies the best arm for each scalarization with high accuracy within a confidence value and a given number of arm pulls as budget. Note that PAC has the goal of identifying the best arm rather than minimizing the regret. The dominance relation for uncertain environments is imported from multi-objective evolutionary optimisation [12] and integrated in the MAB paradigm, where the unbiased estimator is traditionally a mean reward vector and a confidence vector represent the uncertain and stochastic nature of the environment. In the literature [7], the Hoeffding and Bernstein races use uncertainty for the single objective evolutionary direct policy search.

An uniform distribution of the weight vectors does not guaranty a good coverage of the Pareto front. Furthermore, several scalarization functions can identify the same optimal arm, and there could be unidentified Pareto optimal arms. Optimising in stochastic and uncertain environments is costly because numerous arms pulls (or samples) are required for a reasonable accuracy in prediction. Thus, from a computational perspective, it is important to have a *minimal set of scalarization functions with a good coverage of the Pareto front.*

We propose two linear scalarization based Pareto front identification algorithms with efficient exploration / exploitation mechanisms based on topological properties of the Pareto front. In Section 4, we design an efficient linear scalarization based MOMAB algorithm that identifies Pareto fronts with convex shape. *Weight hyper-rectangles decomposition* (WHD) is a Pareto front identification algorithm that groups the weight vectors optimising the same arm from the Pareto front into *weight D-dimensional hyperrectangles*, or simpler weight D-rectangles, where D is the number of objectives.

In general, the Pareto fronts generated with real applications have non-convex shapes, e.g. the wet clutch [10]. Linear scalarization, and implicitly the algorithms that use linear scalarization, has limitations in identifying the entire Pareto front of non-convex shapes. In Section 5, we iteratively decompose Pareto fronts of any shape into convex shapes [8] using topological techniques. The initial algorithm from [8] is designed for two objectives, but similar hierarchical decomposition algorithms from computational geometry can deal with three objectives [2]. The original decomposition algorithm groups the Pareto optimal solutions in convex sub-shapes. We consider that the problem of convex hull decomposition is solved with specific methods from computational geometry, our focus being on the stochastic process. *Hierarchical PAC* identifies Pareto fronts of any shape with a given confidence value.

Our bi-objective stochastic environment is a wet clutch [10] that is a system with one input characterised by a hard non-linearity, when the piston of the clutch gets in contact with the friction plates, and stochastic output because of the variations in environmental temperature. In Section 6, we compare these scalarization based MOMAB algorithms on a bi-objective environments generated with the wet clutch control problem. Section 7 concludes the paper.

2   Multi-Objective Multi-Armed Bandits Problem

Consider an initial set of arms $\mathcal{I}$ with cardinality K, where $K \geq 2$ and let the vector reward space be defined as a D-rectangle $[0, 1]^D$. When arm i is played, a random vector of rewards is received, one component per objective. The random vectors have a stationary distribution with support in $[0, 1]^D$ so that we can apply the Hoeffding inequality. At time steps $t_1, t_2, \ldots$, the corresponding reward vectors $\mathbf{X}_i^{t_1}$, $\mathbf{X}_i^{t_2}, \ldots$ are independently and identically distributed according to an unknown law with unknown expectation vector $\boldsymbol{\mu}_i = (\mu_i^1, \ldots, \mu_i^D)$. Reward values obtained from different arms are also assumed to be independent.

A *policy* π is an algorithm that chooses the next arm to play based on the list of past plays and obtained reward vectors. Let $T_i(N)$ be the number of times a suboptimal arm i has been played by π during the first N plays. The expected reward vectors are computed by averaging the empirical reward vectors observed over the time. The mean of an arm i is estimated to $\widehat{\boldsymbol{\mu}_i}(N) = \sum_{s=1}^{T_i(N)} X_i(s)/T_i(N)$, where $X_i(s)$ is the sampled value s for arm i.

Dominance Relations. In the general case, a reward vector can be better than another reward vector in one objective, and worse in another objective. There are several dominance relations that order vectors in multi-objective optimisation. *Pareto dominance relation* is the natural order for these environments allowing ordering the reward vectors directly in the multi-objective reward space. Let the *Pareto optimal set of arms* $\mathcal{I}^*$, or the Pareto front, be the set of arms whose reward vectors are non-dominated by all the other arms. All arms in the Pareto optimal set $\mathcal{I}^*$ are assumed to be equally important.

We define the Pareto dominance relation as a variant of the Pareto dominance relation for uncertain environments [12]. An arm ℓ dominates another arm i, $\boldsymbol{\mu}_\ell \succ_\epsilon \boldsymbol{\mu}_i$, iff $\exists j$ for which $\mu_i^j + \frac{\epsilon}{2} < \mu_\ell^j - \frac{\epsilon}{2}$, and for all other objectives o, $o \neq j$, we have that $\mu_i^o + \frac{\epsilon}{2} \leq \mu_\ell^o - \frac{\epsilon}{2}$. Pareto front contains all the non-dominated arms ℓ, for which $\nexists i \, \forall j$ such that $\mu_\ell^j < \mu_i^j - \epsilon$. We also assume that the non-dominant arms from $\mathcal{I}$ are good approximations of their true means. Thus, we have $\|\widehat{\mu}_\ell^j - \mu_\ell^j\| < \epsilon$ in all objectives j, and for all arms $\ell \in \mathcal{I}$.

Scalarization functions transform the reward vectors into scalar rewards using e.g. linear or non-linear weighted sums. The linear scalarization function weighs each value of the reward vector and the result is the sum of these weighted values. More formally, a linear scalarized reward is $f_\omega(\mu_i) = \omega \cdot \mu_i = \sum_{k=1}^D \omega^k \cdot \mu_i^k, \forall i$, where $\omega = (\omega^1, \ldots, \omega^D)$ is a set of predefined weight vectors. A valid weight vector has $\sum_{j=1}^D \omega^j = 1$.

The estimated mean value of the scalarized reward vectors is equal to the scalarized value of the estimated mean reward vectors, $f_\omega(\widehat{\boldsymbol{\mu}}_i) = \widehat{f_\omega}(\boldsymbol{\mu}_i)$. In the sequel, for all weight vectors ω, we have that $f_\omega(\widehat{\boldsymbol{\mu}}_i + \boldsymbol{\epsilon}) = f_\omega(\widehat{\boldsymbol{\mu}}_i) + \epsilon$, where $\boldsymbol{\epsilon} = (\epsilon, \ldots, \epsilon)$ and $f_\omega(\boldsymbol{\epsilon}) = \epsilon$. We will use these two properties in the next section.

3 A Baseline Scalarized Pareto Front Identification Algorithm

In the following, we propose a baseline algorithm to identify the Pareto front in stochastic environments. We assume a tolerance error δ and we want to find the Pareto front with probability at least $1 - \delta$. An arm i is optimal for a given scalarization function f_ω iff

$$f_\omega(\widehat{\boldsymbol{\mu}}_i) > \max_{\ell \in \mathcal{I}} f_\omega(\widehat{\boldsymbol{\mu}}_\ell) - \epsilon$$

Algorithm 1. The scalarized PAC algorithm sPAC($\epsilon, \delta, \mathcal{W}$)

---

1 **for** *all arms* $k = \{1, \ldots, K\}$ **do**
2   Pull each arm k for $\ell = \frac{1}{(\epsilon/2)^2} \log\left(\frac{2|\mathcal{W}|K}{\delta}\right)$ times ;
3   Compute the expected mean reward vectors $\widehat{\boldsymbol{\mu}}_k$
4 $\mathcal{I}^* \leftarrow \emptyset$;
5 **for** *all weight vectors* $\omega \in \mathcal{W}$ **do**
6   Select an optimal arm i^* for function f_ω ;
7   Add arm i^* to the Pareto front $\mathcal{I}^* \leftarrow \mathcal{I}^* \cup \{i^*\}$;
8   Delete dominated arms from $\mathcal{I}^*$
9 **return** $\mathcal{I}^*$

The pseudo-code for the sPAC($\epsilon, \delta, \mathcal{W}$) is given in Algorithm 1. sPAC extends the PAC (probably approximatively correct) [6] algorithm to reward vectors. We assume a fixed number of weight vectors $\mathcal{W} \leftarrow \{\omega_1, \ldots, \omega_{|\mathcal{W}|}\}$. The probability that the expected reward vector of an arm i, $\boldsymbol{\mu}_i$, is not the same as its estimated reward vector, $\hat{\boldsymbol{\mu}}_i$, is bounded with the confidence value $\epsilon > 0$ and a small error probability $\delta > 0$. We want to bound the probability of the event $f_\omega(\widehat{\boldsymbol{\mu}}_i) - f_\omega(\boldsymbol{\mu}_i) > \varepsilon$, for any scalarization function f_ω.

In Algorithm 1, each arm is pulled for an equal and fixed number of times $\frac{1}{(\epsilon/2)^2} \log\left(\frac{2|\mathcal{W}|K}{\delta}\right)$. Thus, the budget for this algorithm is fixed $N = \frac{K}{(\epsilon/2)^2} \log\left(\frac{2|\mathcal{W}|K}{\delta}\right)$. For each weight vector, ω, we identify the optimal arms using the N arm pulls. The output for this algorithm is a reunion of the optimal set of arms for each scalarization function f_ω. If an optimal arm is not already in the Pareto front, then that arm is added to the Pareto front $\mathcal{I}^*$. To maintain the Pareto front $\mathcal{I}^*$, the dominated arms are deleted from $\mathcal{I}^*$.

Analysis. By definition, given f_ω, a policy is (ϵ, δ) - correct iff the arm i selected after n trials is

$$P\left(f_\omega(\widehat{\boldsymbol{\mu}}_i) > \max_{\ell \in \mathcal{I}} f_\omega(\widehat{\boldsymbol{\mu}}_\ell) - \epsilon\right) \geq 1 - \delta$$

Our goal is to identify the entire Pareto front with a given confidence ϵ and a small tolerance error δ. Thus, by definition, an algorithm is $(\epsilon, \delta, \mathcal{W})$ correct iff

$$\sum_{\omega \in \mathcal{W}} P\left(f_\omega(\widehat{\boldsymbol{\mu}}_i) > \max_{\ell \in \mathcal{I}} f_\omega(\widehat{\boldsymbol{\mu}}_\ell) - \epsilon\right) \geq 1 - \delta$$

Theorem 1. *sPAC(ϵ, δ, $\mathcal{W}$), cf Algorithm 1, is (ϵ, δ) correct and it has the sample complexity of $\mathcal{O}\left(\frac{K}{\epsilon^2} \cdot \ln\left(\frac{2|\mathcal{W}|K}{\delta}\right)\right)$.*

Proof. Let k be an arm for which $f_\omega(\hat{\mu}_k) < f_\omega(\hat{\mu}_i) - \epsilon$ for any Pareto optimal arm i given the function f_ω. For all objectives j, we want to bound the event $f_\omega(\hat{\mu}_k) > f_\omega(\hat{\mu}_i)$, when $f_\omega(\hat{\mu}_k) > f_\omega(\mu_k) + \epsilon/2$ and $f_\omega(\hat{\mu}_i) < f_\omega(\mu_i) - \epsilon/2$. Then, for any ω,

$$\mathbb{P}(f_\omega(\hat{\mu}_k) - f_\omega(\hat{\mu}_i) > \epsilon) = \mathbb{P}(f_\omega(\hat{\mu}_k) > f_\omega(\mu_k) + \epsilon/2 \text{ or } f_\omega(\hat{\mu}_i) < f_\omega(\mu_i) - \epsilon/2) \leq$$

$$\mathbb{P}(f_\omega(\hat{\mu}_k) > f_\omega(\mu_k) + \epsilon/2) + \mathbb{P}(f_\omega(\hat{\mu}_i) < f_\omega(\mu_i) - \epsilon/2) \leq 4 \cdot e^{-2(\epsilon/2)^2 \ell} = \frac{\delta}{n}$$

where we have used the Hoeffding inequality. If $\ell = \frac{4}{\epsilon^2} \cdot \ln\left(\frac{2|\mathcal{W}|K}{\delta}\right)$ and we sum over K the arms and $\mathcal{W}$ scalarization functions, then the error probability for Algorithm 1 is bounded by δ.   □

Remark. Our *scalarized PAC* algorithm considers that the larger the set of weight vectors, the higher, thus better, the confidence is in the identified Pareto front. If the confidence value ϵ is inverse proportional with the number of used weights $|\mathcal{W}|$, $\epsilon = \frac{1}{|\mathcal{W}|}$, then a larger number of weight vectors means a more accurate identification of Pareto front. A smaller number of weight vectors means a larger, thus worst, confidence value in the identification of Pareto front.

The Wet Clutch Setting. In the following, we exemplify the usage of the proposed Pareto front identification algorithm for a realistic bi-objective example.

Example 1. Wet clutches are typically used in power transmissions of off-road vehicles, which operate under strongly varying environmental conditions. The validation experiments are carried out on a dedicated test bench, where an electro-motor drives a flywheel via a torque converter and two mechanical transmissions. The goal is to learn by minimising simultaneously: 1) the optimal current profile to the electro-hydraulic valve, which controls the pressure of the oil to the clutch, and 2) the engagement time. The output data is stochastic because the behaviour of the machine varies with the surrounding temperature that cannot be exactly controlled, and it has a non-convex Pareto shape.

In Figure 1 a), we show 50 points generated with this application, each point representing a trial of the machine and the jerk time obtained in the given time. The initial problem is a minimisation problem transformed into a maximisation problem by normalising each objective with values between 0 and 1, and then transforming it into a maximisation problem for each of the two objectives. The Pareto optimal set of arms contains 16 optimal reward vectors, and it is about one-third from the total number of arms, i.e. 16/50. Note that the shape of the Pareto front makes it a mixture of convex and non-convex regions.

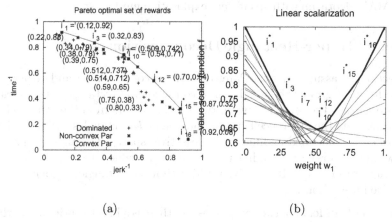

(a)   (b)

Fig. 1. a) The 7-th coordinates on the convex coverage of the Pareto front that is delimited by line segments. The rest of the 9-th coordinates belong to the arms on the non-convex Pareto front and are located in the interior of the line segments. b) The linear scalarization values and the corresponding Pareto optimal arms identified for each weight vector.

To identify the Pareto front with sPAC($\epsilon, \delta, \mathcal{W}$), cf Algorithm 1, we consider 81 weight vectors that are uniform randomly spread in the weight vector space. The step between two weight values is 0.0125 and the confidence value $\epsilon = 0.0$. Only 7 Pareto optimal arms from a total of 16 arms are identified with 81 linear scalarization functions because these 7 arms are on the convex coverage of the Pareto front. The rest of 9 Pareto optimal arms are NOT on the convex coverage and they cannot be identified with the algorithm using linear scalarization functions.

Figure 1 b) shows that the distribution of the weight vectors that identify the 7 Pareto optimal arms are unevenly distributed in the weight vector space. The weight vectors from the first, $\omega_1 = 0.0$, to the twenty-second, $\omega_{22} = 0.275$, weight vectors optimise the first arm $i_1^* = (0.12, 0.92)$. The arm $i_3^* = (0.32, 0.83)$ is found with only four weights, i.e. from the 23-th weight $\omega_{23} = 0.2875$ to $\omega_{28} = 0.325$. The seventh arm $i_7^* = (0.509, 0.742)$ is found with eight weight vectors, from $\omega_{29} = 0.3375$ to $\omega_{41} = 0.5$, whereas the tenth arm $i_{10}^* = (0.54, 0.71)$ is found only with one weight $\omega_{42} = 0.5125$. The twelfth arm $i_{12}^* = (0.70, 0.54)$ is identified with only three weight vectors, from $\omega_{43} = 0.525$ to $\omega_{45} = 0.55$, and the arm $i_{15}^* = (0.87, 0.32)$ is identified with twenty-one weight vectors from $\omega_{46} = 0.5625$ to $\omega_{67} = 0.825$. The arm $i_{16}^* = (0.92, 0.08)$ is identified with fourteen weights from $\omega_{68} = 0.8375$ to $\omega_{81} = 1.0$.

Remark. Note that the weight vectors that identify the same Pareto optimal arms are grouped in disjunct subsets. Therefore, if we found the extreme weight values that identify the same Pareto optimal arm, then we know that any weights between the two weight values identify the same optimal arm. For example, the Pareto optimal arm i_1^* is identified by any of the weight vectors with the first value in the interval $\omega_1 \in [0.0, 0.275]$. In the next section, we use this property to design a scalarized MOMAB algorithm with an efficient exploration mechanism.

4   Weight Hyper-Rectangle Decomposition

In this section, we assume that the Pareto front has a convex shape, and we propose an algorithm that groups the weights into contiguous D-rectangles. This means that we do not need to search between two weight vectors from the same weight D-rectangle. Note the difference between the weight D-rectangles and reward D-rectangles that are defined in different D-objectives spaces. The algorithm starts with D weight vectors corresponding to the D extreme values of a weight D rectangle with support in $[0, 1]^D$. It iteratively adds weight D-rectangles to identify new arms on the Pareto front.

Weight D-rectangles. For each weight vector, there is a Pareto optimal arm that optimises its inner product. Consider $\boldsymbol{\omega}$ a weight vector, and two reward vectors $\boldsymbol{\mu}_i$ and $\boldsymbol{\mu}_h$. If $f_{\boldsymbol{\omega}}(\boldsymbol{\mu}_i) > f_{\boldsymbol{\omega}}(\boldsymbol{\mu}_h)$, then $\sum_{j=1}^{D} \omega^j \cdot (\mu_i^j - \mu_h^j) > 0$. This means that $\exists j$ such that $\mu_i^j > \mu_h^j$, and thus $\boldsymbol{\mu}_i$ is non-dominated by $\boldsymbol{\mu}_h$. Thus,

$$\forall \boldsymbol{\omega} \in \boldsymbol{W}, \ \exists! \, i^* \in \mathcal{I}^* \text{ such that } i^* \to \arg\max_{i \in \mathcal{I}} \boldsymbol{\omega} \cdot \boldsymbol{\mu}_i$$

with $\boldsymbol{W}$ the *weight vector space*. A trivial weight D-rectangle contains a single weight vector. We need a set of scalarization functions, or a set of weight vectors, to generate a variety of arms on the Pareto front.

The *weight D-rectangle* of an arm i^* on the Pareto front, $i^* \in \mathcal{I}^*$, is defined as

$$H_{i^*} = \{\boldsymbol{\omega} \in \boldsymbol{W} \mid i^* \leftarrow \arg\max_{i \in \mathcal{I}} \boldsymbol{\omega} \cdot \boldsymbol{\mu}_i\}$$

This means that if two weight vectors, ω_1 and ω_2, belong to the same weight D-rectangle, H_{i^*}, then $f_{\boldsymbol{\omega}_1}$ and $f_{\boldsymbol{\omega}_2}$ identify the same Pareto optimal arm, i^*.

Let the Pareto front be convex and the weight D-rectangles be contiguous. If ω_1 and ω_2 belong to the same weight D-rectangle, $\boldsymbol{\omega}_1 \in H_{i_1^*}$ and $\boldsymbol{\omega}_2 \in H_{i_1^*}$, then all weight vectors ω_3 between two weight vectors, i.e. for all objectives j, $\omega_3^j \in [\min\{\omega_1^j, \omega_2^j\}, \max\{\omega_1^j, \omega_2^j\}]$, also belong to the same weight D-rectangle, $\boldsymbol{\omega}_3 \in H_{i_1^*}$.

Two weight vectors, $\boldsymbol{\omega}_1 \in H_{i_1^*}$ and $\boldsymbol{\omega}_2 \in H_{i_2^*}$, with $i_1^* \neq i_2^*$, are considered *adjacent* iff, exists an objective $j \in \{1, \ldots, D\}$ such that the distance between the two weight vectors is minimal compared with the other weight vectors in the same or different weight D rectangles. More formally, consider the weight D-rectangle defined by the Cartesian product of the two weight values, $\times_{j=1}^{D}[\min\{\omega_1^j, \omega_2^j\}, \max\{\omega_1^j, \omega_2^j\}]$. These two weight vectors $\boldsymbol{\omega}_1$ and $\boldsymbol{\omega}_2$ are adjacent iff the weight D-rectangle does not contain another weight vector. Thus, $\omega_1 \in H_{i_1^*}$ and $\omega_2 \in H_{i_2^*}$ are adjacent, iff they optimize different arms $i_1^* \neq i_2^*$ in different D-rectangles and the difference between the two weight vectors is minimal in at least one objective. Then, one of the three conditions holds, $\forall j$: 1) $|\omega_1^j - \omega_2^j| \leq \min\{|\omega_1^j - \omega_3^j|, |\omega_3^j - \omega_2^j|\}$, for all $\omega_3 \in H_{i_3^*}$, and $i_3^* \neq i_1^*$ and $i_3^* \neq i_2^*$, 2) $|\omega_1^j - \omega_2^j| \leq |\omega_1^j - \omega_3^j|$, for all $\omega_3 \in H_{i_2^*}$, and 3) $|\omega_1^j - \omega_2^j| \leq |\omega_3^j - \omega_2^j|$, for all $\omega_3 \in H_{i_1^*}$.

Algorithm 2. Weight D-rectangle decomposition $\mathrm{WHD}(\epsilon, \delta, \mathcal{W}_{WHD})$

---

1 Initialise the Pareto front $\mathcal{I}^* \leftarrow \emptyset$; Initialise the weight D-rectangle $\mathcal{H} \leftarrow \emptyset$;
2 Initialise the list of candidate weight vectors $\mathcal{W}_{WHD} \leftarrow \mathcal{W}_1$;
3 $t \rightarrow 1$;
4 **while** *the stopping criteria is not met* **do**
5 $\quad$ Pull each arm for $\ell_t \leftarrow \frac{4}{\epsilon^2} \cdot \ln \frac{2K|\mathcal{W}_t|}{\delta}$ times ;
6 $\quad$ **for** *all weight vectors* $\omega \in \mathcal{W}_t$ **do**
7 $\quad\quad$ Compute the optimal arm i^* for the scalarization function f_ω ;
8 $\quad\quad$ **if** *arm i^* is not in the current set of arms* $i^* \notin \mathcal{I}^*$ **then**
9 $\quad\quad\quad$ Update the current set of arms $\mathcal{I}^* \leftarrow \mathcal{I}^* \cup \{i^*\}$;
10 $\quad\quad\quad$ Update the weight D-rectangle $\mathcal{H}_{i^*} \leftarrow \mathcal{H}_{i^*} \cup \{\omega\}$
11 $\quad\quad$ **else**
12 $\quad\quad\quad$ Add the weight vector $H_{i^*} \leftarrow H_{i^*} \cup \{\omega\}$
13 $\quad$ Set $\mathcal{W}_t$ to be the candidate weight vectors generated between adjacent D-rectangles ;
14 $\quad$ $t \leftarrow t + 1$
15 return $\mathcal{I}^*$;

The Algorithm. The pseudo-code for the *weight D-rectangle decomposition* (WHD) is given in Algorithm 2. The WHD algorithm starts with the initialisation of the Pareto front to the empty set, $\mathcal{I}^* \leftarrow \emptyset$, and the list of weight D-rectangles, $\mathcal{H}$, is also initialized to the empty set. At first, for our initial set of candidate weight vectors $\mathcal{W}_{WHD} \leftarrow \mathcal{W}_1$, we consider a fixed set of weight vectors, like the D extreme weight vectors, i.e. with one objective set to 1 and all the other objectives set to 0. Each iteration t, all arms are pulled $\ell_t \leftarrow \frac{4}{\epsilon^2} \cdot \ln \frac{2K|\mathcal{W}_t|}{\delta}$ times. All weight vectors are selected at random from the list of candidate weight vectors in order to identify the corresponding Pareto optimal arm, i^*. If this optimal arm i^* is already in the Pareto front $\mathcal{I}^*$, then the weight vector ω is added to the corresponding weight D-rectangle, $H_{i^*} \leftarrow H_{i^*} \cup \{\omega\}$. Otherwise, if $i^* \notin \mathcal{I}^*$ is a novel Pareto optima arm, then i^* is added to the Pareto front $\mathcal{I}^* \leftarrow \mathcal{I}^* \cup \{i^*\}$, and a weight D-rectangle is initialised for this optimal arm, $H_{i^*} \leftarrow \{\omega\}$.

The list of weight D-rectangles is also updated with new weight D-rectangle, H_{i^*}, generated between two adjacent weight D-rectangles. To generate a new reward vector that could identify a new Pareto optimal arm, we select uniformly at random two adjacent weight vectors in two different weight D-rectangles. Thus, $\omega_1 \in H_{i_1^*}$ and $\omega_2 \in H_{i_2^*}$, with $i_1^* \neq i_2^*$, and we generate a third weight vector, ω_3, such that, $\forall j, \omega_3^j = (\omega_1^j + \omega_2^j)/2$. The new weight ω_3 is included in the list of candidate weights, $\mathcal{W}$, to later evaluate its optimal arm. The algorithm stops when the list of weight vectors is empty, i.e. $\mathcal{W} \neq \emptyset$, or a fix number of arm pulls, the budget, is reached.

We assume that the Pareto optimal arms can be identified within a bounded confidence value ϵ. Thus, two weight vectors could assigned to two different weight D-rectangles even when their reward values are closer than a given confidence value.

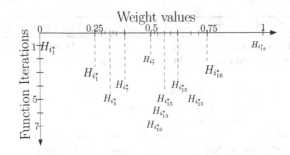

Fig. 2. The weight D-rectangle decomposition algorithm WHD needs 7 iterations (see the vertical coordinates) and only 18 weight vectors (see the horizontal coordinates) to find the convex coverage of the wet clutch simulation.

Analysis. The arms pulls for each iteration is computed such that each of the iterations is (ϵ, δ) correct, and WHD is also (ϵ, δ) correct meaning that the probability of selecting the wrong arm for all scalarization functions is bounded by a small error δ. The number of arms pulls for this algorithm is the sum of all arm pulls in each iteration. Thus, the total budget for this algorithm is

$$N = \frac{K}{(\epsilon/2)^2} \cdot \sum_{t=1}^{L} \ln \frac{2K|\mathcal{W}_t|}{\delta} = \frac{K}{(\epsilon/2)^2} \cdot \ln \left(\frac{2K}{\delta}\right)^L \cdot |\mathcal{W}_1| \dots |\mathcal{W}_L|$$

$$= \frac{LK}{(\epsilon/2)^2} \cdot \ln \frac{2K}{\delta} \cdot \sqrt[L]{|\mathcal{W}_1| \dots |\mathcal{W}_L|} \leq \frac{LK}{(\epsilon/2)^2} \cdot \ln \frac{2K}{\delta} \cdot \frac{|\mathcal{W}|}{L}$$

where $\mathcal{W}$ is the reunion of weight vectors generated $\mathcal{W} = \cup_{t=1}^{L} \mathcal{W}_t$, and L is the number of iterations in WHD. Last inequality is the arithmetical - geometrical mean and the equality holds when the set of weights is (approximatively) equal. Note that each arm is pulled for at most $\approx \frac{L}{(\epsilon/2)^2} \cdot \ln \frac{2K}{\delta} \cdot \frac{|\mathcal{W}|}{L}$ times and the algorithm attains its maximum efficiency in the number of arms pulls when the sets of weights D-rectangles are small.

Theorem 2. *WHD, cf Algorithm 2, is (ϵ, δ) correct and it has the sample complexity of* $\mathcal{O}\left(\frac{LK}{\epsilon^2} \cdot \ln \left(\frac{2K}{\delta} \frac{|\mathcal{W}|}{L}\right)\right)$.

Proof. Thus, using the same rational as in the proof of Theorem 1, WHD algorithm is (ϵ, δ) correct.   □

Note that when compared with the baseline algorithm sPAC, the WHD algorithm has a larger budget given the same amount of weight vectors. WHD stops faster because it generates less weight vectors than with sPAC.

The Wet Clutch Example. In Figure 2, we show that WHD, cf Algorithm 2, finds the convex coverage of the Pareto front with less arm pulls than PAC, cf Algorithm 1. In the first iteration, the two extreme weight vectors, $(0, 1)$ and $(1, 0)$, generate two different weight D-rectangles, $H_{i_1^*}$ and $H_{i_{16}^*}$. In the second iteration, a weight vector in middle of the two extreme vectors, $(0.5, 0.5)$, generates another weight D-rectangle $H_{i_7^*}$. The third iteration, with the weight vectors $(0.25, 0.75)$

and $(0.75, 0.25)$, however, will not generate any new weight D-rectangle. In each of the next iterations, i.e. from the fourth through seventh iteration, a new weight D-rectangle is generated, that are, in the order of generation, $H_{i^*_{15}}$, $H_{i^*_3}$, $H_{i^*_{13}}$ and $H_{i^*_{10}}$. Note that the wideness of the weight D-rectangles is variable. Some of the weight D-rectangles are as narrow as a a single weight vector, i.e. $H_{i^*_{13}}$ and $H_{i^*_{10}}$, or as broad as four weight vectors for $H_{i^*_1}$ and $H_{i^*_{15}}$. The difference with the baseline algorithm is that the weight D-rectangles contains less weight vectors which is why WHD is more efficient than PAC. In total, WHD takes only 7 iterations and 18 weight vectors to find the convex coverage of the Pareto front. Note that the minimum number of iterations vary with the shape of the Pareto optimal set and the distribution of the weight D-rectangles.

Remark. A known problem with linear scalarization is its incapacity to find, using any set of weight vectors, all the points on a non-convex Pareto front [3]. Thus, the scalarized MOMAB algorithms, cf. Algorithm 1 and Algorithm 2, *cannot* identify the entire non-convex Pareto front using any set of weights $\mathcal{W}$. The following algorithm has the goal of identifying all the arms for any shape of the Pareto front by hierarchical decomposing the shape into convex subcomponents of the Pareto front.

5   Hierarchical PAC Algorithm

In this section, we propose an hierarchical shape decomposition algorithm, we call it the hierarchical PAC (hPAC) algorithm with the pseudo-code given in Algorithm 3. hPAC is a recursive procedure call that starts with the initial set of arms $\mathcal{I}$. The first call of the hPAC algorithm has as input the entire set of arms and it returns the whole Pareto front $\mathcal{I}^*$,

$$\mathcal{I}^* \leftarrow \mathsf{hPAC}(\epsilon, \delta, \mathcal{W}_{hPAC}, \mathcal{I})$$

where $\mathcal{I}_1 \leftarrow \mathcal{I}$. We denote the arms assigned to a reward D-rectangle with $\mathcal{I}_d$. The output for each function call is a (sub)set of the Pareto front for the corresponding reward D-rectangle, $\mathcal{I}^*_d$. Each iteration d has two stages: 1) first stage identifies arms on the convex coverage of the Pareto front, $\mathcal{I}^*_d$, and 2) second stage iteratively explores the disjunct convex reward subspaces, or reward D-rectangles, defined by the arms in $\mathcal{I}^*_d$. The search continues until there is no arm to explore or the edge of the reward D rectangle $\mathcal{I}_d$ is smaller than a given fixed confidence value ϵ. Thus, Algorithm 3 always stops.

In the first stage of the d iteration, hPAC, cf Algorithm 3, identifies a subset of the Pareto front $\mathcal{I}^*_d$ from the current set of arms $\mathcal{I}_d$ using one of two algorithms proposed above, cf. Algorithm 1 or Algorithm 2. It is fair to assume that, in one iteration, only a subset of the Pareto front is identified because of the linearity of the scalarization functions and the possible non-convexity of the Pareto front. For sampling efficiency of the algorithm, we can either assume a small number of weight vectors in a simple and straightforward PAC algorithm, cf Algorithm 1, or a larger number of weight vectors and the more efficient but sophisticated WHD algorithm, cf Algorithm 2.

Each arm from $\mathcal{I}_d$ is pulled for a fixed number of times, ℓ_d, to identify its mean reward vector with the confidence value ϵ and error tolerance δ. A minimal set of

Algorithm 3. Hierarchical PAC algorithm $\mathsf{hPAC}(\epsilon, \delta, \mathcal{W}, \mathcal{I}_d)$

---

1 Initialise the subset of the Pareto front $\mathcal{I}_d^* \leftarrow \emptyset$;
2 **for** *all arms $i \in \mathcal{I}_d$* **do**
3   Play arm i for $\ell_d = \frac{4}{\epsilon^2} \ln \frac{4|\mathcal{W}||\mathcal{I}_d|}{\delta}$ times ;
4   Assign $\hat{\mu}_i$ to be the average reward vector of the arm i

5 **for** *all weight vectors $\omega \in \mathcal{W}$* **do**
6   Identify the optimal arm i_ω^* for the scalarization function f_ω ;
7   Merge current Pareto front with the optimal arm, $\mathcal{I}_d^* \leftarrow \mathcal{I}_d^* \cup i_\omega^*$;
8   Remove dominated arms from $\mathcal{I}_d^*$

9 **for** *all adjacent pairs of optimal arms* **do**
10   Select the corresponding subset of arms $\mathcal{I}_d$ within the generated D-rectangle ;
11   **if** *the reward D- rectangle $\mathcal{I}_d$ is not empty* **then**
12         $\mathcal{I}_{d+1}^* \leftarrow \mathsf{hPAC}(\epsilon, \delta, \mathcal{W}, \mathcal{I}_{d+1})$;
13   Merge the two Pareto fronts $\mathcal{I}_d^* \leftarrow \mathcal{I}_d^* \cup \mathcal{I}_{d+1}^*$

14 return $\mathcal{I}_d^*$

---

weight vectors contains only: 1) the extreme weight vectors with support in $[0,1]^D$ with only one weight value set on 1 and the rest of weight values set to 0, and 2) the weight vectors resulted from the combination of either equal weight values or 0s. For each of the corresponding scalarization functions f_ω, a Pareto optimal arm is selected, i^*, and merged with the current Pareto front $\mathcal{I}_d^*$.

In the second stage of d iteration, hPAC decomposes the current reward D-rectangle into smaller reward D-rectangles in order to recursively call hPAC on the arms in these smaller reward D-rectangles. We consider all pairs of optimal arms for which there is no other Pareto optimal arm contained between the two optimal arms. Let i_1^* and i_2^* be two Pareto optimal arms, where i_1^* and $i_2^* \in \mathcal{I}_d^*$. A reward D-rectangle is the Cartesian product $\times_{j=1}^D [\min\{i_1^{*j}, i_2^{*j}\}, \max\{i_1^{*j}, i_2^{*j}\}[$ of D subintervals, one per objective j. We get a set of disjoint reward D-rectangles. Each arm in $\mathcal{I}_d$ is assigned to exactly one of the reward D-rectangles defined by adjacent Pareto optimal arms. A reward D-rectangle can contain more than one arm. It is straightforward to show that if an arm i is not assigned to any of these reward D-rectangles, then the arm i is dominated by the current Pareto optimal set of arms $\mathcal{I}_d^*$ and, therefore, deleted. On each of these reward D-rectangles, $\mathcal{I}_d$, the hierarchical PAC algorithm is called recursively. The current Pareto front $\mathcal{I}_d^*$ is updated with the arms in the newly identified Pareto front, $\mathsf{hPAC}(\epsilon, \delta, \mathcal{W}, \mathcal{I}_d)$. Note that the arms identified as non-dominated by the hPAC algorithm can be dominated by the other arms in $\mathcal{I}_d^*$, therefore the dominated arms should be deleted.

Analysis. Each iteration, hPAC assigns the mean reward vectors within a confidence value ϵ and the probability of erroneously deleting an optimal arm with the tolerance error δ. The total number of times a single arm i is pulled depends on the number of iterations where i is pulled, L_i, and on the number of arms in the corresponding reward D-rectangles. Thus, an arm i is pulled for at most $\frac{L_i}{(\epsilon/2)^2}$ ·

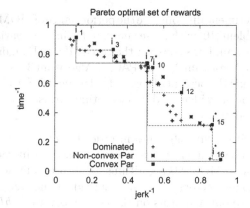

Fig. 3. Pareto optimal arms on the convex front are generated first, and then Pareto optimal arms on the non-convex front are generated

$\ln \frac{2|\mathcal{W}|}{\delta} \cdot \frac{|\mathcal{I}_i|}{L_i}$ times, where $|\mathcal{I}_i| = \sum_{d=1}^{L_i} |\mathcal{I}_d|$ and we have considered the arithmetical - geometrical mean as before. The sample complexity of the algorithm is bounded by $N \leq \sum_{i=1}^{K} \frac{L_i}{(\epsilon/2)^2} \cdot \ln \frac{2|\mathcal{W}|}{\delta} \cdot \frac{|\mathcal{I}_i|}{L_i}$.

Note that the arms that are the most difficult to identify are pulled longest and, thus, the sample complexity of hPAC depends on the distribution of the Pareto front. If the solutions on the Pareto front are uniform randomly distributed, then the number of iterations is smaller than for non-uniform Pareto fronts. Furthermore, the algorithm is more efficient than a simple PAC, cf Algorithm 1, also because the suboptimal arms are discarded. The suboptimal arms that are further away from the Pareto front are discarded at first, and the arms that are closer to the Pareto front are discarded the last. In the experimental section, we show that, even in a practical setting, for the wet clutch example, this algorithm is the most efficient algorithm presented in this paper.

The Wet Clutch Example. In Figure 3, we show the reward subspaces and their succession (different type of lines) in generation. At first, the Pareto optimal arms i_1^*, i_7^* and i_{16}^* are identified as the optimal arms for the weight vectors $(1, 0), (0.5, 0.5)$ and $(0, 1)$, respectively. There are two reward D-rectangles generated between i_1^* and i_7^* and between i_7^* and i_{16}^*. The first reward D-rectangle contains 7 arms and the Pareto optimal arms identified for the three weight vectors are i_2^*, i_3^* and i_5^*. It requires another two iterations to identify also the other two Pareto optimal arms i_4^* and i_6^* not on the convex coverage from the reward D rectangle between i_1^* and i_7^*.

There are 22 arms between the Pareto optimal arms i_7^* and i_{16}^*. In the second iteration in this reward D-rectangle, we identify other three Pareto optimal arms i_8^*, i_{12}^* and i_{15}^*. Another three iterations are required to identify the rest of Pareto optimal arms. The reward D-rectangle between i_{15}^* and i_{16}^* is empty. Note the variable size of these reward subspaces, as well as the different distance between non-convex Pareto optimal arms.

6   Numerical Simulations

In this section, we experimentally compare the performance of MOMAB algorithms on the wet clutch problem [10] with noise generated using a bi-variate normal distribution around the mean. Let's consider the three MOMAB algorithms sPAC, cf Algorithm 1, WHD, cf Algorithm 2, and hPAC, cf Algorithm 3, as before. For each algorithm, we measure: i) the percent of correctly identified Pareto optimal arms, ii) the percentage of erroneously identified Pareto optimal arms, iii) the instantaneous Pareto regret as defined in [3,5].

Settings. Each algorithm runs 30 times in order to collect the means and the standard deviations. The number of weight vectors used is different for the three linear scalarization based MOMAB algorithms. sPAC uses a fix set of weight vectors, i.e. 11 or 81 weight vectors, uniformly spread in the weight vector space to identify the 7 Pareto optimal arms on the convex coverage. $hPAC$ and WHD use each iteration a fixed set of weight vectors. One weight set is $\mathcal{W}_{hPAC} = \mathcal{W}_{WHD} = \{(0,1),(0.5,0.5),(1,0)\}$ meaning that weight vectors are always generated at the half distance between two adjacent weight D-rectangles. Another

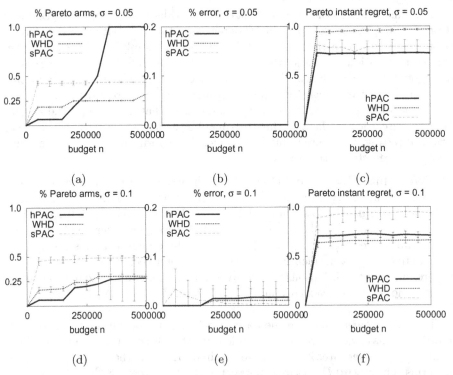

Fig. 4. The performance of sPAC, WHD, and hPAC algorithms for a relatively small number of weight vectors, $|\mathcal{W}_{sPAC}| = 11$ and, for each iteration, $|\mathcal{W}_{hPAC}| = |\mathcal{W}_{WHD}| = 3$, when (top) the variance around the mean is $\sigma = 0.05$, and (bottom) $\sigma = 0.1$

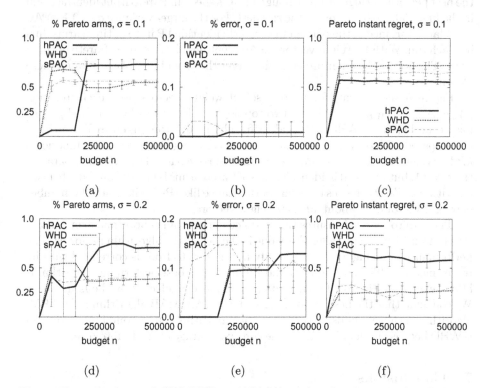

Fig. 5. The performance of sPAC, WHD, and hPAC algorithms for a large number of weight vectors, $|\mathcal{W}_{sPAC}| = 81$ and, each iteration, $|\mathcal{W}_{hPAC}| = |\mathcal{W}_{WHD}| = 11$, when (top) $\sigma = 0.1$, and (bottom) $\sigma = 0.2$

weight set is the 11 weight vectors used also by sPAC, meaning $\mathcal{W}_{hPAC} = \mathcal{W}_{WHD} = \{(0,1),(0.1,0.9),\ldots,(1,0)\}$.

In order to test the robustness of the proposed algorithms at environmental noise, we consider the uncertainty value $\epsilon = \{0.1\}$ and different variance for the normal distribution around the mean, i.e. $\sigma = \{0.05, 0.1, 0.2\}$. Further, we set $\delta = 0.1$, and the budget for each algorithm can be calculated for each confidence value ϵ. The total budget for sPAC when $\epsilon = 0.1$ is $K \cdot \ell \approx 2 \cdot 10^5$. The budget increases considerable for a slight decrease in $\epsilon = 0.05$, i.e. the corresponding sample complexity is $\approx 8 \cdot 10^5$. The arms pulls for the initial iteration in WHD is $\approx 1.5 \cdot 10^5$ when $\epsilon = 0.1$ and the sample complexity increases to $\approx 6.1 \cdot 10^5$ when $\epsilon = 0.05$. We set $N = 5 * 10^5$, which for WHD with $\epsilon = 0.1$ means the generation of about 40-50 weight vectors.

Results. Figure 4 compares the performance of the three MOMAB algorithms when $\epsilon = 0.1$ and variance in the normal distribution generating the noise in the environment is either small $\sigma = 0.05$ (on the top) or $\sigma = 0.1$ (on the bottom), and a small number of weight vectors. For least noisy environments, $\sigma = 0.05$, hPAC is the best performing algorithm although hPAC's speed in Pareto front identification in the first 250.000 iterations is the smallest. For the larger variance $\sigma = 0.1$, hPAC has a large variance in the percent of correctly identified Pareto optimal arms during each run. With 11 weight vectors, sPAC identifies all the convex Pareto optimal arms. The noisier the environments, the larger the probability of error for all these algorithms.

Figure 5 shows that, for a large set of weight vectors used each iteration $|\mathcal{W}_{hPAC}| = |\mathcal{W}_{WHD}| = 11$, the two topological algorithms have a good performance, i.e. hPAC and WHD. $hPAC$ identifies about 90% from the entire Pareto front as compared with only 50% identified with sPAC using 81 scalarization functions. $hPAC$ is identifying the most Pareto optimal arms even for a large variance parameter $\sigma = 0.1$, but its speed in identification of Pareto arms is lower than the other two algorithms. WHD achieves the same performance like sPAC with a smaller number of weight vectors, i.e. about $40 - 50$ weight vectors.

The Pareto regret measures the distance between an suboptimal arm and the Pareto front, thus the smaller this value is the better the algorithm performs. The performance of an algorithm in identifying the Pareto front is correlated with its regret measure. Thus in Figure 4 and Figure 5, the best performing algorithm is hPAC, except in Figure 5 f) where its performance is affected by the large noise. We conclude that the best performing scalarized MOMAB algorithm is the hPAC algorithm because it finds all the Pareto optimal arms. The second best algorithm is WHD because it identifies faster the convex coverage of the Pareto front.

7 Conclusions

In this paper we propose three $MOMAB$ algorithms that identify the Pareto front of stochastic environments. The baseline scalarized Pareto front identification algorithm sPAC uses a fixed set of weight vectors and a fixed number of arm pulls. Weight hyperrectangle decomposition (WHD) groups the weight vectors that optimise the same Pareto optimal arm into weight D-rectangles. WHD iteratively searches new weight vectors in new weight hyperrectangles in order to identify new Pareto optimal arms. Hierarchical PAC (hPAC) identifies Pareto fronts of any shape by iteratively decomposing the current reward hyperrectangle into several reward rectangles containing convex Pareto sub-fronts. We compute the sample complexity of these algorithms, meaning the minimal number of arms necessarily to identify the Pareto front with a give confidence value. The experimental section discusses the behaviour of the proposed linear scalarization based MOMAB algorithms on an example from control theory. The two topological based MOMAB algorithms, i.e. hPAC and WHD, improve the efficiency of the baseline algorithm, sPAC. The advantage of weight hyperrectangle decomposition is that the convex coverage of the Pareto front is identified regardless of its distribution. The advantage of hierarchical PAC is that the entire Pareto front is identified, regardless of its shape. We conclude that the proposed scalarization based MOMAB algorithms are potentially useful tools in optimising in uncertain environments.

Acknowledgments. Madalina M. Drugan was supported by FWO project G.087814N "Multi-criteria RL".

References

1. Auer, P., Cesa-Bianchi, N., Fischer, P.: Finite time analysis of the multiarmed bandit problem. Machine Learning **47**(2/3), 235–256 (2002)
2. de Berg, M., van Kreveld, M., Overmars, M., Schwarzkopf, O.: Computational Geometry. Springer (2000)
3. Drugan, M.M., Nowe, A.: Designing multi-objective multi-armed bandits: a study. In: Proc of International Joint Conference of Neural Networks (IJCNN) (2013)
4. Drugan, M.M., Nowe, A.: Scalarization based pareto optimal set of arms identification algorithms. In: Proc of International Joint Conference of Neural Networks (IJCNN) (2014)
5. Drugan, M.M., Nowe, A., Manderick, B.: Pareto upper confidence bounds algorithms: an empirical study. In: Proc of IEEE Symposium on Adaptive Dynamic Programming and Reinforcement Learning (ADPRL) (2014)
6. Even-Dar, E., Mannor, S., Mansour, Y.: Action elimination and stopping conditions for the multi-armed bandit and reinforcement learning problems. J. of Machine Learning Research **7**, 1079–1105 (2006)
7. Heidrich-Meisner, V., Igel, C.: Hoeffding and bernstein races for selecting policies in evolutionary direct policy search. In: Proc of International Conference on Machine Learning (ICML), p. 51 (2009)
8. Kim, I.Y., Weck, O.L.: Adaptive weighted sum method for multiobjective optimization: a new method for pareto front generation. Structural and Multidisciplinary Optimization **31**(2), 105–116 (2006)
9. Roijers, D., Scharpff, J., Spaan, M., Oliehoek, F., De Weerdt, M., Whiteson, S.: Bounded approximations for linear multi-objective planning under uncertainty. In: ICAPS 2014: Proceedings of the Twenty-Fourth International Conference on Automated Planning and Scheduling (2014)
10. Van Vaerenbergh, K., Rodriguez, A., Gagliolo, M., Vrancx, P., Nowe, A., Stoev, J., Goossens, S., Pinte, G., Symens, W.: Improving wet clutch engagement with reinforcement learning. In: Proc of International Joint Conference of Neural Networks (IJCNN) (2012)
11. Van Moffaert, K., Drugan, M.M., Nowé, A.: Hypervolume-based multi-objective reinforcement learning. In: Purshouse, R.C., Fleming, P.J., Fonseca, C.M., Greco, S., Shaw, J. (eds.) EMO 2013. LNCS, vol. 7811, pp. 352–366. Springer, Heidelberg (2013)
12. Voß, T., Trautmann, H., Igel, C.: New uncertainty handling strategies in multi-objective evolutionary optimization. In: Schaefer, R., Cotta, C., Kołodziej, J., Rudolph, G. (eds.) PPSN XI, Part II. LNCS, vol. 6239, pp. 260–269. Springer, Heidelberg (2010)
13. Yahyaa, S.Q., Drugan, M.M., Manderick, B.: The scalarized multi-objective multi-armed bandit problem: an empirical study of its exploration vs. exploitation trade-off. In: Proc of International Joint Conference on Neural Networks (IJCNN), pp. 2290–2297 (2014)

Elite Accumulative Sampling Strategies for Noisy Multi-objective Optimisation

Jonathan E. Fieldsend$^{(\boxtimes)}$

Computer Science, University of Exeter, Exeter, UK
J.E.Fieldsend@exeter.ac.uk

Abstract. When designing evolutionary algorithms one of the key concerns is the balance between expending function evaluations on exploration versus exploitation. When the optimisation problem experiences observational noise, there is also a trade-off with respect to accuracy refinement – as improving the estimate of a design's performance typically is at the cost of additional function reevaluations. Empirically the most effective resampling approach developed so far is accumulative resampling of the elite set. In this approach elite members are regularly reevaluated, meaning they progressively accumulate reevaluations over time. This results in their approximated objective values having greater fidelity, meaning non-dominated solutions are more likely to be correctly identified. Here we examine four different approaches to accumulative resampling of elite members, embedded within a differential evolution algorithm. Comparing results on 40 variants of the unconstrained IEEE CEC'09 multi-objective test problems, we find that at low noise levels a low fixed resample rate is usually sufficient, however for larger noise magnitudes progressively raising the number of minimum resamples of elite members based on detecting estimated front oscillation tends to improve performance.

Keywords: Pareto optimality · Differential evolution · Uncertainty · Noise

1 Introduction

Many real-world optimisation problems experience noise which corrupts the observed quality values associated with a design. This may be due to, e.g., sensor/measurement error or environmental variation during the evaluation of a built design in embodied optimisation, or due to the stochastic nature of the software simulation being optimised (repeated evaluations leading to slightly different criterion values). Early multi-objective optimisation work raised the issue of noise affecting an evolutionary optimiser [15], but practical work developing evolutionary multi-objective algorithms in this area did not commence in ernest until nearly a decade later. There now exist a wide range of 'noise-tolerant' algorithms, designed specifically for multi-objective optimisation problems with observational noise, e.g. [1–6,8–10,12,13,16–18,20,23,24,26,27,30], and recent

© Springer International Publishing Switzerland 2015
A. Gaspar-Cunha et al. (Eds.): EMO 2015, Part II, LNCS 9019, pp. 172–186, 2015.
DOI: 10.1007/978-3-319-15892-1_12

work has explored the situation where the objective functions themselves are inherently uncertain [29].

The vast majority of noise-tolerant optimisers include some form of resampling (repeated function reevaluation) of designs, in order to improve the estimate of their associated objective values. This is required as noise will mean that poor solutions with 'favourable' noise will be seen as better than they should be, and likewise good solutions that experience detrimental noise will be seen as relatively worse. This has the effect of corrupting fitness assignment and ranking, polluting any elite sets that may be maintained, and generally degrading optimiser performance. Furthermore, it has been observed in a number of studies that as the estimated non-dominated set converges the main driver for updating this set tends to be noise rather than improvements in the designs themselves (see e.g. [10,12,13]). This can seriously impede the ability of an algorithm to locate the Pareto front within the tolerance of the noise width(s).

There are many different approaches taken to resampling in the field, from static approaches, where a fixed number of resamples are taken for each design assessed, through to dynamic approaches based on, for example, reducing the standard error to within an acceptable bound. The reader is directed toward recent work by Siegmund et al. [25] for a full categorisation.

In the work presented here we are solely concerned with *accumulative* sampling approaches (see e.g. [10,20]). These require only a maximum likelihood estimator function being available, $\text{est}(\cdot)$, which takes a set of reevaluated objective vectors associated with a design and provides the best estimate of the underlying noise-free objectives. This differs from a large number of other sampling approaches which rely on the noise experienced being Gaussian, and often utilise variance and standard error estimates [2,8,17,23,25,26].

Accumulative sampling approaches leverage the observation that increasing the number of samples will increase the fidelity of the derived objective vector estimate *irrespective* of the noise density experienced, as long as the estimator is unbiased. That is, at the limit of infinite resamples the estimator will return the noise-free objective vector. Furthermore, even in the case where an estimator converges to the noise-free objective vector plus a bias, dominance-based optimisation can still be performed effectively, as adding a constant does not affect the relative Pareto ranking of solutions [10]. In accumulative resampling of elite members, where the number of reevaluations per member is not limited, there needs to be decision regarding how many function evaluations should be expended on reevaluating elite members. In [20] the elite set is fixed in size, and each generation the entire elite set is resampled once, with a corresponding number of brand new designs also evaluated. In [10] each iteration of the algorithm results in a single new design, and the elite member with the fewest reevaluations is reevaluated a single additional time, with the last 5% of a run entirely devoted to reevaluations. As such, both [20] and [10] split their allocated function evaluations roughly equally between new proposals and previously evaluated proposals. Experiments at the end of [10] however indicate that an equal balance between reevaluations and new design evaluations is not optimal for all problems. Here we examine the use of adaptive reevaluation approaches for elite member accumulative resampling, including methods that ensure the number of effective reevaluations per member increases over time along with methods to increase the reevaluation rate if convergence is impeded by noise.

The rest of the paper is structured as follows. In Sec. 2 the multi-objective optimisation problem with observational noise is defined, along with basic resampling definitions. In Sec. 3 the properties of elite accumulative resampling are discussed, and the proposed adaptive methods described. In Sec. 4 the different approaches are compared empirically on the unconstrained problems of the IEEE CEC'09 multi-objective test suite, modified with additive noise with a range of magnitudes. Sec. 5 contains the paper conclusion and discussion.

2   Multi-objective Optimisation with Noise

Without loss of generality the multi-objective optimisation problem seeks to simultaneously minimise D objectives: $f_d(\mathbf{x})$, $d = 1, \ldots, D$ where each objective depends upon a vector $\mathbf{x} = (x_1, \ldots, x_p, \ldots, x_P)$ of P parameters or decision variables. The parameters may also be subject to equality and inequality constraints which, for simplicity, we assume can be evaluated precisely. The multi-objective optimisation problem may thus be expressed as: minimise $\mathbf{f}(\mathbf{x}) = (f_1(\mathbf{x}), \ldots, f_D(\mathbf{x}))$, subject to the constraints which define $\mathcal{X} \in \mathbb{R}^P$, the feasible search space. When there are multiple competing objectives, solutions may exist for which performance on one objective cannot be improved without degrading performance on at least one other. Such solutions are said to be *Pareto optimal*. The set of all Pareto optimal solutions is said to form the *Pareto set*, whose image in objective space is known as the *Pareto front*.

A decision vector $\mathbf{x}$ is said to *dominate* another $\mathbf{x}'$ iff $f_d(\mathbf{x}) \leq f_d(\mathbf{x}')$ $\forall d = 1, \ldots, D$ and $\mathbf{f}(\mathbf{x}) \neq \mathbf{f}(\mathbf{x}')$. This is often denoted as $\mathbf{x} \prec \mathbf{x}'$. Pareto dominance is a key comparator used in a wide range of evolutionary optimisers – either directly in their fitness assignment and ranking schemes, or as a means to identify their final Pareto set estimate. Elitist multi-objective optimisers generally maintain a mutually non-dominating set A (often called an archive) of solutions which form their estimated Pareto set at any stage in their optimisation. This may be *active* (providing input into the optimisation process) or a *passive* record of the best solutions ever encountered during the optimisation [28]. In a noisy optimisation problem we cannot directly access $\mathbf{f}(\mathbf{x})$, instead we have access to $\mathbf{y}$, which are the criteria contaminated by *observational* noise $\boldsymbol{\epsilon}$. Here we are concerned with additive noise:

$$y_d = f_d(\mathbf{x}) + \epsilon_d. \tag{1}$$

With n repeated reevaluations at a design location $\mathbf{x}$ we obtain a set of noise contaminated objective vectors $Y(\mathbf{x}) = \{\mathbf{y}_i\}_{i=1}^n$, which, in conjunction with an unbiased maximum likelihood estimator will provide us with an estimate of the noise-free evaluation of $\mathbf{f}(\mathbf{x})$: $\widehat{\mathbf{f}(\mathbf{x})} = \mathtt{est}(Y(\mathbf{x}))$. For instance, if the noise was

Gaussian then est$(\cdot)$ would be the mean function, whereas if the noise was Laplacian it would be the median function.

In the noisy situation, we no longer have certainty that one solution dominates another (or that they are mutually non-dominating), as the ϵ experienced by each solution may be of a value sufficient to reverse the ordering of solutions on one or more objective criteria. However, as n increases, our approximation to $\mathbf{f}(\mathbf{x})$ improves (in general this accuracy improves proportionally to $\sqrt{n}$). In order to ensure the exploitation of elite members uses accurately labelled designs, recent noise-tolerant optimisers have focused on resampling elite members preferentially [10,20].

3   Adaptive Accumulative Sampling

Depending on the problem and the noise experienced, the update dynamics of the elite set may vary considerably. If members are regularly leaving the elite set, and new members regularly entering it, then the number of reevaluations per elite member may be in effect quite low – even when reevaluating an elite member for each new solution evaluated. Alternatively, if the membership of A changes relatively irregularly, then the n per elite member may be very large. Neither of these situations may be ideal in practice, as in the first instance the elite members may fail to accumulate sufficient resamples to mitigate the noise when in proximity to the Pareto front, and in the second case some of these function evaluations may be better expended on new designs.

One side-effect of reevaluating previously evaluated solutions is that the estimated Pareto front can *oscillate*. This is distinct from the oscillating/retreating front issue derived from truncating elite sets in noiseless problems [11,14], as its root cause is due to the (estimated) objective location of previously elite solutions moving, rather than the direct exclusion of solutions that are known with certainty to be non-dominated. This therefore affects even unbounded elite sets in the noisy case. An illustrative example is provided in Fig. 1. Here the differential evolution for multi-objective optimisation (DEMO) algorithm [21] is applied to noisy variants of the IEEE CEC'09 UP1 problem, with varying levels of additive observational Gaussian noise. The population in DEMO is maintained using non-dominated sorting, so, subject to the population limit being sufficient to contain the number of non-dominated solutions encountered at any time point, on first glance the population *should* contain the best performing solutions found so far (as it will only discard dominated solutions). However, this maintenance approach is not sufficient in the noisy case with reevaluations – as solutions may be discarded which would later be determined as non-dominated due to reevaluations of elite members degrading their estimated performance ('exposing' the previously dominated solutions). In Fig. 1 we expend one elite member reevaluation for every new design evaluated. The left panel of Fig. 1 shows the number of times the non-dominated subset of the DEMO search population did not contain the non-dominated subset of *all* designs visited so far (based on their est$(\cdot)$). A secondary elite archive is maintained separately from the DEMO population,

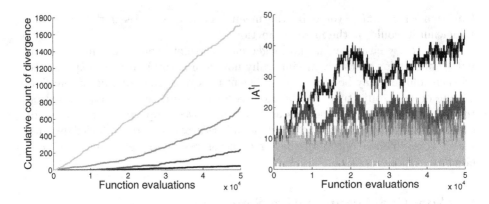

Fig. 1. Empirical oscillation of estimated elite members with Gaussian observational noise on the UP1 problem. Noise standard deviations $\sigma = \{0.01, 0.1, 1.0, 10.0\}$ used for four different runs of the DEMO algorithm using elite reevaluations. *Left:* cumulative number of times the search population in DEMO needed to be updated using a secondary tracking archive (black through to light grey indicates low noise through to high noise). *Right:* Corresponding archive size, $|A^t|$. Note that $|A^t|$ never exceeds 100 – the size of the non-dominated sorting truncated search algorithm in DEMO – but the search population regularly discards solutions which later become non-dominated (and have to be fed back in again) due to the reevaluation of the noisy solutions.

using the techniques described in [7], and is merged in with the search population whenever they are detected to have diverged (redirecting the DEMO population back to the estimated optimal regions of design space). This recalibration can be seen to be regularly required even when the search population membership is much larger than that of the elite set (an order of magnitude bigger in the highest noise case).

We now propose a number of *adaptive* schemes for incremental accumulative sampling, and discuss the reasoning behind them. Each approach treats the optimisation as an incremental process. At each time step t either a new design is evaluated, or a previously evaluated elite solution is *reevaluated*. In both situations the membership of A^t (the estimated elite set at time t) may be altered, as in the first instance the new design may enter A^t, and in the second instance reevaluation may cause the previously elite member to be dominated and/or move to a position in objective space such that solutions that were previously identified as dominated should now (re)enter the elite set.

3.1 Fixed Resamples per Generation

The baseline approach evaluates one new design per algorithm iteration, and reevaluates a single member of the elite set. The reevaluated elite member is

---

Algorithm 1 Resample rate fixed over time.

Require: A^{t-1} Elite non-dominated members identified at previous time step
Require: X^{t-1} Other previously evaluated designs (dominated at $t-1$)
  1: $A^t := A^{t-1}$, $X^t := X^{t-1}$
  2: Propose and evaluate new design, update A^t and X^t
  3: Reevaluate the member of A^t with fewest resamples, update A^{t+1} and X^{t+1}

---

that with the fewest reevaluations contributing to its estimated objective values.[1] The evaluation of a new design may cause a change in the elite set or a change in the set containing all previously evaluated dominated solutions at a time step (X^t). The reevaluation of a solution may also cause multiple changes in *both* sets, as it can mean the removal of elements from A^t to insert into X^t (if the reevaluated solution has moved to a dominated locations, or to a location that now dominates members of A^t), and the addition of elements from X^t to A^t (where designs that were previous dominated are now categorised as non-dominated due to the reevaluated solution moving to an objective space location which no longer covers them). A basic outline is presented in Alg. 1.

3.2 Increasing Resample Rate, Based on Detecting False Convergence

As mentioned above, one of the key issues with noisy optimisation problems is that as an algorithm converges, there is a tendency for noise to drive the search process over improving performance on the underlying criteria. One way of detecting this is to compare the state of the best elite set estimate at one time step, A^t, with that of an earlier time step, e.g. A^{t-m}. If the performance if A^t is *worse* than that of A^{t-m} then (assuming A has not been truncated) this can only be because reevaluations of members of A in the m intervening time steps has meant that their predicted locations through $\mathsf{est}(\cdot)$ have worsened, and that this shift *backwards* of A^t has not been compensated for by finding other designs which provide equivalent or better predicted performance to those in A^{t-m}. In other words, the noise experienced made A^{t-m} seem better than it was, and we have not found any solutions (or reevaluated any) in the intervening m time steps to compensate for this over-estimate. In order to mitigate this, the number of reevaluations are *increased*, making it harder for rogue reevaluations to unduly influence the performance assessment (as outliers should be more quickly diluted with subsequent reevaluations).

    Alg. 2 outlines this approach using the binary $\epsilon+$ indicator (other indicators could also be used, see [19] for a discussion of different indicators and their properties). If the additive ϵ required to make the A^{t-m} set dominate the A^t set is lower that the value required to make the A^t set dominate the A^{t-m} set then the number of reevaluations per iteration is increased. Here the objective values

---

[1] Selection could instead be based on the largest standard error for situations where *a priori* noise density information allows this to be calculated.

Algorithm 2 Increasing resample rate based on convergence assessment.

Require: k	Current resample number
Require: m	Convergence time window
Require: A^{t-m}	Elite non-dominated members identified at time step $t - m$
Require: A^{t-1}	Elite non-dominated members identified at previous time step
Require: X^{t-1}	Other previously evaluated designs (dominated at $t - 1$)

1: $A^t := A^{t-1}$, $X^t := X^{t-1}$
2: Propose and evaluate new design, update A^t and X^t
3: **if** number of time steps since last check meets or exceeds m **then**
4:    **if** $I_{\epsilon+}(A^t, A^{t-m}) > I_{\epsilon+}(A^{t-m}, A^t)$ **then**
5:       $k := k + 1$
6:    **end if**
7: **end if**
8: **for** $i = 1, \ldots, k$ **do**
9:    Reevaluate the member of A^{t+i-1} with fewest resamples, update A^{t+i} and X^{t+i}
10: **end for**

are normalised by the bounds of the minimum bounding box containing A^t and A^{t-m}, and the `est(·)` used are those calculated for the designs at the respective time steps. Rather than compare A^t with A^{t-m} at every time step this is done every m time steps as a minimum (lines 3-7). This allows the increment of k (the number of reevaluations per iteration) time to have an effect before the sets are compared once more.

3.3 Increasing Minimum Revaluation Number, Based on Detecting False Convergence

An alternative approach to increasing the absolute number of reevaluations each iteration, is to increase the minimum number of reevaluations that archive members must have accrued. This approach means the balance of function evaluations expended on reevaluations versus new designs can alter back and forth from one iteration to the next. For instance, if the minimum number of reevaluations per elite member was $k = 10$, after reevaluating a single archive member with the fewest reevaluations (Alg. 3 line 8), if all elite members had at least k reevaluations then no further reevaluations would be taken. On the other hand, if there were elite members with fewer than k reevaluations, then the loop on lines 10-13 may be processed many times before the minimum number of reevaluations condition was satisfied. Note that the check to increase k (lines 3-7) is only undertaken in situations where the elite archive meets the condition that all members have at least k reevaluations each.

3.4 Increasing Average Resamples per Elite Member

As the optimiser progresses we would like to say that the confidence we have in our elite set (our estimate of the Pareto set) increases rather than decreases or stagnates. With accumulative sampling the way to achieve this is to ensure that

---

Algorithm 3 Increasing minimum number of reevaluations for elite members, based on convergence assessment.

Require: m Convergence time window
Require: A^{t-m} Elite non-dominated members identified at time step $t - m$
Require: A^{t-1} Elite non-dominated members identified at previous time step
Require: X^{t-1} Other previously evaluated designs (dominated at $t - 1$)
 1: $A^t := A^{t-1}$, $X^t := X^{t-1}$
 2: Propose and evaluate new design, update A^t and X^t
 3: **if** number of time steps since last check meets or exceeds m **then**
 4:    **if** $I_{\epsilon+}(A^t, A^{t-m}) > I_{\epsilon+}(A^{t-m}, A^t)$ **then**
 5:       $k := k + 1$
 6:    **end if**
 7: **end if**
 8: Reevaluate the member of A^t with fewest resamples, update A^{t+1} and X^{t+1}
 9: $i := 1$
10: **while** member of A^{t+i} with fewest resamples has fewer than k reevaluations **do**
11:    Reeval. the member of A^{t+i} with fewest resamples, update A^{t+i+1} and X^{t+i+1}
12:    $i := i + 1$
13: **end while**

Algorithm 4 Increasing resamples per elite member over time.

Require: A^{t-1} Elite non-dominated members identified at previous time step
Require: X^{t-1} Other previously evaluated designs (dominated at $t - 1$)
Require: α Average number of resamples of A across all previous time steps
 1: $A^t := A^{t-1}$, $X^t := X^{t-1}$
 2: Propose and evaluate new design, determine A^t and X^t
 3: Reevaluate the member of A^t with fewest resamples, update A^t and X^t
 4: $i := 1$
 5: **while** `mean_num_resamp`$(A^t) \leq \alpha$ **do**
 6:    Reevaluate member of A^{t+i-1} with fewest resamples, update A^{t+i} and X^{t+i}
 7:    $i := i + 1$
 8: **end while**

---

the number of reevaluations per set member is always increasing. Comparing one generation directly to the next can be a brittle approach as it may force a relatively large increase in resamples each time step (for instance if $|A| = 1$ for any stretch of time the imposition of an extra reevaluation each iteration can be putative if the set grows in size later during an optimisation). Instead we examine a less stringent average approach here, as outlined in Alg. 4. Here the average number of resamples of solutions in A^t is compared to the average across elite members of all previous time steps, and repeated reevaluations are taken if the current average is lower (lines 4-6). This is in addition to the extra reevaluation taken each generation as standard (line 3), which acts to steadily increase the lower bound on this minimum.

Algorithm 5 Incremental differential evolution candidate creation, **p**.

Require: Z^{t-1} DEMO population at previous time step
1: **p** := copy_random_member(Z^{t-1})
2: $\{\mathbf{a}, \mathbf{b}, \mathbf{c}\}$:= copy_random_members($Z^{t-1} \setminus \{\mathbf{p}\}$)
3: **for** $i := 1, \ldots, |\mathbf{p}|$ **do**
4:    **if** rand() < cross_prob **then**
5:      $p_i := a_i +$ differential_weight $\times (b_i - c_i)$
6:    **end if**
7: **end for**

---

4 Empirical Results

We now compare Algs. 1-4 empirically. We use the DEMO algorithm [21] to generate a new design at each iteration prior to elite set member reevaluation(s). We modify the original DEMO in two ways to use in the noisy optimisation context. Firstly, in order to make the algorithm incremental a single new candidate design is generated from the DEMO population at each time step rather than doubling the population size (before its reduction via ranking and crowding). This is achieved at each algorithm iteration by selecting one of the DEMO population at random to be the base parent (see Alg. 5, line 1). Secondly, due to the noisy environment, there is no guarantee that non-dominated solutions preserved by ranking and crowding truncation at one time step in the DEMO population will be non-dominated on reevaluation (as illustrated in Sec. 3). To mitigate this, a separate elite set A^t is maintained using the data structures introduced in [7],[2] and whenever the DEMO search population, Z^t, does not contain A^t, the omitted members are combined with Z^t prior to DEMO's truncation operator being applied. This was found to significantly improve the performance of DEMO in the noisy domain in our preliminary experimentation.

Further details describing DEMO may be found in the original work [21]. We use a DEMO population of 100 in all experiments, a probability of crossover of 0.9 and differential weight of 0.5. The external archive $|A^t|$ is unbounded, and is updated at each time step using the data structure from [7] to ensure it contains the best estimate of the Pareto set. The algorithm variants are evaluated on the IEEE CEC'09 test suite[3] [31]. We use the unconstrained (bounded) problems from the suite, UP1-10, with the standard number of design parameters, and modify the objective values with independent additive Gaussian noise with standard deviations of $\sigma = \{0.01, 0.1, 1.0, 10.0\}$ (making 40 test problem variants in total). In (1) therefore $\epsilon_d \sim \mathcal{N}(0, \sigma^2)$. We run each algorithm 30 times, for a total of 300,000 function evaluations, and record the generational distance (GD) and inverse generational distance (IGD) every 500 function evaluations using the A^t at that time point. The noise-free reevaluation of the stored A^t is used – the corresponding non-dominated set in its mapping to the noise-free space

---

[2] MATLAB code from https://github.com/fieldsend/.
[3] MATLAB code from http://web.mysites.ntu.edu.sg/epnsugan/PublicSite/Shared Documents/Forms/AllItems.aspx.

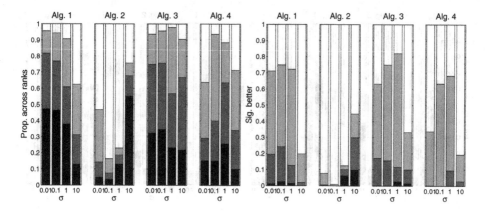

Fig. 2. IGD results, initial 10% of run. The left block of bar charts shows the proportion of time, across problems and polled every 500 function evaluations, where a resampling technique led to the best results (black), the second best (dark grey), third best (light grey) and worst (white), for each noise level. The right set of bar charts shows the corresponding *significance* assessments – black indicates the proportion where the approach is significantly better than all three other sampling approaches, dark grey significantly better than two, light grey significantly better than one and white not significantly better than any others.

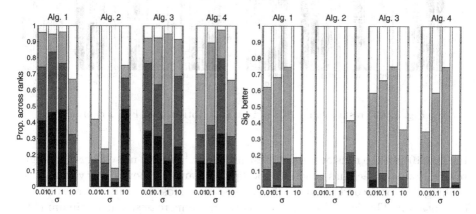

Fig. 3. GD results, initial 10% of run. Description as in Fig. 2 caption.

is extracted for the calculation of the quality measures. We utilise the modified versions of the GD and IGD quality measures which are not susceptible to variation in set size (see [22]). We set the convergence check parameter $m = 100$ for all experiments.

Figs. 2 and 3 give the relative IGD and GD performance for the four reevaluation routines embedded in DEMO, at each of the noise levels, averaged across

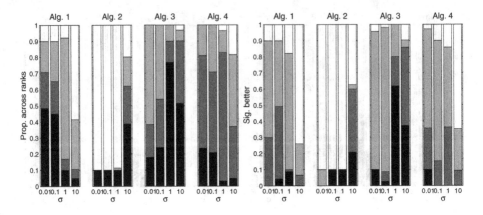

Fig. 4. IGD results, final 10% of run. Description as in Fig. 2 caption.

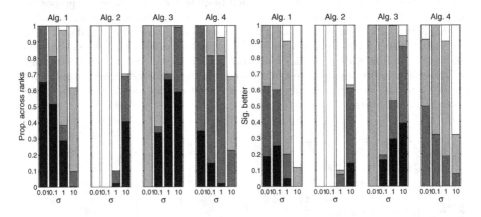

Fig. 5. GD results, final 10% of run. Description as in Fig. 2 caption.

the first 10% of the runs over the 10 test problems.[4] For low noise levels the single revaluation approach has generally good performance for both quality measures, ranking first or second roughly 80% of the time across the initial stages of the optimisation. This is seen to drop off as the noise increases. The minimum elite member reevaluation approach (Alg. 3) performs fairly consistently across the noise levels, and performs better than the steadily increasing reevaluations approach. Interestingly the approach which increases the number of reevaluations at each iteration (if oscillation is detected) tends to perform worst, except for the largest noise level, where its relative performance jumps up.

Figs. 4 and 5 provide the combined results for the final 10% of the runs. The general trends are as for the first 10% but the relative decline (and rise) of the reevaluation approaches as the noise level increases is more pronounced. The

---

[4] 5% statistical significance is assessed using paired Wilcoxon signed ranks tests, with each strategy compared to each of the other competitors.

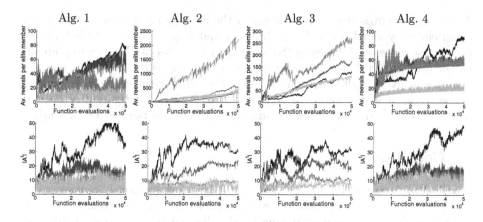

Fig. 6. Average number of resamples per elite member elite on a single run on UP1 (top) and size of A^t (bottom) for each update algorithm. Black through to light grey indicates low noise through to high noise $\sigma = 0.01, 0.1, 1.0, 10.0$.

single revaluation approach degrades more steeply as the noise level rises, such that between $\sigma = 0.1$ and $\sigma = 1.0$ the minimum elite member reevaluation approach (Alg. 3) replaces it as being the preferred approach. Indeed, this approach is best or second best for IGD 90% of the time for the highest two noise values.

4.1   Population Dynamics

We can explore some of the different behaviours of the resampling regimes by examining the elite population dynamics over time, which also lends insight as to *why* particular approaches seem better suited to different noise regimes. Fig. 6 shows the average number of reevaluations per member of A^t as a run progresses through to 50,000 function evaluations on the UP1 problem for each of the accumulative sampling schemes, for the four different noise magnitudes. Some immediate differences become apparent. When noise is high the standard approach of fixing the ratio of resamples to new design evaluations throughout the run can lead to fronts with large oscillations in the average number of resamples per member (repeatedly jumping across the range of 2-15 in a few iterations). On the other hand, increasing the number of reevaluations progressively each time the front is detected as oscillating (Alg. 2) leads to rapidly increasing average number of reevaluations, meaning a relatively small proportion of expended evaluations are on new designs. Alg. 3 (setting a minimum number of resamples for archive members, and increasing this if convergence issues are detected) can be seen to balance these properties, with the average number of resamples increasing steadily in all noise regimes, but not drastically, and the variation relatively small from one iteration to the next. Correspondingly, although the size of A^t varies over time, it is seen to have lower amplitude on its high frequency oscillations (indicating lower churn of elite members with this approach). Alg.

4 similarly removes the wild variation in the average number of reevaluations experienced by members of A^t which Alg. 1 and Alg. 2 are susceptible to, however as the noise level increases the lower bound on this can be seen to plateau, rather than steadily increase as in Alg. 3. This decay is due Alg. 4 comparing the current average reevaluations per member of A^t with *all* previous archive averages, alternatively a moving window approach should mitigate this (though obviously the window size then becomes an additional parameter beyond m).

5   Conclusion

Accumulative resampling of the elite population has previously been seen to provide state-of-the-art performance when embedded in noisy multi-objective optimisers. The management regime for deciding what proportion of function evaluations to expend on accumulative elite reevaluations rather than new designs has not however received much previous attention. Here we have compared four alternative accumulative resampling regimes, which we have embedded in an iterative version of the popular DEMO algorithm, and analysed their performance on 40 variants of the unconstrained CEC'09 test problems. When the noise is level is low (with widths up to 10% of the range of the Pareto front), then having an equal balance of new designs versus elite reevaluations provides relatively good results both at the early stages of optimisation, and also toward the end. For larger noise widths however the balanced approach is not optimal. Due to the estimated front oscillation there is a frequent churn of the elite set membership, meaning the number of reevaluations accrued by members tends to be low, and does not markedly increase as search progresses. This therefore nullifies the benefits of accumulation, as the elite solutions do not progressively get more accurate as time progresses under the standard regime. In these situations the detection of oscillation, and the increase in the minimum acceptable number of reevaluations per elite member in response, is seen to provide consistently good results. We look forward to being able to tackle highly noisy multi-objective problems which these alternative reevaluation regimes would now seem to facilitate (the work presented here including noise widths up to 1000% of the range of the Pareto front).

Further areas of research include examining the use of other indicators to detect oscillations, using different sized time windows in the detection process, allowing the sample rate to *decrease* as well as rise, and investigating switching regimes between maintenance algorithms.

Acknowledgments. The author would like to express his thanks to Prof. Richard Everson for his useful comments whilst drafting this paper, and to Prof. Jürgen Branke and Dr Robin Purshouse for their useful discussions on this topic. MATLAB code for the work presented here is available from https://github.com/fieldsend.

References

1. Basseur, M., Zitzler, E.: A preliminary study on handling uncertainty in indicator-based multiobjective optimization. In: Rothlauf, F., et al. (eds.) EvoWorkshops 2006. LNCS, vol. 3907, pp. 727–739. Springer, Heidelberg (2006)
2. Büche, D., Stoll, P., Dornberger, R., Koumoutsakos, P.: Multiobjective evolutionary algorithm for optimization of noisy combustion processes. IEEE Transactions on Systems, Man, and Cybernetics - Part C: Applications and Reviews 32(4), 460–473 (2002)
3. Bui, L., Abbass, H., Essam, D.: Fitness inheritance for noisy evolutionary multi-objective optimization. In: Proceeding of the Genetic and Evolutionary Computation Conference, pp. 779–785 (2005)
4. Das, S., Konar, A., Chakraborty, U.K.: Improved differential evolution algorithms for handling noisy optimization problems. In: IEEE Congress on Evolutionary Computation, vol. 2, pp. 1691–1698. IEEE (2005)
5. Di Pietro, A., While, L., Barone, L.: Applying Evolutionary Algorithms to Problems with Noisy, Time-consuming Fitness Functions. In: IEEE Congress on Evolutionary Computation, vol. 2, pp. 1254–1261. IEEE (2004)
6. Eskandari, H., Geiger, C.D.: Evolutionary multiobjective optimization in noisy problem environments. Journal of Heuristics 15, 559–595 (2009)
7. Fieldsend, J.E., Everson, R.M.: Efficiently identifying pareto solutions when objective values change. In: Proceedings of the 2014 Conference on Genetic and Evolutionary Computation, pp. 605–612. ACM (2014)
8. Fieldsend, J.E., Everson, R.M.: Multi-objective optimisation in the presence of uncertainty. In: IEEE Congress on Evolutionary Computation, pp. 243–250 (2005)
9. Fieldsend, J.E., Everson, R.M.: On the efficient use of uncertainty when performing expensive ROC optimisation. In: IEEE Congress on Evolutionary Computation, pp. 3984–3991 (2008)
10. Fieldsend, J.E., Everson, R.M.: The Rolling Tide Evolutionary Algorithm: A Multi-Objective Optimiser for Noisy Optimisation Problems. IEEE Transactions on Evolutionary Computation (in press). http://dx.doi.org/10.1109/TEVC.2014.2304415
11. Fieldsend, J.E., Everson, R.M., Singh, S.: Using unconstrained elite archives for multi-objective optimisation. IEEE Transactions on Evolutionary Computation 7, 305–323 (2001)
12. Goh, C.-K., Tan, K.C.: An investigation on noisy environments in evolutionary multiobjective optimization. IEEE Transactions on Evolutionary Computation 11(3), 354–381 (2007)
13. Goh, C.-K., Tan, K.C.: Evolutionary Multi-objective Optimization in Uncertain Environments. SCI, vol. 186. Springer, Heidelberg (2009)
14. Hanne, T.: On the convergence of multi objective evolutionary algorithms. European Journal of Operational Research 117, 553–564 (1999)
15. Horn, J., Nafpliotis, N.: Multiobjective Optimization Using the Niched Pareto Genetic Algorithm. Technical Report 93005, Illinois Genetic Algorithms Laboratory, University of Illinois at Urbana-Champaign (1993)
16. Hughes, E.J.: Constraint handling with uncertain and noisy multi-objective evolution. In: IEEE Congress on Evolutionary Computation, pp. 963–970 (2001)
17. Hughes, E.J.: Evolutionary multi-objective ranking with uncertainty and noise. In: Zitzler, E., Deb, K., Thiele, L., Coello Coello, C.A., Corne, D.W. (eds.) EMO 2001. LNCS, vol. 1993, pp. 329–343. Springer, Heidelberg (2001)

18. Jin, Y., Branke, J.: Evolutionary optimization in uncertain environments-a survey. IEEE Transactions on Evolutionary Computation 9(3), 303–317 (2005)
19. Knowles, J.D., Thiele, L., Zitzler, E.: A tutorial on the performance assessment of stochastic multiobjective optimizers. Technical Report 214, Computer Engineering and Networks Laboratory, ETH Zurich, Switzerland (2006)
20. Park, T., Ryu, K.R.: Accumulative sampling for noisy evolutionary multi-objective optimization. In: Proceeding of the Genetic and Evolutionary Computation Conference, pp. 793–800 (2011)
21. Robič, T., Filipič, B.: DEMO: Differential evolution for multiobjective optimization. In: Coello Coello, C.A., Hernández Aguirre, A., Zitzler, E. (eds.) EMO 2005. LNCS, vol. 3410, pp. 520–533. Springer, Heidelberg (2005)
22. Schutze, O., Esquivel, X., Lara, A., Coello Coello, C.A.: Using the Averaged Hausdorff Distance as a Performance Measure in Evolutionary Multiobjective Optimization. IEEE Transactions on Evolutionary Computation 16(4), 504–522 (2012)
23. Shim, V.A., Tan, K.C., Chia, J.Y., Al Mamun, A.: Multi-objective Optimization with Estimation of Distribution Algorithm in a Noisy Environment. Evolutionary Computation 21(1), 149–177 (2013)
24. Siegmund, F.: Sequential sampling in noisy multi-objective evolutionary optimization. Master's thesis, University of Skövde, School of Humanities and Informatics, Sweden (2009)
25. Siegmund, F., Ng, A., Deb, K.: A comparative study of dynamic resampling strategies for guided evolutionary multi-objective optimization. In: IEEE Congress on Evolutionary Computation (CEC), pp. 1826–1835. IEEE (2013)
26. Syberfeldt, A., Ng, A., John, R.I., Moore, P.: Evolutionary optimisation of noisy multi-objective problems using confidence-based dynamic resampling. European Journal of Operational Research 204, 533–544 (2010)
27. Teich, J.: Pareto-front exploration with uncertain objectives. In: Zitzler, E., Deb, K., Thiele, L., Coello Coello, C.A., Corne, D.W. (eds.) EMO 2001. LNCS, vol. 1993, pp. 314–328. Springer, Heidelberg (2001)
28. van Veldhuizen, D., Lamont, G.: Multiobjective Evolutionary Algorithms: Analyzing the State-of-the-Art. Evolutionary Computation 8(2), 125–147 (2000)
29. Villa, C., Lozinguez, E., Labayrade, R.: Multi-objective optimization under uncertain objectives: application to engineering design problem. In: Purshouse, R.C., Fleming, P.J., Fonseca, C.M., Greco, S., Shaw, J. (eds.) EMO 2013. LNCS, vol. 7811, pp. 796–810. Springer, Heidelberg (2013)
30. Yang, S., Ong, Y.S., Jin, Y.: Evolutionary computation in dynamic and uncertain environments. SCI, vol. 51. Springer, Heidelberg (2007)
31. Zhang, Q., Zhou, A., Zhao, S., Suganthan, P.N., Liu, W., Tiwari, S.: Multiobjective optimization Test Instances for the CEC 2009 Special Session and Competition. Technical Report CES-487, School of Computer Science and Electronic Engineering, University of Essex, UK, April 2009

Guideline Identification for Optimization Under Uncertainty Through the Optimization of a Boomerang Trajectory

Mariapia Marchi[✉], Enrico Rigoni, Rosario Russo, and Alberto Clarich

ESTECO S.p.A., Area Science Park, Loc. Padriciano 99, 34149 Trieste, Italy
{marchi,rigoni,russo,clarich}@esteco.com
http://www.esteco.com

Abstract. Optimization under uncertainty (OUU) is a very important task for practitioners of engineering design optimization. In fact real–world problems are often affected by uncertainties of different kind. The search for robust optimal solutions is intrinsically multiobjective, being formulated as the search for the optimal performance while minimizing its variance. Thus, OUU should garner interest in the evolutionary multi-objective optimization community. It is a challenging topic, because, for instance, engineers have to deal with large scale or highly–constrained problems. The first issue affects the feasibility of the optimization itself, whereas the second affects the reliability of an optimal solution. In this paper, we address the OUU problem to validate a number of best practices through the application to a benchmark problem: the optimization of a boomerang launch parameters. To reduce the computational cost, we consider variable screening as a preliminary step before performing a stochastic optimization. For the latter we use a method recently proposed by the authors, which combines robustness and reliability assessments within a single optimization run.

Keywords: Uncertainty quantification · Polynomial chaos expansion · Evolutionary multiobjective optimization · Robust and reliability–based design optimization · Screening analysis

1 Introduction

Real–world application problems of engineering design optimization are often affected by uncertainties of different kind, such as unknown or changing environmental conditions, variability in material or geometrical properties, model assumptions, etc. Thus, the discipline of optimization under uncertainty (OUU) has been the subject of growing attention. OUU presents several facets and challenges. Below we list some of them.

First there is the problem of uncertainty quantification (UQ): quantifying the *a priori* unknown uncertainty of system responses to uncertain inputs.

Second, the best solutions found with a deterministic approach are not guaranteed to be robust in the presence of uncertainty, as output values might vary

© Springer International Publishing Switzerland 2015
A. Gaspar-Cunha et al. (Eds.): EMO 2015, Part II, LNCS 9019, pp. 187–201, 2015.
DOI: 10.1007/978-3-319-15892-1_13

strongly for small input parameter variations. Moreover, in case of constrained optimization problems they often lie on the boudary between the feasible and unfeasible domain and are likely to violate one or more constraints. Thus, OUU must look for robust and reliable solutions, stable against input parameter variations and with a negligible probability of violating some pre–defined criteria, referred to as *limit state functions* (LSFs). This can be achieved with robust design optimization (RDO) (see e.g. Refs. [1,2]) and reliability–based design optimization (RBDO) (see e.g. Refs. [3–5]).

Third, the search for robust optimal solutions is intrinsically multiobjective, as it is commonly translated in the search for the optimal performance while minimizing its variance. This is a strong contact point between OUU and evolutionary multiobjective optimization (EMO), which is naturally an effective instrument for solving multiobjective optimization problems. Moreover Ref. [3–5] show the importance of using evolutionary algorithms to perform RBDO for both the multiobjective and single objective problems (due to their capability of reaching reliable global solutions).

Four, in real–world use cases OUU has to deal with large scale problems, which combined with the use of computationally expensive function solvers, might pose severe limits to the applicability of OUU.

Five, there are different types of uncertainties: random or probabilistic (due to an inherent variability of physical systems and consequently irreducible) and epistemic or imprecise uncertainties (due to a lack of knowledge or information and, in principle, reducible at some stage of the modeling activity) (see Refs. [6,7]). Here, we focus only on the first type, even though the procedure described in this paper could be applicable to the latter as well.

Most of these challenges are reviewed in Refs. [6,7]. Specific reviews of evolutionary optimization approaches to OUU can be found e.g. in Refs. [2,8,9]. Many papers were devoted to the single–objective case, with new heuristics for determining the fitness function or its optimization in noisy environments, or the use of hierarchical strategies for high performing explorations of the landscape, or the use of surrogate models to reduce the computational cost. Other works highlighted the multiobjective nature of the robust optimisation problem and extended single–objective methods to multi–objective OUU problems, where robustness could be determined by analysing the solution behavior in its neighborhood. Uncertainties effects on the Pareto dominance were studied and a few different choices for robust objectives and constraints were explored. However OUU deserves further investigation in the EMO field.

In this paper, we address the first four challenges of OUU with a practical intent, i.e. with the aim of validating a number of best practices, which can be useful in everyday engineering design, through the application to an example: the optimization of a boomerang.

Though a toy model, this problem is a good representation of realistic design optimization problems as shown in Ref. [10], where an optimization of geometric and launch parameters was performed to determine at the same time optimal shapes and trajectories, together with aerodynamic analyses and the determination of response surface models for the extrapolation of aerodynamic coefficients.

For the present work, for simplicity we confined our investigations to the optimization of the boomerang launch parameters. We started from a deterministic optimization. In order to reduce the computational cost of the stochastic optimization, we performed a preliminary variable screening to detect the most important uncertain variables to be considered as stochastic in the OUU.

For the OUU we used a method recently proposed by the authors, which combines robustness and reliability assessments within a single optimization run [11–13]. We used the polynomial chaos expansion (PCE) method to assess the robustness of the optimization outcomes. The PCE was also used as a stochastic response surface (SRSM) to quickly determine a sample cumulative distribution function (CDF) for estimating the reliability of output properties. Particular attention was given to the convergency check of the UQ parameters, such as sample sizes, PCE degree, etc.

All the simulations shown below were done with the modeFRONTIER[14] (mF) integration platform for multiobjective and multidisciplinary optimization.

In Section 2 we summarize the methods used in this work: the UQ and OUU techniques in 2.1 and the screening analysis in 2.2. In Section 3 we briefly introduce the chosen test case and illustrate the optimization problem. Sections 3.1 and 3.2 respectively describe the deterministic optimization performed on the test case, and the screening analysis and UQ checks preliminary to the stochastic optimization related in Section 3.3. The list of procedural practices is summarized in the Conclusions as a guideline for designers and practitioners. Comments on possible extensions are also provided.

2   Methods

2.1   Uncertainty Quantification and Optimization Under Uncertainty

Probabilistic uncertain input parameters are modeled by random input variables following certain probability density functions (PDFs), which represent the probability of occurrence of an event. Because of the input stochasticity, the system response is also stochastic, but its PDF is not known *a priori*. Given that objectives and constraints in optimization problems are often defined in terms of output variables, UQ becomes very important. Distribution moments, like mean and variance, can be estimated with many techniques, e.g. Monte Carlo or Latin Hypercube Sampling (LHS)[15], or the more efficient PCE[16,17]. The latter is accurate whilst usually requiring a smaller number of function evaluations than sampling techniques, at least in the case of small–medium problem dimensions. This is a crucial advantage, since computational time is one of the major bottlenecks in design optimization processes.

We report below the key ideas of an OUU method recently introduced by the authors. For a more detailed description, see Refs. [11–13] and references therein. Our method relies on the PCE expansion of an output function Y of stochastic input variables $\mathbf{X} = (X_1, ..., X_d)$ on a basis of polynomials ψ_i orthogonal w.r.t. the input variable PDFs. In numerical applications the PCE is truncated to a certain chaos order or polynomial degree k: $Y \simeq \sum_{i=0}^{k} a_i \psi_i(\mathbf{X})$. Thanks to the orthogonality condition, the mean and variance of Y are respectively given by $\mu_Y = a_0$ and $\sigma_Y^2 = \sum_{i=1}^{k} a_i^2 ||\psi_i||^2$. We determine the coefficients a_i via a least–square regression procedure on a sample of N points. The sample can be arbitrarily chosen, with the exception of its minimum size N_{min}, which must be equal to the number of unknown parameters in the PCE, i.e. $N \geq N_{min} = \frac{(k+d)!}{k!d!}$ (in the single–variate case the number of coefficients equals $k + 1$, but in the multivariate case it also depends on the stochastic input space dimension d).

Once the PCE coefficients have been determined, we can use the PCE as a SRSM to approximate the function Y to find its CDF and determine percentile values. The CDF is obtained by evaluating (and sorting) the PCE responses on a LHS set of size N_{perc}. In this way we avoid calls of the real function solvers, sensibly reducing the computational cost w.r.t. a pure sampling approach in real–world applications. Percentile values can be used as probabilistic (or chance) constraints to be optimized according to a given reliability threshold in the so–called performance–measure approach (for a recent overview on RBDO methods, see e.g. Ref. [3,5]).

The illustrated methods have been implemented within mF. The UQ flow is *nested into an optimization process*. The statistical properties (mean, variance, percentiles) computed in the nested UQ flow can be used as objectives or constraints in the main optimization flow, simultaneously enabling both RDO and RBDO. This technique has been benchmarked in Refs. [11–13] on several test–cases with particular attention to the optimal choice of the sample sizes and k for the percentile determination.

Our method does not depend on the optimization algorithm used, which can be chosen on the basis of the optimization problem and other necessities. However, as mentioned in the introduction, EMO algorithms are quite beneficial due to their robustness and ability of finding accurate Pareto fronts.

2.2   Screening Analysis

In the literature various variable screening techniques exist. We consider here the Smoothing Spline–ANOVA (SS–ANOVA) method, a statistical modeling algorithm based on a function decomposition similar to the classical analysis of variance (ANOVA) decomposition and the associated notions of main and interaction effects (higher–order interactions are typically excluded from the analysis). In this context *smoothing* means nonparametric function estimation in presence of stochastic data. In fact, SS–ANOVA belongs to the family of nonparametric or semiparametric models, or, to be specific, to smoothing methods, suitable for regression with noisy data, given the assumption of Gaussian–type responses. For a detailed explanation of the subject, see e.g. Refs. [18,19]. The key elements of this approach are provided below.

In order to determine the regression model f for a given stochastic dataset, SS–ANOVA considers two functionals: $[L(f)]$ defined as the minus log likelihood of the model f given the data (this functional is related to the the goodness of the fit), and $[J(f)]$ defined as a quadratic roughness functional (a measure of the roughness/smoothness of the model f). The regression problem solution can be stated as a constrained minimization problem

$$\min \; L(f), \quad \text{subject to} \quad J(f) \leq \rho, \qquad (1)$$

where the minimization of L guarantees a good fit to the data, while the soft (i.e., inequality) constraint on J – limiting the admissible roughness below a threshold ρ – prevents overfitting. By applying the Lagrange method, Eq. (1) can be rewritten as: $\min \; L(f) + \frac{\lambda}{2}J(f)$, where λ is a Lagrange multiplier. The regression procedure through this minimization is called *general penalized likelihood method* (because J represents a penalty on the roughness), *penalty smoothing*, or *smoothing method with roughness penalty*. The value of the smoothing parameter λ controls the trade–off in the model f between smoothness and fidelity to the data: the larger the λ, the smoother the model. Smaller values imply rougher functions, though with a better agreement to the data. Other smoothing parameters are contained in the expression of $J(f)$. All these parameters are determined through a data-driven process.

These techniques have been implemented in the modeFRONTIER software. SS-ANOVA is a suitable screening technique for detecting important variables in a given dataset. In fact, its results are easily interpretable, as the importance of each term (main effects and interactions) is proportional to the percentage of its contribution to the global variance of the statistical model. This measure can be displayed in an effect bar chart to compare the relative importance of the terms, as shown on Fig. 5 of Section 3.2.

3   Boomerang Test Case

The boomerang problem is an interesting non–standard application, firstly considered as a benchmark in Ref. [10], to which the reader might refer for more details. Here we only summarize some key-points. The main purpose of that work was to design an easy–to–launch boomerang, able to draw a smooth returning trajectory, satisfying at the same time a minimum flight distance constraint. The design process consisted in the geometric optimization of the boomerang shape (built with a CAD software tool) and the optimization of the boomerang launch parameters. The boomerang trajectory was determined by a dynamical

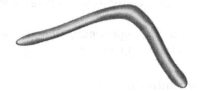

Fig. 1. One of the optimal boomerang geometries found in Ref. [10]

model implemented in MATLAB[20] coupled with a computational fluid dynamic model (CFD), which provided the forces and torques necessary for the trajectory integration. Dedicated response surface models were built for each geometric configuration on the CFD datasets to approximate the aerodynamic coefficients called by the MATLAB scripts. The reasons for this choice are explained in Ref. [10]. mF was used as an integration platform to orchestrate the interaction among the different software tools involved and to drive the overall optimization process, by exploiting its optimization algorithm libraries.

In this work we focused on one subproblem: the optimization of the boomerang launch parameters (see also the description of Fig. 19 in Ref. [10]). We started from one of the optimal designs selected in Ref. [10] (its shape is shown in Fig. 1) and used the response surfaces built in that work for MATLAB computations, which we treated as a black box providing system response values to our input parameters.

The input variables and parameters are summarized in Table 1. Velocity and spin refer to the boomerang initial translational velocity and spin. Tilt is the angle between the boomerang initial rotational axes and the vertical plane (a zero degree tilt corresponds to a vertical plane of rotation). Aim is the angle between the boomerang initial translational velocity and the horizontal plane.

Table 1. Input variables and constant parameters for the boomerang launch

Input variables	Range of variation
Velocity	$[15-30]\ m$
Spin	$[4-10]\ Hz$
Aim	$[0-30]°$
Tilt	$[0-50]°$
Constant parameters	**Value**
Initial height	$2\ m$
Density	$0.5\ Kg/m^3$
Moment of inertia	$0.0011263\ Kg \cdot m^2$
Volume	$0.0590711\ m^3$

Monitored output quantities are listed below:

– Accuracy: the difference between the position from which the boomerang is launched and the final position where the boomerang returns;
– Range: the maximum distance reached by the boomerang during its flight;
– Energy: the (translational plus rotational) energy necessary to launch the boomerang;
– ie: a discrete–valued output, representing the stopping condition for the MATLAB integration of motion equation.

3.1   Deterministic Optimization

Fig. 2 shows the optimization workflow (similar to Fig. 19 of Ref. [10]), with input variables, constant parameters and support file nodes at the top and monitored output variables, objectives and costraints at the bottom. A central script node (EasyDriver) handles the calls to the MATLAB scripts for the evaluation of the boomerang trajectory on the basis of aerodynamic and mass properties as explained in Ref. [10].

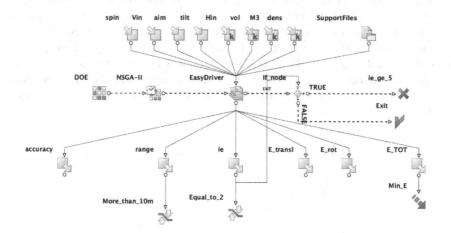

Fig. 2. Workflow for the deterministic optimization

The deterministic optimization problem was single–objective and consisted in the minimization of the total energy, with a constraint on the range spanned by the boomerang trajectory (range>10 m) and an equality constraint on the stopping condition for the MATLAB integration corresponding to a return distance of the boomerang equal to 1 m (ie=2). As we found that several designs (mostly at the beginning of the optimization) provided irregular or unsuccesful trajectories, we decided to insert an "if-node" into our workflow so that such designs would be labelled as error designs, which stabilized the optimization (especially in the case of OUU shown below, because error designs would distort the UQ assessments).

Although the problem was single objective, the optimization was done with NSGA-II[21] to avoid the risk of getting stuck in a local optimum region. We ran 100 generations with an initial population of 50 random individuals (created by a random DOE algorithm), crossover and mutation probabilities equal to 0.9 and 0.25 respectively, a distribution index of 2, and automatic scaling for mutation probability. The first hundred evaluated designs were either error designs (violating the condition imposed in the workflow) or unfeasible due to the violation of the constraint on the stopping criterion ie (boomerangs that do not come back or land at a greater distance than 1 m) or the minimum range constraint. Only after the first 300 designs, the trajectory became more stable and eventually feasible designs were found. In the long run, most of the designs were feasible. The unfeasible ones were distributed evenly between those violating the ie condition (but yielding designs very close to the return accuracy of 1 m) and the constraint on the range.

Fig. 3 shows the history chart of the objective during the deterministic optimization. Only the feasible design values of the total energy are shown. A good convergence was achieved towards the end of the 5000 evaluations.

In Fig. 4 we show the trajectory of the best design found. This design corresponds to an initial velocity slightly greater than 15 m/s, an initial spin of approximately 4.123 Hz, an aim angle of 10.433° and an almost zero tilt angle. This set should make launching the boomerang relatively easy. These parameters yield a total energy of 3.5116 J, most of which is due to the boomerang translational energy, while the rotational energy is approximately 5% of the total energy. The maximum range spanned during the flight is slightly greater than 10 m.

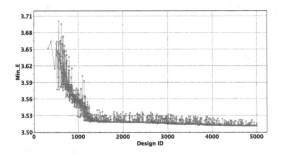

Fig. 3. History chart of the minimization of the total energy (only feasible designs are shown)

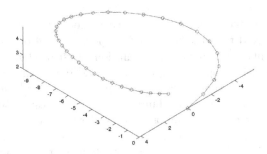

Fig. 4. Trajectory performed by the boomerang with deterministically optimized launch parameters

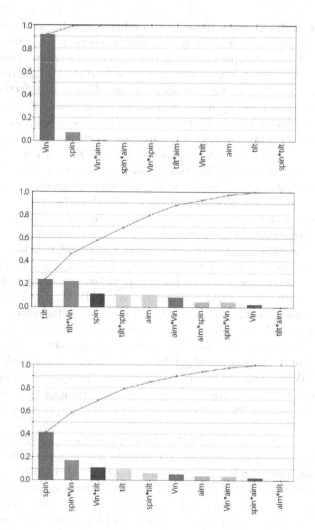

Fig. 5. Input variable main and interaction effects on the boomerang trajectory total energy (top chart), accuracy (middle chart), and maximum range (bottom chart)

3.2   Screening Analysis and Uncertainty Quantification Checks

We begin this section with the analysis of input variable main and interaction effects on output variables to identify which inputs, if stochastic, affect the outputs most. We performed a screening analysis with the SS-ANOVA tool on a database consisting of the first 800 designs evaluated during the deterministic optimization. We found that the total energy of the boomerang trajectory depends almost uniquely on the initial velocity and spin (main effects) (see Fig. 5, top chart). The accuracy appears to depend on several factors, like the tilt angle and its interaction with the initial velocity and spin, the spin itself, etc. (see

Fig. 5, middle chart). None of these is predominant, which is consistent with the complex mechanisms determining a succesful boomerang trajectory. We found similar trends for the stopping criterion too. Finally, the effects determining the range are shown on the bottom chart on Fig. 5. Also in this case there are multiple non–negligible factors, but the predominant effect can be attributed to the spin immediately followed by its interaction with the initial velocity.

From this analysis, the boomerang initial velocity and spin, followed by the tilt angle, appear to be the most important variables for the system response. We chose the first two as stochastic variables for the OUU, while considering the tilt as a deterministic variable like the aim angle, because we wanted to keep the number of stochastic variables as low as possible to reduce the computational effort of the calculations. Moreover, the tilt angle has the biggest effect on the accuracy, which is determined by the stopping criterion ie on which we decided to keep the deterministic constraint ie=2.

The identification of only two important stochastic variables out of four inputs brings about a consistent computational cost reduction in UQ assessments if PCE is used. In fact, for $d = 4$ stochastic inputs and chaos degree $k = 3$ (or 4), $N_{min} = 15$ (or 70), while for $d = 2$ and $k = 3$ (or 4) $N_{min} = 10$ (or 35). This way, approximately two thirds (or four fifths) function evaluations can be spared.

After detecting the most important variables, we performed some UQ checks on the optimal deterministic solution in order to single out the best UQ method for the robust optimization and check the convergence of the sample size and/or PCE degree. We assumed the initial velocity and the spin to be normally distributed around the nominal design values with a distribution standard deviation equal to approximately 3% of the variable range (i.e. respectively 0.5 m/s and 0.2 Hz).

First we did UQ by means of LHS. We generated samples of 13, 20, 50, 100, 200 and 500 designs and computed the sample mean and standard deviation. Then, we considered the PCE technique and estimated the mean and variance with a third order ($k = 3$) polynomial expansion built on samples consisting of 13, 20, 50, 100 designs, and a fourth order ($k = 4$) expansion on samples consisting of 20, 30, 50, 100 designs. The UQ check results are summarized in Fig. 6. By examining the top and middle charts, we can notice that the $k = 3$ and $k = 4$ PCE data converge much faster than the LHS estimates even for the smallest sample sizes, with the exception of the energy standard deviation estimated from the third–order PCE. The PCE data appear to be well converged also in the polynomial degree, since there is no noticeable difference between the third and fourth order results (on the scale of the figures at least). Only the estimates on the accuracy (i.e. return distance) deviate from these trends, as shown on the bottom charts on Fig. 6, where the PCE estimates appear to have bigger oscillations than the LHS outcomes. The reason for this is that the samples contain several designs with different input values but identical accuracy values (that is 1 m) because of the constraint on ie, and in such cases the PCE cannot accurately fit the data.

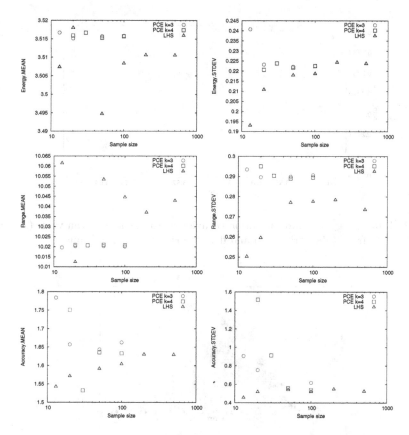

Fig. 6. Comparison of LHS and (third and fourth order) PCE estimates for the mean value (left charts) and standard deviation (right charts) of the boomerang trajectory total energy (top), maximum range (middle) and accuracy (bottom) computed on samples of different sizes. A logarithmic scale is used on the abscissa axes.

3.3   Stochastic Optimization

Based on the results of Section 3.2, we performed an OUU of the boomerang trajectory problem with stochastic spin and velocity as in Section 3.2 and the UQ nested loop based on the PCE method, with a third order expansion and a PCE sample size of 20 designs.

We considered *two robust objectives*, i.e. the minimization of the mean value and standard deviation of the total energy, *a probabilistic constraint* (to optimize the reliability) on the maximum range, such that only the 0.1% of designs should fail achieving a range greater than 10 m (0.1-th percentile of the maximum range > 10 m), and the deterministic constraint ie=2. As an optimizer, we used again NSGA-II with the same parameters as in Section 3.1, starting from an initial population of 50 designs composed by the best solution and other feasible designs found during the deterministic optimization. This way, the algorithm was able to

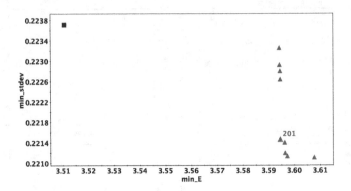

Fig. 7. Total energy standard deviation (y–axis) vs. total energy mean value (x–axis): OUU Pareto front (filled triangles) and deterministic result (filled square)

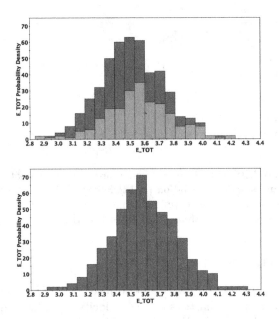

Fig. 8. Histograms on samples of 500 designs centered on the best deterministic (top chart) and one of the stochastic (bottom chart) solutions. The light–colored part of the deterministic histograms denotes designs violating the maximum range constraint.

find some feasible solutions from the very beginning. To compare results with the deterministic run, we considered the same number of real function evaluations, corresponding to 5 generations in this case (for each individual a sample of 20 designs is evaluated in the nested UQ assessment). For the percentile computation we generated samples of 1000 designs, evaluated with the PCE as a SRSM.

Fig. 7 shows the stochastic Pareto front found with the OUU (filled triangles) and the best deterministic solution (filled square). It appears that robustness and reliability have been achieved at the expense of energy. To further inspect the effect of the OUU on the Pareto front, we made UQ assessments on samples of 500 designs around selected solutions. As an example, in Fig. 8 we compare the results for the total energy of the best deterministic design (top chart) and a compromise solution (denoted with the label 201 in Fig. 7) in the central region of the Pareto front (bottom chart of Fig. 8) (the main results illustrated below do not depend on this choice). The OUU design has a probability density which is slightly shifted towards higher energy values and more peaked around the mean value (as expected by looking at Fig. 7). Most importantly, the deterministic solution has a high probability of violating the threshold constraint on the range spanned by the boomerang (the light–colored part of the histograms in the top chart of Fig. 8 denotes the unfeasible designs w.r.t the constraint on the maximum range), while no design out of the 500 hundred drawn around the solution found with the OUU violates that constraint.

4 Conclusions

In this paper we addressed the process of optimization under uncertainty through the application to the optimization of a boomerang trajectory. Starting from a deterministic optimization to obtain a reference solution, we performed a screening analysis to identify the most important input variables influencing output variations. Those variables were used as stochastic variables for the OUU. We compared two methods for UQ, namely the LHS and PCE techniques and performed convergency checks to determine the sample size and the PCE degree. Then we made an OUU looking for robust (minimum mean performance values and minimum standard deviation) and reliable (probabilistic constraints were considered) solutions. The solutions found were compared with the reference deterministic optimum and a further UQ assessment was performed to show the reliability of the stochastic solutions.

We could summarize the following general guidelines for OUU:

1. Increase the knowledge on the examined problem (e.g. with preliminary deterministic optimization, or at least with a preliminary exploration campaign where some designs are evaluated).
2. Identify the most important variables for the problem.
3. Check the parameter convergence of the UQ techniques and choose the method providing the most accurate results with the minimum computational effort, if possible.
4. Perform the OUU and analyze output results.

The second step is very important for high–dimensional problems and real–world applications where evaluating an output might be very expensive. In this paper our goal was to reduce the number of stochastic input variables in order to reduce the minimum number of sample points for the determination of the PCE coefficients (although the size of the problem was not large). An alternative approach would be to identify the subset of the most important variables through screening and use it to build accurate surrogate models and perform the optimization (deterministic or stochastic) while keeping the less important variables fixed at their central values.

The third step is also important in that the choice of the UQ method might affect the computational effort required by OUU, while the convergency checks are necessary to estimate the accuracy of the results. Unfortunately, in real–world applications engineers often need to make a compromise between accuracy of the results and the feasibility of the simulations in a reasonable amount of time. Here we have considered only the UQ and OUU techniques implemented in mF, but the applicatbility of these considerations can be easily extended.

Even though we started from a single–objective optimization problem, the guidelines obtained can be generalized to multiobjective optimization problems, where mean performances could be treated as objectives, and their standard deviations could be used as further objectives or constraints to improve the robustness of the Pareto front solutions. Additional probabilistic constraints could be used to improve the reliability as well. The first part, introduces other problems close to the EMO community interests, such as the many–objective optimization, but this goes beyond our present purposes.

References

1. Branke, J.: Creating robust solutions by means of evolutionary algorithms. In: Eiben, A.E., Bäck, T., Schoenauer, M., Schwefel, H.-P. (eds.) PPSN 1998. LNCS, vol. 1498, pp. 119–128. Springer, Heidelberg (1998)
2. Deb, K., Gupta, H.: Introducing robustness in multi–objective optimization. Evolutionary Computation 14, 463–494 (2006)
3. Deb, K., Padmanabhan, D., Gupta, S., Mall, A.K.: Reliability–based multi-objective optimization using evolutionary algorithms. In: Obayashi, S., Deb, K., Poloni, C., Hiroyasu, T., Murata, T. (eds.) EMO 2007. LNCS, vol. 4403, pp. 66–80. Springer, Heidelberg (2007)
4. Daum, D.A., Deb, K., Branke, J.: Reliability–based optimization for multiple constraints with evolutionary algorithms. In: 2007 IEEE Congress on Evolutionary Computation (CEC 2007), pp. 911–918 (2007)
5. Deb, K., Gupta, S., Daum, D., Branke, J., Mall, A.K., Padmanabhan, D.: Reliability–Based Optimization Using Evolutionary Algorithms. IEEE Transactions on Evolutionary Computation 13, 1054–1074 (2009)
6. Schuëller, G.I., Jensen, H.A.: Computational Methods in Optimization considering Uncertainties – An Overview. Comput. Methods Appl. Mech. Engrg. 198, 2–13 (2008)
7. Beyer, H.-G., Sendhoff, B.: Robust Optimization – A comprehensive Survey. Comput. Methods Appl. Mech. Engrg. 196(33–34), 3190–3218 (2007)
8. Jin, Y., Branke, J.: Evolutionary Optimization in Uncertain Environments – A Survey. IEEE Transactions on Evolutionary Computation 9(3), 1–15 (2005)

9. Barrico, C., Antunes, C.H.: An Evolutionary Approach for Assessing the Degree of Robustness of Solutions to Multi–Objective Models. Evolutionary Computation in Dynamic and Uncertain Environments Studies in Computational Intelligence **51**, 565–582 (2007)
10. Russo, R., Clarich A., Nobile, E., Poloni, C.: Optimization of a boomerang shape using modeFRONTIER. In: 14th AIAA/ISSMO Multidisciplinary Analysis and Optimization Conference, September 2012
11. Clarich, A., Russo, R., Marchi, M., Rigoni, E.: Reliability–based design optimization applying polynomial chaos expansion: theory and applications. In: 10th World Congress on Structural and Multidisciplinary Optimization, Orlando, Florida, USA, pp. 19–24, May 2013
12. Marchi, M., Rigoni, E., Russo, R., Clarich, A.: Percentile via polynomial chaos expansion: bridging robust optimization and reliability. In: International Conference on EVOLVE 2013, A Bridge Between Probability, Set Oriented Numerics, and Evolutionary Computation, Extended Abstract Proceedings, July 10-13, Leiden, NL (2013). ISBN 978-2-87971-118-8, ISSN 2222-9434
13. Marchi, M., Rizzian, L., Rigoni, E., Russo, R., Clarich, A.: Combining robustness and reliability with polynomial chaos techniques in multiobjective optimization problems: use of percentiles. In: Cunha, A., Caetano, E., Ribeiro, P., Muller, G. (eds.) Proceedings of the 9th International Conference on Structural Dynamics, EURODYN 2014, Porto, Portugal, June 30–July 2 (2014)
14. www.esteco.com
15. McKay, M.D., Conover, W.J., Beckman, R.J.: A Comparison of Three Methods for Selecting Values of Input Variables in the Analysis of Output from a Computer Code. Technometrics **21**(2), 239–245 (1979)
16. Wiener, N.: The homogeneous chaos. Amer. J. Math. **60**, 897–936 (1938)
17. Xiu, D., Karniadakis, G.E.: The Wiener–Askey polynomial chaos for stochastic differential equations. SIAM J. Sci. Comput. **24**, 619–644 (2002)
18. Gu, C.: Smoothing Spline ANOVA Models. Springer, New York (2002)
19. Wahba, G.: Spline Models for Observational Data. SIAM, Philadelphia (1990)
20. http://www.mathworks.it/products/matlab/
21. Deb, K., Pratap, A., Agarwal, S., Meyarivan, T.: A fast and elitist multiobjective genetic algorithm: NSGA–II. IEEE Transactions on Evolutionary Computation **6**, 182–197 (2002)

Mulit-Criterion Decision Making

Using Indifference Information in Robust Ordinal Regression

Juergen Branke[1], Salvatore Corrente[2], Salvatore Greco[2,3](✉),
and Walter J. Gutjahr[4]

[1] Warwick Business School, The University of Warwick, Coventry CV4 7AL, UK
juergen.branke@wbs.ac.uk
[2] Department of Economics and Business, University of Catania, Corso Italia,
55, 95129 Catania, Italy
salvatore.corrente@unict.it
[3] Portsmouth Business School,
Centre of Operations Research and Logistics (CORL), University of Portsmouth,
Richmond Building, Portland Street, Portsmouth PO1 3DE, UK
salgreco@unict.it
[4] Department of Statistics and Operations Research,
University of Vienna, 1090 Wien, Austria
walter.gutjahr@univie.ac.at

Abstract. In this paper, we propose an extension to Robust Ordinal Regression allowing it to take into account also preference information from questions about indifference between real and fictitious alternatives. In particular, we allow the decision maker to suggest a new alternative that is different from the existing alternatives, but equally preferable. As shown by several experiments in psychology of the decisions, choosing between alternatives is different from matching two alternatives since the two aspects involve two different reasoning strategies. Consequently, by including this type of preference information one can represent more faithfully the DM's preferences. Such information about indifference should narrow down the set of compatible value functions much more quickly than standard pairwise comparisons, and a first simple example at least indicates that this intuition seems to be correct.

1 Introduction

Multiple Criteria Decision Aiding (MCDA) (see [8,9]) aims to recommend the Decision Maker (DM) a decision that best fits her/his preferences when a plurality of criteria has to be taken into consideration. Typically, in MCDA, a set of alternatives $A = \{a_1, \ldots, a_n\}$ is described in terms of performances with respect to a coherent family of criteria $G = \{g_1, \ldots, g_m\}$ [24]. Without loss of generality, each criterion $g_j \in G$ can be considered as a real-valued function $g_j : A \rightarrow \mathcal{I}_j \subseteq \mathbb{R}$, such that for any $a, b \in A, g_j(a) \geq g_j(b)$ means that a is at least as good as b with respect to criterion g_j.

Given two alternatives $a, b \in A$ and considering their performances with respect to the m criteria belonging to G, very often a will be better than b for

© Springer International Publishing Switzerland 2015
A. Gaspar-Cunha et al. (Eds.): EMO 2015, Part II, LNCS 9019, pp. 205–217, 2015.
DOI: 10.1007/978-3-319-15892-1_14

some of the criteria while b will be better than a for the remaining criteria. For this reason, in order to cope with any multiple criteria decision problem, we need to aggregate the performances of the alternatives taking into account the preferences of the DM. The three most well known aggregation models are the following:

- MAVT - Multi-Attribute Value Theory (see [7,19]) assigning to each alternative $a \in A$ a real number representative of its desirability ,
- outranking methods (see [10,12,24]) building some outranking preference relations S on A, such that for any $a, b \in A$, aSb means that a is at least as good as b,
- decision rule models using a set of "if..., then..." decision rules induced from the DM's preference information through Dominance-based Rough Set Approach (DRSA, see [14,15,26,27]).

Such MCDA models have recently been integrated into Evolutionary Multiobjective Optimization (EMO) as a means to interact with the DM and focus the search to the part of the Pareto front most preferred by the DM [1,3].

In this paper we consider the first model and we take into consideration a value function $U : \prod_{j=1}^{m} \mathcal{I}_j \rightarrow \mathbb{R}$ such that for any $a, b \in A$, a is at least as good as b ($a \succsim b$) if $U(g_1(a), \ldots, g_m(a)) \geq U(g_1(b), \ldots, g_m(b))$. The simplest form of the value function is the additive form, defined as: $U(g_1(a), \ldots, g_m(a)) = \sum_{j=1}^{m} u_j(g_j(a))$, where $u_j(g_j(a))$ are non-decreasing functions of their arguments. In the following, for simplicity of notation, we shall use $U(a)$ instead of $U(g_1(a), \ldots, g_m(a))$ for all $a \in A$.

Application of any decision model requires the definition of its parameters which can be obtained by asking them directly to the DM or inferring them from preference information given by the DM. This second approach seems more practical because the DM can have some difficulty in realizing the exact meaning of the parameters in the preference model and, moreover, their direct elicitation requires a strong cognitive effort from the DM. The typical preference information considered in this case is the pairwise comparisons between alternatives on which the DM feels sufficiently confident. In this paper we propose a different type of preference information expressed in terms of indifference between two alternatives. More precisely, supposing that the DM declares that an alternative a is preferred to another alternative b, we ask the DM to indicate another alternative b^+, obtained by improving b on some criteria, so that alternative b^+ is indifferent to a. Another possible way to get a preference information in terms of indifference is the following. Supposing again that alternative a is preferred to alternative b, one can ask the DM to indicate an alternative a^-, obtained by deteriorating a on some criteria, so that a^- is indifferent to b. Yet another possibility is to consider an alternative a^{+-}, obtained from alternative a improving its performances on some criteria and deteriorating its performances on other criteria, so that a and a^{+-} are indifferent. The main advantage we expect from this type of preference information is that it should reduce the space of compatible value functions much more than usual information supplied in terms of preference pairwise comparisons. Indeed, from the mathematical point of view, the new preference information should be translated by equality constraints that, in case of a value function representing perfectly the preference of the DM, drastically will reduce the space of compatible value functions. In the following, to take into account a certain imprecision in the DM's preferences, we model the indifference information by imposing that the difference of the utilities of two indifferent alternatives, in absolute value, should be no greater than an indifference threshold. Anyway, even if we do not use equality constraints, the space of compatible value function is strongly reduced, especially if the considered indifference threshold is sufficiently small. Moreover, as proved by several experiments in psychology of the decisions [25, 29], choosing between two alternatives is different from matching two alternatives since the two aspects involve two different reasoning strategies. Consequently, putting together usual preference information in terms of pairwise preference of one alternative over another with the new type of preference information we are introducing, permits to build a utility function representing the DMs preferences in a more faithful way. We think that this is beneficial also for the elicitation of preference information within the EMO algorithms, as it should allow a faster convergence of the interactive EMO algorithm to the part of the Pareto front most preferred by the DM.

The paper is organized as follows. In the next section we recall the basic concepts of ordinal regression and robust ordinal regression. In the third section we introduce the new type of preference information. In the fourth section we present a didactic example. Conclusions and perspective for future research are collected in the last section.

2 Ordinal Regression and Robust Ordinal Regression

2.1 Ordinal Regression

Each decision model requires the specification of some parameters. For example, using MAVT, the parameters are related to the formulation of the marginal value functions $u_j(g_j(a)), j = 1, \ldots, m$. Since, as explained previously, the indirect preference information is more applied in practice, within MCDA, many methods have been proposed to determine the parameters characterizing the considered decision model inducing the values of such parameters from some holistic preference comparisons of alternatives given by the DM. This indirect preference elicitation is the base of the ordinal regression paradigm.

The most well-known ordinal regression methodology is the UTA (UTilités Additives) method proposed by Jacquet-Lagrèze and Siskos [17], which aims at inferring one or more additive value functions from a given complete ranking of alternatives belonging to a reference set $A^R \subseteq A$. The method considers a piecewise additive value function $U(g_1(a), \ldots, g_m(a)) = \sum_{j=1}^m u_j(g_j(a))$ having marginal value functions $u_j(\cdot), j = 1, \ldots, m$, being piecewise-linear, with a predefined number of linear pieces. UTA uses linear programming to determine an additive value function compatible with the preference information provided by the DM. Technically, in order to check if there exists at least one additive function compatible with the preferences provided by the DM, one has to solve the following linear programming problem:

$$\varepsilon^* = \max \ \varepsilon, \text{ s.t.}$$

$$\left.\begin{array}{l} U(a^*) \geq U(b^*) + \varepsilon \text{ if } a^* \succ b^*, \text{ with } a^*, b^* \in A^R, \\ U(a^*) = U(b^*) \text{ if } a^* \sim b^*, \text{ with } a^*, b^* \in A^R, \\ \displaystyle\sum_{j=1}^{m} u_j(\beta_j) = 1, \quad u_j(\alpha_j) = 0, \ j = 1, \ldots, m, \quad \Big\} \\ u_j(g_j(a)) \geq u_j(g_j(b)) \text{ if } g_j(a) \geq g_j(b), \forall a, b \in A, j = 1, \ldots, m, \end{array}\right\} \ \begin{array}{c} \\ \Big\} E \end{array} \right\} E^{A^R}$$

where

- β_j and α_j are the best and the worst considered values of criterion $g_j, j = 1, \ldots, m,$
- $\succ$ and $\sim$ are the asymmetric and the symmetric part of the binary relation $\succsim$ representing the DM's preference information, i.e., $a^* \succ b^*$ means that a^* is preferred to b^* while $a^* \sim b^*$ means that a^* and b^* are indifferent,
- here, as always in the following, ε is considered without any constraint on the sign.

If the set of constraints E^{A^R} is feasible and $\varepsilon^* > 0$, then there exists at least one additive value function compatible with the DM's preferences. If there is no compatible value function, i.e., if the preferences of the DM cannot be represented by an additive value function with pre-defined number of linear pieces, [17] suggests either to increase the number of linear pieces in some marginal value functions, or to select the utility function U that gets the sum of deviation errors close to minimum and minimizes the number of ranking errors in the sense of Kendall or Spearman distance.

The ordinal regression paradigm has been applied within the two main MCDA approaches, that is those using a value function as preference model [4,17,18,23,28], and those using an outranking relation as preference model [21,22].

2.2 Robust Ordinal Regression

Usually, from among many sets of parameters of a preference model representing the preference information given by the DM, only one specific set is selected and used to work out a recommendation.

Since the selection of one of these sets of parameters compatible with the preference information given by the DM is rather arbitrary, *Robust Ordinal Regression* (ROR; [5,6,16]) proposes to take into account simultaneously all of them, in order to obtain a recommendation in terms of necessary and possible consequences of applying all the compatible preference models on the considered set of alternatives; the *necessary* weak preference relation holds for any two alternatives $a, b \in A$ ($a \succsim^N b$) if and only if a is at least as good as b for all compatible preference models, while the *possible* weak preference relation holds for this pair $(a \succsim^P b)$ if and only if a is at least as good as b for at least one compatible preference model.

Although UTA^{GMS} [16] is the first method applying the ROR concepts, in the following, we shall describe the GRIP method [11] being its generalization. Then, we shall mention the other applications of the ROR that have been published later in several papers.

2.3   GRIP

In the UTA^{GMS} method [16], which initiated the stream of further developments in ROR, the ranking of reference alternatives does not need to be complete as in the original UTA method [17]. Instead, the DM may provide pairwise comparisons just for those reference alternatives (s)he really wants to compare. Precisely, the DM is expected to provide a partial preorder $\succsim$ on A^R. Obviously, one may also refer to the relations of strict preference $\succ$ or indifference $\sim$.

The transition from a reference preorder to a value function is done in the following way: for $a^*, b^* \in A^R$,

$$\left.\begin{array}{l} U(a^*) \geq U(b^*) + \varepsilon, \quad \text{if} \quad a^* \succ b^*, \\[2mm] U(a^*) = U(b^*), \quad \text{if} \quad a^* \sim b^*, \end{array}\right\} E_1$$

where ε is a (generally small) positive value.

Observe that $a^* \sim b^*$ can be represented as follows:

$$|U(a^*) - U(b^*)| \leq \delta, \tag{1}$$

i.e.

$$\left.\begin{array}{l} U(a^*) - U(b^*) \leq \delta, \\[2mm] U(b^*) - U(a^*) \leq \delta, \end{array}\right\} \tag{2}$$

where δ is a non-negative indifference threshold considered to take into account imprecision in the preference information.

Observe that the case $\delta = 0$ collapses to the constraints expressed as equality, i.e. $U(a^*) = U(b^*)$. It is apparent that if the indifference constraints are expressed in terms of equality, one can get a more precise inference of the utility function U (e.g. in case U is expressed as weighted sum and there are only two criteria, then a single indifference comparison formulated in terms of equality is enough to determine univocally the utility function). However, observe that this greater precision can be misleading because a certain imprecision is always implicit in the preference information given by the DM.

In some decision making situations, the DM is willing to provide more information than a partial preorder on a set of reference alternatives, such as "a^* is preferred to b^* at least as much as c^* is preferred to d^*". The information related to the intensity of preference is also accounted for by the GRIP method [11]. It may refer to the comprehensive comparison of pairs of reference alternatives on all criteria or on a particular criterion only. Precisely, in the holistic case, the DM may provide a partial preorder $\succsim^*$ on $A^R \times A^R$, whose meaning is: for $a^*, b^*, c^*, d^* \in A^R$,

$$(a^*, b^*) \succsim^* (c^*, d^*) \Leftrightarrow a^* \text{ is preferred to } b^* \text{ at least as much as } c^* \text{ is preferred to } d^*.$$

When referring to a particular criterion $g_j \in G$, rather than to all criteria jointly, the meaning of the expected partial preorder $\succsim^*_j$ on $A^R \times A^R$ is the following: for $a^*, b^*, c^*, d^* \in A^R$,

$$(a^*, b^*) \succsim^*_j (c^*, d^*) \Leftrightarrow a^* \text{ is preferred to } b^* \text{ at least as much as } c^* \text{ is preferred to } d^* \text{ on criterion } g_j.$$

In both cases, the DM is allowed to refer to the strict preference and indifference relations rather than to weak preference only. The transition from the partial preorder expressing intensity of preference to a value function is the following: for $a^*, b^*, c^*, d^* \in A^R$,

$$\left.\begin{array}{l} U(a^*) - U(b^*) \geq U(c^*) - U(d^*) + \varepsilon, \text{ if } (a^*, b^*) \succ (c^*, d^*), \\[4pt] U(a^*) - U(b^*) = U(c^*) - U(d^*), \text{ if } (a^*, b^*) \sim (c^*, d^*), \\[4pt] u_j(a^*) - u_j(b^*) \geq u_j(c^*) - u_j(d^*) + \varepsilon, \text{ if } (a^*, b^*) \succ_j (c^*, d^*) \text{ for } g_j \in G, \\[4pt] u_j(a^*) - u_j(b^*) = u_j(c^*) - u_j(d^*), \text{ if } (a^*, b^*) \sim_j (c^*, d^*) \text{ for } g_j \in G. \end{array}\right\} E_2$$

In order to check if there exists at least one model compatible with the preferences of the DM we solve the following linear programming problem:

$$\varepsilon^* = \max \varepsilon \text{ s.t.}$$
$$E \cup E_1 \cup E_2 = E^{DM} \tag{3}$$

If the set of constraints E^{DM} is feasible and $\varepsilon^* > 0$, then there exists at least one additive value function compatible with the preference information provided by the DM, otherwise no additive value function is compatible with the provided information. In this case, the analyst can decide to check for the cause of the incompatibility [20] or can continue the decision aiding process accepting the incompatibility.

Denoting by $\mathcal{U}_{A^R}$ the set of value functions compatible with the preference information provided by the DM, in the GRIP method three necessary and three possible preference relations can be defined:

- $a \succsim^N b$ iff $U(a) \geq U(b)$ for all $U \in \mathcal{U}_{A^R}$, with $a, b \in A$,
- $a \succsim^P b$ iff $U(a) \geq U(b)$ for at least one $U \in \mathcal{U}_{A^R}$, with $a, b \in A$,
- $(a, b) \succsim^{*N} (c, d)$ iff $U(a) - U(b) \geq U(c) - U(d)$ for all $U \in \mathcal{U}_{A^R}$, with $a, b, c, d \in A$,
- $(a, b) \succsim^{*P} (c, d)$ iff $U(a) - U(b) \geq U(c) - U(d)$ for at least one $U \in \mathcal{U}_{A^R}$, with $a, b \in A$,

- $(a,b) \mathrel{\succsim}_j^{*N} (c,d)$ iff $u_j(a) - u_j(b) \geq u_j(c) - u_j(d)$ for all $U \in \mathcal{U}_{AR}$, with $a,b,c,d \in A$, $g_j \in G$,
- $(a,b) \mathrel{\succsim}_j^{*P} (c,d)$ iff $u_j(a) - u_j(b) \geq u_j(c) - u_j(d)$ for at least one $U \in \mathcal{U}_{AR}$, with $a,b \in A$, $g_j \in G$.

Given alternatives $a,b,c,d \in A$, and the sets of constraints

$$\left. \begin{array}{c} U(b) \geq U(a) + \varepsilon \\ E^{DM} \end{array} \right\} E^N(a,b), \qquad \left. \begin{array}{c} U(a) \geq U(b) \\ E^{DM} \end{array} \right\} E^P(a,b),$$

$$\left. \begin{array}{c} U(c) - U(d) \geq U(a) - U(b) + \varepsilon \\ E^{DM} \end{array} \right\} E^N(a,b,c,d),$$

$$\left. \begin{array}{c} U(a) - U(b) \geq U(c) - U(d) \\ E^{DM} \end{array} \right\} E^P(a,b,c,d),$$

$$\left. \begin{array}{c} u_j(c) - u_j(d) \geq u_j(a) - u_j(b) + \varepsilon \\ E^{DM} \end{array} \right\} E_j^N(a,b,c,d),$$

$$\left. \begin{array}{c} u_j(a) - u_j(b) \geq u_j(c) - u_j(d) \\ E^{DM} \end{array} \right\} E_j^P(a,b,c,d),$$

we get that:

- $a \mathrel{\succsim}^N b$ iff $E^N(a,b)$ is infeasible or if $E^N(a,b)$ is feasible and $\varepsilon^N(a,b) \leq 0$, where $\varepsilon^N(a,b) = \max \varepsilon$, s.t. $E^N(a,b)$;
- $a \mathrel{\succsim}^P b$ iff $E^P(a,b)$ is feasible and $\varepsilon^P(a,b) > 0$, where $\varepsilon^P(a,b) = \max \varepsilon$, s.t. $E^P(a,b)$;
- $(a,b) \mathrel{\succsim}^{*N} (c,d)$ iff $E^N(a,b,c,d)$ is infeasible or if $E^N(a,b,c,d)$ is feasible and $\varepsilon^N(a,b,c,d) \leq 0$, where $\varepsilon^N(a,b,c,d) = \max \varepsilon$, s.t. $E^N(a,b,c,d)$;
- $(a,b) \mathrel{\succsim}^{*P} (c,d)$ iff $E^P(a,b,c,d)$ is feasible and $\varepsilon^P(a,b,c,d) > 0$, where $\varepsilon^P(a,b,c,d) = \max \varepsilon$, s.t. $E^P(a,b,c,d)$;
- $(a,b) \mathrel{\succsim}_j^{*N} (c,d)$ iff $E_j^N(a,b,c,d)$ is infeasible or if $E_j^N(a,b,c,d)$ is feasible and $\varepsilon_j^N(a,b,c,d) \leq 0$, where $\varepsilon_j^N(a,b,c,d) = \max \varepsilon$, s.t. $E_j^N(a,b,c,d)$;
- $(a,b) \mathrel{\succsim}_j^{*P} (c,d)$ iff $E_j^P(a,b,c,d)$ is feasible and $\varepsilon_j^P(a,b,c,d) > 0$, where $\varepsilon_j^P(a,b,c,d) = \max \varepsilon$, s.t. $E_j^P(a,b,c,d)$;

As to properties of $\mathrel{\succsim}^N$ and $\mathrel{\succsim}^P$ on A, let us remind after [16] that:

- $\mathrel{\succsim}^N$ is a partial preorder on A,
- $\mathrel{\succsim}^N \subseteq \mathrel{\succsim}^P$,
- $a \mathrel{\succsim}^N b$ and $b \mathrel{\succsim}^P c \Rightarrow a \mathrel{\succsim}^P c$, $\forall a,b,c \in A$,
- $a \mathrel{\succsim}^P b$ and $b \mathrel{\succsim}^N c \Rightarrow a \mathrel{\succsim}^P c$, $\forall a,b,c \in A$,
- $a \mathrel{\succsim}^N b$ or $b \mathrel{\succsim}^P a$, $\forall a,b \in A$.

The above properties are the minimal ones characterizing $\mathrel{\succsim}^N$ and $\mathrel{\succsim}^P$ [13]. Other interesting properties of $\mathrel{\succsim}^N$ and $\mathrel{\succsim}^P$ are the following [16]:

- $\mathrel{\succsim}^P$ is strongly complete and negatively transitive,
- $\mathrel{\succ}^P$ is complete, irreflexive and transitive.

3   Preference Information in Terms of Pairwise Indifference Comparisons

In this section we introduce a new type of preference information expressed in terms of indifference between alternatives. Of course, this new type of preference information is supposed to be added to the type of preference information already considered within GRIP and, more in general, within the ROR methods. Even more, as explained in the following, the new type of preference information is very often based on some preference information expressed in terms of strict preference already considered within ROR.

We shall present three typical types of preference information expressed in terms of indifference pairwise comparisons:

- suppose that the DM has already declared that a^* is preferred to b^*. In this case one can ask the DM to indicate a new alternative b_H^{*+} obtained from b^* improving the performances on criteria from $H \subseteq G$ such that a^* is indifferent to b_H^{*+}. This preference information will be represented as follows:

$$|U(a^*) - U(b_H^{*+})| \leq \delta, \tag{4}$$

i.e.

$$\left. \begin{array}{l} U(a^*) - U(b_H^{*+}) \leq \delta, \\ U(b_H^{*+}) - U(a^*) \leq \delta \end{array} \right\} E(a^*, b_H^{*+})$$

where δ is a non-negative indifference threshold considered to take into account imprecision in the preference information;

- suppose again that the DM has already declared that a^* is preferred to b^* and let us ask the DM to indicate a new alternative a_K^{*-} obtained from a^* deteriorating the performances on criteria from $K \subseteq G$ such that a_K^{*-} is indifferent with b^*. This preference information will be represented as follows:

$$|U(a_K^{*-}) - U(b^*)| \leq \delta, \tag{5}$$

i.e.

$$\left. \begin{array}{l} U(a_K^{*-}) - U(b^*) \leq \delta, \\ U(b^*) - U(a_K^{*-}) \leq \delta \end{array} \right\} E(a_K^{*-}, b^*);$$

- let us consider a reference alternative a^* and let us ask the DM to indicate a new alternative $a_{H,K}^{*+-}$ obtained from a^* by improving the performances on criteria from H and deteriorating the performances on criteria from K with $H, K \subseteq G, H \cap K = \emptyset$ such that $a_{H,K}^{*+-}$ is indifferent to a^*. This preference information will be represented as follows:

$$|U(a_{H,K}^{*+-}) - U(a^*)| \leq \delta, \tag{6}$$

i.e.

$$\left. \begin{array}{l} U(a_{H,K}^{*+-}) - U(a^*) \leq \delta, \\ U(a^*) - U(a_{H,K}^{*+-}) \leq \delta \end{array} \right\} E(a^*, a_{H,K}^{*+-}).$$

Besides the above three typical types, other preference information expressed in terms of indifference pairwise comparisons can be the following: supposing again that a^* is preferred to b^*,

- $a^{*+-}_{H,K}$ is indifferent with b^*, with the related constraint denoted by $E(a^{*+-}_{H,K}, b^*)$,
- a^* is indifferent with $b^{*+-}_{R,S}$, with the related constraint denoted by $E(a^*, b^{*+-}_{R,S})$,
- $a^{*+-}_{H,K}$ is indifferent with $b^{*+-}_{R,S}$, with the related constraint denoted by $E(a^{*+-}_{H,K}, b^{*+-}_{R,S})$,

with $H, K, R, S \subseteq G, H \cap K = \emptyset, R \cap S = \emptyset$.

ROR methodology proceeds as explained before, simply adding constraints $E(a^*, b^{*+}_H), E(a^{*-}_K, b^*), E(a^*, a^{*+-}_{H,K}), E(a^{*+-}_{H,K}, b^*), E(a^*, b^{*+-}_{R,S})$ and $E(a^{*+-}_{H,K}, b^{*+-}_{R,S})$ to set of constraints E^{DM}.

Let us observe that the new type of preference information is translated by inequalities such as the classical preference information as introduced in all ROR methods and, therefore, the recommendations obtained by the new model can be considered appropriate and consistent for the decision problem at hand. Moreover, as already observed in [25,29], choosing between two alternatives is different from matching two alternatives since the two aspects involve two different reasoning strategies. Consequently, we think that putting together these types of preference information can represent more faithfully the DM's preferences.

4   Didactic Example

In order to illustrate the proposed methodology, in this section we shall provide a didactic example. Let us suppose that 8 alternatives are evaluated on 4 criteria that should be maximized. The evaluations of the alternatives on the considered criteria are shown in Table 1 and, for the sake of simplicity, we shall suppose that the evaluation criteria can assume 5 discrete values only $(1, \ldots, 5)$.

Let us observe that the dominance relation on the set of alternatives A is empty because no alternative dominates another alternative.

In a first moment, let us suppose that the DM provides the following preference information:

$$a \succ f, \quad c \succ h, \quad b \succ e, \quad c \succ d, \quad d \succ f, \quad e \succ h$$

Using this preference information, we get the following necessary preference relation:

$$\succsim^N = \{(a,f), (b,e), (b,h), (c,d), (c,f), (c,h), (d,f), (e,h)\} \cup \{(x,x) : x \in A\}.$$

Table 1. Alternatives' evaluations

Alternative / Criterion	g_1	g_2	g_3	g_4
a	5	5	1	5
b	5	3	5	1
c	4	1	5	5
d	4	4	4	2
e	4	4	2	4
f	5	2	3	2
h	4	2	3	4
l	5	5	3	1

Let us suppose now that dealing with the same decision problem the DM provides the following preference information:

- $a \succ f, \quad c \succ h, \quad b \succ e,$
- $a \sim f^+_{\{2,4\}},$
- $c^-_{\{1,3\}} \sim h,$
- $e \sim e^{+-}_{\{3\},\{4\}},$

where $f^+_{\{2,4\}} = (5,4,3,4)$, $c^-_{\{1,3\}} = (3,1,4,5)$ and $e^{+-}_{\{3\},\{4\}} = (4,4,4,2) = d$.
After considering the new set of constraints

$$
\left.
\begin{aligned}
U(a) &\geq U(f) + \varepsilon, \\
U(c) &\geq U(h) + \varepsilon, \\
U(b) &\geq U(e) + \varepsilon, \\
U(a) - U(f^+_{\{2,4\}}) &\leq \delta, \\
U(f^+_{\{2,4\}}) - U(a) &\leq \delta, \\
U(c^-_{\{1,3\}}) - U(h) &\leq \delta, \\
U(h) - U(c^-_{\{1,3\}}) &\leq \delta, \\
U(e^{+-}_{\{3\},\{4\}}) - U(e) &\leq \delta, \\
U(e) - U(e^{+-}_{\{3\},\{4\}}) &\leq \delta,
\end{aligned}
\right\} E^{DM}_*
$$

translating the preference information provided by the DM where $\delta = 10^{-4}$, we solve the linear programming problems shown in Section 3, getting the following necessary preference information:

$$\succsim^N = \{(a,e),(a,f),(a,h),(b,d),(b,e),(c,h),(d,f),(d,b),(d,e)\} \cup \{(x,x):x \in A\}.$$

For example, by solving the optimization problem

$$\varepsilon^* = \max \varepsilon, \text{ s.t.}$$

$$\left.\begin{array}{c} U(e) \geq U(a) + \varepsilon, \\ E_*^{DM} \cup E \end{array}\right\}$$

where E is the set of normalization and monotonicity constraints defined in Section 2, we get $\varepsilon^* = 0$ and, consequently, $a \succsim^N e$.

Observe that in this second case we get a slightly richer preference relation in terms of pairs of alternatives from A for which necessary preference holds (9 non-trivial pairs in this second case vs 8 non-trivial pairs in the first case), with a smaller cognitive effort in terms of number of alternatives from A considered in the preference information (6 - a, b, c, e, f, h - in the second case vs 7 - a, b, c, d, e, f, h - in the first case).

5   Conclusions

In this paper we introduced new types of preference information in Robust Ordinal Regression. More precisely we considered pairwise indifference comparisons between real or fictitious alternatives. We believe that this new type of preference information could permit to get a more precise induction of the DM value function with a smaller cognitive effort. Moreover, the introduction of the new type of preference information makes the obtained value function more faithful because, according to the evidence of a certain number of experiments in Psychology of the decision, choosing between two alternatives (corresponding to the usual preference information) is different from matching two alternatives (corresponding to the new type of preference information) since the two aspects involve two different reasoning strategies. The results of a very first didactic example presented in this paper seem promising, but a lot of work remains to be done. In particular we envisage the following perspectives for the future research:

- we have to measure the advantages in terms of smaller cognitive effort and better results of the MCDA procedure offered by the new type of preference information;
- we have to discuss how to manage the selection of criteria to be modified in order to get indifference in the considered pairwise comparisons between alternatives;
- we have to verify how beneficial can be the use of the new type of preference information in EMO procedures based on preferences, especially those procedures based on ROR (e.g. [1–3].

Acknowledgments. This work has been partly funded by the "Programma Operativo Nazionale" Ricerca & Competitivitá "2007-2013" within the project "PON04a2 E SINERGREEN-RES-NOVAE".

References

1. Branke, J., Greco, S., Słowiński, R., Zielniewicz, P.: Interactive evolutionary multiobjective optimization using robust ordinal regression. In: Ehrgott, M., Fonseca, C.M., Gandibleux, X., Hao, J.-K., Sevaux, M. (eds.) EMO 2009. LNCS, vol. 5467, pp. 554–568. Springer, Heidelberg (2009)
2. Branke, J., Greco, S., Słowiński, R., Zielniewicz, P.: Interactive evolutionary multiobjective optimization driven by robust ordinal regression. Bulletin of the Polish Academy of Sciences - Technical Sciences 58(3), 347–358 (2010)
3. Branke, J., Greco, S., Słowiński, R., Zielniewicz, P.: Learning value functions in interactive evolutionary multiobjective optimization. IEEE Transactions on Evolutionary Computation (forthcoming)
4. Charnes, A., Cooper, W.W., Ferguson, R.O.: Optimal estimation of executive compensation by linear programming. Management Science 1(2), 138–151 (1955)
5. Corrente, S., Greco, S., Kadziński, M., Słowiński, R.: Robust ordinal regression in preference learning and ranking. Machine Learning 93, 381–422 (2013)
6. Corrente, S., Greco, S., Kadziński, M., Słowiński, R.: Robust ordinal regression. Wiley Enciclopedia of Operational Research (forthcoming)
7. Dyer, J.S.: MAUT – multiattribute utility theory. In: Figueira, J., Greco, S., Ehrgott, M. (eds.) Multiple Criteria Decision Analysis: State of the Art Surveys, pp. 265–292. Springer, Berlin (2005)
8. Ehrgott, M., Figueira, J., Greco, S. (eds.): Trends in Multiple Criteria Decision Analysis. Springer, Berlin (2010)
9. Figueira, J., Greco, S., Ehrgott, M. (eds.): Multiple Criteria Decision Analysis: State of the Art Surveys. Springer, Berlin (2005)
10. Figueira, J., Greco, S., Roy, B., Słowiński, R.: An overview of ELECTRE methods and their recent extensions. Journal of Multicriteria Decision Analysis 20(1–2), 61–65 (2013)
11. Figueira, J., Greco, S., Słowiński, R.: Building a set of additive value functions representing a reference preorder and intensities of preference: GRIP method. European Journal of Operational Research 195(2), 460–486 (2009)
12. Figueira, J., Mousseau, V., Roy, B.: ELECTRE methods. In: Figueira, J., Greco, S., Ehrgott, M. (eds.) Multiple Criteria Decision Analysis: State of the Art Surveys, pp. 133–153. Springer, Berlin (2005)
13. Giarlotta, A., Greco, S.: Necessary and possible preference structures. Journal of Mathematical Economics 49(2), 163–172 (2013)
14. Greco, S., Matarazzo, B., Słowiński, R.: Rough sets theory for multicriteria decision analysis. European Journal of Operational Research 129(1), 1–47 (2001)
15. Greco, S., Matarazzo, B., Słowiński, R.: Decision rule approach. In: Figueira, J., Greco, S., Ehrgott, M. (eds.) Multiple Criteria Decision Analysis: State of the Art Surveys, pp. 507–562. Springer, Berlin (2005)
16. Greco, S., Mousseau, V., Słowiński, R.: Ordinal regression revisited: multiple criteria ranking using a set of additive value functions. European Journal of Operational Research 191(2), 416–436 (2008)
17. Jacquet-Lagrèze, E., Siskos, J.: Assessing a set of additive utility functions for multicriteria decision-making, the UTA method. European Journal of Operational Research 10(2), 151–164 (1982)
18. Jacquet-Lagrèze, E., Siskos, Y.: Preference disaggregation: 20 years of MCDA experience. European Journal of Operational Research 130(2), 233–245 (2001)

19. Keeney, R.L., Raiffa, H.: Decisions with multiple objectives: Preferences and value tradeoffs. J. Wiley, New York (1993)
20. Mousseau, V., Figueira, J., Dias, L., Gomes da Silva, C., Climaco, J.: Resolving inconsistencies among constraints on the parameters of an MCDA model. European Journal of Operational Research 147(1), 72–93 (2003)
21. Mousseau, V., Słowiński, R.: Inferring an ELECTRE TRI model from assignment examples. Journal of Global Optimization 12(2), 157–174 (1998)
22. Mousseau, V., Słowiński, R., Zielniewicz, P.: A user–oriented implementation of the ELECTRE-TRI method integrating preference elicitation support. Computers & Operations Research 27(7–8), 757–777 (2000)
23. Pekelman, D., Sen, S.K.: Mathematical programming models for the determination of attribute weights. Management Science 20(8), 1217–1229 (1974)
24. Roy, B.: Multicriteria Methodology for Decision Aiding. Kluwer Academic, Dordrecht (1996)
25. Slovic, P.: Choice between equally valued alternatives. Journal of Experimental Psychology: Human Perception and Performance 1(3), 280 (1975)
26. Słowiński, R., Greco, S., Matarazzo, B.: Rough sets in decision making. In: Meyers, R.A. (ed.) Encyclopedia of Complexity and Systems Science, pp. 7753–7786. Springer, New York (2009)
27. Słowiński, R., Greco, S., Matarazzo, B.: Rough Set and Rule-based Multicriteria Decision Aiding. Pesquisa Operacional 32(2), 213–269 (2012)
28. Srinivasan, V., Shocker, A.D.: Estimating the weights for multiple attributes in a composite criterion using pairwise judgments. Psychometrika 38(4), 473–493 (1973)
29. Tversky, A., Sattath, S., Slovic, P.: Contingent weighting in judgment and choice. Psychological Review 95(3), 371 (1988)

A Multi-objective Genetic Algorithm for Inferring Inter-criteria Parameters for Water Supply Consensus

Pavel A. Álvarez[1](✉), Danielle C. Morais[2], Juan C. Leyva[1], and Adiel T. Almeida[2]

[1] Department of Economic Science, University of Occident, 80020 Culiacan, Sin, Mexico
{pavel.alvarez,juan.leyva}@udo.mx
[2] Department of Production Engineering, Federal University of Pernambuco,
Recife, PE 50740-530, Brazil
daniellemorais@yahoo.com.br, almeidaatd@gmail.com

Abstract. This work is based on a disaggregation approach for the ELECTRE III method for the group decision-making. We provide a procedure in which the group is supported for modifying the parameters of outranking methods in an iterative and interactive process. In this work, we provide an application of the procedure through evaluating eight municipal districts for Water Company to invest in projects of water supply. An inferring parameters model performed by NSGA-II obtains marginal information from decision makers with more disagreement, which supports the stage of parameters modification in correspondence with the preferences of the whole group (collective ranking). This paper shows how the inferring parameters model may be used to help the decision makers with different interest to iteratively reach an agreement on how to rank cities at a time, reflecting the preferences at the individual level and at the collective level.

Keywords: Inferring parameters · Preference disaggregation analysis · Outranking method · Genetic algorithm · NSGA-II · Group decision-making/GDSS

1 Introduction

The multicriteria group decision-making (GDM) process is a difficult task when the group tries to reach a consensus. When the members of the group can generate their own individual result and latter obtaining a collective solution, it is common to find that some individual results present a significant difference from the collective solution. This problem is very common in consensus schemas (see Parreiras et al. [1]) because the collective preference is generated from divergent individual solutions. Then the collective solution can reflect some individual preferences better than others. This consensus schema is related to an individual procedure called parallel coordination mode of work [2]. In the parallel mode, the members of the group work in an iterative way before reaching a consensus. It means they generate repeatedly the individual solution in this procedure.

The problem with the above approach is that DMs can obtain divergent results with respect to the global solution. In this case a tool to support the stage of parameters

© Springer International Publishing Switzerland 2015
A. Gaspar-Cunha et al. (Eds.): EMO 2015, Part II, LNCS 9019, pp. 218–233, 2015.
DOI: 10.1007/978-3-319-15892-1_15

modification is needed. A common technique to support the DM with complex models is based on preference-disaggregation analysis (PDA). This is based on previous information given by the DM.

In this paper we apply an outranking method based for inferring inter-criteria parameters in the stage of definition of parameters to support DMs in a group decision-making approach. This model includes DM's marginal information to propose parameters for the outranking method, which generate individual solutions closer to the collective solution. The inferring model is exploited with a genetic algorithm, which takes as input DM's original parameters and shared preference between individual and collective solution. The output of the genetic algorithms is a set of parameters (w, q, p, v), which can generate a ranking (individual solution) more similar to the group ranking (collective solution).

This paper aims the application of a group decision procedure for the analysis of destination of resources for water supply, in order to prioritize the city in which the project will be implanted, in agreement with specific criteria [3]. In this procedure a group of four Decision Makers (DMs), which represent different interests, are supported by an inferring inter-criteria parameter model for outranking methods in the stage of parameters modification. This model proposes DMs inter-criteria parameters for ranking municipal districts. This paper describes how the set of criteria was developed, the use of an appropriate Multicriteria Decision Aids method (ELECTRE III, a ranking method proposed by Roy [4]), and the application of the method using the SADGAGE software [5], which support a multicriteria Group Decision Making (GDM) to reach a consensus collective solution.

The paper is organized as follows. The multicriteria method group-based chosen to provide the decision support is presented in Section 2. Section 3 shows the results of applying the proposed methodology for prioritization of municipal districts to implement water supply. Finally, Section 4 draws the main conclusions of this study.

2   MCDA Method for Group of Decision Makers

A variety of studies have been developed using group decision-making based in individual solutions approach. Alencar et al. [6] developed a multicriteria group decision model where a collective evaluation is undertaken from individual preferences. Morais and Almeida [7] propose a method based on the generation of individual rankings permitting the members be involved in the process to choose an alternative in-group for the water resources problem. Silva et al. [8] used a group decision approach for aggregating the individual result to obtain a global ranking. The problem with the above approach is that DMs can obtain divergent results with respect to the global solution. In this case a tool to support the stage of parameters modification is needed.

2.1   A Consensus Procedure for Reaching a Higher Agreement

Reaching a consensus in the group decision-making problem is an interactive and iterative procedure. Fig. 1 illustrates the schema of a group decision-making procedure where the DMs work in an individual manner. In each iteration the DMs construct their individual preferential models and individual ranking. With a group preferential aggregation method, which is based on individual's results, a temporal collective ranking is generated. Two methods based on individual's preferences have been recognized in the scientific literature [9, 10].

In each iteration we calculate two consensus parameters, a consensus measure and a proximity measure. The first parameter guides the consensus process, and the second parameter supports the group discussion phase of the consensus process. The problem addressed is how to find the individual positions converge and, therefore, how to support the decision makers in obtaining and agreeing with a specific solution. To accomplish this goal, a consensus level α required for that solution is fixed in advance ($\alpha \in [0,1], \alpha > 0.5$). When the consensus measure reaches this level, then the decision-making session is finished and the solution is obtained. If that scenario does not occur, then the decision-makers' preferences must be modified. This modification is accomplished in a group discussion session in which we use a proximity measure to propose a feedback process based on simple rules, which supports the decision makers in changing their preferences. Here is when the inferring parameter tool is used to support the parameter modification stage, because the complexity for the DM.

The proposed consensus model compares the positions of the alternatives based on the individual solutions and the group's temporary solution. Based on the consensus level and the offset of the individual solutions, the model gives feedback suggesting the direction in which the individual decision makers should change their preferences, based in an inferring parameter model. The all procedure of the consensus for this group multi-criteria ranking problem is presented in Fig. 1.

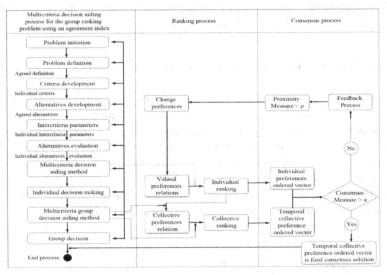

Fig. 1. Group multicriteria decision aid procedure for reaching higher agreement [5]

Consensus and Proximity Measures

The consensus procedure considers the position-weighted measure (proximity) $wP_A^{k,G}$ developed by Leyva and Alvarez [11] for expressing the differences in the ranking discrepancies between the group temporary ranking and the individual rankings (see [12]). The measure considers the rankings of a set of alternatives from the most important to the least important.

1. We use a multi-criteria decision making method (e.g., Electre III, Promethee II) to obtain the individual rankings R_k for each decision maker; then, we use a group multi-criteria decision making method (Electre for groups, Promethee for groups) to obtain a collective ranking of alternatives R_G.

2. We calculate the proximity of the k-th decision maker's individual solution to the collective temporary solution, called $wP_A^{k,G}$.

3. The global consensus measure, called C_A, is calculated by the aggregation of the above consensus degrees for each decision maker, using the following expression:

$$C_A = \sum_{k=1}^{n} \frac{wP_A^{k,G}}{n}$$

where n is the number of decision makers in the group.

Feedback Process

When the consensus measure C_A has not reached the consensus level required, then the decision makers' rankings must be modified. The proximity measures $wP_A^{k,G}$ is used to build a feedback process so that decision makers can adjust their preferences in order to achieve closer preferences between them. This feedback mechanism (adapted from that of Herrera-Viedma, Herrera [13]) will be applied when the consensus level is not satisfactory and will be finished when a satisfactory consensus level is reached.

In the feedback process, every DM has the opportunity to make some preference modification, however in this procedure the DMs with more disagreement with the global preference must to be in priority to getting closer individual preference with the global preference. In order to meet this goal, the procedure regards the following. "if the proximity of the k-th decision maker's individual solution to the collective temporary solution is less than a predefined threshold $\rho \in [0,1]$ then the decision maker has to change their preferences, and it will be carried out in the following way":

1. Compute, for each member k, the agreement level $wP_A^{k,G}$.

2. Compute the global consensus measure C_A

3. If the global consensus measure C_A exceeds the predefined threshold α, the procedure is stopped because a consensus order has been reached. In the opposite case ($C_A < \alpha$), then we move to step 4 (the feedback process);

4. In this step, the members can exchange information and discuss and modify their ranking to reach a consensus order. For this purpose, we proceed as follows:

4.1. Identify the members k whose proximity measure $wP_A^{k,G}$ is less than a predetermined threshold ρ ; then, these members must change their preferences (w, q, p, v) supported by the inferring parameter tool.

4.2. The inferring parameter tool suggests different sets of new parameters (w, q, p, v) for the DM. Each k member selects a set that matches better his/her preferences and the individual ranking R_k is modified. Then, the aggregation procedure is again performed to produce a new collective temporary ranking R_G. Return to step 1.

2.2   A Genetic Algorithm for Inferring Inter-criteria Parameters

We identify two works in the disaggregation methods for group approach. Bregar, Gyorkos [14] infer w, λ parameters. Damart, Dias [15] propose a methodology for group, which cooperatively develop a common multicriteria evaluation model to sort actions. Covantes et al. [16] developed a inferring model for the sorting THESEUS outranking method. Fernandez et al. [17] developed the most recent work for the ranking problem in a outranking method for inferring parameters. However those methods are not developed for consensus schemes, when DMs generate their individual solution before the collective solution.

In a group decision-making process, the methods and tools of decision-making support should include procedures that allow group members to include their preferences, which are reflected in a collective decision. The inferring parameters tool is based on a model that considers both individual and group preferences to propose parameters, which obtain individual results matching the collective preference.

The model for this problem includes individual and group information to propose parameters for the outranking method, which generate individual solutions closer to the collective solution. The inferring model is exploited with a genetic algorithm, which takes as input DM's marginal preference (original inter-criteria parameters) and, similar preference between individual and collective solution. The output of the genetic algorithms is a set of parameters (w, q, p, v), which can generate a ranking (individual solution) more similar to the group ranking (collective solution).

The inferring parameters model is exploited with a genetic algorithm because the facility of converging and obtaining the Pareto frontier since first running. The complexity of multiobjective optimization problems can be handled with evolutionary algorithm because computer efficiency of these methods is less sensitive to problem size in comparison to the traditional techniques [18]. The next sections present the brief information of genetic algorithm (GA) and the proposed model.

Genetic Algorithm

In this section, we present a multiobjective evolutionary algorithm (MOEA) based on a posterior articulation of preferences, which are able to detect the set of inter-criteria parameters that one DM should consider to change their evaluation in a particular subset of criteria $\{g_{k_1}, g_{k_2}, ..., g_{k_i}\}$, with the purpose of reducing the pairwise disagreements between the d-th order and the collective temporary order. The algorithm borrows fundamental elements from NSGA II [19]. In the following we present in further detail the fundamental aspects of the algorithms.

A solution is encoded as a set of pseudo-criteria in an m-ary string. For each criterion defined by the DM corresponds the w, q, p and v pseudo-criteria. The representation of each criterion in the m-ary string corresponds to 4 real values (see Fig 2.). The algorithm borrows fundamental elements from NSGA II [19]. The GA's value parameters used for the model application: 40 individuals for population size, 10,000 generations, 0.90 crossover index and 0.4 mutation index. Although the NSGA II can find the Pareto front in the firsts runs, 10000 generations were needed to reach a diversity of parameters proposals for the DM. When the algorithm was ran with 5,000 generation or less, the results show few different proposals. The complexity of the problem requires higher exploration on the continuous search space.

	w_1	q_1	p_1	v_1	w_2	q_2	p_2	v_2	...	w_n	q_n	p_n	v_n
Individual $\tilde{p} =$	0.2	10	20	35	0.1	1	3	5	...	0.25	2.5	3.5	4.2

Fig. 2. Individual representation of the GA for inferring inter-criteria parameters

Objective Functions

Basic definitions.

Definition 1. Let $A = \{a_1, a_2, ..., a_m\}$ be a set of m alternatives and $G = \{g_1, g_2, ..., g_n\}$ be a set of n decision criteria defined on A. Without loss of generality, we can consider the first t criteria $G_o = \{g_1, g_2, ..., g_t\}$ as objective criteria and the rest $n-t$ criteria $G_s = \{g_{t+1}, g_{t+2}, ..., g_n\}$ as subjective criteria.

Definition 2. Let K^d be the set of pairwise disagreements between the d-th order O^d and the collective temporary order O^G defined as:

$$K^d(O^d, O^G) = \begin{cases} (a_i, a_j) \in A \times A \mid i < j, (O^d(a_i) < O^d(a_j) \wedge O^G(a_i) > O^G(a_j)) \vee \\ (O^d(a_i) > O^d(a_j) \wedge O^G(a_i) < O^G(a_j)) \end{cases}$$

The first complement of K^d, $K^{d,C}_{first}$ is the set of pairwise agreements between the first d-th order O^d and the collective temporary order O^G and is defined as:

$$K^{d,C}_{first}(O^d, O^G) = \begin{cases} (a_i, a_j) \in A \times A \mid i < j, (O^d(a_i) < O^d(a_j) \wedge O^G(a_i) < O^G(a_j)) \vee \\ (O^d(a_i) > O^d(a_j) \wedge O^G(a_i) > O^G(a_j)) \end{cases}$$

Note that $\left| K^d \right| + \left| K^{d,C}_{first} \right| = (m-1)!$

A Multiobjective Evolutionary Algorithm for Identification of Inter-criteria Parameters

We use a value encoding scheme for representing a potential solution.

Let $\tilde{p} = p_1 p_2 \cdots p_{4n}$ be the schematic representation of an individual's chromosome.

$\tilde{p} \in \prod_{i=1}^{4n} C_i$, where C_i is the set of values that p_i can takes. This set of values is dependent of the problem.

In this section we describe the objective functions f_1 , f_2 and f_3 . The fitness of an individual is calculated according to a given fitness procedure. The approach for defining individual's fitness involves the non dominated solutions in a similar form of NSGA II [19]. We define the objective function f_1 of an individual $\tilde{p}$ as follows:

Let $\tilde{p} = p_1 p_2 \cdots p_{4n}$ be the schematic representation of an individual's chromosome. Let $\tilde{p}_R = r_1 r_2 \cdots r_{4n}$ be a reference individual representing the original inter-criteria parameters, i.e.:

$r_1 = w_1$, $r_2 = q_1$, $r_3 = p_1$, $r_4 = v_1$, $r_5 = w_2$, $r_6 = q_2$, ..., $r_{4n-2} = q_n$, $r_{4n-1} = p_n$, ...,

$r_{4n} = v_n$. Then:

$$f_1(\tilde{p}) = \left| \{ (p_i, r_i) \mid p_i \neq r_i i = 1, 2, ..., 4n \} \right|$$

$f_1(\tilde{p})$ is the number of modified inter-criteria parameters. Note that the quality of solution increases with decreasing f_1 score. With this objective function we want to preserve, as much as possible, the original inter-criteria parameters.

The objective function f_2 of an individual $\tilde{p}$ measures the amount of pairwise disagreements between the *d-th* order O^d and the collective temporary order O^G and we chose to define it as:

$$f_2(\tilde{p}) = \left| K^d \right|$$

$f_2(\tilde{p})$ is the number of pairwise disagreements between the *d-th* order and the collective temporary order.

With this objective function we want to reduce the pairwise disagreements between the *d-th* order and the collective temporary order in order to increase the value of the proximity measure $wP_A^{d,G}$.

The objective function f_3 of an individual $\tilde{p}$ measures the amount of pairwise agreements between the *d-th* order O^d and the collective temporary order O^G and we chose to define it as:

$$f_3(\tilde{p}) = \left| K_{first}^{d,C} \right|$$

$f_3(\tilde{p})$ is the number of pairwise agreements between the *d-th* order and the collective temporary order.

With this objective function we want to maximize the original pairwise agreements between the *d-th* order and the collective temporary order in order to preserve the rationality and values system of the DM.

We are interested in:

Individuals whose objective function f_1 value is close to zero. This assures us that the ordering represented by the individual is almost equal to the original set of inter-criteria parameters; this is one characteristic always appreciated for all rational decision maker.

Individuals whose objective function f_2 value is close to zero. This objective improves the feasibility and reduces the disagreements between two orders.

Individuals whose objective function f_3 is closet to $(m-1)!$ This assure us that the ordering represented by the individual increases the value of the proximity measure $wP_A^{d,G}$.

Then, we use the genetic algorithm as evolutionary search for solving the multiobjective combinatorial problem:

$$Min(f_1(\vec{p})),Min(f_2(\vec{p})),Max(f_3(\vec{p}))$$

$$Subject\ to$$

$$\vec{p} \in \prod_{i=1}^{4n} C_i \tag{1}$$

where C_i is the set of values that p_i can takes.

3   Proposed Model Application for Destination of Resources for Water Supply in a Group Decision Making Approach

The model proposed to deal with individual preference desegregation to generate a collective ranking derived by individual results was applied for destination of resources for water supply. The same problem of water supply was deal before by Morais and Almeida [3], however the procedure performed by the DMs in [3], the all group defined the parameters in agreement without generating individual results. In the present work the water supply problem is deal with different approach proposed to support the DMs to generate a ranking agreement derived from individual results. In this sense, a brief description of the water supply problem is described.

The implementation of the plan for water supply can be developed only in one city at time. Eight cities are competing for the water supply resource and four DMs (each representing different interest) should prioritize them in order to know the sequence, which the resource would be used. The DMs should reach an agreement of the collective ranking generated.

In this study were considered four interest groups that act as decision-makers (DM) as follows:

DM 1: Water Company, which that traditionally acts as the responsible of implantation of the system.

DM2: Environmental Agency, which is responsible to guarantee that the environment is protected with the project designed.

DM3: Local Groups, which are representatives from local community. These people can actively influence the decisions to support their own interests.

DM4: Financial groups, that are entities tied to the government of the state with enough power to influence the decisions. In terms of financial aspects, bankers can impose pressure in favor of their own interests. Some of the players are: Caixa Economica Federal (Federal Bank) and The Interamerican Development Bank.

It should be noticed that each decision group has an opinion and a specific interest. So, conflicts can be generated due to a series of factors. It is clear that what is needed is a simplified methodology for the choice of the place for the implantation of a water supply system, by evaluating several aspects.

Some meetings were arranged with these interest groups. In these meetings they had the opportunity to express the points and list the factors they perceived sufficiently important to be taken into account in the selection process and defined four criteria to analysis the districts:

Cost of Investment: Each district requires different amount of resource for the water supply project. This criterion specifies the investment amount.

Population: It is the number of persons shall be beneficed by the project. It is expected the greatest number of people is beneficed by the project.

Quality of life: This criterion represents the sanitary and hygienic conditions of the population. The Life Conditions Index (LCI) is used to estimates sanitation conditions. For example, the LCI for Brazil is .723, while that for Pernambuco state is .616 [20]. Thus, a city with the lowest LCI has higher priority to be attended with a water supply project with respect a city with higher LCI.

Tourism: The criterion is measuring an economic aspect of the project. This evaluation was taken by the actors related with the condition of the city in respect to the tourism. The values attributed to each verbal concept are 0.00 (weak), 0.33 (regular), 0.67 (good) and 1.00 (very good).

3.1   Results

The evaluation was treated as a the multicriteria ranking problem and performed with SADGAGE system [5] hosted in http://mcdss.udo.mx/sadgage. Four DMs are representing different group's interests for the prioritization on destination of resource for water supply. The DMs are evaluating 8 municipal districts for enterprise to invest in using the funds available. Table 1 shows code of each municipal district and four criteria codes for evaluation. Table 2 shows the performance matrix of municipal districts for each criterion.

Table 1. Alternatives and criteria for water supply problem

Alternatives		Criteria	
A	Alianca	C1	Cost of investment
B	Moreno	C2	Population
C	Ouricuri	C3	Quality of life
D	Passira	C4	Tourism
E	Pocao		
F	Porto Galinhas		
G	Toritama		
H	Trindade		

Table 2. Performance of the alternatives

	C1	C2	C3	C4
A	589,176	37,188	0.476	0.00
B	1,548,354	45,481	0.600	0.33
C	2,053,485	56,623	0.443	0.00
D	804,270	29,131	0.474	0.33
E	2,191,952	11,177	0.478	0.67
F	5,181,246	10,995	0.500	1.00
G	2,135,702	21,794	0.600	0.00
H	1,457,073	21,919	0.496	0.00

Table 3. Parameters: weights, indifference and preference thresholds in iteration 1

		C1 Min	C2 Max	C3 Min	C4 Max
DM1	w	0.30	0.25	0.25	0.20
	q	500000	10000	0.05	0.33
	p	1000000	20000	0.1	0.67
DM2	w	0.10	0.30	0.20	0.40
	q	480000	0	0.025	0
	p	1000000	0	0.6	0.33
DM3	w	0.10	0.30	0.40	0.20
	q	300000	9000	0.05	0
	p	600000	125000	0.07	0.33
DM4	w	0.40	0.15	0.30	0.15
	q	300000	10000	0.01	0.33
	p	600000	20000	0.03	0.67

Table 3 presents the criteria selection, directions of the criteria, weight and thresholds regarding every criterion selected by the DM. The next stage is constructing the preferential model of every DM. This stage can be performed with different outranking methods. For this example, the ELECTRE III method was used and a MOEA [21] for exploits the preferential model generating a raking of alternatives for every DM. Those methods are embedded in SADGAGE system.

At this stage we have the individual ranking of every DM shown in Table 4. The collective ranking is shown in the last column of Table 4. This is a temporal ranking that reflects the group preferences in iteration 1. The group preferences were generated by an aggregation approach for groups which was strongly based on ELECTRE and for simplicity we call ELECTRE-GD [9]. In Table 4 we have a row showing the similarity (proximity) between the individuals' rankings and the collective ranking. This value was calculated by the proximity index proposed by Leyva and Alvarez [11].

In iteration 1, the rankings of the DM2 and DM3 with proximity value 0.536 and 0.660, respectively, present greater difference to the collective ranking than DM1 (0.907) and DM4 (0.886). The similarity values obtained in this iteration gain a consensus level (CA) of 0.747. For this procedure a required level of consensus is $\alpha = 0.8$ (CA > α). In this case the consensus level is not reached (0.747 < α), it means that some DMs must change their preferences through a modifying parameters stage.

Table 4. Individuals and group ranking in iteration 1

Position	DM1	DM2	DM3	DM4	Collective
1	D	C	C	A	A
2	A	E	D	C	D
3	C	F	A	D	C
4	H	B	E	H	E
5	B	D	F	B	B
6	E	A	B	E	F
7	F	H	H	F	H
8	G	G	G	G	G
λ	0.55	0.597	0.6	0.55	
Disagrees	5	10	9	5	
Proximity	0.907	0.536	0.660	0.886	
		Consensus level (C_A) 0. 747			

Disagrees: the number of differences between two rankings seen in pairwise format

To improve the consensus level a greater similarity value is required for some DM. In the analysis of destination of resource for water supply, the DMs with low proximity level ($\rho = 0.7$) were required to interact in preference modifying procedure to obtain a solution, which does not present strong preference chances to support. Then, DM2 and DM3 with similarity level of 0.536 and 0.660, respectively were involved in the procedure. For the 4.1 feedback stage of the consensus process the DM2 and DM3 are the k members with $wP_A^{k,G} < \rho$ in iteration 1. To support this stage we use the inferring parameter model of the Section 3.2.2 that use as input a set of similar preference between individual and group solutions. The model concerns number of parameters' changes ($f1$), number of disagreements ($f2$) and agrees of the first ranking that still remaining ($f3$); those objectives are evaluated in the new parameters proposal (See Equation (1)). Regarding with the 4.2 feedback stage of the consensus process where the DM2 and DM3 need to change their individual preference (inter-criteria parameters), the output of the inferring parameter tool is shown in the Table 5 and Table 6, respectively. The new parameters suggested for the DM2 and DM3 implies some changes in the weight, indifference and preference thresholds.

The column 2, 3 and 4 is the evaluation of the $f1, f2, f3$, respectively. Column 5 is the epsilon value used for the new proposal o parameters. The values of the new parameters are limited by an epsilon value. The new parameter is in a neighborhood of $p_i = r_i \pm (r_i * \varepsilon)$. The column 6 presents the proposed parameters and column 7 the possible ranking will be generated. The last column is the proximity index that new ranking presents with the collective ranking of iteration 1.

In iteration 2, the new inter-criteria parameters suggested by inferring model in Table 5 and Table 6 are showed to the DM2 and DM3, respectively to support the modifying parameters stage.

The DM2 found that proposal 4 and 5 generate the same ranking (see Table 5) and the alternative C remains in high position. However proposal 5 requires stronger preference changes in weights for criterion 3 and 4. The proposal 4 requires smooth preference changes on weights for every criterion. DM2 selected proposal 4 because remains C in high position, the changes are easier to support and generates acceptable proximity level.

Table 5. Inter-criteria parameters proposals for improving the DM2' agreement

No.	f_1	f_2	f_3	ε	Proposal	Ranking	Proximity
1	5	2/28	18/19	0.10	q1=270000, p1=660000, q3=0.0465381, p3=0.0746381, p4=0.321273,	D ≻ C ≻ A ≻ B ≻ F ≻ E ≻ H ≻ G	0.854
*2	4	2/28	18/19	0.50	q1=208183, p1=480797, q3=0.0326906, p3=0.0943696,	C ≻ D ≻ A ≻ B ≻ E ≻ H ≻ F ≻ G	0.825
3	4	2/28	18/19	0.50	q1=450000, p1=479315, q3=0.0326906, p3=0.0943696,	D ≻ C ≻ A ≻ B ≻ E ≻ H ≻ F ≻ G	0.871
4	8	1/28	19/19	0.75	q1=185422, p1=646962, w2=0.196, p2=14131, q3=0.0421931, p3=0.0846173, w4=0.304, p4=0.0825,	D ≻ E ≻ A ≻ F ≻ C ≻ B ≻ H ≻ G	0.782
5	7	2/28	18/19	0.75	q1=274227, p1=1050000, q2=3093, p2=21875, q3=0.0421931, p3=0.0846173, p4=0.0825,	C ≻ A ≻ D ≻ H ≻ E ≻ B ≻ F ≻ G	0.842
6	6	2/28	18/19	1.00	q1=334210, q2=0, p2=25000, q3=0.0153812, p3=0.118739, p4=0.241184,	C ≻ D ≻ A ≻ B ≻ E ≻ F ≻ H ≻ G	0.832

*: proposal selected by the DM2

Table 6. Inter-criteria parameters proposals for improving the DM3' agreement

No.	f_1	f_2	f_3	ε	Proposal	Ranking	Proximity
1	4	5/28	16/18	0.10	p1=1029870, q3=0.0266334, p3=0.561145, p4=0.299604,	D ≻ E ≻ B ≻ C ≻ F ≻ A ≻ H ≻ G	0.643
2	6	3/28	18/18	0.50	w1=0.057, q1=240000, p1=829692, w2=0.343, q3=0.0331671, p3=0.42281,	E ≻ D ≻ B ≻ C ≻ A ≻ F ≻ H ≻ G	0.657
3	8	2/28	18/18	0.50	w1=0.057, q1=240000, p1=500075, w2=0.343, w3=0.247, q3=0.0331671, p3=0.42281, w4=0.353,	B ≻ D ≻ C ≻ A ≻ E ≻ H ≻ G ≻ F	0.677
*4	8	2/28	18/18	0.50	w1=0.057, q1=264983, p1=1046210, w2=0.343, w3=0.247, q3=0.0331671, p3=0.42281, w4=0.353,	D ≻ C ≻ B ≻ A ≻ E ≻ H ≻ F ≻ G	0.8
5	6	2/28	18/18	0.75	q1=490411, p1=935514, w3=0.293, q3=0.0362182, p3=0.265269, w4=0.307,	D ≻ C ≻ B ≻ A ≻ E ≻ H ≻ F ≻ G	0.8
6	6	2/28	18/18	0.75	q1=148032, p1=935514, w3=0.293, q3=0.0362182, p3=0.265269, w4=0.307,	D ≻ B ≻ A ≻ C ≻ E ≻ H ≻ F ≻ G	0.802
7	8	1/28	18/18	1.00	w1=0.2, q1=105087, p1=597152, w2=0.193, w3=0.398, q3=0.0222516, p3=0.15778, w4=0.209,	D ≻ A ≻ C ≻ B ≻ H ≻ E ≻ F ≻ G	0.914

*: proposal selected by the DM3

DM3 was interested on proposal 5's result (see Table 6) because remains better his/her most preferred alternatives (A, C, D). However, this proposal requires to change 7 parameters. Instead DM3 selected proposal 2 because remains A, C and D on high positions and has good proximity level with less number of preference changes (4 parameters).

The proposal selected by DM2 and DM3 are used to generate new individual's preference models and rankings, the DM1 and DM4 remain their previous ranking (see Table 7). With the rankings showed in Table 7 a new collective ranking is generated (see column 6). With the new temporal solution, the proximity level for DM1 and DM4 remain high proximity levels, 0.946 and 0.857, respectively.

The individual solution from DM2 and DM3 reflect improved proximity level 0.850 and 0.854, respectively. The number of disagreements where reduced to 3 for DM2 and 4 for DM3. This new proximity between rankings generates a better consensus level $(C_A=0.877)$ greater than required $(C_A > \alpha)$. In this model application, the procedure finished with the consensus level obtained by iteration 2.

Table 7. Individuals and group ranking in iteration 2

Position	DM1	*DM2	*DM3	DM4	+Collective
1	D	D	C	A	D
2	A	C	D	C	A
3	C	B	A	D	C
4	H	A	H	H	E
5	B	E	E	B	B
6	E	H	B	E	H
7	F	F	F	F	F
8	G	G	G	G	G
λ	0.55	0.600	0.600	0.55	
Disagrees	3	3	4	5	
Proximity	0.946	0.850	0.854	0.857	
Consensus level (C_A) 0.877					

* DM with new ranking   + New collective ranking

The procedure developed with the proposed model helped the participants to obtain a collective solution (iteration 2), which reflect their individual preference better than the first collective solution. In the first collective solution of the Table 4 the DM2 and DM3 presented more disagreement with the collective solution. In this situation the collective solution did not reflect their individual preference of DM2 and DM3 as much as for DM1 and DM4. The collective solution in iteration 1 (Table 4) presents the alternatives A and D in first and second order respectively. However, three individual solutions show D better ranked than A (DM1, DM2, DM3). The above has a negative impact in the DM's agreement and as consequence in the consensus level. We can see how this situation is avoided with the collective ranking in the second iterations. Here alternatives D is ranked in higher positions (DM1, DM2 and DM3' ranking). Thus the collective solution shows D higher than A (Table 7), now every DM shows better agreement and it is supported by the number of disagrees reduced in every DM's ranking (see Table 7). The new collective solution obtained in the iteration 2 shows more proximity level for each DM and as consequence better consensus level (0.877). In different situation, the proximity level obtained for each DM in iterations can be reduced for some DM's and increment for other. The goal is get equilibrium in the agreement context for this been reflected in the consensus level. The final results must include a collective solution reflecting every DM's preferences.

The application of the model showed above presents the interaction process of the DMs in 2 iterations. This procedure shows how a group can be supported in the modifying parameters stage to generate new individual solutions closer to the group solutions. In others words, the DMs can obtain individual preferences closer to the group preferences. The most interesting aspect of the procedure is when the consensus level is reached the collective solution reflects individual preference better than first solutions. The amount of iteration could depend of the problem complexity and the availability of the DMs to change their individual information.

4   Concluding Remarks

In this work, we present a multiobjective optimization model to infer inter-criteria parameters for outranking methods (w, q, p, v) in an interactive process of group decision-making for the multicriteria ranking problem. Here we describe the interactive procedure for group decision-making. In this procedure a multiobjective optimization model and an evolutionary algorithm support the stage of modification of preference by iterative and interactive activities with DMs.

The model was tested by the application in a real problem to evaluate municipal districts for water supply by DMs with different interests. This model provides both the facilitator and DMs a support tool for modifying inter-criteria parameters in an interactive group decision-making process. The inferring parameter model can reduce the facilitator's influence on DMs when they are helped in the modification of parameters. However, this needs further research.

References

1. Parreiras, R.O., Ekel, P.Y., Morais, D.C.: Fuzzy Set Based Consensus Schemes for Multicriteria Group Decision making Applied to Strategic Planning. Group Decision and Negotiation 21(2), 153–183 (2011)
2. Cao, P.P., Burstein, F., Pedro, S.J.: Extending Coordination Theory to the Field of Distributed Group Multiple Criteria Decision-Making. Journal of Decision Systems 13(3), 287–305 (2004)
3. Morais, D.C., Almeida, A.T.: Water supply system decision making using multicriteria analysis. Water S.A. 32(2), 229–235 (2006)
4. Roy, B.: Un algorithme de classements fondé sur une représentation floue des préférences en présence de critères multiples. Cahiers du CERO (Centre d'Etudes de Recherche Opérationnelle) 20(1), 3–24 (1978)
5. Leyva, J.C., Alvarez, P.A.: SADGAGE: A MCGDSS to solve the multicriteria ranking problem in a distributed and asynchronous environment. In: Proceedings of the Fourth International Workshop on Knowledge Discovery, Knowledge Management and Decision Support. Atlantis Press, Mazatlan (2013)

6. Alencar, L.H., Almeida, A.T., Morais, D.C.: A multicriteria group decision model aggregating the preferences of decision-makers based on electre methods. Pesquisa Operacional **30**(3), 687–702 (2010)

7. Morais, D.C., Almeida, A.T.: Group decision making on water resources based on analysis of individual rankings. Omega **40**(1), 42–52 (2012)

8. Silva, V.B.S., Morais, D.C., Almeida, A.T.: A Multicriteria Group Decision Model to Support Watershed Committees in Brazil. Water Resources Management **24**(14), 4075–4091 (2010)

9. Leyva, J.C., Fernández, E.: A new method for group decision support based on ELECTRE III methodology. European Journal of Operational Research **148**(1), 14–27 (2003)

10. Macharis, C., Brans, J.P., Mareschal, B.: The GDSS PROMETHEE procedure - A PROMETHEE-GAIA based procedure for group decision support. Journal of Decision Systems **7**, 283–307 (1998)

11. Leyva López, J.C., Alvarez Carrillo, P.A.: Accentuating the rank positions in an agreement index with reference to a consensus order. International Transactions in Operational Research (2014) doi:10.1111/itor.12146

12. Jabeur, K., Martel, J.-M.: An Agreement Index with Respect to a Consensus Preorder. Group Decision and Negotiation **19**(6), 571–590 (2009)

13. Herrera-Viedma, E., Herrera, F., Chiclana, F.: A consensus model for multiperson decision making with different preference structures. IEEE Transactions on Systems, Man, and Cybernetics - Part A: Systems and Humans **32**(3), 394–402 (2002)

14. Bregar, A., Gyorkos, J., Rozman, I.: An alternative sorting procedure for interactive group decision support based on the pseudo-criterion concept. Journal of Systemics, Cybernetics and Informatics **1**(4), 66–71 (2003)

15. Damart, S., Dias, L.C., Mousseau, V.: Supporting groups in sorting decisions: Methodology and use of a multi-criteria aggregation/disaggregation DSS. Decision Support Systems **43**(4), 1464–1475 (2007)

16. Covantes, E., Fernandez, E., Navarro, J.: Robustness analysis of a MOEA-based elicitation method for outranking model parameters. In: Proceedings of the 2013 10th International Conference on Electrical Engineering, Computing Science and Automatic Control (CCE), pp. 209–214 (2013)

17. Fernandez, E., Navarro, J., Mazcorro, G.: Evolutionary multi-objective optimization for inferring outranking model's parameters under scarce reference information and effects of reinforced preference. Foundations of Computing and Decision Sciences **37**(3) (2012)

18. Michalewicz, Z.: Genetic Algorithms + Data Structures = Evolution Programs. Springer, Heidelberg (1996)

19. Deb, K., Pratap, A., Agrawal, S., Meyarivan, T.: A fast and elitist multiobjective genetic algorithm: NSGA-II. IEEE Transactions on Evolutionary Computation **6**(2), 182–197 (2002)

20. FIDEM-PE, Foundation of Metropolitan Development of the Pernambuco State. Research of Quality of Life. Perfil Municipal. Governo do Estado de Pernambuco, Recife (2000)

21. Leyva, J.C., Aguilera, M.A.: A Multiobjective Evolutionary Algorithm for Deriving Final Ranking from a Fuzzy Outranking Relation **3410**, 235–249 (2005)

Genetic Algorithm Approach for a Class of Multi-criteria, Multi-vehicle Planner of UAVs

Emory Freitas and Jose Reginaldo Hughes Carvalho[(✉)]

Institute of Computing, Federal University of Amazonas, Manaus, Amazonas, Brazil
{emory,reginaldo}@icomp.ufam.edu.br,
http://www.icomp.ufam.edu.br

Abstract. This work presents an solution for the problem of planning the routes of a small fleet of hand-launched fixed-wing UAVs to monitor by vision an area of forest at the minimum possible time, for rescuing support, fire detection, deforestation mitigation, among other important applications in the Amazon rain forest operational scenario. However, time is not the only criteria to be considered. Given that it is virtually impossible to recover a missing airplane in the deep jungle, another relevant objective is to apply as fewer UAVs as possible, to reduce the risks for the whole UAV system. Moreover, another aspect that complicates this specific application is the non-holonomic constraint of the fixed wing airplanes (lack of instantaneous lateral velocity). We had to restrict the changing of heading, and thus, the roll angle, to prevent the drifting of the PoI (point-of-interest) out of the camera plane. As a consequence, long turn radius imply longer route. The authors modeled this problem using the multi vehicle routing problem, combined with Dubins path generator to address the non-holonomicity aspect, in a multicriteria formulation to reach a solution that simultaneously takes into account time, number of UAVs and expended resources. The resulting problem formulation has a computational effort that tends to be extremely high as the number of way-points increases. The authors applied Genetic Algorithm to solve this multi-criteria NP-hard Dubins adapted routing problem. The presented results, applied to a fleet of mini-UAVs, show the effectiveness of the proposed work in providing a satisfactory solution in a very reasonable execution time.

Keywords: Path Planning · Multi Vehicle Routing Problem · Dubins Path

1 Introduction

This work addresses the problem of routing a small fleet of hand-launched UAV to overflight specific areas in the Amazon rain forest at a minimum time. Typical

This work was partially sponsored by Financing Agency for Studies and Projects (Finep), process 01-10 0611-00, the Foundation for Research Support of the State of Amazonas (FAPEAM), processes 01135/2011 (Pronex), 114/2014 (ProTI-Pesquisa), and PROMOBILE/Samsung under the terms of Brazilian federal law 8.248/91.

© Springer International Publishing Switzerland 2015
A. Gaspar-Cunha et al. (Eds.): EMO 2015, Part II, LNCS 9019, pp. 234–248, 2015.
DOI: 10.1007/978-3-319-15892-1_16

missions include, but are not restricted to, fire detection, deforestation mitigation, seek-and-track targets, and rescue support. One would infer that an optimization problem to reach minimum time would promote the maximization of the number of vehicles per mission, in a divide to conquer approach. However, it is virtually impossible to recover a small UAV in the deep jungle in the occurrence of failures. Therefore, besides minimizing the execution time, one has also to consider to apply as fewer airplanes as possible to reduce the risks for the fleet as a whole. These objectives are clearly conflicting, and the resulting multi-criteria multi-vehicle non-holonomic combinatorial optimization problem constitutes the scope of this paper.

An aircraft flight plan is usually defined by a set of way-points in geodesic coordinates. Considering a fleet of UAV, the flight planning phase may be seen as the vehicles routing problem (VRP) [13], known to be NP-Hard. The problem will have to take into account the non-holonomic constraints of the airplane. An well-known approach is to approximate an airplane in cruiser flight mode as a Dubins Car (maximum steering angle and no backward capability) and consider the Dubins curve [1] when computing the total path length.

One can find solutions to the VRP with Dubins curve in the literature, mostly for the single vehicle problem (the travelling salesman problem-TSP). In [2], the authors address the single vehicle DVRP, and their methodology is subsequently improved by [10] and [11]. In [15] the authors propose a solution based on Genetic Algorithms to the single-vehicle problem with Dubins constraints. One can find approaches for a fleet of vehicles, however, they do not consider non-holonomic constraints [13]. As far as the authors are concerned, none of the studied approaches proposes a solution for a multi-criteria, multi-vehicle with Dubins curve formulation.

In the multi-objective optimization field of research, the Genetic Algorithms (GA) is a suited tool when considering the complexity of an NP-hard problem [7][8]. The proposal is to provide to each airplane a good set (measured by the Fitness Function) of way-points so that the combined routes will reach the overall mission goals. In other words, the proposed methodology seeks to simultaneously minimize mission execution time, total flight distance and number of applied UAV, while satisfying the operational constraints of the aircraft (flight autonomy, radio link range, etc). The results were very promising, and a very detailed discussion about the different combinations of these criteria showed that the proposed methodology is a useful decision-making supporting tool.

The main contributions of this papers are: *i)* a novel formulation of the routing problem of a fleet of Dubins vehicles, considering the simultaneous minimization of multiple mission objectives; *ii)* a solution proposal based on GA contextualized to a fleet of mini UAVs; *iii)* a three-step methodology to validate the algorithm.

This paper is organized as follows: Section 2 formulates the problem, while providing the fundamental concepts and the related works. Section 3 details the proposed algorithm to address the routing problem, showing the steps towards the solution. Section 4 will presents the experimental results. The last sections the authors conclude the paper.

2   The Multi-objective Dubins Vehicle Routing Problem (MDVRP)

The Vehicle Routing Problem (VRP) was first introduced by [13]. It is a NP-hard combinatorial optimization problem that, in summary, consists on providing an optimal route for a fleet of vehicles to attend a set of clients with known demands. Although the problem supports multiples depots, usually, it assumes that there is one coincident start and end points. In this context, each vehicle begins the route from a given point, attends the clients and returns to the same point. The overall goal is to visit every and each client just once in a way that the associated cost for all vehicles is minimized. This problem is clearly a generalization of the well-known traveling salesman problem (TSP) for the case where there are more than one seller, or, in other words, the TSP is basically the VRP with just one seller. Accordingly, let VRP be defined as an undirected graph $G(V, E)$, where $V = \{0, 1, ..., N\}$ is the set of nodes, and $E = \{(i, j) : i < j\}$ the set of edges. Vertex V_0 represents the starting point for a fleet of m vehicles, while the remaining nodes are related to the clients position (e.g. cities, locations). A non-negative cost (distance or time of traveling) c_{ij} is defined for each element in E. The VRP usually has the objective to minimize the total traveled distance.

The classic VRP does not consider maneuvering constraints, an intrinsic characteristic of all non-holonomic vehicles (those that can not move sideways). In most cases a car-like vehicle can perform the point-to-point route, providing enough space to maneuver. However, for a fixed-wing airplane, although space might not be a problem, maneuvering is an issue as it can not move backward, and reducing the speed to perform a short radius curve without drifting may lead to stall. The Dubins curve [1] is a method to find the minimum path from two starting and ending configurations, while considering the maneuvering constraints of a non-holonomic vehicle with a maximum steering angle and with no backward capabilities. These characteristics are a valid approximation of the cruiser flight of a fixed-wing aircraft (Eq. 1). Therefore, in this work the VRP is adapted to consider the Dubins curve in the planned path.

We apply the algorithm using the same adaptation proposed in [10]. The Dubins path is defined by an initial configuration (x_i, y_i, θ_i) and a final desired configuration (x_f, y_f, θ_f), where, in this case, (x, y) indicates latitude and longitude, respectively, and θ is the airplane heading angle. To differentiate from the classic formulation, the VRP with Dubins vehicles will be denoted as DVRP.

$$\begin{bmatrix} x \\ y \\ \theta \end{bmatrix} = \begin{bmatrix} v\cos(\theta) \\ v\sin(\theta) \\ \frac{v}{\rho}u \end{bmatrix}, \tag{1}$$

v being the longitudinal velocity, ρ the minimum ray of curvature, and u the control input. The velocity is normalized and considered constant, and thus, time and distance depend on the edge cost c_{ij}. For a fixed-wing airplane, edge cost is not just the straight distance between two consecutive clients, but the distance defined by the local Dubins arc $D(c_i, c_j)$.

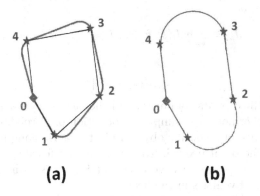

(a)   (b)

Fig. 1. Two Dubins approaches: Left: The AA algorithm; Right: The BAA algorithm

Depending on the heading angle set at two consecutive vertexes, the length of the resulting Dubin arc may be much higher than the direct distance. A common approach is to use the Dubins Alternating Algorithm (AA) [9], which sets the orientation between vertices V_i and V_{i+1} to be given by the line that connects them. This strategy forces the UAV to head to the next vertex as soon as it leaves the current one, however it might promote high accelerations due to the aggressive changing of orientation. Another variant of Dubins, Best Alternating Algorithm (BAA)[10], proposes to use as much line segments as possible in order to minimize the number of arcs. The BAA states that if the i index of the vertex V_i is even, its orientation is based on the orientation of vertex V_{i+1}, otherwise, it will be based on the Vertex V_{i-1}, forcing the alternation of straight lines curves. Figure 1 illustrates the two approaches.

There are a sort of cost functions that can be minimized in the VRP, most of them are linear combinations of the edge costs c_{ij}, and thus, related to the minimization of the traveled distance. In this paper the authors will work with a multi-objective formulation, combining criteria that capture different aspects of the mission that should be optimized simultaneously. In general, given a vector of n objectives $J(\cdot) = [(J_1(\cdot), J_2(\cdot), \cdots J_n(\cdot)]$ evaluated over the set S of all feasible solutions $s \in S$. A multi-objective problem may be stated as:

$$\min_{s} \mathcal{V}(J(s))$$
$$s \in S$$

where the Value Function $\mathcal{V}(\cdot)$ incorporates the decision maker preferences about the Pareto Optimal solutions [12].

Among all existing methodologies to define $\mathcal{V}(\cdot)$, either implicit or explicit, we will use the no-preferred approach based on the difference from the vector $J(s)$ to the ideal vector $J_u = [J_{u1}, J_{u2}, \cdots J_{un}]$ measured by the normalized

Euclidean (l_2) norm, or

$$J_{ui} = \min_{s \in S} J_i(s) \text{ for all } i = 1, \ldots, n, \tag{2}$$

with

$$\mathcal{V}(s) = \|(J(s) - J_u\|^2. \tag{3}$$

Such $\mathcal{V}(\cdot)$ is justified by its simplicity, and the lack of a clear priority among the objectives. Furthermore, the simple introduction of a weighted vector w to modulate the influence of a specific objective in the norm computation will act as a prioritization factor, if needed. To define the set feasible solutions S, let $G(V, E)$ be a graph composed by V vertices and E_{ij} edges with cost c_{ij}, and M be the total number of vehicles in the fleet.

Then, we formulated the the multi-objective problem as follows:

$$\min_{r_k \in G(V,E)} \|(J(r_k) - J_u\|^2$$

$$s.t.$$

$$\sum_{j=1}^{N} \sum_{k=1}^{N} x_{0jk} \leq K \tag{4}$$

$$\sum_{j=1}^{N} x_{0jk} = \sum_{i=1}^{N} x_{i0k} \leq 1 \ (k \in U) \tag{5}$$

$$\sum_{\substack{j=0 \\ j \neq i}}^{N} \sum_{k=1}^{K} x_{ijk} = 1 \ (i \in T) \tag{6}$$

$$\sum_{\substack{i=0 \\ i \neq j}}^{N} \sum_{k=1}^{K} x_{ijk} = 1 \ (j \in T) \tag{7}$$

$$r_k \leq D_k \ (k \in U) \tag{8}$$

where
- $U = \{1, 2, ..., K\}$ is the set of all UAVs;
- $T = \{1, 2, ..., N\}$ is the set of all targets (visiting way-points);
- $V = T \cup \{0\}$ aggregates the starting point 0 (or the depot in the VRP) to the set of visiting way-points;
- Dk is the maximum distance U_k can fly ($k \in U$);
- x_{ijk} are the decision variables, given by:

$$x_{ijk} = \begin{cases} 1 & \text{if } U_k \text{ travels from target } i \in V \text{ to } j \in V, i \neq j \\ 0 & \text{otherwise} \end{cases}$$

and

$$r_k = \sum_{i=1}^{N} \sum_{j=1}^{N} x_{ijk} c_{ij}$$

is the length of the k-th route of the set of K valid routes in graph $G(V, E)$. Constraint (4) ensures that there are no more than K routes out of the base. Constraint (5) guarantees that every UAV that leaves the base also returns to the base at the end of the tour. Constraints (6) and (7) ensure that all and each target is visited only once. Constraint (8) ensures that the maximum distance traveled by each UAV will not be exceeded.

In this work, we will apply Genetic Algorithm to solve this multi-criteria DVRP optimization problem, named here as MDVRP.

3   The MDVRP via Genetic Algorithm

Basically, a Genetic Algorithm (GA) is an interactive procedure that maintains a population of individual (in our case, an individual is the joint routes for the UAV fleet to overfly all giving way-points only once) which represents candidates of solutions for a given problem. GA is a powerful tool to solve high complex optimization problem and high complex formulation. The method is based on a mathematic function to generate a diverse population of individuals, either by changing its values or attributes or by combining two individuals into a new one. This new individual has a different configuration, and the Dubins path changed. GA measures the adaptation of each individuals to the problem by a fitness function. The best adapted individual is assigned as the solution of the problem. The GA algorithm derived for this MDVRP problem was based on [4], [5], [6], and is explained in the sequence.

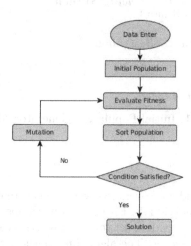

Fig. 2. Algorithm GA

3.1 Initial Population

An important factor for a fast convergence in any GA algorithm is the diversification of the initial population. However, diversification implies computational cost. Given this trade-off between good initial solutions and added cost, the authors proposed an implementation based on the interactive improvement of the solution. Given a randomly generated initial solution for a fleet composed by one UAV, the algorithm consists in improving it interactively by a mutation function. The solutions are recorded and then the best one, according fitness values, is selected and added to the Initial Population set. This Local Search technique, when applied along with GA is known as Hybrid Algorithm [14].

initialization;
l ← desired maximum list lenght;
n ← number of tweaks desired to sample the gradient;
S ← some initial candidate solution Best **repeat**
    if $Lenght(L) > l$ **then**
      | Remove oldest element from L
    end
    R ← Tweak(Copy(S)) BAA(R) **for** $n - 1 times$ **do**
      W ← Tweak(Copy(S)) BAA(W) **if**
      $W \notin Land(Quality(W) > Quality(R) or R \in L)$ **then**
        | R ← W
      end
    end
    if $R \notin LandQuality(R) > Quality(S)$ **then**
      | S ← R Enqueue R into L
    end
    if $Quality(S) > Quality(Best)$ **then**
      | Best ← S
    end
    return Best
until *we have run out of time*;

Algorithm 1. Initial Population Generator Algorithm

3.2 Elitism

GA utilizes mutations and crossovers to diverse the current population, generating and eliminate new individuals. Elitism is a mechanism to preserve the best current solutions for the next data set (or generation), avoiding they are deleted during mutation and crossover. The algorithm to select the best individuals is very simple. One set what percentage $b\%$ of the current population will compose the elite set. Then, the algorithm moves the top $b\%$ of best adapted individuals according to its fitness value to the next Population P_{i+1}.

3.3   Mutation and Crossover

Mutation is a diversification by changing characteristics of one or more individuals of the current population, while crossover is a diversification of the next population creating new individuals utilizing two or more individuals of the current population. After these steps, AA or BAA algorithm is used to generate a new initial and final configuration at two consecutive way-points to evaluate the new route distance based on the Dubin arc length. In this paper, the authors are using a probability factor to decide between crossover and mutation operation, described in [16].

Mutation. Mutation is the step used by GAs to create new solutions on population, producing leap in solution area to avoid local and global maxima. Inversion and Swap are example of techniques to avoid local maximum, insertion and displacement, on the other hand, avoid global maximum. In this work the authors defined three Mutation operators.

- Swap: two way-points c_1 and c_2 of any route individual I are randomly chosen and have their visiting order interchanged. The mutation pair do not need to belong to the same route. Swap is defined as: $\{c_1, c_2 \in S \mid c_1 \neq c_2\}$;
- Insertion: an way-point is randomly selected and inserted in another position of a randomly selected route. The probability to create a new route with only one way-point is $\frac{1}{2*V}$, where V is the number of UAVs in the current solution;
- Inversion: a route segment of an individual is selected and the visiting order is inverted;
- Displacement: a route segment of an individual is selected, and it is inserted in another position (keeping the order), either in the same individual or in a different one. This operator is similar to the crossiover operator (detailed next), with the difference that the insertion is random, while crossover performs the insertion such that to minimize the total cost after the operator. The probability to create new routes by this operator is not clear. Recent approaches utilizes the probability of $\frac{1}{2*V}$, where V is the number of vehicles in the current solution.

Crossover. As mentioned before, this operator is responsible to generate one or more new individuals (or route solutions) by combining two or more from the current population. That is, given an individual I_i and another individual I_j, one can obtain a set of child solution $C = C_i, ..., C_n$. The fitness value for the child solution may be better or worse when compared to its parents values. In this work this comparison is not performed, and all children will be used to form the next population. As one of the problems objective is to minimize the number of UAVs, the crossover algorithm follows some steps to generate new individuals:

- Find the largest route between the parents;
- Eliminate the way-points already visited;

- Select new routes: New routes are inserted on different routes to generate a new one. When a route of a given UAV has already way-points (i.e. it is already being used), the first way-point of the inserted route is positioned just after the closest way-point of its current set.

3.4   Fitness: A Multi-objective Formulation

The Fitness function is the equivalent of the Value Function of Eq. (3). In GA, the Fitness function evaluates how adapted (or fitted) an individual is to the problem. The fitness is measured by each individual of the current population and its value reflects the quality of the current solution, and the best fitted solution will be assigned as the problem solution. Therefore, the selection of the Fitness function became critical to the success of the GA in providing a satisfactory solution for the MDVR Problem. In this work we will define three criteria:

- J_1: Total mission flight distance, defined as the sum of the lengths of all routes in a giving solution. This criterion is the most used in the routing problem and it is related with the resources spent to accomplish the mission;
- J_2: Mission execution time. For a multiple-vehicle formulation, the mission finishes when the last vehicle returns to the basis. Therefore, the mission execution time is approximated to the longest route (considering all vehicles flying as the same constant speed). The minimization of the execution time is a key objective when time is critical, like in rescuing missions, surveillance, fire detection, among others;
- J_3: The number of vehicles launched to perform the mission. In some real scenarios, like in the Amazon rainforest, it is important to use as fewer vehicles as possible for at least two reasons: *i)* to reduce the risks for the fleet, as in the deep jungle it is virtually impossible to recover a missing mini UAV; *ii)* to reduce the readiness time for the next mission. In having part of the fleet ready to launch, a second mission could be performed before the UAVs used in the previous missions are ready to fly again.

Their respective mathematical representations are given by:

$$J_1 = \sum_{i=1}^{N}\sum_{j=1}^{N}\sum_{k=1}^{K} r_{ijk}, \ J_2 = \max_{k \in K} \sum_{i=1}^{N}\sum_{j=1}^{N} r_{ijk}, \ J_3 = \sum_{j=1}^{N}\sum_{k=1}^{K} x_{0jk} \qquad (9)$$

As mentioned in Section 2, the ideal vector J_u is a non-reachable objective vector composed by the minimum of each criterion individually, as if it was the only one to be optimized (Eq. (2)). This ideal vector will be set as reference to evaluate how far the current solution is from it. The squared of the Euclidean metric will be used in the Fitness function. To avoid dimensional issues, we normalized the distance by its respective ideal value. We notice that there is a clear conflict between J_1 and J_2 and between J_2 and J_3. The conflicting nature between J_1 and J_2 can be understood by considering that, to minimize the mission execution time, less way-points will be applied to each vehicles, and then, more vehicles will have to be launched to visit all way-points, and the total route length will increase. Concerning J_2 and J_3, as the number of applied UAVs decreases, more way-points will be assigned to the remaining vehicles, increasing the mission execution time. Finally, given that, in this application, launching and arrival points are the same, both J_1 and J_3 promote the minimization of the number of vehicles, not being clearly conflicting. Furthermore, given that the number of UAVs in the fleet is just four vehicles, we did not to include the criteria J_3 explicitly in the fitness function, but rather, we considered it after solving the bi-objective problem formed by J_1 and J_2 (see next Section for more details). It is also worth to mention that, under the assumption of a constant cruiser speed, the bi-objective formulation of J_1 vs. J_2 is correlated to the classical time vs. energy problem. Thus, the Fitness function $Fit(P_i)$ for the MDVRP is given by:

$$Fit(P_i) = \|J_d\|^2 = w_1 J_{d1}^2 + w_2 J_{d2}^2 \tag{10}$$

where,

$$J_{d1} = (J_1 - J_{u1})/J_{u1} \tag{11}$$

is the normalized deviation from the value J_{u1}, the route that ideally uses less resources.

$$J_{d2} = (J_2 - J_{u2})/J_{u2} \tag{12}$$

is the normalized deviation of J_2, the mission execution time (the time to cover the longest route) to the ideal value J_{u2} (time to visit the furthest client). And, w_1 and w_2 are weights to reflect the relative importance between the criteria. We will use the convex combination $w_2 = (1 - w_1), w_1 \in [0, 1]$

4   Experiments

This work is part of the mission planner subsystem of an unmanned aircraft system based on hand-launched fixed-wing UAVs carrying a digital camera pointed to the ground. The following three incremental steps methodology, also detailed in the next sections, were used to validate the results:

1. Validate the core of the GA method: The GA implementation, and its diversification functions, was validated by solving the classical VRP using the benchmark presented in (Augerat, 1995);
2. Validate the BAA adaptation: The GA implementation was then used to solve BAA adaptation of the DVRP for the same set of clients . As the arcs will increase the route length, the solution of the VRP are expected to be lower bound references for the BAA-DVRP;
3. Validate the MDRVP formulation: The MDRVP was then solved of the multiple criteria problem, combining the criteria in two, generating three two-dimensional trade-off curves (J_1 vs. J_2, J_2 vs. J_3, and J_1 vs. J_3). The method will be validated based on the Pareto-optimality of the solutions that formed the trade-off curve. Additionally, the trade-off curves will be useful to analyze the cross-influence between the criteria.

Table 1. Experiments results of the GA algorithm applied to the instance A-n32-k5, and the average total traveling cost with its iteration

Elistim	Swap	Inversion	Insertion	Displacement	Average Total Cost	Iteration
0.5	0.05	0.05	0.15	0.2	501.80	905.06
0.25	0.15	0.15	0.25	0.3	474.06	817.60
0.25	0.2	0.2	0.3	0.15	475.20	709.76
0.5	0.2	0.2	0.3	0.15	466.63	800.16
0.5	0.3	0.3	0.15	0.2	514.73	831.63
0.5	0.15	0.15	0.2	0.15	484.10	872.80

4.1   Classic VRP Results

The first step is to solve the classical VRP to find a lower bound to the next Step. An well-known instance used to validate is the A-n32-k5, or Set A of (Augerat, 1995). This instance has 32 vertexes, with values of demand ranging from 1 to 24. The vehicle capacity is 100. The solution of this instance is such that the minimum number of vehicles is 5 and the total cost is 784. The GA algorithm was executed using: Population of 500 individuals, 50% elitism, the mutation applied 0.2%, 0.2%, 0.4% of swap, inversion and insertion, respectively. Table 1 shows the progression of operators values. For each setup the algorithm was executed 30 times, and the average values can be anso seen on Table 1. The random parameters are first set as related work and then increase to show the behavior of algorithm in relation these values. As the GA algorithm could find the optimal solution for this classical problem in all 30 trials, the authors considered the GA implementation validated.

4.2   Classic VRP as the Lower Bound for the DVRP

In this step, the cost function of the classic VRP was modified to consider the Dubins curve, using the same method as proposed in [10]. The so-called *BAA* was executed using the same instance, and respecting the same 5 vehicles. After several executions, all solutions were higher than the Classic VRP, as expected. The solution was in average 712.23 or 20% higher than the lower bound.

4.3   Applying the MDVRP to an Actual Scenario

To validate the planner in an actual scenario, we considered the Forest Reserve Adolpho Ducke (Figure 3), an environmental protected area nearby Manaus, the two million inhabitants capital of the Brazilian State of Amazonas. The deforestation pressure around the almost perfect 10 km x 10 km square is so intense that its dark green highlights in the satellite imagery as if it was a consequence of digital mosaicking process, but it is not. In this context, a closed and regular monitoring is fundamental to establish a protection line around the reserve. Although located in the surrounding area of Manaus, the ground access to the limits of the reserve is difficult, specially the north and east sides. Thus, aerial monitoring is mandatory.

Fig. 3. Adolph Ducke reserve. Left: Manaus and the so-called meeting of the waters. The arrow points to the reserve location, northeastern Manaus; Center: Close view at the Reserve surrounding area. The arrow points to north. Right: Missions Way-Points to monitor the limits of the Reserve.

It is important no mention that the authors are evaluating the performance of the route planner. The aspects related to how the UAVs actually executed the plan were considered out of the scope of this paper, and it will be addressed in a future article. Micropilot©autopilot is the main airplane onboard computer, connected via radio to the ground station, an Intel i7 notebook with Micropilot©companion software, Horizon©, installed. The experiments were conducted on the ground station. A map of the area is uploaded to Horizon and the way-points are collected using the software proper function. The selected set of way points is input to the MDVRP that optimizes different routes, according to the algorithm described in the previous section. The resulting routes will be sent back to Horizon©and them uploaded to the autopilot hardware onboard the respective aircraft. The mission considered here is to monitor the limits of the reserve by overflying 16 selected spots, marked as point-of-interests (PoI) distant hundreds of meters from each other. including the center of the reserve (Figure 3). We set the Dubins curvature radius as 500 m, which allows turns with very small roll angles, keeping the camera pointing to the ground. We estimated J_{u1} and J_{u2} by solving its respective mono objective VRP, resulting on $J_{u1} = 57700m$ and $J_{u2} = 28003m$.

We solved MDVRP for $J_1 vs. J_2$ using the Fitness function as given by Eq. (10), varying $w_1 = 0$ to $w_1 = 1$ in a step of 0.1. The algorithm was executed 15 times for each value of w_1. Figure 4 presents the $J_1 x J_2$ solutions, where the small red circles represents the non-dominated solution, small green circle the dominated solution, and the crossed red circle marks the position of the ideal solution. Due to the limitation of space only a summary of the results was presented. In Table 2 one can verify that all frequent solution are non-dominated. In fact, after all 110 times the MDVRP was executed for different values of w_1 up to 75% of the given solutions were non-dominated, considering

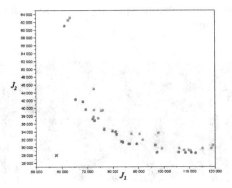

Fig. 4. Non-dominated (red) and dominated (green) solutions of the $J_1 \times J_2$ bi-criteria problem. The red "x" points to the ideal solution.

a population of 500 individuals in the GA method. Interesting to mention that, after increasing the population to 750 individuals (the limit of the computational capacity of our notebook), we could not see a significant improvement on the rate of non-dominated solution.

By analyzing Table 2 the decision maker can use J_3 implicitly to decide among the best fitted solutions. The $\|J_A\|_2$ column presents values considering no preference between J_1 and J_2 ($w_1 = 0.5$), and in this case, given that there is just an small difference from line 1 and 4, and the solution of line 4 uses one aircraft less, it will be the preferable flight plan when the minimization of number of UAV (J_3) applies. In column $\|J_B\|_2$ J_1 is more important ($w_1 = 0.8$), and all solutions lead to the use of 2 UAVs, being the line 7 the best fitted one. Similar analysis can be done by any combination of preference among the criteria without having to recalculate the MDVRP, providing that a good set of non-dominant solution was found. Figure 5 shows the best Dubin path (in terms of $\|J\|_2$ for a given number of UAV, from 1 to 4.

Table 2. Most frequent solutions for J_1 vs. J_2

#	$\|J_A\|_2$ $w_1 = 0.5$	$\|J_B\|_2$ $w_1 = 0.8$	J_1 total lenght	J_2 longer route	J_3 UAVs	freq.
1	0.3997	0.3087	76446.9	34521.5	3	18
2	0.4614	0.4011	83320.2	31514.8	3	25
3	0.4692	0.4092	83869.9	31363.7	3	7
4	0.4096	0.2733	72812.7	36821.7	2	6
5	0.4144	0.2674	72057.9	37283.6	2	7
6	0.4633	0.2591	69289.5	39694.6	2	7
7	0.5265	0.2567	65322.9	42274.5	2	13
8	1.1819	0.5305	61060.5	61060.5	1	6

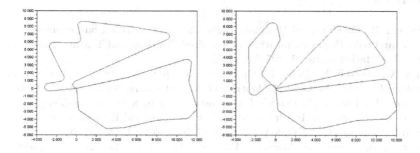

Fig. 5. The routes found by MDVRP for 2 and 3 UAVs

5   Conclusions

This work addressed the problem of finding optimal routes for a fleet of mini UAVs in monitoring missions over environmental protected areas. This hand-launched fixed-wing UAV is being developed by a Brazilian network of researchers and engineers. The main purpose of the present results is to implement the mission planner subsystem. A typical mission has to deal with multiple conflicting objectives, such as to reduce the time to execute the mission, while minimizing the applied resources. To use as fewer UAVs as possible in the mission is also an important criteria. This scenarios leaded to a multi-criteria formulation for the problem.

The method was validated in three steps: first, the GA implementation was compared with a well-known benchmark; second, the VRP adaptation with Dubins curves was validated comparing the results with the classic VRP, and, finally, the multiple objective formulation was validated comparing the results of three objectives organized into three pairs of bi-criteria problems. The proposed methodology passed all three steps.

The application of GA reduced the complexity of the implementation of the algorithm, while provided solutions that, in most cases, were optimal. The bi-criteria formulation promoted a good learning process about the mission profile. Changing the importance weight of one criterion when compared to the other could give to the authors important insights of how to configure the mission for better achieve the overall goals.

The results were promising, as the optimal routes were very consistent with the non-holonomic characteristics of the airplane. The idea of constraint the steering-angle of the lateral movement is not a limitation of the method, as it can be modulated to fit to a better operational condition. A future work is to consider no overlapping routes (to avoid collision).

References

1. Dubins, L.E.: On Curves of Minimal Lenght with a a Constraint on Coverage Curvature, and with Prescribe Initital and Terminal Positions and Tagents. American Journal of Mathematics **79**, 497–516 (1957)
2. Savla, K., Frazzoli, E., Bullo, F.: On the point-to-point and traveling salesperson problems for dubins' vehicle. In: American Control Conference, pp. 786–791 (2005)
3. Machado, P., Tavares, J., Pereira, F.B., Costa, E.: Vehicle routing problem: doing it the evolutionary way. In: Proceedings of the Genetic and Evolutionary Computation Conference (GECCO 2002), July 9–13, New York, USA (2002)
4. Pereira, F.B., Tavares, J., Machado, P., Costa, E.: GVR: a new genetic representation for the vehicle routing problem. In: O'Neill, M., Sutcliffe, R.F.E., Ryan, C., Eaton, M., Griffith, N.J.L. (eds.) AICS 2002. LNCS (LNAI), vol. 2464, pp. 95–102. Springer, Heidelberg (2002)
5. Tavares, J., Pereira, F.B., Machado, P., Costa, E.: Crossover and diversity a study about GVR. In: Proceedings of the Analysis and Design of Representations and Operators (ADoRo-2003) a Bird-of-a-Feather Workshop at the 2003 Genetic and Evolutionary Computation Conference (GECCO 2003), July 12–16, Chicago, Illinois, USA (2003)
6. Potvin, J.-Y.: State-of-the art review - evolutionary algorithms for vehicle routing. INFORMS Journal on Computing **21**(4), 518–548 (2009)
7. Pohl, A., Lamont, G.: Multi-objective uav mission planning using evolutionary computation. In: Em Winter Simulation Conference, WSC 2008, pp. 1268–1279 (2008)
8. Deb, K.: Multi-Objective Optimization using Evolutionary Algorithms. Wiley (2002)
9. Isaacs, J.T., Klein, D.J., Hespanha, J.P.: Algorithms for the traveling salesman problem with neighborhoods involving a dubins vehicle. In: Em Proc. of the: Amer. Contr. Conf. (2011)
10. Guimaraes Macharet, D., Alves Neto, A., da Camara Neto, V.F., Campos, M.F.M.: Data gathering tour optimization for dubins vehicles. In: The IEEE Congress on Evolutionary Computation, (CeC 2012), Brisbane, Australia (2012)
11. Obermeyer, K.J.: Path planning for a UAV performing reconnaissance of static ground targets in terrain. In: Em AIAA Conf. on Guidance, Navigation and Control, Chicago, IL, USA (2009)
12. Carvalho, J.R.H., Ferreira, P.A.V.: Multiple-Criteria Control: A Convex Programming Approach, (extended version). Automatica **31**(7), 1025–1029 (1995)
13. Dantzig, G.B., Ramser, J.H.: The truck dispatching problem. Management Science **6**, 80–91 (1959)
14. de Oliveira, H.C.B., Alexandrino, J.L., de Souza, M.M.: Memetic and genetic algorithms: A comparison among different approaches to solve vehicle routing problem with time windows. In: International Conference non Hybrid Intelligent Systems, 55 (2006)
15. Yu, X., Hung, J.Y.: A genetic algorithm for the dubins traveling salesman problem. In: IEEE International Symposium on Industrial Electronics (ISIE 2012), Hangzhou, pp. 1256–1261 (2012)
16. Yan, X.S., Li, H., et al.: A fast evolutionary algorithm for combinatorial optimization problems. In: Proceedings of the Fourth International Conference on Machine Learning and Cybernetics, pp. 3288–3292. IEEE Press (2005)

An Interactive Evolutionary Multiobjective Optimization Method: Interactive WASF-GA

Ana B. Ruiz[1]([⊠]), Mariano Luque[1], Kaisa Miettinen[2], and Rubén Saborido[3]

[1] Department of Applied Economics (Mathematics),
Universidad de Málaga, Málaga, Spain
{abruiz,mluque}@uma.es

[2] Department of Mathematical Information Technology,
University of Jyvaskyla, Jyvaskyla, Finland
kaisa.miettinen@jyu.fi

[3] Polytechnique Montréal Researchers in Software Engineering,
École Polytechnique de Montréal, Montreal, Canada
ruben.saborido-infantes@polymtl.ca

Abstract. In this paper, we describe an interactive evolutionary algorithm called *Interactive WASF-GA* to solve multiobjective optimization problems. This algorithm is based on a preference-based evolutionary multiobjective optimization algorithm called WASF-GA. In Interactive WASF-GA, a decision maker provides preference information at each iteration simply as a reference point consisting of desirable objective function values and the number of solutions to be compared. Using this information, the desired number of solutions is generated to represent the region of interest of the Pareto optimal front associated to the reference point given. Interactive WASF-GA implies a much lower computational cost than the original WASF-GA because it generates a small number of solutions. This speeds up the convergence of the algorithm, making it suitable for many decision-making problems. Its efficiency and usefulness is demonstrated with a five-objective optimization problem.

Keywords: Multiobjective programming · Pareto optimal solutions · Reference point approach · Interactive methods · Evolutionary algorithms

1 Introduction

Many real-world applications arising in e.g. engineering involve solving multiobjective optimization problems where several conflicting objectives must be optimized over a set of feasible solutions. In many occasions, these problems can be complex to solve because they deal with different types of functions (nonlinear, nondifferentiable, discontinuous, etc.) and different types of variables (continuous, integer, binary, etc.). They may even involve black-box functions, whose computational cost can be high.

Commonly, there is no solution where all the objectives can reach their individual optima and we look for so-called *Pareto optimal solutions*. These solutions

© Springer International Publishing Switzerland 2015
A. Gaspar-Cunha et al. (Eds.): EMO 2015, Part II, LNCS 9019, pp. 249–263, 2015.
DOI: 10.1007/978-3-319-15892-1_17

are defined as solutions where an improvement of any objective always implies a sacrifice in at least one of the others. The set of Pareto optimal solutions is called the *Pareto optimal set* and its image in the objective space is known as the *Pareto optimal front*. A *decision maker (DM)*, a person who is interested in solving the problem, decides which Pareto optimal solution best satisfies his/her preferences and this solution is commonly known as *the most preferred solution*.

There exists a great amount of methods to deal with multiobjective optimization problems in the literature. On the one hand, interactive Multiple Criteria Decision Making (MCDM) methods are widely used due to the gradual incorporation of the DM's preferences into the solution process in order to generate one or a small set of Pareto optimal solutions according to these preferences [13,16]. On the other hand, during the last decades, Evolutionary Multiobjective Optimization (EMO) algorithms have become very popular for solving different types of problems [1,2]. Their main aim is the approximation of the whole Pareto optimal front. However, although knowing the ranges of the objectives functions and the conflict degree among them can be of great help for having a good knowledge of the problem itself, the task of identifying a single preferred Pareto optimal solution that pleases the DM may not be easy. Also, approximating the whole Pareto optimal front may be impossible in e.g. large scale or computationally complex problems. These difficulties can be managed by considering an interactive method that uses tools from an EMO algorithm. To be more precise, one can incorporate preference information into EMO algorithms to overcome various (computational and cognitive) challenges [8].

Some interactive EMO methods have been proposed in the literature, including the following ones. The Reference-Point-Based NSGA-II (R-NSGA-II) proposed in [5] modifies NSGA-II [4] as follows. According to one or several reference points given by a DM, the crowding distance used in NSGA-II is replaced by a preference distance, which equally emphasizes objective vectors that are close to any of the reference points with respect to the Euclidean distance. In [21] an interactive EMO method called the Preference Based Evolutionary Algorithm (PBEA) was proposed, which modifies the EMO algorithm IBEA [27]. PBEA allows the DM to interactively give reference points, with which the binary quality indicator of IBEA (which measures the minimal distance by which an individual needs to be improved in each objective to become nondominated) is redefined using an achievement scalarizing function [24] from MCDM. A Preference-based Interactive Evolutionary (PIE) algorithm was proposed in [18]. Starting from a solution selected from a randomly generated population or from a reference point, PIE progressively improves the objective function values by minimizing an achievement scalarizing function [15] at each iteration using a single-objective evolutionary algorithm. The DM guides interactively the algorithm by deciding from which solution, at which distance from the Pareto optimal front and in which direction the search for the next solutions is continued. iMOEA/D [9] is an interactive version of the well-known MOEA/D method [25], where a set of solutions is shown to the DM at intermediate generations, who must choose one of them. Then, the search is guided to the neighbourhood of the selected solution by relocating the weight vectors which determine the search directions.

In [23], an interactive EMO method called iPICEA-G , which is based on the PICEA-G algorithm (Preference-Inspired Co-Evolutionary Algorithm) [22], was proposed. In this method, the DM's preferences can be given either as a search direction or as a reference point. In the former case, the DM has to indicate the importance (s)he gives to each objective function and an angle between 0 and $\Pi/2$ which determine the search range. This kind of information may be difficult to provide for the DM. In the case of a reference point, all objectives are given the same importance and the search range is set according to the number of objective functions. In [19], an interactive evolutionary algorithm was suggested which tries to find the most preferred solution with a limited number iterations expecting DM's involvement. The preference information is given by choosing a desirable solution among a set of solutions.

Whatever algorithm is used, the final purpose of solving any multiobjective optimization problem is that the DM can find her/his most preferred solution. Thus, once a set of solutions that approximates the Pareto optimal front is found, we cannot overlook the decision making phase in which the DM must make an adequate decision to choose the final solution. Obviously, the DM plays an active role in the process and an interactive method is supposed to be appealing and acceptable to her/him because (s)he is involved in the process. However, it is important to consider several issues. On the one hand, for decision making purposes, only a few solutions must be analysed by the DM in order not to overwhelm her/him. Comparing too many solutions may be difficult in the presence of a high number of objectives. On the other hand, asking preference information in a format as simple as possible is very important since it makes the interactive process more meaningful. Besides, if the DM feels that the solutions obtained reflect well enough her/his wishes, and they are improved progressively, (s)he is more motivated and it is more likely that (s)he wants to keep on iterating until the most preferred solution is found.

Based on this, in this paper, we concentrate on the decision making phase and the interaction with the DM necessary for solving any multiobjective optimization problem. Taking into account the previous ideas, we propose an interactive EMO method that generates a small set of solutions at each iteration and which needs from the DM preference information which is not cognitively demanding. The interactive method proposed is called *Interactive WASF-GA* and it is an interactive version of a preference-based EMO algorithm called WASF-GA [17]. At each iteration of Interactive WASF-GA, the DM indicates the number of solutions to be compared and a reference point containing aspiration levels, that is, objective function values that are desirable. According to this, WASF-GA is executed iteratively to generate the desired number of solutions in the region of interest defined by the given reference point. However, we do not only propose an interactive algorithm, but we also suggest a user interface aimed at enhancing the interaction with DM when solving a problem with Interactive WASF-GA.

The rest of this paper is organized as follows. In Section 2, we introduce the main concepts and notations used, including a brief overview of the WASF-GA algorithm. Interactive WASF-GA is motivated and described in Section 3, where we also carry out a comparative analysis with respect to other interactive EMO algorithms. In Section 4, a computational implementation is described, showing the graphical user interface proposed and the solution process of a five-objective optimization problem. Finally, conclusions are drawn in Section 5.

2   Formulation and Background Concepts

2.1   Concepts and Notation

A general *multiobjective optimization problem* is defined by

$$
\begin{aligned}
\text{minimize} \quad & \{f_1(\mathbf{x}), \dots, f_k(\mathbf{x})\} \\
\text{subject to} \quad & \mathbf{x} \in S,
\end{aligned}
\tag{1}
$$

where $f_i : S \to \mathbf{R}$, for $i = 1, \dots, k$ ($k \geq 2$) are the *objective functions* that we wish to optimize (to minimize in our case) simultaneously. The decision variables $\mathbf{x} = (x_1, \dots, x_n)^T$ are referred to as *solutions* or *decision vectors* and they belong to $S \subset \mathbf{R}^n$, called the feasible set. The images of the solutions $\mathbf{f}(\mathbf{x}) = (f_1(\mathbf{x}), \dots, f_k(\mathbf{x}))^T$ are called *objective vectors*. The image of the feasible set in the *objective space* $\mathbf{R}^k$ is called the *feasible objective region* $Z = \mathbf{f}(S)$.

Since, in the presence of conflicting objective functions, it is not possible to find a solution where all the objectives can reach their individual optima, there exist solutions that are mathematically incomparable. In these solutions, no objective function can be improved without deteriorating at least one of the others. A solution $\mathbf{x} \in S$ is said to be *Pareto optimal* if there does not exist another $\mathbf{x}' \in S$ such that $f_i(\mathbf{x}') \leq f_i(\mathbf{x})$ for all $i = 1, \dots, k$ and $f_j(\mathbf{x}') < f_j(\mathbf{x})$ for at least one index j. The corresponding objective vector $\mathbf{f}(\mathbf{x})$ is called a *Pareto optimal objective vector*. The set of all Pareto optimal solutions is called a *Pareto optimal set*, denoted by E, and the set of all Pareto optimal objective vectors is called a *Pareto optimal front*, denoted by $\mathbf{f}(E)$.

Given two objective vectors $\mathbf{z}, \mathbf{z}' \in Z$, we say that $\mathbf{z}$ *dominates* $\mathbf{z}'$ if and only if $z_i \leq z_i'$ for all $i = 1, \dots, k$, with at least one strict inequality. In the context of EMO algorithms, we refer to a *nondominated set* as a set of solutions whose objective vectors are not dominated by the objective vector corresponding to any other solution in the set.

The *ideal objective vector* and the *nadir objective vector* are defined, respectively, as $\mathbf{z}^\star = (z_1^\star, \dots, z_k^\star)^T$ such that $z_i^\star = \min_{\mathbf{x} \in E} f_i(\mathbf{x})$ ($i = 1, \dots, k$), and as $\mathbf{z}^{nad} = (z_1^{nad}, \dots, z_k^{nad})^T$ such that $z_i^{nad} = \max_{\mathbf{x} \in E} f_i(\mathbf{x})$ ($i = 1, \dots, k$). That is, the ideal and the nadir values are, respectively, the best and the worst values that each objective function can achieve in the Pareto optimal front (that is, they define lower and upper bounds for the objective functions). While the ideal objective vector can be easily obtained, the nadir objective vector is, in general, more difficult to calculate and typically we need to settle for approximations [3,20]. In what follows, we assume that the Pareto optimal front is bounded and that there are available estimations of the ranges of the objective function values.

From the mathematical point of view, all Pareto optimal solutions can be regarded as equally desirable and we need information about the preferences of a DM to identify one as the final solution to be implemented [13]. A natural way to express preferences consists of specifying desirable objective function values, which constitute the components of the so-called reference point. A *reference point* is given by $\mathbf{q} = (q_1, \ldots, q_k)^T$, where q_i is an aspiration level for the objective function f_i provided by the DM, for all $i = 1, \ldots, k$. Usually, $\mathbf{q}$ is said to be *achievable* for (1) if $\mathbf{q} \in Z + \mathbb{R}_+^k$ (where $\mathbb{R}_+^k = \{\mathbf{y} \in \mathbb{R}^k \mid y_i \geq 0 \text{ for } i = 1, \ldots, k\}$), that is, if either $\mathbf{q} \in Z$ or $\mathbf{q}$ is dominated by some Pareto optimal objective vector. Otherwise, the reference point is said to be *unachievable*, that is, for an unachievable reference point, all components cannot be achieved simultaneously (in some situations, a reference point is unachievable because some components cannot be achieved although other ones can be attained).

Once a reference point is given, a so-called *achievement (scalarizing) function* (ASF) [24] can be minimized over the feasible set in order to find the Pareto optimal solution which best fits the reference point. These functions combine the original objective functions with the preferences of the DM into a scalar valued function. For an overview about ASFs, see [15].

2.2 WASF-GA Algorithm

As previously mentioned, the interactive method we proposed is based on the preference-based EMO algorithm called WASF-GA [17]. This algorithm tries to approximate the region of interest of the Pareto optimal front defined by a reference point $\mathbf{q}$ given by a DM. In [17], the *region of interest of the Pareto optimal front associated to* $\mathbf{q}$ is defined as followed. When $\mathbf{q}$ is achievable, the region of interest is the subset of Pareto optimal objective vectors $\mathbf{f}(\mathbf{x})$, with $\mathbf{x} \in E$, which verify that $f_i(\mathbf{x}) \leq q_i$, for every $i = 1, \ldots, k$. On the other hand, if $\mathbf{q}$ is unachievable, the region of interest is formed by the Pareto optimal objective vectors $\mathbf{f}(\mathbf{x})$, with $\mathbf{x} \in E$, which verify that $f_i(\mathbf{x}) \geq q_i$, for every $i = 1, \ldots, k$. Therefore, in the achievable case, this region of interest contains all the Pareto optimal solutions which dominate the reference point and, thus, which are the most interesting solutions for the DM. In the unachievable case, the region of interest is formed by the Pareto optimal solutions which are dominated by the reference point. In this case, solutions lying in this region are likely to be more appealing for the DM than the ones outside it because, at them, the objective function values differ from the aspiration values as little as possible, although they do not improve any of them. The solutions outside this region may improve some of the aspiration levels (and not all of them) but at the expense of a sacrifice in the rest of reference levels, what may not be so attractive for the DM.

To approximate the region of interest, WASF-GA maintains a diverse set of nondominated solutions by considering, on the one hand, a predefined set of weight vectors in the weight vector space $(0,1)^k$ (let us consider N_μ vectors of weights) and, on the other hand, by minimizing at each generation the ASF proposed by Wierzbicki in [24] for the reference point given. Roughly speaking, at each generation of WASF-GA, parents and offspring are classified into several fronts.

This classification is done according to the values that each individual takes on the ASF, for the reference point and for each of the weight vectors in the set. To be more precise, the first front is formed by the solutions which reach the lowest value of the ASF for each of the N_μ weight vectors; the second front is constituted by the individuals with the next lowest value of the ASF for each of the N_μ weight vectors, and so on until every individual has been included into some front. Afterwards, the solutions which are passed to the next generation are those in the lower level fronts until completing the new population. The solutions selected can be considered as the best individuals at the current generation for minimizing the ASF with respect to the weight vectors considered. The outcome of WASF-GA is the first front of the last generation, which has N_μ individuals. From the practical point of view, the region of interest is approximated by projecting the reference point onto the Pareto optimal front in different ways, by using the set of projection directions (or search directions) defined by the inverses of the N_μ weight vectors considered.

Figure 1 gives a graphical idea of the working procedure of WASF-GA in a biobjective optimization problem. The region of interest in the Pareto optimal front has been highlighted with a bold line in both cases, and the arrows represent the projection directions determined by a set of weight vectors. It can be seen that, by varying the weight vectors and by emphasizing at each generation those individuals which minimize the ASF for each weight vector, the region of interest can be approximated by projecting the reference point onto the Pareto optimal front using several projection directions, for both unachievable and achievable reference points. For more details about WASF-GA, see [17].

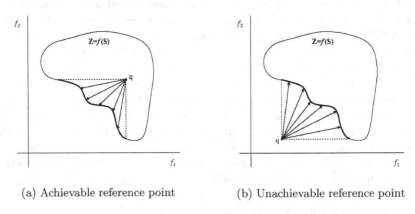

(a) Achievable reference point   (b) Unachievable reference point

Fig. 1. Idea of the working procedure of WASF-GA

3   Interactive WASF-GA

Based on the success of interactive MCDM methods, we propose a new interactive method using a preference-based EMO algorithm. In Interactive WASF-GA, the preference information indicated by the DM at each iteration it are aspiration levels for the objective functions, which determine a reference point denoted by q^{it}, and the number of solutions (s)he wants to compare, denoted by N_S^{it}. The set of new solutions is generated by applying the WASF-GA algorithm using as many weight vectors as the number of solutions indicated by the DM, that is, $N_\mu = N_S^{it}$ at each iteration. Let us denote by $\mu^{it,j}$ the weights vectors used at iteration it, for $j = 1, \ldots, N_S^{it}$.

3.1 Algorithm of Interactive WASF-GA

The steps of the Interactive WASF-GA are the following ones:

Step 1. Initialization. Initialize $it = 1$.

Step 2. Preference information I. If $it > 1$ and the DM wants to generate new solutions using the previous reference point, set $q^{it} = q^{it-1}$. Otherwise, ask the DM to specify a reference point q and set $q^{it} = q$.

Step 3. Preference information II. Ask the DM how many solutions (s)he would like to see, N_S^{it}. If $it > 1$ and $N_S^{it} = N_S^{it-1}$, set $\mu^{it,j} = \mu^{it-1,j}$ for every $j = 1, \ldots, N_S^{it}$ and go to Step 5. Otherwise, continue.

Step 4. Generation of the weight vectors. Following the procedure described in [17], generate N_S^{it} weight vectors, denoted by $\mu^{it,j}$ for $j = 1, \ldots, N_S^{it}$.

Step 5. Generation of solutions. Generate N_S^{it} solutions with the WASF-GA algorithm using the set of weight vectors $\mu^{it,j}$ for $j = 1, \ldots, N_S^{it}$, and show the solutions to the DM.

Step 6. Termination rule. Ask the DM to select the most preferred of the N_S^{it} solutions and denote it by x^{it}. If the DM wishes to *Stop*, the solution process concludes with x^{it} as the final solution and $f(x^{it})$ as a final objective vector. Otherwise, set $it = it + 1$ and go to Step 2.

Next, let us make some remarks about some aspects of the algorithm:

- When $it = 1$, the DM must give a reference point in Step 2 because no reference point was provided previously.
- The number of solutions to be shown to the DM can be changed at each iteration in Step 3, but (s)he can alternatively maintain the same number along several iterations. In that case, the same set of weight vectors can be used through these iterations and only the reference point changes.
- When the DM decides to generate new solutions using the same reference point, only the N_S^{it} weight vectors must be recalculated. Given that the procedure described in [17] generates an initial large number of vectors from which the weight vectors needed are selected, the new N_S^{it} weight vectors can be obtained using again the same initial vectors to reduce the computational effort. Furthermore, the weight vectors that were already used at the previous

iteration must be internally removed in order to assure that different solutions are provided to the DM.

- The final population generated at one iteration can be used as the initial population at the next iteration, which allows to accelerate the speed of the solution process. This increases the convergence speed of the algorithm since the initial population is already close to the Pareto optimal front.
- In order to guarantee at least local Pareto optimality of the final solution, the last solution chosen by the DM can be locally improved by minimizing the ASF proposed in [24] using the objective function values achieved by this solution as the reference point, with some local optimization method.

3.2   Comparative Analysis

In order to compare Interactive WASF-GA with some of the reference point-based interactive EMO algorithms mentioned in Section 1, we present Table 1, which summarizes the main features of each algorithm. Next, we detail the information given on each column and, as an example, we explain this information for the Reference-Point-Based NSGA-II algorithm [5]. In this algorithm, at each iteration, the DM must specify one or several reference points, which is indicated in the *'Preference information'* column. At each iteration with preferences, the outcome population shown to the DM consists of individuals in the first nondominated front of the last generation. This is indicated in the *'Solutions shown to the DM'* column. The *'Computational cost'* column contains the complexity of the basic operations of each algorithm in one iteration, considering their worst cases. In this column, k represents the number of objective functions and N is the population size used. For the Reference-Point-Based NSGA-II algorithm, we have used only one reference point and we have taken into account the computational cost needed for carrying out the nondominated sorting procedure, the preference distance assignment, the preference distance sorting and the ϵ-based selection strategy (see [5]). Finally, if the algorithm needs to set any additional parameters to be executed, they are indicated in the last column. In the example considered, a value for ϵ is necessary to compute the niching operator.

Let us now analyse Table 1. Firstly, from the cognitive point of view, the preference information required from the DM in Interactive WASF-GA is very simple compared to some of the other methods. For example, in the PIE algorithm or in iPICEA-G, the DM is asked for the percentage of distance to the (unknown) Pareto optimal front or for a search angle, respectively. This type of information may not be easy to understand by the DM. Secondly, as it can be seen in the third column, the only method that generates exactly the number of solutions the DM wants to see is Interactive WASF-GA. Except from PIE (which generates one solution at each iteration) and i-MOEA/D, the rest of methods show the solutions generated at the last generation. Consequently, the number of solutions shown to the DM may be too high for making a fair comparison and cannot be known beforehand. Besides, Interactive WASF-GA shows nondominated solutions which approximate the region of interest, instead of showing nondominated solutions generated at intermediate generations, as in i-MOEA/D. This may be seen as a

Table 1. Comparison of several methods

At each iteration	Preference information	Solutions shown to the DM	Computational cost	Additional parameters
Interactive WASF-GA	A reference point and the number of solutions to be compared (N_S)	The N_S solutions generated at the last generation	$O(k \cdot N \cdot N_S)$	No
Reference-Point-Based NSGA-II [5]	One or several reference points	First nondominated front of the last generation	$O(k \cdot N^2)$	To control the extent of solutions, an ϵ-clearing idea is used in the niching operator
PBEA [21]	A reference point	Population of the last generation	$O(k \cdot N^2)$	The extent of solutions is controlled by an operator δ
PIE [18]	Preferential weights, a reference point and the distance to the Pareto optimal front	A solution at the distance indicated to the Pareto optimal front	$O(k \cdot N)$	If the DM wants to investigate solutions previously obtained, (s)he must indicate the number of solutions to be shown
iMOEA/D [9]	The number of solutions to be shown and choosing one solution among a set of solutions	Solutions at intermediate generations	$O(k \cdot N^2)$	The number of iterations to be taken and a reduction factor of the preferred region
iPECEA-G [23]	A reference point or a search direction with a search angle	Population of the last generation	$O(k \cdot N^2)$	Search angle to control the extent of solutions

strength of our algorithm, since the solutions found at intermediate generations may be still far from the region of interest and may not give a good idea about the real trade-offs among the objectives in this region. Thirdly, regarding the 'Computational cost' column, the algorithm proposed has a much lower computational cost than those needed by other algorithms given that the number of solutions the DM wants to compare in Interactive WASF-GA, denoted by N_S, is expected to be much lower than the population size. The only method with a lower computational cost than Interactive WASF-GA is PIE because this method internally solves a single-objective (scalarized) optimization problem with a single-objective algorithm instead of solving the multiobjective optimization problem itself. Finally, it is worthy to mention that Interactive WASF-GA does not need to set any additional parameter during the solution process, while the other algorithms do require some (see last column). In most of them, these additional parameters control the extent of solutions in the region approximated in the Pareto optimal front and, consequently, they affect the outcome of the algorithm.

The previous analysis highlights that, in comparison with some of the state-of-art interactive EMO algorithms, Interactive WASF-GA requires very simple preference information from the DM and it is able to generate exactly the number of solutions the DM wants to see in the region of interest. Besides, its computational cost is quite limited and it does not need to set any additional parameter during the solution process. As shortcomings of Interactive WASF-GA, we can say that, on the one hand, the distribution of the N_S weight vectors influences the distribution of the solutions generated and, thus, special emphasis must be taken for using weights which produce well-distributed projection directions [17]. This may be overcome by producing a large number of weight vectors (e.g., 100 or more) and then using the k-means clustering [12] to select the N_S most representative ones.

On the other hand, one may think that it may be not assured that the DM is shown exactly N_S solutions because minimizing the ASF using different weight vectors does not assure to generate different Pareto optimal solutions (for example, in problems with discontinuous Pareto optimal fronts). In order to avoid such a situation, more than N_S weight vectors may be used in WASF-GA, e.g. we can use $N_S^* = 2 \cdot N_S$ vectors. In this way, more solutions are generated in the region of interest and, afterwards, the set of solutions obtained may be filtered using e.g. the k-means clustering in order to get the N_S most representative solutions. This procedure, which is also used in [11], increases the computational cost and it must only be applied in case this situation is internally detected.

4   Computational Implementation

In this section, we demonstrate the computational implementation created for Interactive WASF-GA, which is in a preliminary development phase. It has been developed in Java by using jMetal [7], a Java-based framework for multiobjective optimization. In order to check the performance of the method proposed, we have introduced into the platform several test problems from the ZDT, DTLZ and WFG families [6,10,26], respectively by now, for which the number of objectives can vary between 2 and 6. Of course, this implementation must be further improved so that other multiobjective optimization problems considered.

The main menu can be seen in Figure 2, where we consider the DTLZ2 test problem with 5 objective functions. The information is organized as follows:

- **Algorithm's Configuration.** There are three parameters in this box: (a) the number of solutions the DM would like to compare at the current iteration; (b) the population size and (c) the number of generations, the latter two being technical parameters. In the implementation, default values are recommended for these technical parameter for each problem although they can be modified if so desired. For example, if the DM thinks that the solutions are not good enough and (s)he wishes to obtain solutions closer to the Pareto optimal front, one can allow more generations to be carried out.
- **Problem's Configuration.** In this box, the multiobjective optimization problem to be solved is selected.
- **Reference Point.** Approximations of the ideal and the nadir values are provided to the DM in order to let her/him know the ranges of the objective functions. By clicking on each slider and moving it, the DM can set the aspiration level for each objective, and the corresponding numerical values are shown in the *Value* column.
- **Solution Process.** To generate the N_S^{it} solutions the DM wishes to compare, (s)he must click the *Start* button. If (s)he decides to take a new iteration by

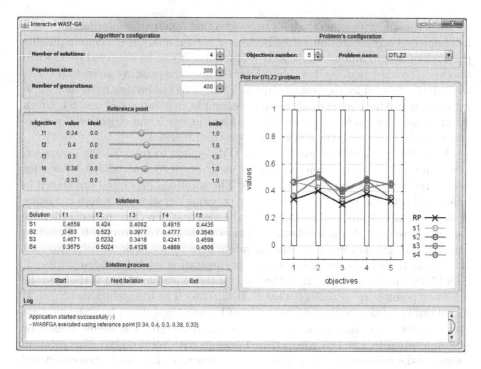

Fig. 2. Interface of Interactive WASF-GA - DTLZ2 problem, iteration 1

changing some preference information (the reference point and/or the number of solutions to be generated), (s)he must click the *Next Iteration* button to generate new solutions.

- **Solutions.** Here, the objective values of the N_S^{it} solutions obtained for the current reference point are shown.
- **Plot for the Problem.** The objective vectors of the solutions found and the reference point are shown graphically in order to ease the comparison among them. For bi-objective optimization problems, they are plotted in $\mathbf{R}^2$, and also the Pareto optimal front is shown if it is known. For multiobjective optimization problems with three or more objective functions, we use a *value path* [14] representation to shown the solutions obtained, as can be see in Figure 2. The reference point (labelled as RP) and each solution are plot by lines that go across different columns which represent the objective function values they reach. The lower and upper ends of each column represent the total values range of each objective function, that is, its ideal and nadir values, respectively.
- **Log.** The *Log* box indicates if there has been any error during the execution.

In what follows, we illustrate the performance of Interactive WASF-GA with the DTLZ2 problem with 5 objectives. Let us assume that the preference information given by the DM at the first iteration is the one shown in Figure 2, that is, he set the first reference point as $\mathbf{q}^1 = (0.34, 0.4, 0.3, 0.38, 0.33)$ and wanted to generate four solutions ($N_S^1 = 4$). Analysing the value path and the objective values of the solutions generated, it can be seen that none of the solutions obtained has improved any aspiration value. Besides, a careful analysis of them highlights the conflict degree among the objective functions. It can be observed that, when a solution reaches an objective value closer to the corresponding aspiration value, the values achieved by the rest of objective functions are further from their aspiration values. This can be easily seen, for example, in solutions S2 and S4, which attain values close to the aspiration levels for objective 5 and objective 1, respectively, at the expense of the rest of the objective functions.

According to the above analysis, the DM decided to relax all the aspiration levels and he set the new reference point as $\mathbf{q}^2 = (0.36, 0.42, 0.33, 0.4, 0.36)$ for generating four new solutions ($N_S^2 = 4$). The solutions generated are shown in image and table (a) of Figure 3. As at the previous iteration, no solution reaches or improves any aspiration level, but it can be seen that now the ranges of objective values achieved by all the solutions are closer to their aspiration levels. Based on this, the DM wished to have another iteration in order to check the solutions that could be obtained if he maintained the same aspiration levels for the objectives 1, 2 and 5 and he relaxed a bit more the ones for the objectives 3 and 4. He fixed the reference point as $\mathbf{q}^3 = (0.36, 0.42, 0.38, 0.45, 0.36)$. The four solutions ($N_S^3 = 4$) found can be seen in image and table (b) of Figure 3 and it can be observed that now they are even closer to the reference point. Although the reference point was still unattainable, the DM was satisfied enough with solution S2. This solution improved the values achieved for the objective functions 1, 2 and 5 when compared to the ones reached by most of the solutions at the previous iteration and, at the same time, it attained the second best values for the objectives 3 and 4. After three iterations the DM found the most preferred solution and was convinced of its goodness.

With this example, we have shown the behaviour of Interactive WASF-GA and the user interface proposed. If the DM changes the reference point, we have seen that the solutions generated are different from the ones previously produced. And if the DM indicates a higher or a smaller number of solutions, more or less solutions are produced accordingly. We have not computationally compared our algorithm with other interactive methods because a quantitative assessment of interactive approaches is very difficult in practice when interacting with a DM. Furthermore, traditional comparative tables which evaluate the performance of EMO algorithms after several independent runs are not meaningful for assessing Interactive WASF-GA because we focus on the DM's interaction and the decision making phase, and not only on the approximation of the Pareto optimal front.

5   Conclusions

In this paper, a new interactive evolutionary algorithm has been proposed for solving multiobjective optimization problems. The new algorithm is called Interactive WASF-GA and it is based on the preference-based EMO algorithm WASF-GA. At each iteration of Interactive WASF-GA, very easy to understand preference information is asked to the DM: just a reference point (containing desirable objective

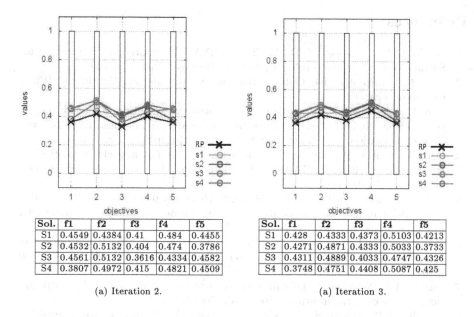

Sol.	f1	f2	f3	f4	f5
S1	0.4549	0.4384	0.41	0.484	0.4455
S2	0.4532	0.5132	0.404	0.474	0.3786
S3	0.4561	0.5132	0.3616	0.4334	0.4582
S4	0.3807	0.4972	0.415	0.4821	0.4509

(a) Iteration 2.

Sol.	f1	f2	f3	f4	f5
S1	0.428	0.4333	0.4373	0.5103	0.4213
S2	0.4271	0.4871	0.4333	0.5033	0.3733
S3	0.4311	0.4889	0.4033	0.4747	0.4326
S4	0.3748	0.4751	0.4408	0.5087	0.425

(a) Iteration 3.

Fig. 3. Solution process of the DTLZ2 problem

function values) and the number of solutions the DM wishes to compare. According to this information, a set with this number of solutions is generated in order to represent the region of interest of the Pareto optimal front defined by the reference point given. Subsequently, the DM analyses the solutions found and decides either to stop or to carry out a new iteration by redefining the preference information.

While the original WASF-GA approximates the region of interest with a high number of nondominated solutions, the interactive version only needs to generate few representative nondominated solutions. This fact allows to accelerate the solution process and reduces the computational cost needed. Besides, as the DM just compares a small number of solutions in the region of interest in order to find her/his most preferred solution, the solution process is not very demanding and requires a low cognitive effort. Furthermore, Interactive WASF-GA is able to generate as many solutions as the DM indicates, and this is a strength in comparison to other interactive EMO algorithms, which provide the DM with sets of solutions that may be too large to be compared. Also, it is noteworthy that Interactive WASF-GA does not need to set any additional parameter.

We have demonstrated the applicability of Interactive WASF-GA with a five-objective optimization problem which has shown how our algorithm can be used for reaching a solution interesting for the DM. Next, we plan to apply the algorithm proposed to real-life multiobjective optimization problems.

Acknowledgements. This research was partly supported by the Spanish Ministry of Innovation and Science (MTM2010-14992) and by the Andalusia Regional Ministry of Innovation, Science and Enterprises (PAI groups SEJ-445 and SEJ-532).

References

1. Coello, C.A.C., Lamont, G.B., Veldhuizen, D.A.V.: Evolutionary Algorithms for Solving Multi-Objective Problems, 2nd edn. Springer, New York (2007)
2. Deb, K.: Multi-objective Optimization using Evolutionary Algorithms. Wiley, Chichester (2001)
3. Deb, K., Miettinen, K., Chaudhuri, S.: Towards an estimation of nadir objective vector using a hybrid of evolutionary and local search approaches. IEEE Transactions on Evolutionary Computation 14(6), 821–841 (2010)
4. Deb, K., Pratap, A., Agarwal, S., Meyarivan, T.: A fast and elitist multiobjective genetic algorithm: NSGA-II. IEEE Transactions on Evolutionary Computation 6(2), 182–197 (2002)
5. Deb, K., Sundar, J., Ubay, B., Chaudhuri, S.: Reference point based multi-objective optimization using evolutionary algorithm. International Journal of Computational Intelligence Research 2(6), 273–286 (2006)
6. Deb, K., Thiele, L., Laumanns, M., Zitzler, E.: Scalable multi-objective optimization test problems. In: Congress on Evolutionary Computation, CEC-2002, pp. 825–830 (2002)
7. Durillo, J.J., Nebro, A.J.: jMetal: A java framework for multi-objective optimization. Advances in Engineering Software 42, 760–771 (2011)
8. Figueira, J.R., Greco, S., Mousseau, V., Słowiński, R.: Interactive multiobjective optimization using a set of additive value functions. In: Branke, J., Deb, K., Miettinen, K., Słowiński, R. (eds.) Multiobjective Optimization. LNCS, vol. 5252, pp. 97–119. Springer, Heidelberg (2008)
9. Gong, M., Liu, F., Zhang, W., Jiao, L., Zhang, Q.: Interactive MOEA/D for multi-objective decision making. In: 13th Annual Conference on Genetic and Evolutionary Computation, GECCO 2011, pp. 721–728 (2011)
10. Huband, S., Hingston, P., Barone, L., While, L.: A review of multiobjective test problems and a scalable test problem toolkit. IEEE Transactions on Evolutionary Computation 10(5), 477–506 (2006)
11. Luque, M., Ruiz, F., Steuer, R.E.: Modified interactive Chebyshev algorithm (MICA) for convex multiobjective programming. European Journal of Operational Research 204(3), 557–564 (2010)
12. MacQueen, J.B.: Some methods for classification and analysis of multivariate observations. In: 5-th Berkeley Symposium on Mathematical Statistics and Probability, vol. 1, pp. 281–297. University of California Pressley, Berkeley (1967)
13. Miettinen, K.: Nonlinear Multiobjective Optimization. Kluwer Academic Publishers, Boston (1999)
14. Miettinen, K.: Survey of methods to visualize alternatives in multiple criteria decision making problems. OR Spectrum 36(1), 3–37 (2014)
15. Miettinen, K., Mäkelä, M.M.: On scalarizing functions in multiobjective optimization. OR Spectrum 24(2), 193–213 (2002)
16. Miettinen, K., Ruiz, F., Wierzbicki, A.P.: Introduction to multiobjective optimization: interactive approaches. In: Branke, J., Deb, K., Miettinen, K., Słowiński, R. (eds.) Multiobjective Optimization. LNCS, vol. 5252, pp. 27–57. Springer, Heidelberg (2008)
17. Ruiz, A.B., Saborido, R., Luque, M.: A preference-based evolutionary algorithm for multiobjective optimization: The weighting achievement scalarizing function genetic algorithm. Journal of Global Optimization (2014, in press). doi:10.1007/s10898-014-0214-y

18. Sindhya, K., Ruiz, A.B., Miettinen, K.: A preference based interactive evolutionary algorithm for multi-objective optimization: PIE. In: Takahashi, R.H.C., Deb, K., Wanner, E.F., Greco, S. (eds.) EMO 2011. LNCS, vol. 6576, pp. 212–225. Springer, Heidelberg (2011)

19. Sinha, A., Korhonen, P., Wallenius, J., Deb, K.: An interactive evolutionary multiobjective optimization algorithm with a limited number of decision maker calls. European Journal of Operational Research **233**(3), 674–688 (2014)

20. Szczepanski, M., Wierzbicki, A.P.: Application of multiple crieterion evolutionary algorithm to vector optimization, decision support and reference point approaches. Journal of Telecommunications and Information Technology **3**(3), 16–33 (2003)

21. Thiele, L., Miettinen, K., Korhonen, P., Molina, J.: A preference-based evolutionary algorithm for multi-objective optimization. Evolutionary Computation **17**(3), 411–436 (2009)

22. Wang, R., Purshouse, R.C., Fleming, P.J.: Preference-inspired coevolutionary algorithms for many-objective optimization. IEEE Transactions on Evolutionary Computation **17**(4), 474–494 (2013)

23. Wang, R., Purshouse, R.C., Fleming, P.J.: "Whatever works best for you"- a new method for a priori and progressive multi-objective optimisation. In: Purshouse, R.C., Fleming, P.J., Fonseca, C.M., Greco, S., Shaw, J. (eds.) EMO 2013. LNCS, vol. 7811, pp. 337–351. Springer, Heidelberg (2013)

24. Wierzbicki, A.P.: The use of reference objectives in multiobjective optimization. In: Fandel, G., Gal, T. (eds.) Multiple Criteria Decision Making, Theory and Applications, pp. 468–486. Springer (1980)

25. Zhang, Q., Li, H.: MOEA/D: A multiobjective evolutionary algorithm based on decomposition. IEEE Transactions on Evolutionary Computation **11**(6), 712–731 (2007)

26. Zitzler, E., Deb, K., Thiele, L.: Comparison of multiobjective evolutionary algorithms: Empirical results. Evolutionary Computation **8**(2), 173–195 (2000)

27. Zitzler, E., Künzli, S.: Indicator-based selection in multiobjective search. In: Yao, X., Burke, E.K., Lozano, J.A., Smith, J., Merelo-Guervós, J.J., Bullinaria, J.A., Rowe, J.E., Tiño, P., Kabán, A., Schwefel, H.-P. (eds.) PPSN VIII. LNCS, vol. 3242, pp. 832–842. Springer, Heidelberg (2004)

On Generalizing Lipschitz Global Methods for Multiobjective Optimization

Alberto Lovison[1]([✉]) and Markus E. Hartikainen[2]

[1] Department of Mathematics, University of Padova, Padua, Italy
[2] Department of Mathematical Information Technology, University of Jyvaskyla,
P.O. Box 35, FI-40014 Jyvaskyla, Finland
lovison@math.unipd.it

Abstract. Lipschitz global methods for single-objective optimization can represent the optimal solutions with desired accuracy. In this paper, we highlight some directions on how the Lipschitz global methods can be extended as faithfully as possible to multiobjective optimization problems. In particular, we present a multiobjective version of the Pijavskiĭ-Schubert algorithm.

Keywords: Global and Lipschitz optimization · Multiobjective optimization · Multiple criteria decision making

1 Introduction

Exact global search methods are a well known class of algorithms belonging to the single-objective optimization literature.[1] These methods usually demonstrate appreciable speed of convergence and furthermore guarantee that the global optimum of the function under exam is approximated with arbitrary precision in a finite time, providing some constraints on the functions at hand. A well known example of these methods is the Pijavskiĭ-Schubert algorithm [18,19], which is quickly reviewed in Section 2.

Unfortunately, it appears that in the available multiobjective literature there has not been so much attention dedicated to the complete or deterministic methods for global search. Nevertheless, at least in the single objective case and when limited computational resources are available, global deterministic methods have proven their effectiveness and are known and widely employed. Now and then we have witnessed the attempt of producing adaptations of some of these methods for the multiobjective case. However, at least to the knowledge of the authors, most of those adaptations follow one of the following schemes:

1. the method uses a scalarization of the multiobjective problem to a single objective optimization problem and then applies the global algorithm to the scalarization, or

---

[1] These methods are referred to also as *complete* or *deterministic*, possibly referring to more specific features [16].

A. Gaspar-Cunha et al. (Eds.): EMO 2015, Part II, LNCS 9019, pp. 264–278, 2015.
DOI: 10.1007/978-3-319-15892-1_18

2. the method translates the underlying idea of the global method in the multiobjective format, but then applies a non deterministic method to produce the Pareto set.

In both cases, we encounter the following problems contrasting with the global and exact character desired:

1. the method cannot guarantee a systematic covering of the Pareto set, or
2. the method operates at some point some non deterministic choice.

Well-known and widely used methods belonging to the latter class are the evolutionary multiobjective optimization methods. To partially overcome the first problem some method try to realize a systematic covering of the space of parameters. However, this could not lead to a correspondingly systematic covering of the Pareto set, especially in non convex cases. Therefore we do not consider this approach as genuinely multiobjective and we would prefer to tackle the multiobjective nature of the problem directly. We believe that the best strategy for approximating the Pareto set is adopting a set-wise approach. That is, instead of having a single point converging to a Pareto optimum at a time and then repeating this for a number of points, it is better to make converge multiple points at the same time towards the whole Pareto set. This set-wise concept of convergence is already adopted by evolutionary multiobjective optimization methods, and it is in contrast to point-wise convergence followed by most scalarization methods.

We restate our claim about the nonexistence of exact global methods in a more positive sense, by presenting a pair of methods both attempting to adhere to the most possible extent to deterministic methods and to guarantee a complete representation of the Pareto set, at least requesting some regularity conditions on the functions at hand. The first method [4] uses the Karush-Kuhn-Tucker conditions to write a non negative auxiliary function whose zero set contains the set of Pareto optima. The zero set of such a function is approximated by using an associated ordinary differential equation and suitable iteration schemes obtained from a discretization of it. Set-wise convergence with respect to the Hausdorff distance is obtained if suitable regularity conditions are met. The set obtained is the set of subcritical points, which strictly contains the set of Pareto optima. The approximation obtained consists in a collection of hypercubes which covers the subcritical set. The second method [12] uses a qualitatively similar approach obtained from the Smale's first and second order conditions [20, 21]. For the special case of two functions in two variables, it is possible to write a multiobjective extension of the Pijavskiĭ-Shubert algorithm, i.e., it is possible to guarantee the convergence to the Pareto set in global sense with respect to the Hausdorff distance.

In both cases, the set obtained is a strict superset of the set of Pareto optima, corresponding to first order conditions of optimality, i.e., extensions of the notion of critical point for a single function, and furthermore, the application is limited to low dimensional examples. Therefore effective and straightforward approximation methods for the Pareto set are still missing, at least in the authors' knowledge.

In this paper, our main scope is to present a Lipschitz global optimization algorithm for multiple objectives, namely an extension of the Pijavskiĭ-Shubert method which does not make use of auxiliary functions and that approximates the set of global Pareto optima within a desired tolerance measured according to the Hausdorff distance (see Section 2 for details). This method is a first step in the direction proposed in [12] where the possibility of defining exact and global strategies was outlined. This method produces an approximation of the Pareto set consisting in a covering composed by arbitrarily small hypercubes. In perspective, the method can be combined with surface tracing methods to generate a faithful geometric surrogate of the Pareto set, as in the methods [5–7,11].

2   Pijavskiĭ-Shubert

The Pijavskiĭ-Shubert algorithm [18,19] (from now on noted as the *P-S algorithm*) is a 1-dimensional globally convergent method assuming that a global Lipschitz constant is known in the domain of the search process. At each step of the process, there is a finite number of points in the domain where the function has been evaluated. Those points are taken as the extrema of a collection of subintervals. For every subinterval a lower bound of the unknown function is determined on the basis of the Lipschitz constant, on the subinterval width and the values of the function at the extrema. The subinterval with the lowest estimate is chosen for further sampling and subdivision, by taking the point where the lower bound is predicted to be located.

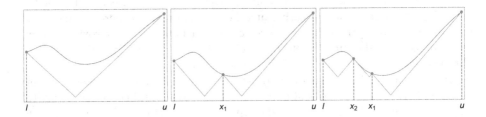

Fig. 1. Workings of the Pijavskiĭ-Shubert algorithm. The domain is divided in subintervals and for each subinterval a lower bound is computed on the basis of the global Lipschitz constant for the function in study. The interval with the lowest lower bound is then divided exactly at the position of the lower bound.

This method allows for detecting subintervals where the global minimum cannot be located, discarding them from further analysis. Indeed if the lower bound corresponding to a subinterval is higher than one of the already computed values, it is impossible that the global optimum would be contained into the interval.

This method can be extended to higher dimensions although the computational complexity rises exponentially [15]. Nevertheless the method has been the starting point for several efficient global algorithms, such as DIRECT [8] and Lipschitz Global Optimization (LGO) [17] and many more [9,10][2].

3 Extending P-S Algorithm to Multiple Objectives

In [12], tessellation of the 2D domain by means of equilateral triangles and an auxiliary scalar function was used for deciding if a triangle could contain a portion of the Pareto set or not. However in the method presented here, we will try to define an approach valid for higher dimensions and we will avoid any scalarization or auxiliary function. In particular we will estimate a vector lower bound for every hyper interval, i.e., for each one of the available objective function. This estimate will be on the lines of the scalar method, i.e., based on the Lipschitz constant and on the hyper interval diameter. Then we will not combine the single objective lower bounds in a unique scalar indicator but we will keep the vector as it is and compare and rank different intervals on the basis of Pareto dominance. More precisely, we will partition the set of hyper intervals in two classes, the *discarded* and the *candidates* for further division. To decide if an hyper interval should be discarded we will check if the estimated lower bound dominates one of the already computed points. In that case there cannot exist a point inside the hyper interval belonging to the Pareto set, so we are warranted to discard it. All other hyper intervals will be selected for further division in the subsequent iterations.

This will produce several candidate intervals for each iteration, but this does not constitute a problem, because it is typical for multiobjective methods and it occurs also for some scalar global optimization methods like [8]. A detailed formal description of the algorithm proposed is given in Algorithm 1.

4 Global Convergence of Deterministic Algorithms

We recall some definitions from [12] about global convergence of multiobjective algorithms, starting from the standard scalar case $m = 1$.

4.1 Algorithms and Global Convergence in Scalar Optimization

Let $f : D \to \mathbb{R}$ be a Lipschitz continuous function, where W can be the n-hypercube $[0,1]^n$ for simplicity or a smooth n-dimensional compact manifold.

---

[2] In a more general view, we notice that a Lipschitz constant represent a proxy for the complexity of a function. As a result they can be used for optimization as well as for other purposes, e.g., for function approximation (see [14]).

Algorithm 1. multi-Pijavskĭ-Shubert (mPS) method

1: Let the domain (decision space) be a hyper-rectangle $D := [0,1]^n$, possibly after a suitable normalization
2: Let the unknown function $f : D \to \mathbb{R}^m$ be globally Lipschitz continuos with constants different for every component $L_1, \ldots, L_m$.
3: Evaluate f on the corners of D
4: Initialize the set of evaluated points as $E := \left\{ (q,v) \mid q \in \{\text{corners of } D\}, v = f(q) \right\}$
5: Initialize the set of active subintervals as $S := D$
6: Set $nIter \in \mathbb{N}$ as the maximum number of iterations
7: **for** $i = 1$ to $nIter$ **do**
8:    **for all** $I \in S$ **do**
9:       Divide the hyper interval I in the 2^n subintervals obtained by halving all dimensions of the original hyper interval. Remove the hyper interval I from the list S and add the subintervals to the list.
10:      Evaluate f on all the midpoints p of the k-faces of the hyper interval I, for all $1 \leqslant k \leqslant n$, i.e., all the corners of the subintervals. Add $(p, f(p))$ to the list E.
11:    **end for**
12:    Associate to each interval I in S the vectors $v_{I,\iota} = (f_j(q_\iota) + L_j \text{diam} I)_{\iota,j}$, where q_ι is a corner of I and $j = 1, \ldots, m$.
13:    **for all** $I \in S$ **do**
14:       **if** all the upper bounds in the corners $(f_j(q_\iota) + L_j \text{diam} I)_j$ are dominated by some vector of values v in E **then**
15:          discard the interval I from S
16:       **end if**
17:    **end for**
18: **end for**

– Let us denote by $f^\star$ the absolute, or *global*, maximum value of the function f, $x^\star$ being a point in W realizing the maximum. In other words, $x^\star$ is an *optimum*, while $X_f^\star$ is the set of all optima:

$$f^\star = \max_{x \in W} f(x), \qquad X_f^\star := \left\{ x^\star \in W \mid f(x^\star) = f^\star \right\}. \tag{1}$$

– An *algorithm* is a finite sequence of well–defined instructions, which, when running on a function f, produces the sample sequence

$$X_f := \{x_1, \ldots, x_k, \ldots\} \subseteq W.$$

In particular the function f is assumed to be actually computed in the point x_k at the kth step of the algorithm.

– We denote by X_f the full infinite sequence produced by an algorithm when given a function f, by $X_{f,k}$ or X_k the partial k–sequence. $\overline{X_f}$ is the closure of X_f while $X_f' = \overline{X_f} \setminus X_f$, is the set of limit points of X_f.

Assume we are not in the trivial case $X_f^\star \cap X_k \neq \emptyset$ for any finite k.

– An algorithm *sees* the global minimum of the function f if $X_f^\star \cap X_f' \neq \emptyset$.

- An algorithm *localizes* the global minimum if $X'_f = X^\star_f$ (or *weakly localizes* if $X'_f \subseteq X^\star_f$).

It seems useful to give further precise description of a class of algorithms for detecting structured subsets rather than scatters of points.

- A *set–wise sequential algorithm* is a deterministic algorithm which, besides the sample sequence $X_f = \{x_1, x_2, \dots\}$ where actually the function f has to be evaluated, generates a sequence of subsets $\{S_1, \dots, S_k, \dots\}$, $S_k \subseteq W$, intended to give an approximation of the Pareto set.
- Notice that more or less explicitly, any multiobjective optimization strategy is a set-wise sequential algorithm. If not specified in a different way, the sequence of sets approximating the Pareto set is given by the non dominated sets of the partial sequences:

$$S_k := nd(\{x1, \dots, x_k\}).$$

- It is a common belief that in typical cases the Pareto set is a finite collection of smooth manifolds with edges and corners. Such objects have interesting properties and are called *stratified sets* (see [13] for a discussion with the point of view of multiobjective optimization). Simplicial methods like [1,6,7, 11,12] at each new iteration produce a simplicial complex as approximation of the Pareto optimal set. These methods have the fundamental property of offering a parametric representation of the Pareto set, which appears as a very useful tool for exploring the available solutions during the decision process.

4.2   Convergence in Multiobjective Optimization

To be convergent, an algorithm should produce a sequence $S_1, S_2, \dots$, converging in some sense to the set of optima θ_{op}. Expressing a crude translation of the concepts of *seeing* and *localizing* the optima is poorly useful, because apart from degenerate cases, the set of Pareto optimal values does not consist in a single (vector) value $f^\star \in \mathbb{R}^m$. In the generic case, the set of Pareto optimal values is infinite, as well as, of course, the set of Pareto optima θ_{op}. Limits have to be considered in a set–wise sense, and therefore we need a concept of distance between sets.

- Let $A, B \subseteq W$. The *Hausdorff distance* between A and B is defined as

$$d_{\mathcal{H}}(A, B) := \max \left\{ \max_{x \in A} \min_{y \in B} d(x, y), \max_{y \in B} \min_{x \in A} d(x, y) \right\}. \tag{2}$$

- We say that a set-wise sequential algorithm $\mathcal{A}$ *sees* the set of global Pareto optima θ_{op} if

$$\lim_{k \to \infty} \min_{x \in S_k, y \in \theta_{op}} d(x, y) = 0. \quad (\mathcal{A} \text{ sees } \theta_{op}) \tag{3}$$

(In a sense the limit set $\lim_k S_k \cap \theta_{op} \neq \emptyset$, i.e., at least the Pareto set generated by the algorithm touches a portion of the global Pareto set, i.e., it generalizes the statement $X'_f \cap X^\star_f \neq \emptyset$.)

- We say that $\mathcal{A}$ *weakly localizes* the set of global Pareto optima θ_{op} if

$$\lim_{k \to \infty} \max_{t \in \theta_{op}} d(t, S_k) = 0. \qquad (\mathcal{A} \text{ weakly localizes } \theta_{op}) \qquad (4)$$

(In a sense the limit set will contain all portions of the global Pareto set $\theta_{op} \subseteq \lim_k S_k$. The limit set is possibly larger than the Pareto set.)
- We say that $\mathcal{A}$ *strictly localizes* the global Pareto optima θ_{op} if

$$\lim_{k \to \infty} d_{\mathcal{H}}(S_k, \theta_{op}) = 0, \qquad (\mathcal{A} \text{ strictly localizes } \theta_{op}) \qquad (5)$$

i.e., the Pareto set generated by the algorithm *coincides* with the true Pareto set.
- Dealing with algorithms which merely see the global optimum, or that localize non strictly the set of Pareto optima seems not completely satisfactory from the global multiobjective optimization point of view. For instance, an algorithm optimizing only to one component of the vector function f would give a non dominated point, and it would *see* the Pareto optimum.

4.3 Convergence for mPS

The convergence proof for the mPS algorithm 1 is twofold. Let us consider a point x that it is not in the sequence of sampled points. Clearly this sequence could be a dense subset of W in principle, but it has zero Lebesgue measure.

Proposition 1. *prop:conv1 Let $x \in W$ be a Pareto optimal point. Then*

1. *for every iteration step $k \in \mathbb{N}$ there exists a cell C_k in S_k containing x,*
2. $\lim_{k \to \infty} \operatorname{diam} C_k = 0$.

Proof. et us assume that x is Pareto optimal and that at the step $k + 1 > 0$ there is no cell in S_k containing x. Assume that at step k there was a cell C_k containing x, thus that cell must have been discarded at the $k + 1$ step. So there must exist a point p_t in the sequence of evaluated points such that $f_j(p_t) > f_j(q_r) + L_j \operatorname{diam}(C_k)$ for all j and all vertices q_r in the cell C_k. But because of the Lipschitz property, $f_j(x) < f_j(q_r) + L_j \|x - q_r\| < \max_{q_r} f_j(q_r) + L_j \operatorname{diam}(C_k) < f_j(p_t)$, so x is dominated and not a Pareto optimum. A contradiction.   □

Proposition 2. *prop:conv2 Let $x \in W$ be not Pareto optimal point. Then there exists $k \in \mathbb{N}$ such that for every $k' \geqslant k$ there is no cell C in $S_{k'}$ containing x.*

Proof. f x is not Pareto, let d the minimum distance from a Pareto optimum p dominating x and let $\ell = \min_{j=1,\dots,m} f_j(p) - f_j(x) > 0$. Let $\bar{L} = \max_j L_j$. Assume that there exists for every $k \in \mathbb{N}$ a cell C_k in S_k that contains x. As $k \to \infty$ the cell size $\operatorname{diam} C_k \to 0$, so let $\tilde{k}$ such that $\operatorname{diam} C_{\tilde{k}} < \frac{\ell}{2 \max_{j=1,\dots,m} L_j}$ and consider any of the vertices y of $C_{\tilde{k}}$ and any of the vertices q of the cell $V_{\tilde{k}}$ containing p. Note that such a cell exists for every k because of the preceding

Proposition, and that at the same iteration the cells in S_k have all the same size, and let $d = \mathrm{diam}C_{\tilde{k}} = \mathrm{diam}V_{\tilde{k}}$. We have

$$f_j(q) - f_j(y) > f_j(p) - L_j d - (f_j(x) + L_j d) = f_j(p) - f_j(x) - 2L_j d > \ell - 2\left(\max L_j\right)d, \quad (6)$$

so, if

$$\mathrm{diam}C_{\tilde{k}} < \frac{\ell}{2\max L_j}, \quad (7)$$

all the vertices of $V_{\tilde{k}}$ dominate the vertices of $C_{\tilde{k}}$, so the cell $C_{\tilde{k}}$ will be discarded at the $\tilde{k} + 1$ step, leading to a contradiction. □

From the above propositions the convergence of Algorithm 1 follows straightforwardly.

Theorem 1. *teo:naiveconv Let $f : D \to \mathbb{R}^m$ globally Lipschitz continuous, with Lipschitz constants $L_1, \ldots, L_m$ and consider the application of Algorithm 1 to f. Consider the sequence of families of sets S_k, where S_k is the active set of intervals at the k-th step. Then Algorithm 1 strictly localizes the Pareto set of f.*

5   Benchmarks

For testing our method, we consider a set of three non degenerate following functions, so the Pareto sets are $m - 1$ dimensional objects, both in the decision and in the objectives spaces, as it is expected for typical cases [13].

5.1   DTLZ2 with Three Decision Variables and 2 Objectives

This function is part of a collection of test functions largely known and used in literature [3]. The function is scalable to any number of decision variables and objectives, but we have used here the the version with three decision variables and two objectives. We have performed two runs of the algorithm which are documented in Figure 2. In panel (a) we report the outcome of three iterations of Algorithm 1, corresponding to a total of 305 function evaluation while in panel (b) we represent the outcome of four iterations (1205 function evaluations). In both panels, the left figure represents the design space, while the right figure is the objective space. Transparent cubes in design space represent the active cells of the algorithm, i.e., the cells which are candidate for further splitting in the subsequent iterations. The same active cells are mapped to the objectives space into generic polygons (light blue regions in the right parts of the panels). This region surrounds the non dominated points and can be considered as an approximation of the Pareto front.

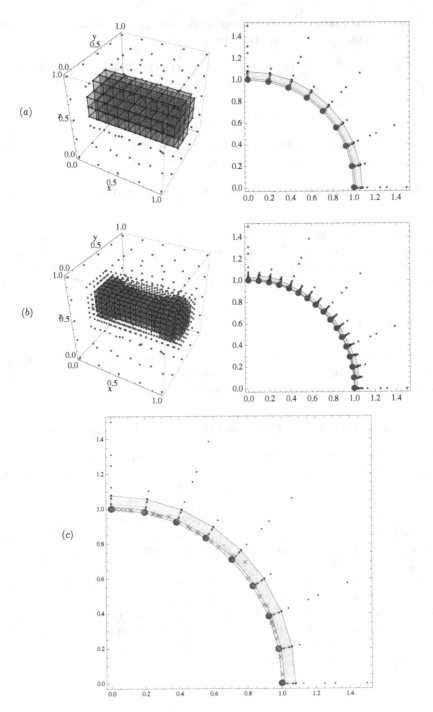

Fig. 2. Test function DTLZ2 with $m = 2$ objectives and $n = 3$ dimensions for the design space. See text for the details.

For comparison with a well known evolutionary strategy we report in panel (c) the outcome of the MOEA/D method [22] with three generations (where the first one is a random sample) corresponding to a total of 303 function evaluations, since the population size is 101. The green crosses correspond to the Pareto non dominated values of the points produced by the algorithm. In the same panel we also represent the outcome of our method applied for three iterations (i.e., 305 function evaluations), marked with red dots.

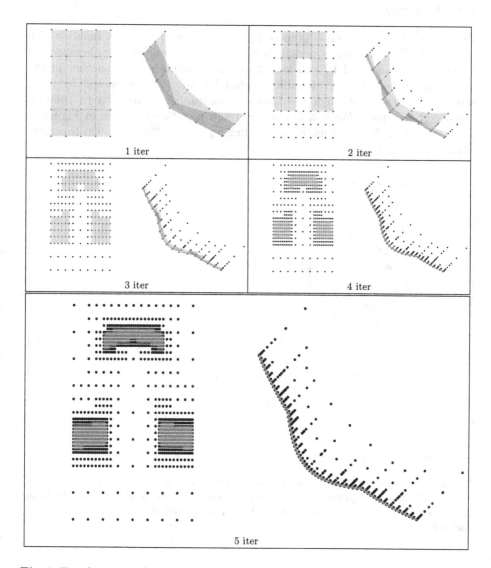

Fig. 3. Test function $L\&H2x2$. The iterations from 1 to 5 of the method are reported, the decision space on the left and the objectives space on the right for each panel.

Just for attempting to compare the outcomes of the two methods, we observe that MOEA/D seems to span more densely the range of Pareto front, although none of those points dominates a point produced by mPS. On the other side, the Pareto set obtained with mPS dominates 14 out of the 44 points in the front corresponding to MOEA/D, i.e., the 31% of the points composing the front, attesting the higher accuracy of the new method.

5.2 *L&H2x2*

The *L&H2x2* is an example proposed in [11] and used as a test function also in [2,7,12]. The example is paradigmatic for the non convex case, because the Pareto set is composed by two local fronts superimposing one another in the objectives space. The corresponding global Pareto set is composed by three separate branches, although we observe a unique connected Pareto front in the objectives space. We test our method and plot the outcomes in Figure 3, going from 1 to 5 iterations, corresponding to 25, 81, 201, 445, 920 functions evaluations.

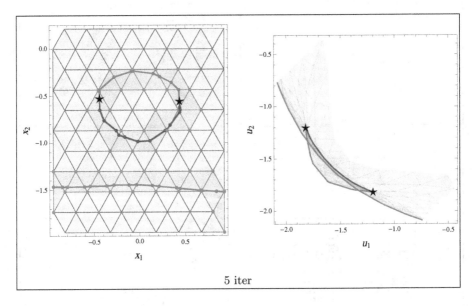

Fig. 4. Outcome of the SiCon method on the test function *L&H2x2*. Left panel: decision space. Right panel: objectives space. The Pareto set. Orange lines represent the local Pareto set while the red line is the part of the Pareto critical set which is not locally optimal.

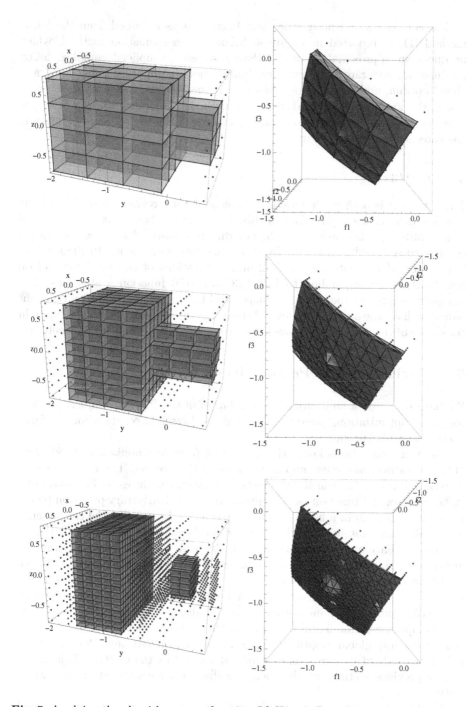

Fig. 5. Applying the algorithg to test function $L\&H3 \times 3$. Decision space is on the left side and the objectives space is on the right side. The number of iterations grows from the top to the bottom.

For comparison purposes, the local Pareto set as obtained from the SiCon method [11] is reported in Figure 4. SiCon is a continuation method, which produces an approximation of the Pareto set as a simplicial complex. SiCon produces an accurate representation but cannot distinguish among local and global optima, and its outcome cannot be refined as easily as with mPS. As a result, the Pareto set obtained with this method (the orange lines in figure) is composed by two connected components while actually the set of global optima has three separate components.

5.3   *L&H*3x3

This function is a three dimensional version of the previous example, for the function definition see [11]. Also in this case there are two superimposing local fronts, although, because of the higher dimensionality of the decision space, the Pareto set results composed by two connected components. In Figure 3 we have reported the outcomes of 1, 2 and 3 iterations of the mPS method on this function, corresponding to 125, 633 and 6156 function evaluations. The left panels represent the points evaluated and the active cells while on the right panels we have the function values and the images of the active cells. The bump in the center of the surface corresponds to the smaller component of the Pareto set.

6   Conclusions and Perspectives

We have proposed a multiobjective translation of the Pijavskĭ-Shubert method for global optimization, assuming that a global Lipschitz constant for the functions at hand is known.

As far as the authors know, this is the first fully deterministic method provably generating convergent approximations to Pareto sets. The convergence is defined in terms of Hausdorff distance between sets, i.e., the exact Pareto set of a sufficiently regular function can be approximated with arbitrary precision (small Hausdorff distance) in a finite number of steps. We have called the convergence of the algorithm intended in this sense *strict localization* of the set of Pareto optima. We have mentioned several methods inspired by the same ideas in the global optimization literature, and we have observed that either they fall in the set of local methods, because they focus on searching for single optimal points or even only critical points, or either they make use of some random choice at some point, missing in some sense an exact localization of the whole Pareto set.

The approximation found by means of the proposed method is sharp, in the sense that only globally optimal points are approximated, and Pareto critical points or local optima are sooner or later discarded by the method. This differs from a previous method which strictly localized the singular set or the Pareto critical set [12].

We have tested the method on three non convex examples and compared the results for one of the cases with a well known evolutionary method, obtaining positive results on the side of the accuracy of the representation. Actually the strategy is a very conservative one, therefore densely distributed representations of the Pareto set can be obtained with a large number of function evaluations. Nevertheless, as there are many efficient generalizations and extensions of the Pijavskĭ-Shubert method, we figure that some of these variants can give valid inspirations for writing new algorithms less computationally demanding and also accessible for higher dimensional problems. We expect that such extensions should be very attractive for experts and practitioners in multiple criteria decision making community.

References

1. Askar, S., Tiwari, A.: Multi-Objective Optimisation problems: a symbolic algorithm for performance measurement of evolutionary computing techniques. In: Ehrgott, M., Fonseca, C.M., Gandibleux, X., Hao, J.-K., Sevaux, M. (eds.) EMO 2009. LNCS, vol. 5467, pp. 169–182. Springer, Heidelberg (2009)
2. Custódio, A.L., Madeira, J.F.A., Vaz, A.I.F., Vicente, L.N.: Direct Multisearch for Multiobjective Optimization. SIAM Journal on Optimimization 21, 1109–1140 (2011)
3. Deb, K., Thiele, L., Laumanns, M., Zitzler, E.: Scalable multi-objective optimization test problems. In: IEEE International Conference on E-Commerce Technology, vol. 1, pp. 825–830 (2002)
4. Dellnitz, M., Schütze, O., Hestermeyer, T.: Covering Pareto Sets by Multilevel Subdivision Techniques. Journal of Optimization Theory and Applications 124, 113–136 (2005)
5. Hartikainen, M., Miettinen, K., Wiecek, M.M.: Constructing a Pareto Front Approximation for Decision Making. Mathematical Methods of Operations Research 73, 209–234 (2011)
6. Hartikainen, M., Miettinen, K., Wiecek, M.M.: PAINT: Pareto Front Interpolation for Nonlinear Multiobjective Optimization. Computational Optimization and Applications 52, 845–867 (2012)
7. Hartikainen, M., Lovison, A.: PAINT-SiCon: Constructing Consistent Parametric Representations of Pareto Sets in Nonconvex Multiobjective Optimization. Journal of Global Optimization (2014, to appear). http://dx.doi.org/10.1007/s10898-014-0232-9
8. Jones, D.R., Perttunen, C.D., Stuckman, B.E.: Lipschitzian Optimization Without the Lipschitz Constant. Journal of Optimization Theory and Applications 79, 157–181 (1993)
9. Kvasov, D.E., Sergeyev, Y.D.: A Univariate Global Search Working with a Set of Lipschitz Constants for the First Derivative. Optimization Letters 3, 303–318 (2009)
10. Lera, D., Sergeyev, Y.D.: Lipschitz and Hölder Global Optimization using Space-Filling Curves. Applied Numerical Mathematics 60, 115–129 (2009)
11. Lovison, A.: Singular Continuation: Generating Piecewise Linear Approximations to Pareto Sets via Global Analysis. SIAM Journal on Optimization 21, 463–490 (2011)
12. Lovison, A.: Global Search Perspectives for Multiobjective Optimization. Journal of Global Optimization 57, 385–398 (2013)
13. Lovison, A., Pecci, F.: Hierarchical Stratification of Pareto Sets. http://arxiv-web3.library.cornell.edu/abs/1407.1755

14. Lovison, A., Rigoni, E.: Adaptive Sampling with a Lipschitz Criterion for Accurate Metamodeling. Communications in Applied and Industrial Mathematics **1**, 110–126 (2010)
15. Mladineo, R.H.: An Algorithm for Finding the Global Maximum of a Multimodal. Multivariate Function. Mathematical Programming **34**, 188–200 (1986)
16. Neumaier, A.: Complete search in continuous global optimization and constraint satisfaction. Acta Numerica **13**, 271–369 (2004)
17. Pintér, J.D.: Nonlinear Optimization with GAMS/LGO. Journal of Global Optimization **38**, 79–101 (2007)
18. Piyavskii, S.: An Algorithm for Finding the Absolute Extremum of a Function. USSR Computational Mathematics and Mathematical Physics **12**, 57–67 (1972)
19. Shubert, B.: A Sequential Method Seeking the Global Maximum of a Function. SIAM Journal on Numerical Analysis **9**, 379–388 (1972)
20. Smale, S.: Global analysis and economics. I. pareto optimum and a generalization of morse theory. In: Dynamical Systems (Proc. Sympos., Univ. Bahia, Salvador, 1971), pp. 531–544. Academic Press, New York (1973)
21. Smale, S.: Optimizing several functions. In: Manifolds-Tokyo 1973 (Proc. Internat. Conf., Tokyo, 1973), pp. 69–75. Univ. Tokyo Press, Tokyo (1975)
22. Zhang, Q., Li, H.: MOEA/D: A Multi-objective Evolutionary Algorithm Based on Decomposition. IEEE Transactions on Evolutionary Computation **11**, 712–731 (2007)

Dealing with Scarce Optimization Time in Complex Logistics Optimization: A Study on the Biobjective Swap-Body Inventory Routing Problem

Sandra Huber[(⊠)] and Martin Josef Geiger

Helmut Schmidt University, Holstenhofweg 85, 22043 Hamburg, Germany
sandra-huber@hsu-hh.de
http://logistik.hsu-hh.de

Abstract. In this paper a biobjective Swap Body Inventory Routing Problem (SB-IRP) is considered: A combination of the Swap Body Vehicle Routing Problem (SB-VRP), which minimizes fixed and variable routing costs, and the Inventory Routing Problem (IRP). The problem is based on the context of the VeRoLog Solver Challenge 2014, where our proposed VeRoLog Solver was ranked third, and our previous work on the IRP. Since we are investigating a multi-period problem, an additional objective function is formulated which includes inventory levels at the customers. Dealing with the allocation of scarce optimization time to the VeRoLog Solver is an essential topic, since an alternating approach of the determination of a replenishment strategy and the routing is considered. We propose an Iterative Variable Neighborhood Search and analyze the allocation of the computational time by extended VeRoLog test instances.

Keywords: Vehicle routing problem · Swap locations · Iterated variable neighborhood search · Multi-objective optimization

1 Introduction

The Vendor Managed Inventory (VMI) concept is an example of a successful cooperation between a customer and a supplier [7]. The supplier is in charge of the inventory levels at the customers and decides when and how much should be delivered. In fact, two supply chain management aspects, namely the inventory management and transportation, are combined to enhance the performance of the supply chain [1]. The customers do not have to cope with inventory control and replenishment orders and the supplier can reduce the distribution cost since deliveries for different customers can be combined [6,7]. Thus, the inventory, and the distribution effort can be smoothened [1]. In this sense, VMI is an example of value adding through logistics [6]. The VMI strategy is based on the solution of a multi-objective SB-IRP that simultaneously includes the minimization of

© Springer International Publishing Switzerland 2015
A. Gaspar-Cunha et al. (Eds.): EMO 2015, Part II, LNCS 9019, pp. 279–294, 2015.
DOI: 10.1007/978-3-319-15892-1_19

inventory levels and routing costs which are truly in conflict to each other. For example, small delivery quantities lead to low inventory levels over time, whereby large delivery quantities give the possibility to minimize the routing costs [15].

This paper introduces swap locations with predefined actions, different customer characteristics and possible vehicle configurations within the IRP. In this problem description of practical relevance, the possibility is given to leave the depot with a semi-trailer attached to the truck and also deliver goods to customers where only a truck is allowed (called truck customers). This distinction of customer properties is made because of e. g. area restrictions [9] and demand characteristics of the customers. The delivery is possible since the vehicle can drive to a swap location, park the semi-trailer and continue with only the truck.

Since the SB-IRP is an extension of the recently investigated Swap Body Vehicle Routing Problem, which was proposed for the VeRoLog Solver Challenge [14], we want to explain our motivation to investigate the SB-IRP. Besides minimizing the routing costs, analyzing a multi-period SB-IRP brings up the question of how to integrate inventory levels at the customers. This is necessary because deliveries can take place before the actual demand and then are stored at the customers. We include the inventory levels by formulating a second objective function.

From a practical point of view, the VeRoLog Solver Challenge set the termination criterion to 600 seconds computing time on a 4-core CPU for the single-period SB-VRP. Since we are now investigating 30 periods, the computational time for the VeRoLog Solver becomes an issue. As a research question, we may ask how to manage the limited computing time (e. g. 600 seconds) in a multi-period model. This is the scope of this article.

The remainder of this paper is organized as follows. In section 2 the SB-IRP is introduced and described. Section 3 presents the Iterative Variable Neighborhood Search for the considered problem. Computational results are analyzed in section 4, followed by the conclusion in section 5.

2   Problem Description of the Swap Body Inventory Routing Problem

We consider a multi-objective and multi-period SB-IRP which is an extension of the SB-VRP. The SB-VRP was initially introduced by the VeRoLog Solver Challenge [14,20] and a further description is also given in [16].

The SB-IRP is classified based on the characterization of [2]. Our problem introduces an one-to-many network where homogenous products are delivered with a heterogeneous fleet over a finite planning horizon, $t = 1, \ldots, 30$. Also the fleet size is unconstrained and the inventory levels are restricted to be non-negative. It is also assumed that the demand q_{it} of customer i in period t ($q_{it} \geq 0$) is deterministic since, from a practical point of view, numerous companies have an idea of their consumption patterns [11]. The SB-VRP is summarized with respect to its main characteristics:

- customer properties,
- asymmetric travel times and distance costs,
- maximum driving time DT^{max} of the vehicle driver per period,
- the possible vehicle configurations and
- swap locations, including the different swap actions.

As illustrated in Figure 1, the vehicle can leave as a truck or as a train. When a truck leaves the depot, only one swap body with capacity Q is attached. Alternatively, a truck and a semi-trailer (train configuration) can depart from the warehouse with two swap bodies. This differentiation is due to mainly two reasons: 1) customer- and 2) demand characteristics. Based on these characteristics, three types of customers are identified: truck-, train- and mandatory train customers. *Truck customers* can exclusively be reached by a truck configuration which is caused by accessibility constraints [9]. The second class of customers are called *train customers* which can either be approached by a truck or by a train vehicle composition. Differently, a *mandatory train customer* must be visited by a truck and a semi-trailer since $q_{it} > Q$.

In Figure 2 the different actions are described when a train configuration moves to a swap location. One main action is the parking of the semi-trailer SB_2 at the selected swap location. Note that every swap body must be picked up before arriving back at the depot. Also a swap action is possible, which allows the driver to change the currently used swap body by parking it at the swap location (SB_1 in the Figure 2) and using SB_2 for further deliveries. The last action is an exchange operation which is e. g. necessary when a truck customer should be visited with SB_2. This might be helpful when not enough goods are in SB_1. Hereby, SB_1 is parked and the truck continues with SB_2. These swap actions are solely possible at the swap locations and not for example at the train customers as proposed in [4,19]. Moreover, no transshipments, respectively load transfers of goods [6,8] are allowed between the swap bodies.

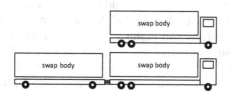

Fig. 1. Example of a truck- and train configuration [14,16]

The objectives of the problem are the minimization of the total costs of the routes and the total inventory levels at the customers for every period. The following components are included in the sum of the routing costs: the truck and the semi-trailer costs. With respect to the truck costs, fixed costs *[Monetary Units (MU)/usage]* occur when a truck is used in the solution. Additionally, the travelled distances by the truck *[MU/km]* and the driver costs *[MU/h]* are

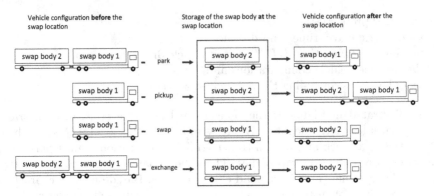

Fig. 2. Description of the possible actions at the swap location [14, 16]

included. Also fixed [$MU/usage$] and variable costs [MU/km] are considered for the semi-trailer, but no driver costs.

The solution should simultaneously determine for every period:

- when and how much to deliver to the customers,
- which vehicle configuration should be used to serve a customer and
- which swap locations and what kind of actions should be applied.

3   Applied Solution Methods

The SB-IRP is $\mathcal{NP}$-hard since the VRP is a special case [5,6]. Based on the complexity, the size of the test instances (number of customers is between approx. 50 and 550), and the limited computing time (600 seconds), a heuristic is applied.

With the aim of solving the SB-IRP, the problem is decomposed in two decision levels: the determination of the delivery periods and the subsequent generation of the tours.

Another idea to solve the SB-IRP could be to apply multilevel programming which aims to incorporate the fact that different point of views exist on several decision hierarchies, such as top management and divisions. Another common feature is that each subordinate level executes its policy after, and in view of, previous decisions [3]. In contrast to this idea, our multi-objective solution approach only assumes one decision maker (DM), here the supplier, whose decisions are not influenced by e. g. the customers. The main advantage of VMI concepts is especially that the supplier has the responsibility of the inventory levels at the customers. The focus of the DM is to investigate the trade-off between the two objectives of the SB-IRP.

3.1   Replenishment Strategy

In order to determine a replenishment plan, a delivery alternative is represented by an n-dimensional vector $\pi = (\pi_1, \ldots, \pi_n)$ where each position describes an integer value. The length of this vector is equivalent to the number of customers which are investigated. Notice that this value shows for how many periods the customer is delivered in a row (delivery periods). E. g., a vector $\pi = (1, 2, 1)$ means that the exact demand of the first and the third customer is serviced every period. The demand of the second customer is fulfilled every second day.

The initial replenishment strategy is computed by identical delivery periods for all customers, starting with 1 and then increasing every value by $+1$. As long as the values can be varied, the vectors are added to an archive of non-dominated solutions. Due to capacity constraints of the vehicles and the stock capacity, the procedure terminates at some point.

Besides, when e. g. one customer has a demand $q_{it} > Q$ (mandatory train customer), delivery periods of two might not be realizable for this customer. To improve the initial replenishment procedure, also maximal delivery periods for every customer are determined. If a position cannot be varied by $+1$, then the maximal delivery period value is used. This procedure is continued until every position in the vector assumes its maximum delivery period value.

The improvement procedure on the replenishment vectors applies a neighborhood search operator which diversifies each position of the vector π by ± 1. Since at most one delivery per day is assumed, values < 1 are infeasible. The maximum number of neighboring solutions is therefore $2n$ [10]. Let us assume that 549 customers are investigated, then one neighborhood consists at most 1098 vectors in the archive. Furthermore, assume that the subsequent computation of the routing costs is 30 seconds for one vector. Then, more than 9 hours (30 second times 1098 vectors) would be needed to investigate only one neighborhood.

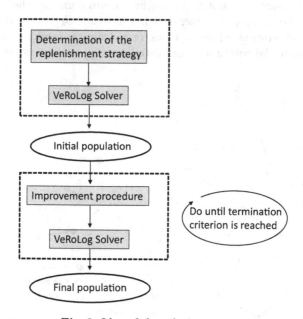

Fig. 3. Idea of the solution process

It becomes quite clear that, in this light, only a subset of the neighborhood can be investigated in a practical experimental setting.

3.2   Generation of the Tours

We recently achieved good results in the VeRoLog Solver Challenge and were ranked third in the competition [13]. Based on our competitive algorithm, the routing for the SB-IRP is solved by this idea.

Exemplarily, the idea of the solution process is shown in Figure 3. The initial replenishment strategies are used as an input for the generation of the tours. Thereby, deliveries for every customer and period are planned, and then the VeRoLog code is used to generate the tours, resulting in the initial population. After that the initial replenishment strategies are varied (see section 3.1) and the VeRoLog Solver is applied to compute the tours. This is done until the termination criterion is reached. The algorithmic idea of the VeRoLog Solver stems from [16]. Thus, only the key points are here summarized.

Various tour segments are representing a tour and the aim of the solution method is to assign customers to the different segments $S_k, k = 1, \ldots, 4$. An illustrative example of the different tour types and segments S_k are depicted in Fig. 4 and described in Table 1.

Construction Heuristic. In order to assign a customer to a route segment S_k, a customer is randomly selected. The priority of S_k is: S_1, S_2, S_3, S_4. If the selected customer does not fit in the first route segment, the second route segment is tested etc. If the customer does not fit in any S_k, a new tour is opened. The vehicle configuration, either a truck or a train, is randomly chosen. The generation of an initial solution is completed once all customers are allocated.

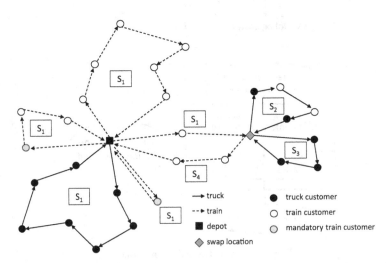

Fig. 4. Illustrative example of tour types and segments [16]

Table 1. Characteristics of the different tour types [4]

Tour types	Description
Pure truck tour	• A tour starts with a truck at the depot and visits 7 truck customers before returning to the depot. • In the tour representation customers are only assigned to S_1.
Pure train tour	• A train visits 6 train customers and arrives back at the depot. • From a representation's point of view, train customers are assigned to S_1 and $S_2 = \emptyset \wedge S_3 = \emptyset \wedge S_4 = \emptyset$.
Pure mandatory train tour	• Only a *mandatory train customer* is visited.
Combined mandatory train tour	• This tour type includes *mandatory train customers* and *train customers*. • The train visits a mandatory train customer followed by two train customers.
Combined tour	• Customers are assigned to every route segment. Thus, a train and a truck vehicle configuration are used. A train services a customer and then continues to the swap location in order to park SB_2. • Then, the truck visits a *truck customer*, 2 *train customers* and another *truck customer* which are assigned to S_2 and applies a swap action at the swap location. • After that, the truck using SB_2 drives to the customers on S_3. Prior to the arrival at the depot, a pickup action is executed and customers of the fourth route segment are visited by using the train.

Improvement Heuristic. The initial route plan **X** is improved by a combination of a Variable Neighborhood Search (VNS) [18] and Iterated Local Search (ILS) [17]. Neighboring alternatives are investigated by a list of several neighborhood operators. When an alternative cannot be improved, which means that a local optimum is obtained, a perturbation move is applied and the search continues based on this alternative. This operator modifies the alternative by randomly removing entire tours and then reconstructing new ones. In the following, let $\mathcal{X}$ denote the set of feasible alternatives.

The VNS makes use of several neighborhood operators:

Intra-Tour-Operator:

1. Intra-Move (INTRA) relocates a selected customer in another S_i in the same tour. A similar operator has been proposed by Van Breedam [21].
2. The classical 2-opt (2-OPT) tries to improve a tour by replacing two arcs with two other arcs.

Inter-Tour-Operator:

1. Two-Inter-Exchange (2-EX) is a move in which the positions of two customers are swapped [12].
2. Inter-Move (INTER) inserts a chosen customer of one tour in another tour.

Problem-Specific-Operator:

1. Change swap locations (CSL) modifies the swap location of every tour.

The acceptance criterion of every operator is the same: a neighboring alternative is accepted iff Tourcost $(X') <$ Tourcost $(X) \wedge \mathbf{X}' \in \mathcal{X}$.

4 Experimental Study

4.1 Test Instances

To evaluate the performance of our solution method, the VeRoLog instances are taken [13]. These real-world instances have been proposed by the EURO Working Group on Vehicle Routing and Logistics Optimization in collaboration with the PTV Group, a company based in Karlsruhe, Germany. The competition was divided into two phases: the qualification and the final phase. At the beginning of the qualification phase, three instances were made public (see small, medium and large1 instance in Table 2). After announcing the teams for the final phase, one more instance was released (large2). In the final phase, another two instances were used to test the VeRoLog Solver (final1 and final2).

The instances are summarized in Table 2. Every instance comes in three variants: a normal-, an all without trailer- and an all with trailer instance. More specifically, the 'normal' instance can contain truck-, train and mandatory train customers. An 'all without trailer' instance only includes truck customers. Train customers and mandatory train customers are used for the 'all with trailer' instance [14].

In order to study the effects of taking delivery quantity decisions into account, we extended the given VeRoLog benchmark instances by the demand data and the stock capacity for every customer. The given instances provided customer's demand data for one period. We simulated the demand for another 29 periods by varying the given demand q_{i1} by at most 20%. Also it has to be taken into account that the swap body capacity is not exceeded. Note that the demand q_{i1}

Table 2. Description of the SB-VRP instances

	# swap locations	# truck customers	# train customers	# mandatory train customer	released
small					
normal	20	15	41	1	
all without trailer	20	57	0	0	Feb 1, 2014
all with trailer	20	0	57	0	
medium					
normal	41	20	186	0	
all without trailer	41	206	0	0	Feb 1, 2014
all with trailer	41	0	206	0	
large1					
normal	99	50	498	0	
all without trailer	99	548	0	0	Feb 1, 2014
all with trailer	99	0	548	0	
large2					
normal	101	50	500	0	
all without trailer	101	550	0	0	May 1, 2014
all with trailer	101	0	550	0	
final1					
normal	102	50	499	0	
all without trailer	102	549	0	0	Jul 1, 2014
all with trailer	102	0	549	0	
final2					
normal	102	50	499	0	
all without trailer	102	549	0	0	Jul 1, 2014
all with trailer	102	0	549	0	

of the SB-VRP and the SB-IRP is the same for the first period. In this sense, the original problem of the Solver Challenge is included in our data's first period. It follows that the provided checker of the VeRoLog Solver Challenge [14] can be used to verify the solutions for each period.

The stock capacity is a hard constraint and it is not allowed to exceed this value at any time. The stock capacity of the customer is set to 10 times q_{i1}.

4.2 Computational Results

Parameter settings
During a tuning phase, several parameters were tested and fixed:

- number of tours that should be removed by the perturbation,
- number of other customers in the vicinity of each customer; this allows the reduction of the neighborhoods and
- number of neighboring swap locations which are tested by the CSL operator.

The algorithm removes two tours, the number of neighboring customers is set to 100, and the number of swap locations which are tested is 15.

Table 3. Hypervolume values for the initial population (single test run)

Instance	0.5 s	1 s	1.5 s	2 s
small				
normal	0.6188	0.6362	0.6327	0.6302
all without trailer	0.5624	0.5215	0.5647	0.5658
all with trailer	0.6668	0.6579	0.6694	0.6682
medium				
normal	0.7063	0.7203	0.7215	0.7201
all without trailer	0.6869	0.7139	0.7178	0.7248
all with trailer	0.7010	0.7199	0.7220	0.7223
large1				
normal	0.7201	0.7621	0.7688	0.7701
all without trailer	0.6703	0.7066	0.7173	0.7208
all with trailer	0.7261	0.7652	0.7733	0.7762
large2				
normal	0.7259	0.7586	0.7658	0.7686
all without trailer	0.6821	0.7188	0.7265	0.7287
all with trailer	0.7237	0.7601	0.7655	0.7704
final1				
normal	0.0877	0.2384	0.2622	0.2765
all without trailer	0.3443	0.7631	0.8218	0.8448
all with trailer	0.0818	0.2293	0.2570	0.2600
final2				
normal	0.7145	0.7477	0.7637	0.7641
all without trailer	0.6627	0.7177	0.7254	0.7285
all with trailer	0.7073	0.7423	0.7516	0.7551

The sequence of the neighborhood operators is for every experiment: 2-EX, CSL, INTRA, 2-OPT and INTER.

Also several other parameters must be investigated, regarding the alternating process of the replenishment strategy and the VeRoLog Solver, and are further explained with respect to:

- How long is the VeRoLog Solver applied for each period?
- What is the termination criterion?
- How large is the subset of solutions when altering the values in π?

Since the allocation of the computational time is studied, 4 variants are tested and analyzed. The VeRoLog Solver is applied to each period for: 0.5, 1, 1.5 or 2 seconds. In order to clarify the running time for one routing plan, an example is briefly explained. Assume that the VeRoLog Solver is applied one second for each period, then it takes approx. 30 seconds to compute one routing plan (since our problem definition includes 30 periods).

The overall termination criterion is composed in the following way: 30 periods times 600 seconds (given by the VeRoLog organizers) times 4 cores, resulting in 72,000 seconds. For comparison reasons with the computational time of the

SB-VRP, the termination criterion is multiplied by four because the experiments of this article only utilize a single core. All experiments were run on an Intel Xeon X5650 2.66 GHz (single CPU core).

In section 3.1, the diversification of the replenishment strategy is described. Since it takes a long time to investigate one complete neighborhood, the size of the subset is reduced to 1% at a time. Thereby, the selection of the vectors is randomly chosen.

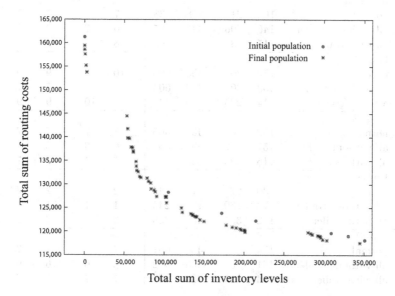

Fig. 5. The initial population (dots) and the final population (stars) of the small-normal instance with an allocation of 2 seconds to the VeRoLog Solver

In Table 3 the hypervolume values of the initial population are presented for the different running times of the VeRoLog Solver over a single test run. Additionally, Figure 5 shows the approximation of the Pareto-Front of the initial and the final population ('small-normal instance' with an allocation of 2 seconds to the VeRoLog Solver).

The hypervolume is commonly used by many researchers to measure the size of the objective space which is dominated by the generated solutions [22]. The hypervolume of a set is measured relative to an anti-optimal point in space [22]. As the worst-possible point, we assume the maximal value of each objective function value plus 10%. Then, the rectangle of each solution with respect to the worst-possible point is computed. The volume is enclosed by the union of the rectangles [23].

The variant with one second never leads to the best hypervolume value. For fourteen out of 18 instances (large1, large2, final1 and final2) the variant with 2 seconds achieves the best hypervolume values. For the 'small- and medium normal', 'small-all without trailer' as well as for the 'medium-all without trailer'

Table 4. CPU time in seconds and cardinality of the initial population

Instance	0.5 s	1 s	1.5 s	2 s	0.5 s	1 s	1.5 s	2 s
small								
normal	145	295	445	595	7	8	7	8
all without trailer	145	295	445	594	5	5	3	4
all with trailer	145	295	445	594	8	8	8	8
medium								
normal	108	218	328	438	7	8	7	7
all without trailer	146	296	445	595	6	7	6	7
all with trailer	115	233	350	468	8	7	7	7
large1								
normal	152	302	453	605	10	9	9	9
all without trailer	153	304	456	604	7	7	7	7
all with trailer	114	225	339	447	9	10	9	9
large2								
normal	152	300	455	605	7	7	7	7
all without trailer	157	302	454	603	6	6	6	6
all with trailer	115	225	339	448	7	7	7	7
final1								
normal	139	277	410	544	2	2	2	2
all without trailer	70	130	187	245	1	1	1	1
all with trailer	152	307	458	605	2	2	2	2
final2								
normal	154	301	453	601	7	7	7	7
all without trailer	151	303	453	602	6	6	6	6
all with trailer	114	225	336	448	7	7	7	7

instances, the variants with 1.0 and 1.5 achieve slightly better values than allocating two seconds to the VeRoLog Solver (see Table 3).

Table 4 gives the CPU times in seconds (left part of the Table) and the cardinality (right part) of the initial population for all instances. Analyzing the CPU times for the different variants, it can be seen that the computational times strictly increase when the computational time for the VeRoLog Solver is increased which has to be expected. Besides, it can be observed that the number of initial solutions varies with respect to the test instances. These results are simply caused by different demand data, vehicle- and stock capacities.

Surprisingly, the cardinality of the set of initial solutions for instance 'final1' is low, although the same procedure is conducted. One main difference between the 'final1' and the 'large1/large2' respectively 'final2' instances are the composition of the demand data (normal), which is the reason for the different results. For example, the 'large1/large2' and 'final2' instances only have a single customer which has a demand $q_{i1} > 800$. In contrast, the 'final1' instance consists of 73 customers with a demand higher or equal to 800, and 69 have a demand greater than or equal to 900. This characteristic leads to difficulties in assigning several customers to one vehicle, since we also combine the demand of several periods.

Table 5. Hypervolume values for the final population (single test run)

Instance	0.5 s	1 s	1.5 s	2 s
small				
normal	0.8176	0.8063	0.7951	0.7783
all without trailer	0.8288	0.8251	0.7962	0.8037
all with trailer	0.8176	0.7925	0.7902	0.7821
medium				
normal	0.7807	0.7666	0.7594	0.7506
all without trailer	0.7776	0.7683	0.7731	0.7556
all with trailer	0.7735	0.772	0.7561	0.7518
large1				
normal	0.7381	0.7758	0.7844	0.7913
all without trailer	0.6994	0.7265	0.7335	0.7407
all with trailer	0.7449	0.7801	0.7914	0.7933
large2				
normal	0.7489	0.7742	0.7776	0.7802
all without trailer	0.7055	0.7365	0.7388	0.7409
all with trailer	0.7444	0.7731	0.7760	0.7778
final1				
normal	0.2953	0.3794	0.3519	0.3369
all without trailer	0.5034	0.9039	0.9633	0.9321
all with trailer	0.2858	0.3549	0.3601	0.3600
final2				
normal	0.7441	0.7678	0.7768	0.7800
all without trailer	0.7002	0.7376	0.7472	0.7530
all with trailer	0.7420	0.7638	0.7735	0.7762

This characteristic results in a higher number of vehicles and therefore higher routing costs.

Since the demand characteristic of this instance could appear to be problematic, one idea to overcome this issue is to adapt the replenishment strategy, e. g. instead of delivering only the exact demand of the customers, the delivery quantities might be splitted to different periods as long as the demand in q_{it} can be satisfied. Another idea could be to modify the procedure for the initial replenishment strategies. At the moment, each position in π is changed by $+1$ at once and if π_i cannot be modified, the maximal delivery period value is assumed. Alternatively, only one position can be changed at a time in order to increase the initial set of solutions.

The hypervolumes for the final population are illustrated in Table 5. For the small and the medium instances, the highest hypervolume is achieved when 0.5 seconds are allocated to the Solver. For instances large1/large2 and final2 the highest hypervolume is reached when 2 seconds are given to the VeRoLog Solver. The allocation of 1 and 1.5 seconds to the Routing Solver lead to the highest hypervolume value for the final1 instances.

Table 6. Number of evaluations of the final population

Instance	0.5 s	1 s	1.5 s	2 s
small				
normal	4963	2442	1620	1213
all without trailer	4962	2442	1621	1212
all with trailer	4999	2449	1623	1213
medium				
normal	5741	2754	2009	1439
all without trailer	4938	2439	1621	1212
all with trailer	5966	3010	2033	1590
large1				
normal	4890	2397	1603	1203
all without trailer	4764	2396	1596	1204
all with trailer	5963	3014	2022	1558
large2				
normal	4683	2400	1597	1202
all without trailer	4812	2373	1603	1201
all with trailer	5959	3015	2027	1491
final1				
normal	4485	2323	1572	1192
all without trailer	4165	2195	1510	1169
all with trailer	4533	2364	1575	1190
final2				
normal	4706	2388	1598	1203
all without trailer	4814	2389	1594	1200
all with trailer	5957	3108	2051	1509

The number of evaluations is illustrated in Table 6. The highest value is reached when half a second is allocated to the Solver and the lowest when two seconds are assigned, which already has been suspected before the experiments. Note that the number of evaluations is higher for the *medium* instance compared with the *small* instance, and we must see into why this happens.

For nine out of 18 instances ('large1/large2' as well as 'final2'), the allocation of two seconds leads to the highest hypervolume and proves to be superior. Although some counterexamples exist, the overall conclusion is rather strong in favor of the assigning of two seconds, since the larger instances with up to 550 customers are more suitable for practical cases.

5   Conclusion

This article has developed an optimization approach for the biobjective SB-IRP. After introducing the problem, which is based on a real-world application, a two-stage solution approach is proposed that alternates between the determination of the replenishment strategy and the VeRoLog Solver. Subsequently, different variants for the allocation of the computational times are conducted and analyzed on the extended VeRoLog test instances.

By comparing the hypervolume values for the different utilization of the computational times, considerable differences became obvious. For the smaller instances (between 57 and 200 customers), the allocation of little computation time to VeRoLog Solver achieves the highest hypervolume values. Contrary for the larger instances (approximately 500 customers) it is better to allocate more computational time to the routing plans, with the exception of the final1 instance, where also the cardinality of the set of initial solutions is small compared to the other instances. Comparing the results with respect to the size of the instances, it seems that, it is more convenient for smaller instances to invest more time in the replenishment strategies than in the routing. On the contrary, for larger instances the routing is more important. A higher hypervolume is achieved when fewer solutions are investigated and more computational time is allocated to the routing.

Although the conducted experiments indicate that the allocation of 2 seconds to VeRoLog Solver is promising for larger instances, future research should be dedicated towards the investigation of further variants, e. g. 3 seconds. Also the allocation of longer computational times to instances greater than 500 customers should be explored.

References

1. Abdollahi, M., Arvan, M., Omidvar, A., Ameri, F.: A simulation optimization approach to apply value at risk analysis on the inventory routing problem with backlogged demand. International Journal of Industrial Engineering Computations **5**(4), 603–620 (2014)
2. Andersson, H., Hoff, A., Christiansen, M., Hasle, G., Løkketangen, A.: Industrial aspects and literature survey: Combined inventory management and routing. Computers & Operations Research **37**(9), 1515–1536 (2010)
3. Bard, J.F.: Practical Bilevel Optimization: Algorithms and Applications. Nonconvex optimization and its applications. Kluwer Academics, Dordrecht (1998)
4. Chao, I.M.: A tabu search method for the truck and trailer routing problem. Computers & Operations Research **29**(1), 33–51 (2002)
5. Coelho, L.C., Cordeau, J.F., Laporte, G.: Thirty years of inventory routing. Transportation Science **48**(1), 1–19 (2014)
6. Coelho, L.C., Cordeau, J.F., Laporte, G.: The inventory-routing problem with transshipment. Computers & Operations Research **39**(11), 2537–2548 (2012)
7. Desaulniers, G., Rakke, J.G., Coelho, L.C.: A branch-price-and-cut algorithm for the inventory-routing problem. Tech. Rep. G-2014-19, Les Cahiers du GERAD, Montrél (Québec), Canada (April 2014)
8. Drexl, M.: Branch-and-price and heuristic column generation for the generalized truck-and-trailer routing problem. Journal of Quantitative Methods for Economics and Business Administration **12**(1), 5–38 (2011)
9. Drexl, M.: Applications of the vehicle routing problem with trailers and transshipments. European Journal of Operational Research **227**(2), 275–283 (2013)

10. Geiger, M.J., Sevaux, M.: On the use of reference points for the biobjective inventory routing problem. In: Proceedings of the 9th Metaheuristics International Conference MIC 2011, pp. 141–149 (2011)
11. Gudehus, T., Kotzab, H.: Comprehensive logistics, 2nd edn. Springer, Heidelberg (2012)
12. Gündüz, H.I.: The Single-Stage Location-Routing Problem with Time Windows. In: Böse, J.W., Hu, H., Jahn, C., Shi, X., Stahlbock, R., Voß, S. (eds.) ICCL 2011. LNCS, vol. 6971, pp. 44–58. Springer, Heidelberg (2011)
13. Heid, W., Hasle, G., Vigo, D.: Verolog solver challenge - results. VeRoLog (EURO Working Group on Vehicle Routing and Logistics Optimization) and PTV Group p. 1 (2014). http://verolog.deis.unibo.it/verolog-solver-challenge (online; accessed August 31, 2014)
14. Heid, W., Hasle, G., Vigo, D.: Verolog solver challenge 2014 - VSC2014 problem description. VeRoLog (EURO Working Group on Vehicle Routing and Logistics Optimization) and PTV Group pp. 1–6 (2014). http://verolog.deis.unibo.it/news-events/general-news/verolog-solver-challenge-2014 (online; accessed May 23, 2014)
15. Huber, S., Geiger, M.J.: Simulation of preference information in an interactive reference point-based method for the bi-objective inventory routing problem. Journal of Multi-Criteria Decision Analysis (2014, accepted)
16. Huber, S., Geiger, M.J.: Swap Body Vehicle Routing Problem: A Heuristic Solution Approach. In: González-Ramírez, R.G., Schulte, F., Voß, S., Ceroni Díaz, J.A. (eds.) ICCL 2014. LNCS, vol. 8760, pp. 16–30. Springer, Heidelberg (2014)
17. Lourenço, H.R., Martin, O.C., Stützle, T.: Iterated local search. In: Glover, F., Kochenberger, G.A. (eds.) Handbook of Metaheuristics, International Series in Operations Research & Management Science, vol. 57, pp. 321–353. Springer (2003)
18. Mladenović, N., Hansen, P.: Variable neighborhood search. Computers & Operations Research 24(11), 1097–1100 (1997)
19. Scheuerer, S.: A tabu search method for the truck and trailer routing problem. Computers & Operations Research 33(4), 894–909 (2006)
20. Schulte, F., Voß, S., Wenzel, P.: Heuristic routing software for planning of combined road transport with swap bodies: A practical case. In: Proceedings of MKWI 2014 - Multikonferenz Wirtschaftsinformatik, 15–28 February 2014, pp. 1513–1524. Paderborn (2014)
21. Van Breedam, A.: Comparing descent heuristics and metaheuristics for the vehicle routing problem. Computers & Operations Research 28(4), 289–315 (2001)
22. While, L., Hingston, P., Barone, L., Huband, S.: A faster algorithm for calculating hypervolume. IEEE Transactions on Evolutionary Computation 10(1), 29–38 (2006)
23. Zitzler, E.: Evolutionary Algorithms for Multiobjective Optimization: Methods and Applications. Doctoral dissertation ETH 13398, Swiss Federal Institute of Technology (ETH), Zürich, Switzerland (1999)

Machine Decision Makers as a Laboratory for Interactive EMO

Manuel López-Ibáñez[1]([✉]) and Joshua Knowles[2]

[1] IRIDIA, Université Libre de Bruxelles (ULB), Brussels, Belgium
manuel.lopez-ibanez@ulb.ac.be
[2] School of Computer Science, University of Manchester, Manchester, UK
j.knowles@manchester.ac.uk

Abstract. A key challenge, perhaps the central challenge, of multi-objective optimization is how to deal with candidate solutions that are ultimately evaluated by the hidden or unknown preferences of a human decision maker (DM) who understands and cares about the optimization problem. Alternative ways of addressing this challenge exist but perhaps the favoured one currently is the interactive approach (proposed in various forms). Here, an evolutionary multi-objective optimization algorithm (EMOA) is controlled by a series of interactions with the DM so that preferences can be elicited and the direction of search controlled. MCDM has a key role to play in designing and evaluating these approaches, particularly in testing them with real DMs, but so far quantitative assessment of interactive EMOAs has been limited. In this paper, we propose a conceptual framework for this problem of quantitative assessment, based on the definition of machine decision makers (machine DMs), made somewhat realistic by the incorporation of various non-idealities. The machine DM proposed here draws from earlier models of DM biases and inconsistencies in the MCDM literature. As a practical illustration of our approach, we use the proposed machine DM to study the performance of an interactive EMOA, and discuss how this framework could help in the evaluation and development of better interactive EMOAs.

Keywords: Machine decision makers · Artificial decision makers · MCDM · Interactive EMO · Performance assessment

1 Introduction

Good introductions to the current state of research in interactive evolutionary multi-objective optimization algorithms (iEMOAS) [3,10,21] indicate that several important issues remain to be tackled when combining methods from multi-criteria decision making (MCDM) and evolutionary multi-objective optimization (EMO). We think that a most fundamental one is that few interactive approaches have been evaluated in a quantitative way, in particular in a way that would reveal how effective and efficient they are at finding a solution the decision-maker (DM) finds satisfactory or preferred. More than this, we think

© Springer International Publishing Switzerland 2015
A. Gaspar-Cunha et al. (Eds.): EMO 2015, Part II, LNCS 9019, pp. 295–309, 2015.
DOI: 10.1007/978-3-319-15892-1_20

that quantitative assessment of methods involving decision-making remains controversial[1], and this controversy has perhaps led to a position where it is standard to propose new methods, and to justify them on theoretical grounds, but not to test them in a way that reveals practical properties. Testing is difficult (and controversial) because of the need for a DM who interacts with the system (and cares about the result), and the fact that DMs are all different [6,27]. Nevertheless, there is ongoing research pursuing the goal of quantitative assessment by means of empirical analysis [7,16]. In one of the first works pursuing a theoretical quantitative analysis, Brockhoff et al. [5] derived bounds on both the runtime and the number of queries to the DM for two different local-search based EMOAs, under the simplifying assumption that the DM would be able to answer a query whenever the EMOA had an incomparable pair of solutions to choose between. A common criticism of such quantitative analysis is that they rely on strongly simplifying assumptions about human DMs.

Simulating human DMs is not an easy task and most work in MCDM has focused on how to model and elicit the DM's preferences. Few works have considered the simulation of human factors and other non-idealities of the decision-making process. Kornbluth [17] incorporates "unsureness" of the DM as a range of values of the utility function for which the DM cannot make a decision. Morgan [19] describes the development of an even more realistic model to simulate expert decision-making in dynamic and time-critical environments (e.g., air combat); yet, the model itself is not fully described and its applicability to other contexts remains unclear. Perhaps the most extensive simulation of DM biases and other non-idealities has been conducted by Stewart [22–24], who studied how the ranking of efficient solutions is affected by these non-idealities. Nonetheless, the usual approach when evaluating interactive EMOAs in the recent literature is to add random noise to the DM's preferences [7,16].

The step which we propose (and begin to investigate here) beyond recent studies is to use *more realistic* machine decision makers, and to perform a more detailed and multi-factorial analysis of the relationships between performance, DM satisfaction (i.e., finding and recommending a solution close to the ideal one the DM would choose), DM biases, and the EMO/MCDM interactive approach.

This paper is structured as follows. Section 2 presents our proposed conceptual framework for quantitative assessment of interactive multi-objective optimization. In particular, we propose the concept of machine DMs as a problem- and preference-independent framework for the simulation of realistic DMs. Section 3 describes a possible instantiation of our framework based on previous work from the EMO and MCDM literature. The goal of this instantiation is not only to serve as an example, but also to show how variations of the human non-idealities can have strong and surprising effects on the behavior of an interactive EMOA. Towards this goal, we present in Section 4 experimental results that confirm these effects. Finally, Section 5 discusses in more depth related work and the

---

[1] This is apparent from several discussions at the Dagstuhl series of seminars concerned with MCDM and EMO [1].

context of our proposal within the ongoing effort to combine MCDM and EMO approaches.

2   A Conceptual Framework of Machine Decision Making in Interactive Multi-objective Optimization

One of the difficulties when comparing interactive EMOAs is that competing algorithms not only differ in their interaction style and the preference models they can handle, but also on the underlying MOEA that ultimately provides the alternatives that are considered by the DM. Thus, it is difficult to assess whether any observable differences are due to one or another aspect, or their precise combination. The other major difficulty, as discussed above, is how to simulate a realistic DM in a way that allows us to understand the influence that human biases and other non-idealities have on the behavior of iEMOAs.

In order to overcome the above difficulties, we propose a conceptual framework for the quantitative analysis of iEMOAs. The architecture of our proposed framework is shown in Figure 1. It is composed of three main modules: a machine decision maker, an interactive module and an EMOA. Traditionally, research on interactive EMOAs has focused on the combination of the two latter components and considered the machine DM as a preference function, perhaps with some added noise.

Let us assume a multi-objective optimization problem with m objectives, and let $z = (z_1, \ldots, z_m) \in \mathbb{R}^m$ represent an objective vector. Moreover, let us assume that there is an ideal preference function $U(z) \in \mathbb{R}$ that must be maximized in order to satisfy the DM. Then, in our proposed framework, the machine DM simulates a *true preference function* $U(\cdot)$, but in addition it simulates several biases that distort the expression of the true preference function. As a result, the interaction module does not have access to the true preference function and instead it interacts (either directly or indirectly) with the resulting *imperfect preference function* $\hat{U}(\cdot)$. Another characteristic of the proposed framework is that the machine DM may also distort the true set of objectives (the *true criteria* $z \in \mathbb{R}^m$) such that they are different from the criteria optimized by the EMOA (the *modeled criteria* $\hat{z} \in \mathbb{R}^{m'}$). The rationale for this is that the way the DM sees the problem is not necessarily the way that the EMOA is able to optimize it. If both non-idealities are present, they are combined such that the imperfect preference function is evaluated on the modeled criteria.

A particular instantiation of each module can be defined by setting their parameters (the parameter layer in Fig. 1). An instantiation of the framework can then be applied to a given preference function and optimization problem, and the effect of different parameter settings can hopefully help us to identify general effects on the performance of various interactive EMOAs. Thus, the main goal of the proposed framework is to enable a factorial analysis, where the effect of the different modules can be analyzed separately and independently from a particular preference function and optimization problem being tackled.

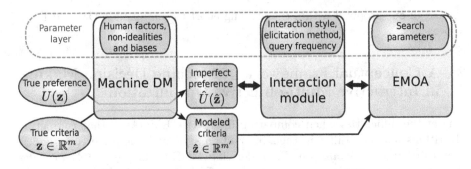

Fig. 1. A conceptual framework for the quantitative assessment of iEMOAs. The framework comprises three modules: the machine decision maker, the EMOA, and the interaction approach, which mediates between the other two. The parameter layer (top) encapsulates the settings of each module to be controlled, which enables experiments in a range of scenarios to be conducted. One goal of this framework is to generalize over the actual DM's preferences (*true preference*) and the actual optimization problem at hand (*true criteria*). Moreover, although the EMOA interacts with the DM via the interaction module, the problem being optimized by the EMOA (*modeled criteria*) is not necessarily identical to the problem seen by the machine DM (*true criteria*).

3   An Example Instantiation of Our Framework

Rather than proposing our own interactive EMOA (iEMOA) here, we chose to recast previous work as an instantiation of our proposed framework. This allows us to re-analyze results from the literature in a new light, while also showing how we can combine advanced iEMOAs with previous studies in MCDM.

The EMO Algorithm. The EMO algorithm used here is BC-EMOA [2], which is a variant of NSGA-II that learns the DM's preference function using support vector ranking (SVR).

The Interaction Module. BC-EMOA offers several alternatives for interaction. Here we consider the one recommended by the authors, which is based on periodically presenting a set of solutions to the DM, who must rank them according to her preferences. This means that, when the interaction module is active, the EMOA algorithm does not have direct access to the preference values.

The Machine DM. The machine DM used in this work follows the simulation of non-idealities proposed by Stewart [22]:

- **Omitted objectives.** From the m objectives considered by the DM, $q < m$ objectives are not known, that is, they are not modeled by the iEMOA. This may be due to a failure to identify relevant objectives when modeling the problem. The $m - q$ objectives ($\hat{z}_k$, $k = 1, \ldots, m - q$) that are known are selected randomly with a probability proportional to their true weights.

- **Mixed objectives.** The objectives modeled might actually correspond to the aggregation of two or more of the objectives internally considered by the DM. The machine DM simulates this "mixing" by making the $m - q$ known objectives a combination of two true objectives such that $\hat{z}_k = (1 - \gamma)z_{c_k} + \gamma z_{c_{k+1}}$, where $\gamma \in [0,1)$ is called the *mixing* parameter, and c_k is the position of objective k in the random selection described above (the position for not selected objectives does not matter).
- **Imperfect preference function.** Instead of the true preference function $U(\boldsymbol{z})$, the machine DM uses a transformed function $\hat{U}(\hat{\boldsymbol{z}})$, which is similar to the true preference function except that
  - the q objectives that are not modeled are always set to the same value,
  - and the addition of Gaussian noise $N(0, \sigma^2)$, where σ is a parameter of the machine DM.

4   Experimental Analysis

4.1   Experimental Setup

In order to assess the impact of the non-idealities and compare our results with those reported in the literature, we use similar parameters for BC-EMOA as in the original paper [2], that is, the algorithm runs for 500 generations, with a population of 100 individuals, and the standard crossover and mutation operators of NSGA-II. After the first 200 generations, the algorithm presents 10 solutions to the machine DM, who must rank them according to her preferences (either the true preference function or the imperfect one). This interaction occurs at most three times, with 20 generations between each interaction.

In the experimental analysis, we compare the best solution returned by three variants of the algorithm:

- **G**, or the *gold standard* variant [2], uses the true utility function $U(\boldsymbol{z})$ directly without any interaction. The MOEA optimizes the true criteria $\boldsymbol{z}$.
- **M**, or the *modeled* variant, uses the imperfect utility function $\hat{U}(\hat{\boldsymbol{z}})$ directly without any interaction. In addition, the MOEA uses the modeled criteria $\hat{\boldsymbol{z}}$ instead of the true criteria $\boldsymbol{z}$. When $q = 0$, $\gamma = 0$ and $\sigma = 0$, then $\hat{U}(\hat{\boldsymbol{z}})$ and $U(\boldsymbol{z})$ are equal, thus **G** and **M** are equivalent.
- **I**, or the *interactive* variant, interacts with the machine DM as described above. The machine DM uses the imperfect utility function to rank solutions and the MOEA optimizes the modeled criteria.

As for the machine DM parameters, we consider all combinations of the following settings: $m - q \in \{0, 1, 2\}$, $\gamma \in \{0.0, 0.05, 0.1, 0.2\}$ and $\sigma \in \{0.0, 0.05, 0.1, 0.2\}$, which are in the range of those considered by Stewart [22].

As a simple benchmark problem suite, we consider DTLZ1, DTLZ2, DTLZ6 and DTLZ7 [11] with $m = 5, 7$ objectives and $n = 2m$ variables. As for the true preference function, although in principle any arbitrary additive value function could be used [22], for simplicity, we consider here a machine DM with a linear scalarizing function and three different sets of randomly generated weights. The application of the machine DM described above to such a preference function is straightforward.

We repeat each run 10 times with different random seeds, but in order to reduce variance, we use the same set of 10 seeds for all the variants compared. Since some of the variants are equivalent, they will produce the same results.

We assess the results according to normalized true utility value $U(\cdot)$ of the most preferred solution found computed as follows. For a given problem and preference function, we compute the maximum and minimum values of the true preference function ever found over all runs of the algorithms, then we normalize the true preference values corresponding to the most preferred solution returned by each run to the interval $[0, 1]$, as $U'(\boldsymbol{z}) = \frac{(U(\boldsymbol{z}) - U_{min})}{(U_{max} - U_{min})}$, such that 0 corresponds to the worst value and 1 to the best one.

4.2   Experimental Results

We present the results in terms of plots (e.g., Figure 2) of the mean U' for each of the three variants **G**, **M** and **I** described above. Each point corresponds to the mean value of the 10 runs for each of the three preference settings. The error bars denote a 95% confidence interval around the mean.

Figure 2 shows the results of omitting objectives (parameter q) without noise or mixing objectives for all problem instances with $m = 5$ (left) and $m = 7$ (right) objectives. When no objective is omitted ($q = 0$), the variants **G** and **M** are equivalent. As soon as we omit one or two objectives, there is a significant drop in the normalized preference value of the solutions found by **M** and **I**, which for $q = 2$ becomes more than two times worse for some problems. Interestingly, the difference between **I** and **M** often decreases for larger q. This is explained by the fact that by dropping objectives, learning the imperfect preference function becomes easier, despite the fact that the imperfect function is further away from the true one.

Figure 3 shows the results of mixing objectives, where the degree of mixing is controlled by parameter γ. The similar values obtained by **G** and **M** suggest that $\gamma = 0.1$ is not high enough to produce a noticeable effect. On the other hand, further increasing γ up to 0.2 produces an effect that is problem-dependent: In the case of DTLZ2, it makes more difficult for the interactive approach **I** to produce as good preferred solution as either **G** or **M**, whereas in the case of DTLZ7 and $m = 5$, the value $\gamma = 0.2$ produces the opposite effect.

Figure 4 shows the effect of adding noise to the utility function via parameter σ. With a sufficient high noise, we can observe that the variant using the imperfect preference (**M**) deteriorates noticeably with respect to the gold variant **G**. However, the effect on the actual interactive variant **I** is limited. In fact, for DTLZ1 and $m = 5$, a high σ does actually help **I** to approximate better the most preferred solution. Except for a few cases (DTLZ2 and DTZL7 with $m = 7$), the noise does not seem strong enough to produce any differences between **G** and **M**, and neither it has a remarkable effect on **I**.

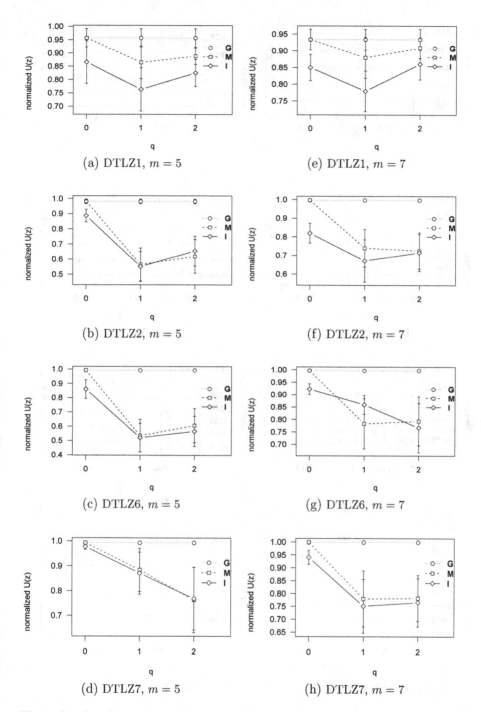

Fig. 2. Results when omitting q out of m objectives without noise ($\sigma = 0.0$) nor mixing of objectives ($\gamma = 0.0$)

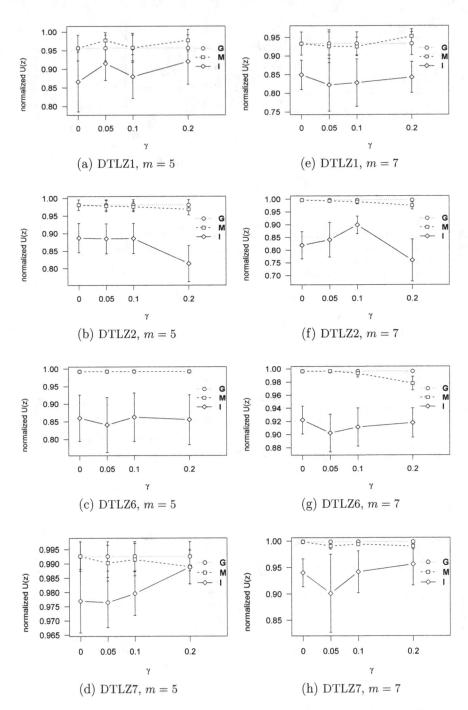

Fig. 3. Results when mixing objectives (parameter γ) without omitting objectives ($q = 0$) and without noise ($\sigma = 0.0$)

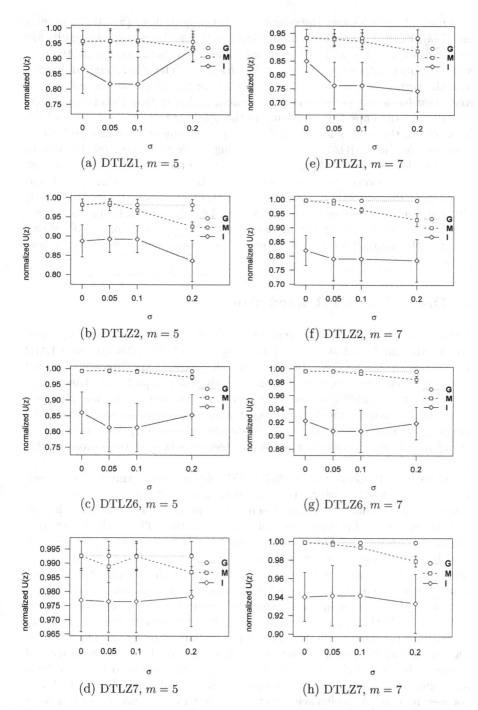

Fig. 4. Results when adding noise to the utility function (parameter σ) without omitting objectives ($q = 0$) and mixing them ($\gamma = 0.0$)

Finally, Figure 5 shows the combined effect of both noise ($\sigma = 0.2$) and mixing of objectives ($\gamma = 0.2$) when omitting $q = 0, 1, 2$ objectives. The results are mostly as expected, that is, the combined effect is a sharp decrease of the normalized utility with respect to the gold variant **G**. The other major difference with the corresponding plots in Fig. 2 is that, in this case, the interactive approach that learns the preferences (**I**) is sometimes better than the approach that directly uses the imperfect preference of the machine DM (**M**), e.g., in DTLZ2. In addition, the results for DTLZ7 when $q = 0$ suggest that this problem is quite easy for BC-EMOA. Yet, when omitting one or two objectives, there is an enormous degradation of the utility value, despite the fact that there is almost no difference between the values obtained by the interactive algorithm **I** and the variant using directly the imperfect preference **M**. Thus, this indicates that, although learning the preference function is relatively easy in this problem, even in the presence of noise or mixing of objectives, the omission of two objectives will lead the algorithm to a completely wrong answer with respect to the true preference.

5 Discussion and Related Work

Although nearly 15 years old, the review by Coello Coello [9] is worthy of attention for the number of attempts at merging methods from MCDM with EMO approaches already proposed by that time. Of course, in the intervening period there has been much further progress in EMO and in preference-based optimization and a proliferation of algorithms and interaction schemes. However, we believe the advancement in methods for the *assessment* of interactive EMO methods (in terms of their ability to satisfy a decision maker, or decision makers of different types) is less clear and there seems to be much less work in this direction.

As reviewed by Coello Coello [9], MCDM splits in broad terms into the French and American schools, and there is in general much difference of opinion about how preferences should be elicited, the benefits and vagaries of different schemes, and any number of difficulties associated with this task. This is all further compounded when one considers how to marry an MCDM approach eliciting and modelling preferences interactively (or otherwise) to an EMO algorithm, which generates a highly non-deterministic trajectory through the candidate solution space. But sidestepping this undoubted complexity, we can ask a simpler question. If a DM had a certain set of preferences and if they were expressible in some fixed way (as in the American school), how could we assess whether a particular modeling of this DM — the outcome of the method of preference elicitation — was successful? We could answer this simple question by agreeing to use not a human, but a machine DM for which we have full access to the underlying preference model. What makes it a *machine DM* simulating a human DM, and not merely a simple preference structure, is that we can also add into it more complex actions like biases, inconsistencies, learning and so forth (but all controllable by us). Using this approach, we can then evaluate how well the DM's

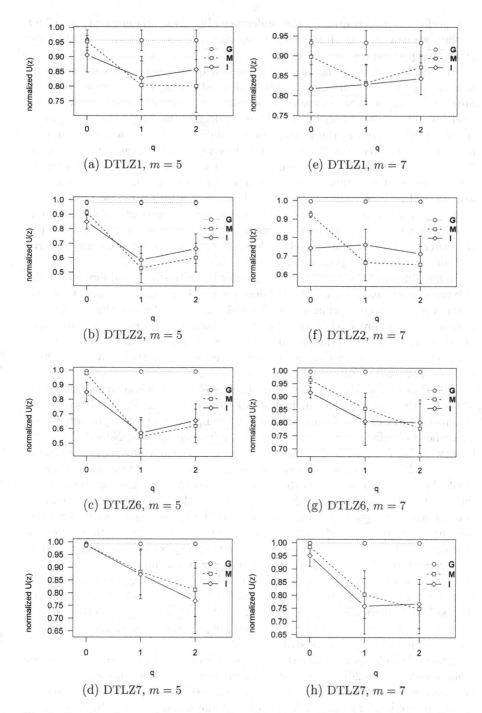

Fig. 5. Results when omitting q out of m objectives with both noise ($\sigma = 0.2$) and mixing of objectives ($\gamma = 0.2$)

true preference is captured by the preference elicitation scheme, or in an EMO system, we can measure how well the search finds a solution that satisfies the DM's true preferences.

We are by no means the first to approach quantitative assessment in this way. Indeed, we were inspired in this work most strongly by previous work by Stewart [22–24], and based several aspects of our approach on it. In particular, modeling several non-idealities of real human decision making, Stewart [22] measured the robustness in the ranking of alternatives obtained when attempting to model the DM as a stable additively-independent (non-linear) value function. Working with simulations involving 7 criteria and 100 nondominated alternatives, a series of sensitivity analyses showed that the 'elicited' value functions worked well in preserving the 'true' ranking of solutions provided that (i) the value function was modeled as piecewise linear with a sufficient number of pieces (4 seemed sufficient for value functions derived from Prospect Theory [15]), (ii) the criteria are close to additively independent, and (iii) not too many criteria are missed out (missing one or two was not very detrimental).

While seminal as a simulation to quantify the robustness and limits of additive value functions, the paper (ibid.) does not consider either the elicitation process per se, or search (optimization). Thus the problem of evaluating iEMOAs is a good deal more involved. We hope that we have remained true to Stewart's aim of using (machine) DMs that mimic real human behaviours in important ways, while showing how we can begin to assess complete interactive EMO algorithms in ways that matter for the DM who would be using them.

The choice of iEMOA used in the study, BC-EMO, is somewhat arbitrary; although, the fact that it can handle up to five different types of preference information means that future analysis can compare the results presented here with those obtained with more complex preference models. Nonetheless, our framework is intended to be able to abstract away from any particular EMO algorithm, DM's preference model and preference elicitation technique. Thus, it would be very interesting to extend our analysis to other leading methods in interactive EMO such as [4,8,12–14,20,28] within the same framework, and this is our longer term goal. In particular, it would be important to study potential mismatch between methods that use reference points and a DM with preferences that do not rely on such, or vice versa. Evaluating very sophisticated approaches, like robust ordinal regression [13] or those based on machine learning [7], would shed light on whether these approaches really work in practice, and under what violations of DM assumptions do they also begin to exhibit poorer performance. And as EMO algorithms become more adept at handling many objectives, it will be increasingly important to put proposals for preference-based many-objective methods such as [28] to more stringent testing.

From a broader perspective still, the problem of eliciting human preferences while searching [25,26] is not limited to mathematical optimization per se. A common scenario where human DMs are exercised on a daily basis are web or database searches for things such as books, holidays, and property. In these cases, human users both seem to learn about or construct their preferences through the interaction, and also learn the querying system. Work in this area (ibid.) seems advanced compared to work on iEMOAs in terms of the interaction systems proposed, and also concerning evaluation of them. We see a bright future for iEMOAs if and only if we can embrace a similar focus on the objective evaluation of our MCDM/EMO hybrids, including realistic and generalized DM models (machine DMs), and a more stringent testing regime that presents a variety of challenges to the working of these methods.

6 Conclusion

We have given an illustration how EMO/MCDM interactive approaches can be evaluated quantitatively using a conceptual framework based on machine DMs incorporating human-like non-idealities. Whilst iEMOAs are becoming more sophisticated, incorporating more advanced methods both from EMO and MCDM, it has remained unclear up to now how we can quantitatively assess the improvement. We proposed and demonstrated a few parameters of machine DMs for assessing iEMOA performance quantitatively. Importantly, the robustness of iEMOAs can be evaluated using this approach and we show with an example that existing iEMOAs are not very robust in the face of these lesser tested non-idealities. In particular, BC-EMOA is very sensitive to the omission of objectives and much less sensitive to their mixing. In fact, we observed that there are non-obvious interactions between various parameters of the machine DM. For instance, under some circumstances, the mixing of objectives may actually help the iEMOA to not get confused by an imperfect preference function. From these foundations, we hope to build a more comprehensive testing facility for the interactive EMOA community.

The experiments reported here are evidently preliminary. We plan in the future to extend the experiments to more complex preference functions. Given the variability of these preliminary results, carrying out a full factorial ANOVA would help to identify the most important factors. Future work will extend the framework of machine DMs to other types of interactions, such as goal programming [24] and aspiration-based techniques [23]. It is currently an open question how to extend the framework to incorporate a wider range of human behaviors and other non-idealities, in particular, the role of learning or evolving preferences by the DM. Finally, the ultimate goal of this framework should be to provide incentives and a way to benchmark interactive EMOAs able to cope with the complex behaviors of human DMs, possibly enabling at some point in the future the automatic design of interactive EMOAs [18].

Acknowledgments. We would like to thank the Research Center Schloss Dagstuhl[2] for hosting the "Learning in Multiobjective Optimization" seminar (id 12041). We also thank the authors of BC-EMOA for providing its source code. Manuel López-Ibáñez acknowledges support from the Belgian F.R.S.-FNRS, of which he is a postdoctoral researcher. Joshua Knowles thanks Theo Stewart for useful discussions.

---

[2] http://www.dagstuhl.de/en

References

1. Auger, A., Brockhoff, D., López-Ibáñez, M., Miettinen, K., Naujoks, B., Rudolph, G.: Which questions should be asked to find the most appropriate method for decision making and problem solving? (Working Group "Algorithm Design Methods"). In: Greco et al. [12], pp. 50–99
2. Battiti, R., Passerini, A.: Brain-computer evolutionary multiobjective optimization: A genetic algorithm adapting to the decision maker. IEEE Transactions on Evolutionary Computation 14(5), 671–687 (2010)
3. Branke, J.: Consideration of partial user preferences in evolutionary multiobjective optimization. In: Branke, J., Deb, K., Miettinen, K., Słowiński, R. (eds.) Multiobjective Optimization. LNCS, vol. 5252, pp. 157–178. Springer, Heidelberg (2008)
4. Branke, J., Greco, S., Słowiński, R., Zielniewicz, P.: Interactive evolutionary multiobjective optimization driven by robust ordinal regression. Bulletin of the Polish Academy of Sciences: Technical Sciences 58(3), 347–358 (2010)
5. Brockhoff, D., López-Ibáñez, M., Naujoks, B., Rudolph, G.: Runtime analysis of simple interactive evolutionary biobjective optimization algorithms. In: Coello, C.A.C., Cutello, V., Deb, K., Forrest, S., Nicosia, G., Pavone, M. (eds.) PPSN 2012, Part I. LNCS, vol. 7491, pp. 123–132. Springer, Heidelberg (2012)
6. Buchanan, J.T.: An experimental evaluation of interactive MCDM methods and the decision making process. Journal of the Operational Research Society 45(9), 1050–1059 (1994)
7. Campigotto, P., Passerini, A.: Adapting to a realistic decision maker: experiments towards a reactive multi-objective optimizer. In: Blum, C., Battiti, R. (eds.) LION 4. LNCS, vol. 6073, pp. 338–341. Springer, Heidelberg (2010)
8. Chaudhuri, S., Deb, K.: An interactive evolutionary multi-objective optimization and decision making procedure. Applied Soft Computing 10(2), 496–511 (2010)
9. Coello Coello, C.A.: Handling preferences in evolutionary multiobjective optimization: a survey. In: Proceedings of the 2000 Congress on Evolutionary Computation (CEC 2000), pp. 30–37. IEEE Press, Piscataway (July 2000)
10. Deb, K., Köksalan, M.: Guest editorial: Special issue on preference-based multiobjective evolutionary algorithms. IEEE Transactions on Evolutionary Computation 14(5), 669–670 (2010)
11. Deb, K., Thiele, L., Laumanns, M., Zitzler, E.: Scalable test problems for evolutionary multiobjective optimization. In: Abraham, A., Jain, L., Goldberg, R. (eds.) Evolutionary Multiobjective Optimization. Advanced Information and Knowledge Processing, pp. 105–145. Springer, London (2005)
12. Greco, S., Knowles, J.D., Miettinen, K., Zitzler, E. (eds.): Learning in Multiobjective Optimization (Dagstuhl Seminar 12041), Dagstuhl Reports, 2(1). Schloss Dagstuhl-Leibniz-Zentrum für Informatik, Germany (2012)
13. Greco, S., Matarazzo, B., Słowiński, R.: Interactive evolutionary multiobjective optimization using dominance-based rough set approach. In: Ishibuchi, H., et al. (eds.) Proceedings of the 2010 Congress on Evolutionary Computation (CEC 2010), pp. 1–8. IEEE Press, Piscataway, NJ (2010)
14. Greenwood, G.W., Hu, X., D'Ambrosio, J.G.: Fitness functions for multiple objective optimization problems: combining preferences with pareto rankings. In: Foundations of Genetic Algorithms (FOGA), pp. 437–455. Morgan Kaufmann Publishers, San Francisco (1996)
15. Kahneman, D., Tversky, A.: Prospect theory: An analysis of decision under risk. Econometrica 47(2), 263–291 (1979)

16. Köksalan, M., Karahan, I.: An interactive territory defining evolutionary algorithm: iTDEA. IEEE Transactions on Evolutionary Computation 14(5), 702–722 (2010)
17. Kornbluth, J.: Sequential multi-criterion decision making. Omega 13(6), 569–574 (1985)
18. López-Ibáñez, M., Stützle, T.: The automatic design of multi-objective ant colony optimization algorithms. IEEE Transactions on Evolutionary Computation 16(6), 861–875 (2012)
19. Morgan, P.D.: Simulation of an adaptive behavior mechanism in an expert decision-maker. IEEE Transactions on Systems, Man, and Cybernetics 23(1), 65–76 (1993)
20. Phelps, S., Köksalan, M.: An interactive evolutionary metaheuristic for multiobjective combinatorial optimization. Management Science 49(12), 1726–1738 (2003)
21. Purshouse, R.C., Deb, K., Mansor, M.M., Mostaghim, S., Wang, R.: A review of hybrid evolutionary multiple criteria decision making methods. COIN Report 2014005, Computational Optimization and Innovation (COIN) Laboratory, University of Michigan, USA (January 2014)
22. Stewart, T.J.: Robustness of additive value function methods in MCDM. Journal of Multi-Criteria Decision Analysis 5(4), 301–309 (1996)
23. Stewart, T.J.: Evaluation and refinement of aspiration-based methods in MCDM. European Journal of Operational Research 113(3), 643–652 (1999)
24. Stewart, T.J.: Goal programming and cognitive biases in decision-making. Journal of the Operational Research Society 56(10), 1166–1175 (2005)
25. Viappiani, P., Faltings, B., Pu, P.: Preference-based search using example-critiquing with suggestions. Journal of Artificial Intelligence Research 27, 465–503 (2006)
26. Viappiani, P., Pu, P., Faltings, B.: Preference-based search with adaptive recommendations. AI Communications 21(2), 155–175 (2008)
27. Wallenius, J.: Comparative evaluation of some interactive approaches to multicriterion optimization. Management Science 21(12), 1387–1396 (1975)
28. Wang, R., Purshouse, R.C., Fleming, P.J.: Preference-inspired coevolutionary algorithms for many-objective optimization. IEEE Transactions on Evolutionary Computation 17(4), 474–494 (2013)

Real World Applications

Aircraft Air Inlet Design Optimization via Surrogate-Assisted Evolutionary Computation

Andre Lombardi[1]([⊠]), Denise Ferrari[2], and Luis Santos[3]

[1] Embraer, São José dos Campos, São Paulo, SP, Brazil
andre.lombardi@embraer.com.br
[2] Instituto Tecnológico de Aeronáutica, São José dos Campos, São Paulo, SP, Brazil
denise@ita.br
[3] Universidade de São Paulo, São Paulo, SP, Brazil
lsantos@ime.usp.br

Abstract. In aviation, the performance impact of auxiliary air inlets used for system ventilation is significant. The flow phenomena and consequently the numerical model, is highly non-linear, leading to a compromise between pressure recovery and drag for a given mass flow condition. This work follows a step-by-step approach which highlights the important issues related to solving such complex optimization problem, using surrogate methods coupled to evolutionary algorithms. Its conclusions can be used as a guideline to similar industrial applications.

Keywords: Design optimization · Genetic algorithm · Surrogate modeling · Air inlet · CFD · Aerodynamics

1 Introduction

Air inlets are employed in aviation to provide external air for several purposes such as cooling flow for heat exchangers, compartment ventilation, electronic systems cooling, and auxiliary power unit (APU) operation. Installation and dimensioning requirements for an air inlet are dictated by the requirements of other connected systems and are dependent of performance-related factors such as ingested air flow rate, total pressure and flow distortion. Larger air inlets placed in regions of high pressure and where the boundary layer is thinner are good candidates, although may increase aircraft drag, which is undesirable. Hence, there is a compromise between improving air inlet performance and deteriorating the overall aircraft performance due to drag production by the inlet. As most commercial aircraft usually have several air inlets, their contribution to the overall aircraft drag is significant, thus the optimization of such components constitutes a great opportunity for global performance improvement.

Reaching such a compromise solution between local air inlet and global aircraft performance, subjected to the variation of geometrical and operational parameters is a hard task due to the highly non-linear and possibly non-monotonical characteristics of the flow phenomena involved, possibly resulting in

© Springer International Publishing Switzerland 2015
A. Gaspar-Cunha et al. (Eds.): EMO 2015, Part II, LNCS 9019, pp. 313–327, 2015.
DOI: 10.1007/978-3-319-15892-1_21

multimodal responses with local maxima and minima that might deceive direct optimization methods.

Experimental studies [1–3] have been conducted in order to determine the influence of air inlet geometrical parameters on performance metrics and drag production. However, these results have limited application because of the simplified set up of the experimental schemes adopted [1–3]. The common industrial practice is to rely on existent correlations based on experimental data [4]. High fidelity calculations on realistic geometries are often obtained using Computational Fluid Dynamics (CFD), which require high performance computers and might take several hours depending on computational model size and complexity. Thus, an intensive use of high fidelity models for optimization might not be practical, especially during aircraft preliminary design, as systems architectures are under development and there is often the need for fast decision making.

This work describes in details the rational combination of advanced techniques to reach a fast and accurate optimal design. Our approach is to investigate sequentially design space sampling strategies, construction of a surrogate model, definition of an optimization problem and its solution using an evolutionary algorithm. As those steps are taken, it will become clear that for complex non-linear problems, as typically found in fluid dynamics applications, important questions arise. Our main contribution is to propose a set of guidelines or, at least, call attention to the potential difficulties and issues one might face while solving similar problems in practical fluid dynamics applications.

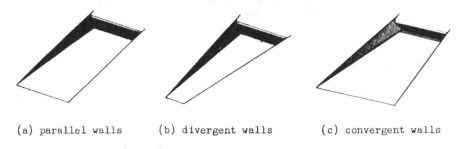

(a) parallel walls   (b) divergent walls   (c) convergent walls

Fig. 1. Simple configurations for submerged air inlets with varying sidewall angles

2   The Physical Problem and Numerical Solution

Air Inlet Performance Parameters

For the sake of simplicity, we consider only straight lip inlets, as seen in Figure 1. The ramp angle induces the flow into inlet throat. The dynamic pressure of the flow is low close to the solid surface so that, for a given pressure recovery, the inlet area must be enlarged in order to overcome this effect. Placing the inlets along the vehicle requires taking into consideration the local pressure distribution [5]. Along the sidewalls forms a pair of vortices that captures higher dynamic pressure from the flowfield and increases the pressure recovery. Figure 2 shows the right side vortex of a parallel and a divergent wall inlet (as obtained from a numerical solution). Both ramp (which translates to throat area) and sidewall angles have important effect on inlet performance, since they are associated with the conversion of dynamic pressure into total pressure at the inlet, providing the necessary pressure difference to allow the influx of cooling air into the internal systems.

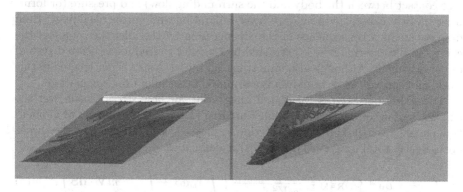

Fig. 2. Streamlines (in red) obtained by CFD simulation, representing flow path in air inlets with parallel walls (on left) and divergent walls (on right)

Mass Flow Ratio

In order to generalize conclusions, previous experimental investigations [1] used a non-dimensional parameter to represent throat area effect which is the mass flow ratio (MFR):

$$\text{MFR} = \frac{\dot{m}_1}{\dot{m}_0}, \tag{1}$$

where $\dot{m}_1 = \rho_1 V_1 A_{throat}$ is the mass flow rate actually passing through the inlet throat, $\dot{m}_0 = \rho_0 V_0 A_{throat}$ is the ideal mass flow rate, ρ is the density and V is the flow velocity. Indexes 0 and 1 represents the freestream and throat conditions, respectively. For incompressible flow, the expression simplifies to $\text{MFR} = V_1/V_0$.

Pressure Recovery

A commonly used inlet performance characteristic is pressure recovery [1,4], which can be calculated as:

$$\eta = \frac{P_T - P_{S\infty}}{P_{T\infty} - P_{S\infty}}, \tag{2}$$

where, P_T is the area averaged total pressure at the air inlet throat, $P_{S\infty}$ is the freestream static pressure, $P_{T\infty}$ is the freestream total pressure. Pressure recovery is a representation of the amount of the freestream dynamic pressure that is recovered by the air inlet.

Drag Coefficient

The cost of having pressure recovery (at a given MFR) is increasing drag of the air inlet. Since energy is being extracted from the flowfield, the resistance of the vehicle will increase. Here, two components of drag are taken into consideration: parasitic drag and ram drag. Parasitic drag is present whenever a body is immerse in a viscous non-steady flow and is caused, mainly, by skin friction drag (due to surface contact between the body and the surrounding flow) and pressure (or form) drag (mostly caused by boundary layer separation and vortex shedding). Ram drag occurs as a consequence of the air ingestion by the inlet, i.e., some of the energy in the external flow is extracted by the inlet to feed its associated system or compartment. Although also important, air outlet design is not treated in this work, and consequently, ram drag is accounted in air inlet drag calculation.

Parasitic drag (D_0) is calculated integrating air flow stress tensor over the exposed surface of the air inlet, while ram drag (D_{RAM}) is the amount of momentum passing over the throat cross-section, expressed as an integral over the throat area. Hence, total drag coefficient (C_D) is defined as:

$$C_D = C_{D0} + C_{DRAM} = \frac{2}{\rho_0 V_0^2 S_{throat}} \left(\int_S \tau_w dS + \int_{throat} \dot{m} \mathbf{V} \cdot \mathbf{dS} \right), \qquad (3)$$

where $\dot{m}$ is the mass flow rate across the inlet throat, $\mathbf{V}$ is the velocity vector, S_{throat} is the inlet throat cross-sectional area, τ_w is the viscous stress on the inlet surface wall (S) and $\rho_0 V_0^2/2$ is the freestream dynamic pressure.

High-fidelity Model

As mentioned previously, pressure recovery and drag coefficient calculations are carried out by CFD, which involves the solution of a set of partial differential equations using a discretized domain (a mesh or grid) that represents the problem geometry. The process here adopted follows a benchmark standard [6,7]. Previous studies [8–10] indicate that the level of accuracy compared to experimental data validate the present process.

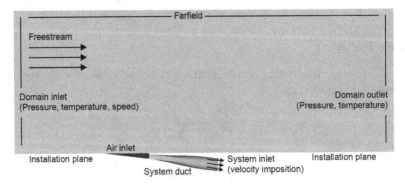

Fig. 3. Scheme of model geometry and boundary conditions.

Geometry

Figure 3 represents the geometric model considered in this work, in which air inlets are placed in a flat plate. Since the air inlets are symmetrical, only half model is simulated. The use of CFD is not limited to flat surfaces, it was a choice for the present study to focus on the development of the process guidelines.

Computational Mesh

After geometry was built, a computational mesh that discretizes the domain formed by the model boundaries was constructed. For this task, structured multi-block grid features of ICEMCFD were used, producing a computational mesh of about 3 million elements with adequate refinement in regions of interest (near wall and across surfaces with higher curvature).For each of the simulated inlet configurations a different mesh was constructed. Since conclusions of this work are based on some level of comparison of results obtained from different meshes, use of structured multi-block meshes in these cases (instead of common unstructured hybrid meshes) is recommended as there is a perfect block topology association between any given two meshes (and thus configurations). As a result, spurious differences in the results due to distinctions on meshes topology are not expected.

CFD Solver

Computational meshes for each configuration were imported into CFD++ v. 14.1.1 which is a commercial CFD solver employed for the simulations and was set to resolve the Reynolds-Averaged Navier-Stokes (RANS) equations with Realizable k-ϵ turbulence model. To each wall (physical or not) in the geometric model/computational mesh, a boundary condition was assigned. For all the studies a single reference flow condition was used (standard atmosphere sea level pressure and temperature at Mach number 0.197). Inlet flow condition is varied to provide corresponding MFR value.

Equations were solved in high performance computers using 32 processors per case. Convergence of each configuration has been obtained after about 2000 iterations, after approximately 8 hours of simulation. This relatively high computational cost motivates the use of sampling and surrogate modeling techniques.

3   The Optimization Problem and Surrogate Modeling

The problem under investigation is the design optimization of an air inlet considering effects of the variation of two parameters, inlet mass flow ratio (x_1) and sidewalls divergence angle (x_2), on two responses, pressure recovery (y_1) and drag coefficient (y_2). A larger optimization problem with respect to number of design variables and/or objective functions (or constraints) might be easily set up. As the results will show, even for this small sized problem, the non-linearity of the flowfield presents interesting issues that need to be discussed. This problem could be treated either as a multi-objective or as a single-objective optimization problem, in which the responses are combined into a single objective function. Both approaches are considered. The responses pressure recovery and drag coefficient, are respectively represented by $y_1 = f_1(\mathbf{x})$ and $y_2 = f_2(\mathbf{x})$, where $\mathbf{x} = (x_1, x_2)^T \in \mathbf{X} \subset \mathbb{R}^2$ corresponds to the air inlet parameters of interest. The optimal design is obtained by maximizing pressure recovery (y_1) and minimizing drag coefficient (y_2).

For multi-objective optimization, both y_1 and y_2 are considered as objective functions. Some operational changes in these functions are required, however, as the adopted algorithm (*nsga2*, contained in *mco* package for R computing environment [11]) works by minimizing objectives. There are two options: $w_1 = -y_1$, and $z_1 = 1/y_1$, where minimization of w_1 or z_1 results in maximization of y_1. For the final drag coefficient objective function the notation $w_2 = z_2 = y_2$ is adopted for symbolic consistency. Therefore, multi-objective optimizations with two set of objective functions will be performed: one with w_1 and w_2 (**mo1**) and another with z_1 and z_2 (**mo2**).

For single-objective optimization, the objective function is defined as $y = f(\mathbf{x}) = f_1(\mathbf{x})/f_2(\mathbf{x})$ so its maximization (*GA* package for R computing environment [12], which instead of *nsga2*, search for maxima in functions) is obtained for maximum pressure recovery (y_1) and minimum drag coefficient (y_2).

Design Space Sampling

Due to the high computational cost of CFD simulations (8 CPU hours per design point), instead of working directly with the objective functions, a surrogate model was be adopted. Several distinct methods can be used to obtain an adequate set of data as input to build the surrogate. In this study, the sampling points were generated using space-filling Latin Hypercube Design (LHD) [13], and cover the two dimensional space of considered parameters: mass flow ratio (x_1, in the interval $[0, 1.5]$) and divergence angle (x_2, in the interval $[0, 10]$).

Construction of LHD was carried out using function *maximinlhs* found in the *lhs* R package. Despite the lower cost of evaluating a surrogate model instead of CFD model, building a surrogate model is not a trivial task. In order to evaluate the trade-off between cost and accuracy of the surrogate model, three sets of training points were sequentially obtained: the first set (**set1**) with only 5 points, is a subset of the second one (**set2**), containing 10 points which, in turn, is a subset of the third set (**set3**), with 21 points (see Figure 4).

Surrogate Model

Response vectors y_1, y_2 and y obtained from CFD simulations are used as input to build surrogate models $\hat{y}_1$, $\hat{y}_2$ and $\hat{y}$ that represent y_1, y_2 and y for any point within the design space. The use of the Kriging method provides an estimate for prediction errors, which is valuable to ascertain the accuracy of the surrogate model ([14]). For this contribution, the *DiceKriging* R package is used to build a Kriging model. Basic parameters for function *km* are the training data, along with surrogate model settings, like covariance function type, set here to be of *Gaussian* type.

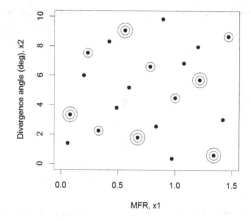

Fig. 4. Scatter plot of points generated by Latin Hypercube Sampling. Original set of 5 points (set1) is represented as the larger markers (3 circles); second set (set2), as midsize markers (2 circles) and third set (set3), as small dots. Sets of data were built in a "nested" manner, so that set1 is contained in set2, and both are contained in set3.

The Kriging surrogate models were assessed by "leave-one-out" cross-validation (CV), in which the i-th Kriging estimate is obtained by iteratively excluding the corresponding point of the training data, which produces $\hat{y}_{-i}$. This procedure is repeated for each point in the training dataset, so that it is possible to compute the global lack-of-fit of the surrogate:

$$\text{RMSE} = \sqrt{\frac{1}{n}\sum_{i=1}^{n} e_{CV}^2(i)}, \tag{4}$$

where $e_{CV}^2(i) = y(\mathbf{x}^{(i)}) - \hat{y}_{-i}(\mathbf{x}^{(i)})$. Reasonable values for this metric depend on the application field. For aeronautical applications, maximum acceptable errors in air inlet pressure recovery predictions ($\hat{y}_1$) are around 10% (so $\text{RMSE}_{y_1} \leq 0.1$), while for drag coefficient (in aircraft scale) predictions, errors up to 0.00005 (half "drag count") might be admitted. Converting drag coefficient to air inlets scale, acceptable errors in $\hat{y}_2$ are below 1.0 ($RMSE_{y_2} \leq 1.0$). Propagating these errors within the objective function y, acceptable values for $RMSE_y$ are around 0.01. Besides global metrics for inaccuracy, cross-validated standard error for the surrogate model can also be calculated for each training point.

Figure 5 shows contours of $\hat{y}_1$ obtained for each subset of points as well as cross-validation for the corresponding surrogate. Figures 6 and 7 show analogous results for $\hat{y}_2$ and $\hat{y}$. Plots on the right in the corresponding figures show a comparison of original responses and cross-validated Kriging estimates. Error bars for two $se_{CV}(i)$ determine approximate 95% confidence intervals for each estimate.

The results obtained from set3 (shown in Figure 5) indicate non-linear behavior of pressure recovery (y_1) as a function of MFR (x_1) and divergence angle (x_2), and three main local maxima can be pointed out close to coordinates $(0.3, 10)$, $(0.5, 0.0)$ and $(1.0, 1.0)$. The literature [1] presents pressure recovery peak ranging

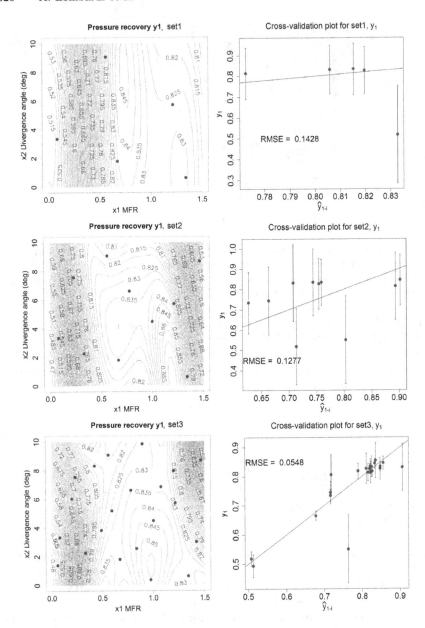

Fig. 5. Contour plot of the Kriging surrogate $\hat{y}_1$ obtained from CFD results for LHS design points in set1, set2 and set3 (left side plot). Training points are represented by the points plotted over the contours. Cross-validation of Kriging surrogate responses, where error bars correspond to 95% confidence interval (right side plot).

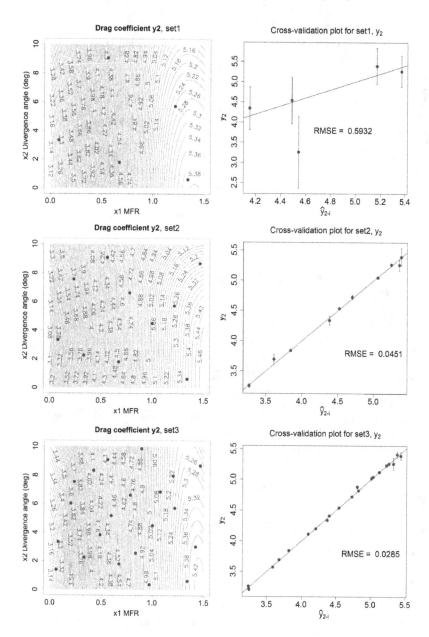

Fig. 6. Contour plot of the Kriging surrogate $\hat{y}_2$ obtained from CFD results for LHS design points in set1, set2 and set3 (left side plot). Training points are represented by the points plotted over the contours. Cross-validation of Kriging surrogate responses, where error bars correspond to 95% confidence interval (right side plot).

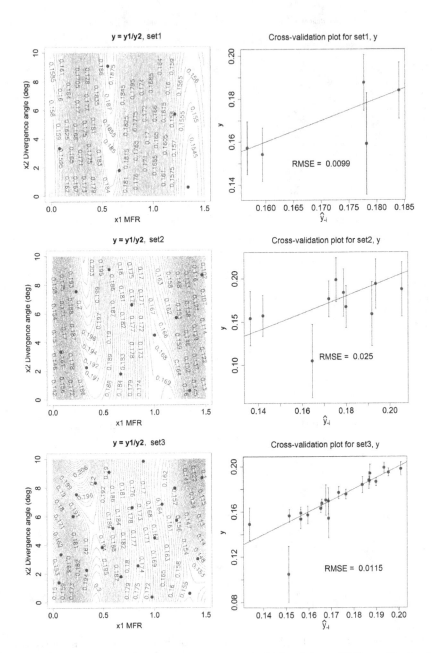

Fig. 7. Contour plot of the Kriging surrogate $\hat{y}$ obtained from CFD results for LHS design points in set1, set2 and set3 (left side plot). Training points are represented by the points plotted over the contours. Cross-validation of Kriging surrogate responses, where error bars correspond to 95% confidence interval (right side plot).

from approximately MFR = 1.0 for air inlets with parallel walls, to MFR = 0.2 for inlets with high wall divergence angle. Thus, the obtained results are in agreement with experimental observations.

Generally speaking, the RMSE results obtained for the surrogates suggest improvements in model prediction accuracy as the number of training points increases. Thus, regarding the assessment of surrogates based solely on RMSE values (see in Figures 5, 6 and 7), it is possible to conclude that: (i) The Kriging model obtained from **set3** is the only surrogate acceptable for predicting y_1 (pressure recovery), as $RMSE_{y_1} \approx 0.05 \leq 0.1$; (ii) Drag coefficient (y_2) is well represented by all surrogates built from any of the training data sets, since except for predictions on very low and very high values of MFR, surrogates obtained from all sets of data are very similar in topology and magnitude; and (iii) Strictly considering RMSE results for response y, only the Kriging model based on **set1** would qualify.

The RMSE measure should not be considered alone to evaluate the quality of the surrogate. Observe, for example, the results of $\hat{y}$ for **set1**: the RMSE outperforms the other two surrogates for the same response, obtained with higher computational cost (**set2** and **set3**). Such incoherent result is actually accidental, since the surrogate estimates for points in **set1** are contained in an narrower range (which reduces RMSE) but present large confidence intervals, which means the estimates are very imprecise. Furthermore, this surrogate is inaccurate. The bottom plot in Figure 7 shows a contour plot of the response surface for y based on **set3**. Close inspection provides insight on approximate location of maxima: $(0.2, 10)$ and $(0.4, 0)$. This feature is completely missed out using surrogate based on **set1**.

In terms of computational cost, surrogates obtained based on **set1** took about 40 hours of high fidelity model simulation. This scales to 80 hours for surrogates calculated from **set2** and then to more than 160 hours if **set3** is used as training dataset. Although considerably less expensive, use of training data sets of sizes comparable to **set1**, or even **set2**, could cause significant impact on prediction precision and accuracy, thus on optimization results.

4   Single and Multiobjective Solutions

As previously stated, optimization is performed using evolutionary computation. As input to the genetic algorithms, instead of CFD simulation results for each individual in the population, the less expensive Kriging surrogate model results were be calculated. For single-objective optimization package GA ([12]) from R statistical software was used. The objective (or fitness) function to be **maximized** is $y = f(\mathbf{x})$. The following set of genetic algorithm parameters were fixed: (i) Population size: 100 individuals; (ii) Number of iterations (generations): 1000; (iii) Crossover probability: 80%; (iv) Mutation probability: 10%; (v) Elitism: 5% (best) individuals.

No convergence criteria was established; the execution log was checked to ensure that at least during the last 50 iterations (generations), fitness value had not changed. The optimal design obtained in this process was then simulated via CFD and compared to surrogate model results.

For multi-objective optimization, the *nsga2* function from R package *mco* was employed with the same settings as before. Genetic algorithm was run for the two sets of objective functions previously presented (mo1: w_1 and w_2; mo2: z_1 and z_2) by minimization (instead of *ga*). Resulting Pareto fronts will be presented next.

Optimization Results

Multi-objective optimization (via *nsga2*) for two different function definition approaches was performed. Figure 8 shows results for both approaches ("mo1" and "mo2"), comparing pareto fronts generated using Kriging surrogates built as previously described. For a given set of training points used, it is noticeable that Pareto fronts obtained from mo1 and mo2 are coincident, though the points are not exactly the same, they lie on the same line, leading to the conclusion that optimization performed for both multi-objective function sets (mo1 and mo2) might be equivalent for the problem discussed in this work.

In addition, Pareto fronts obtained from different training points present considerable differences in topology (including Pareto front points location in design space). However, as the number of training points increases, some convergence is observed since groups of Pareto points become clustered in the design space. This is particularly noticeable comparing Pareto front results obtained for surrogates based on training sets **set2** and **set3**.

Single-objective optimization (via *ga* for y) produced the evolution presented in Figure 9(a), which corresponds to optimization of Kriging surrogate for y based on data in **set3** (21 points). The optimal design is found at coordinates $(0.214, 9.993)$ with fitness function value $\hat{y} = 0.216$, which is represented by a black dot in the bottom plots in Figure 8.

An additional run of the genetic algorithm (also using Kriging surrogate obtained for **set3**) with different values of population (50), crossing-over probability (60%), mutation probability (5%) and elitism (1%) was performed and results are shown in Figure 9(b). Although same fitness and optimal design point were eventually found, genetic algorithm evolution passed through a local maximum at $(0.421, 0.010)$ (see the bottom plot at left side of Figure 7) and took about 400 generations to get closer to the global maximum. This interesting result comes as consequence of the multimodal nature of the studied responses (especially y_1). Figure 4 shows LHS design has not put any point actually very close to the optimal design point obtained via single-objective optimization. For this reason, an additional CFD simulation was performed to evaluate obtained surrogate at this point. Calculated response is $y = 0.215$ which is close enough to surrogate result at this point ($\hat{y} = 0.216$).

Another interesting observation is that single-objective optimization results are contained in the corresponding Pareto fronts for the proposed multi-objective optimization (in Figure 8, single-objective optimization results are represented by the black dot). Additionally, single-objective optimal design is located in the same cluster of optimal designs generated by multi-objective optimization. This

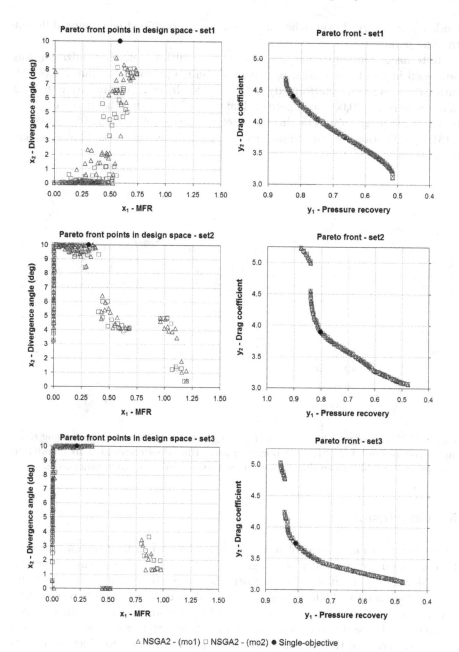

Fig. 8. Multi-objective optimization results using Kriging surrogate data obtained from set1, set2 and set3: location of Pareto front points in the design space (on left side) and plot of Pareto front for single-objective, "mo1" and "mo2" optimization approaches (on right side).

tendency is observed even when using surrogates obtained employing different training points sets.

In terms of computational cost, each of the multi-objective optimization was processed in less than 5 minutes using an average personal computer (200,000 surrogate evaluations performed), while single-objective optimization took 2 minutes to run (100,000 surrogate evaluations). Considering the amount of processing time to generate training data set responses (more than 160 hours for set3), optimization computational cost is negligible for this case. This is typical of industrial applications.

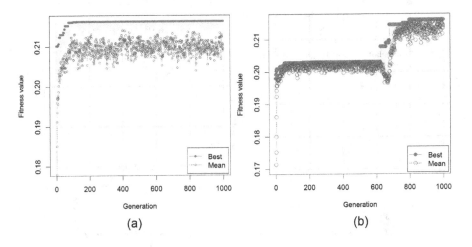

(a)   (b)

Fig. 9. Evolution of population fitness for two genetic algorithm executions using Kriging surrogate built from training points in set3. Blue dots correspond to population average fitness while green dots represent best fitness value for each generation.

5   Conclusion

In this contribution, aspects of aeronautical submerged air inlet design optimization are discussed. Influence of two geometric parameters (divergence angle and mass flow ratio/throat area) on performance responses (pressure recovery and drag coefficient) was modeled using CFD simulations. In order to reduce simulation results fluctuation the same mesh topology was maintained for all analyzed geometries. Three different sets of training points were sampled from design space and resulting Kriging surrogate models were assessed considering acceptable levels of error for responses established based on application experience, leading to verification of an adequate sampling size. Non-linear behavior of the studied phenomena is observed as multi-modal features are presented in the contours of obtained surrogate models. Use of surrogates as means of design evaluation greatly reduced computational time, making optimization process viable for this case. Pareto fronts generated adopting the chosen two sets of multi-objective functions were shown to be coincident. Calculated single-objective optimal design was contained in corresponding Pareto front, though it consistently lied in the same "cluster" of designs. Moreover, genetic algorithms parameters were shown to cause some influence on optimization result, possibly leading to different optimal designs if convergence is not reached. Finally, optimization problem studied in this work – a single incompressible flow case with two objective functions and the variation of only two parameters – presented many aspects and interesting issues under a perspective not usually seen in industrial applications.

References

1. Mossman, E.A., Randall, L.M.: An Experimental Investigation of the Design Variables for NACA Submerged Duct Entrances. NACA research memorandum, National Advisory Committee for Aeronautics (1948)
2. Frick, C.W., Ames Research Center: An Experimental Investigation of NACA Submerged-duct Entrances. Advance confidential report. National Advisory Committee for Aeronautics (1945)
3. Martin, N.J., Holzhauser, C.A., Ames Research Center and United States: National Advisory Committee for Aeronautics. An Experimental Investigation at Large Scale of Several Configurations of an NACA Submerged Air Intake. NACA research memorandum. National Advisory Committee for Aeronautics (1948)
4. ESDU 86002 Drag and pressure recovery characteristics of auxiliary air inlets at subsonic speeds
5. Sacks, A.H., Spreiter, J.R., United States: National Advisory Committee for Aeronautics. Theoretical Investigation of Submerged Inlets at Low Speeds. Technical note. National Advisory Committee for Aeronautics (1951)
6. Rumsey, C.L.: Consistency, verification, and validation of turbulence models for reynolds-averaged navier-stokes applications. In: Proceedings of the Institution of Mechanical Engineers, Part G: Journal of Aerospace Engineering **224**(11), 1211–1218
7. Rumsey, C.L.: Turbulence modeling verification and validation. AIAA Paper 2014–0201 (2014)
8. Hime, L., Perez, C.C., Silva, L.F.F., Ferreira, S.B., Jesus, A.B., Takase, V.L., Vinagre, H.T.M.: A review of the characteristics of submerged air inlets. In: 18th International Congress of Mechanical Engineering - COBEM (2005)
9. Perez, C.C., Ferreira, S.B., Silva, L.F.F., Jesus, A.B., Oliveira, G.L.: Numerical study of the performance improvement of submerged air intakes using vortex generators. In: 25th International Congress of the Aeronautical Sciences (2006)
10. Perez, C.C., Silva, L.F.F., Ferreira, S.B., Jesus, A.B., Oliveira, G.L.: Effective use of vortex generators to improve the performance of submerged air inlets for aircraft. In: 11th Brazilian Congress of Thermal Sciences and Engineering - ENCIT (2006)
11. Deb, K., Pratap, A., Agarwal, S., Meyarivan, T.: A fast elitist multi-objective genetic algorithm: Nsga-ii. IEEE Transactions on Evolutionary Computation **6**, 182–197 (2000)
12. Scrucca, L.: GA: a package for genetic algorithms in R. Journal of Statistical Software **53**, 1–37 (2013)
13. Stein, M.: Large sample properties of simulations using latin hypercube sampling. Technometrics **29**(2), 143–151 (1987)
14. Roustant, O., Ginsbourger, D., Deville, Y., et al.: DiceKriging, DiceOptim: Two R packages for the analysis of computer experiments by kriging-based metamodeling and optimization. Journal of Statistical Software **51**, 1–55 (2012)

Diesel Engine Drive-Cycle Optimization with Liger

Stefanos Giagkiozis$^{(\boxtimes)}$, Robert J. Lygoe, Ioannis Giagkiozis,
and Peter J. Fleming

Department of Automatic Control and Systems Engineering,
The University of Sheffield, Sheffield S1 3JD, UK
{s.giagkiozis,i.giagkiozis,p.fleming}@sheffield.ac.uk, blygoe@ford.co
http://www.sheffield.ac.uk

Abstract. In the current market, engineers are continually required to optimize their designs to realise improved performance whilst meeting ever more stringent regulations and competing for market share. This reality increases the demand for optimization. Due to these, and several other reasons, real-world optimization problems often have a large search space, are non-convex, and have expensive-to-evaluate objective functions that have many conflicting objectives. However, even if these problems are overcome, to select an acceptable solution, the decision making process itself is equally demanding. Some of these difficulties could be alleviated if a tool existed to support the analyst and decision maker throughout the entire process. The aim of this work is to illustrate and share insight gained in using Liger in such a scenario. Liger is an open source integrated optimization environment and its use is described in a case study of involving the calibration of a diesel engine using multi-models. The benefits of using Liger are demonstrated along with the procedure we followed to obtain an optimized engine calibration that complies with performance and regulatory requirements.

Keywords: Diesel engine optimization · Liger · Integrated optimization environment · Multi-objective optimization · Drive-cycle optimization

1 Introduction

Designing a modern day vehicle is a feat that requires many considerations to be taken into account. These considerations are based on constraints imposed by the industry, the customers and government regulations. We could argue that the main objective of the industry is profit from vehicle sales, which requires high appeal to customers and low production cost. Governments are introducing ever more stringent regulations pertaining to the reduction of harmful emissions to humans and the environment. Broadly speaking, although customer requirements are quite varied, they can be split into a small number of groups, such as purchase cost and vehicle safety along with fuel economy, performance,

© Springer International Publishing Switzerland 2015
A. Gaspar-Cunha et al. (Eds.): EMO 2015, Part II, LNCS 9019, pp. 328–342, 2015.
DOI: 10.1007/978-3-319-15892-1_22

design and brand. The design process has to incorporate all these requirements simultaneously.

Diesel engines by design are reliable, safe and their main advantage is fuel economy. They have a high thermal efficiency, which is a result of their high compression ratio, but this also leads to increased nitrogen oxide (NOx) emissions[21]. Modern diesel engines have been vastly improved, resulting in reduced emissions and an overall cleaner engine, while maintaining their high efficiency. More detailed analyses of the different emission control technologies currently applied on diesel engines can be found in [10] and [11].

What is of interest now is how to best utilise the hardware described above, in order to obtain optimal, or at least improved performance. The design parameters of the hardware and the calibration of the software parameters controlling that hardware can be used to alter the engine performance. Therefore, it is preferable to automate this procedure by means of an optimization algorithm (stochastic, gradient-based or some other alternative) to reduce the engineering overhead that would be introduced with empirical calibration. Furthermore, although for non-convex problems a *certificate* of optimality cannot be issued, as is the case for a large number of convex problems, empirical calibration often produces inferior results.

The use of optimization enables the study of a number of additional issues that could not be easily addressed with empirical calibration. One issue is the fact that the number of competing objectives is usually large but is kept to a relatively small number to reduce the complexity for the decision maker (DM). Nevertheless, multi-objective optimization is becoming more and more prevalent, given that it can be applied to different kinds of real-world problems (e.g. [13], [15], [19], [20]). An issue associated with population-based algorithms, is that due to the sheer number of available algorithms it is virtually impossible, even for experts in the field, to select the best algorithm for a given problem. Even if a small subset of algorithms is shortlisted for evaluation, their implementation can be challenging and costly for the practitioner. In [7], an overview of population-based algorithms and the considerations in choosing an appropriate algorithm are given. It is obvious from the conclusions in [7], that it is not a trivial task to make a decision on the algorithm family, implementation or the selection of the configurable parameters.

Some of the problems identified above can be solved by developing tools that make the use of optimization algorithms straightforward and provide tools to assist the DM in exploring the resulting solutions. Such tools are, for example, the OpenMDAO [8], OpenOpt [2], TAO [16] and Liger [6]. Tools such as these can simplify the task of optimisation by incorporating state-of-the-art algorithms into their libraries and providing advanced visualization to assist in the design of experiments. The use of such tools can reduce engineering time while producing as good or better solutions in comparison with algorithms built by practitioners that are potentially non-experts.

In this case study, we are interested in reducing fuel consumption, nitrogen oxide (NOx) emissions and particulate matter (PM) production in a diesel engine drive-cycle. The diesel engine under investigation is equipped with an exhaust gas recirculation (EGR) and turbocharging systems and a common rail injection system. Optimization tasks like this one, are very commonly performed in the industry and we are interested in evaluating the effectiveness of using an integrated optimization environment in performing this task, while we obtain a set of possible settings that can be used on the real engine. Liger [6] is used, in order to complete all of the optimizations; statistical models of the engine, provided by Ford, represent an entire drive-cycle of the engine under investigation. It is not our goal to compare algorithms or other optimization frameworks, but merely demonstrate the use of one such tool, Liger, to solve a practical problem. We also do not claim that Liger is the best available alternative, but we want to stress its strengths and its weaknesses. This could potentially help other developers of similar tools improve their own software.

The remainder of this paper is organised as follows. In Section 2 we describe the diesel engine model and in Section 3 we present the mathematical formulation of our optimization problem. Section 4 provides details on the experiment set up and the use of Liger to complete the experiments. Subsequently, in Section 5 we present and comment on the obtained solutions with Liger. In Section 6 we discuss shortcomings, benefits and future potential in the use of Liger for engineering design problems. Lastly, this work is summarised and concluded in Section 7.

2   Diesel Engine Drive-Cycle Model

Modern diesel engines are comprised of a large number of control variables. Many of the components that are added on the powertrain to reduce emissions, improve efficiency or vehicle performance, can be calibrated in order to improve their performance. However, optimal performance of an isolated component does not guarantee that the collection of components will perform equally well. This, in turn, increases the scope of the optimization problem and increasing its parameter space. Furthermore, the operating conditions of the engine vary in time from *turn-on* to *turn-off*. An operating period from engine turn-on to turn-off is called the drive-cycle. During a drive-cycle the engine is subjected to varying environmental and usage parameters, for example, temperature, humidity, oxygen density, load and speed requirements. The engine performance will not be the same for different points in this parameter space. Since models of the engine are not built from first principles (since such a task would be prohibitively complex and expensive), creating a single model that will describe all those variations in the engine state usually leads to inaccurate models. The alternative scenario is to create models in the neighbourhood of a set of parameter points (operating points) and then combine this set of models to obtain the estimated response of the engine throughout a drive-cycle. Such models are called multi-models.

2.1   Multi-models

Multi-models provide the engine response for different operating points. The collection of local models for all operating points comprise the global model. For example, when fuel consumption is to be evaluated for the engine, the operating point at which the evaluation is required must be provided. If the operating point provided matches one of the local models in the global model for the fuel consumption response, then an output can be evaluated. If the operating point is not in the discrete set of operating points, an output cannot be evaluated, since no interpolation takes place between local models. The layers of the entire engine model, can be seen in Figure 1. Local models provide a local response, based on the local inputs. The local model is chosen by the operating point (global inputs), which in our case is a set of engine speed and brake torque.

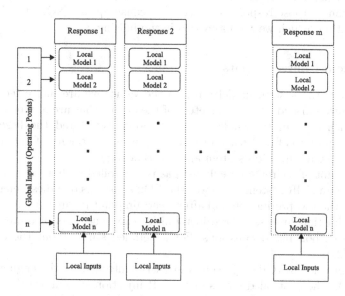

Fig. 1. A general view of how multi-layer models are used to estimate a response, based on a predefined set of operating points

A disadvantage of using multi-models is that a large amount of data is required; more engine physical tests (bench-tests) are required in order to estimate the parameters of the local models. Another disadvantage is that local models cannot be used to interpolate between operating points. This means that if we want to obtain data at an operating point that is not in the discrete set of operating points, we must design another experiment for that operating point and estimate the parameters for that new local model.

2.2 Obtaining the Multi-models

In order to obtain the models for each of the responses that we are interested in, discrete sets of engine speed and brake torque were first defined. These sets are discrete operating points of the engine and represent the drive-cycle profile, on which the vehicle will be tested. Each country has a different drive-cycle, for which the emissions need to be limited. In total, there are 30 operating points in our models. These operating points are split into two test plans, based on the engine operating temperature, namely the hot-fast and warm operating regions. In the hot-fast regions 18 sets of engine speed and brake torque operating points are defined and in the warm region there are 12 such sets.

At each set of speed and torque, experiments were performed and 29 outputs were measured. Finally, using the MATLAB model-based calibration (MBC) toolbox, a local model for all the responses was obtained, for each of the operating points. Amongst those responses are the fuel consumption, NOx emissions and PM production, which we are interested in minimising.

2.3 System Control Inputs

The inputs of the system model are also the values we are interested in calibrating. So we must identify parameters of the engine that are of interest to us, based on their effect on the whole system. As previously stated, the diesel engine we are working on is fitted with an exhaust gas recirculation (EGR) system, a common rail (CR) injection system and a turbocharger.

The amount of recirculated exhaust gas is controlled with a valve, the EGR valve. Using an EGR system can lower the NOx emissions of a diesel engine, but it increases the PM production and after a certain point it can cause instabilities in the combustion process. These instabilities can cause loss of power and an increase in carbon based emissions. For a detailed study about the effects of EGR, see [21].

The common rail (CR) injection system is controlled by the engine control unit (ECU). The controlled states of the CR injection system are the pressure in the CR, the quantity of injected fuel and the timing of the injection. These variables have an effect on the NOx emissions, particulate production and the performance of the engine. The emissions can be controlled by changing the rate of the injected fuel [17]. The amount of PM produced can be controlled by changing the pressure in the common rail. The NOx emissions also depend on the timing and quantity of injected fuel by a pilot injection [14]. This pilot injection is also used to reduce the noise generated by the combustion process.

The turbocharger increases the density of the air inside the cylinder, which increases the overall pressure in the cylinder and the power output. This is controlled using the turbocharger valve and is described as a percentage of actuation. More information on turbocharging and effects of the different techniques of turbocharging can be found in [11].

The controlled states of the hardware described above are the local inputs of our multi-models and can be optimized in order to obtain a better performance.

The hot-fast region models have six inputs, the percentage amount that the EGR and turbocharger valves are from being fully open, the main injection phase, the CR pressure and the pilot injection phase and quantity. The warm region models have two more inputs, the quantity and timing of a second pilot injection. Both the hot-fast and the warm region models have two global inputs, the engine speed and the brake torque, that are used for the switching between the local models, as described previously (Figure 1).

Finally, the range of the inputs is also defined in the models and is given by the convex hull boundary. A sample convex hull boundary for three of the inputs of the system can be seen in Figure 2. All values for the inputs inside the "blue" mass, are valid. Any values outside the constraint are not allowed.

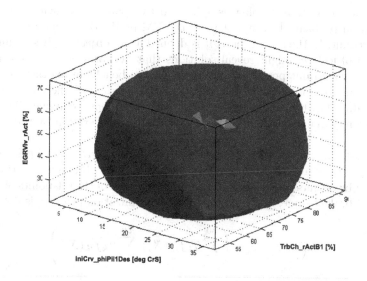

Fig. 2. A convex hull boundary example for three of the system inputs Turbocharger valve actuation (*TrbCh_rActB1 [%]*) on the y-axis, CR pressure (*RailP_pFit [hPa]*) on the x-axis and EGR valve (*EGRVlv_rAct [%]*)) actuation on the z-axis

3   Diesel Engine Parameter Optimization

The models described in Section 2, will be used to define our objective functions, based on which we will measure the performance of the engine model for each of our decision vectors.

As seen in the previous section, the calibration settings have different effects on the performance of the engine and sometimes opposing effects. For example, according to [11], the EGR can reduce NOx emissions, but has been associated with increased PM production. Based on the study in [18], the pilot injection timing and quantity, controlled by the CR injection system, can reduce PM, previously increased by the EGR system, but at the cost of increased combustion noise. It is clear that there is a trade-off relationship between a number of the objective functions.

Because we are dealing with a multi-objective problem, there are many possible solutions that cannot be distinguished in terms of fitness from one another. After the optimization process is performed, a set of solutions is obtained. The feasible solutions are selected from the final population. The decision maker, identifies the trade-offs between the objective functions and, in turn, will make an informed decision about the final design.

3.1   Formulation of the Optimisation Problem

Our engine calibration problem is split into two parts, each being a separate optimization problem. In the first part, the PM production response is defined as a constraint. In the second part, it is defined as an objective. Re-defining the PM as an objective, might reveal solutions that do not satisfy the initial PM production constraint posed, but result in a large benefit in another objective. It should also be noted that the NOx and PM outputs of the engine can be further reduced by the addition of exhaust after-treatment techniques. The after-treatment requirements should also be taken into consideration by the decision maker.

The first part, as stated previously, is to perform a drive-cycle optimization. In order to perform the drive-cycle optimization, the objective functions are multiplied by a time-weight and then summed for each of the set-points of engine speed and brake torque. Mathematically, the optimization problem is defined as follows,

$$\underset{X_j}{\text{minimize}} \quad F = \left(\sum_{k=1}^{30} T_k f_{1k}(X_j), \sum_{k=1}^{30} T_k f_{2k}(X_j) \right)$$

$$\text{subject to} \quad 0 < \sum_{k=1}^{30} T_k f_{3k}(X_j) \le \sum_{k=1}^{30} pT_k,$$

$$g_k(X_j) < 0, \tag{1}$$

$$j = \begin{cases} 1 \text{ , for } 1 \le k \le 18 \\ 2 \text{ , for } 19 \le k \le 30 \end{cases}$$

$$X_1 = (x_1, x_2, x_3, x_4, x_5, x_6),$$

$$X_2 = (x_1, x_2, x_3, x_4, x_5, x_6, x_7, x_8),$$

$$k = 1, 2, \dots, 30.$$

In (1), $F = [F_1, F_2]$ is the objective function vector, where F_1 and F_2 are the drive-cycle responses for the fuel consumption and the NOx emissions, comprised by their respective local responses. The local responses are f_{1k} is the fuel mass consumption for each speed and torque operating point, k, f_{2k} is the NOx emissions response and f_{3k} is the PM response. g_k is the convex hull boundary at point k (described in detail in Section 2, which is a function of the same inputs and effectively defines our search space. In total there are 30 operating points. Because there are two operating regions, the hot-fast and the warm, which are defined for different operating temperatures of the engine, there are two sets of models for each of the responses and the total speed and torque set-points are split in these two categories.

The models for the hot-fast operating region have 6 local inputs, therefore 6 decision variables are defined in the decision vector X_1 and the warm region models have 8 local inputs, defined in the X_2 decision vector. The hot-fast operating region is denoted by $j = 1$ and the warm operating region by, $j = 2$, which consist of 18 and 12 operating points, respectively. Finally, T_k is the time weight vector in seconds and it defines how long each operating point is "active" in the drive-cycle. The local constraint for the PM production is denoted by p. A local constraint is a constraint that must be met at each individual operating point, which in our case was $3.5g/hour$ in the case of the PM production.

The second part of this case study, is to perform another drive-cycle optimization, with the PM production (f_{3k}) defined as an objective. In (1), the objective function vector becomes $F = [F_1, F_2, F_3]$. This can be interpreted as an exploratory optimization, in order to obtain solutions in a wider range for all of our three objectives.

4   Integrated Optimization

Empirically, we know that a large number of optimizations will be needed to solve the problem in (1). Furthermore, more than one iteration of the entire process is performed, to ensure that a satisfactory population of solutions is produced. This process is time consuming but can be greatly simplified by making use of the right tools. Liger, currently has a collection of implementations of several population-based algorithms. Additionally, Liger also offers a collection of visualization tools, that can assist the decision maker. Finally, nodes that allow Liger to work together with external software, such as MATLAB, are provided, in order to further simplify the task.

4.1   Algorithm Choice

Two of the algorithms that have already been implemented and tested on Liger are the GDE3, a version of generalized differential evolution [12] and a non-dominated sorting genetic algorithm implementation (NSGA-II) [1]. Unfortunately, these algorithms are not well equipped to handle constraints. Given that our optimization problems have constraints that need to be satisfied, more tests will be necessary in order to obtain a satisfactory number of feasible solutions.

An alternative approach, that is planned to be implemented in Liger by the developers, is proposed in [4] and further described in [5]. Fonseca and Fleming [4], propose the use of progressive preference articulation (PPA), in order to benefit from the existence of constraints in an optimization problem. This approach allows the decision maker to initially perform the optimizations without any constraints, in order to obtain solutions from the entire search space. The decision maker can then gradually limit the freedom of objective functions, effectively add constraints, to obtain a richer population within these limits.

4.2 Experiment Description

For each part of our problem described previously, a total of 150 tests are performed, with each of the algorithms. Basic constraint handling was implemented using a simple penalty constraint approach. The penalty is assigned to a function that does not perform within the set limits, in order to direct the algorithm towards the feasible region. This is necessary because the optimization library that Liger currently uses is a C++ version of jMetal [3] ported in Qt, which does not support constraint handling.

Both for NSGA-II and GDE3, a population of 100 will be used. The maximum iterations for GDE3 are set to 250; the CR parameter, which controls the crossover operation and the F parameter, which scales the mutation operation, are both set to 0.5. For NSGA-II, the objective evaluations were set to 10000, the crossover probability to 0.9 and the crossover distribution index at 20. These are the default values provided in [12] and [1] and we chose not to change them. The interested reader can find more information about the effects of the parameters above for GDE3 in [12] and for NSGA-II in [1].

4.3 Liger Work-Flow

Designing optimization work-flows in Liger is very intuitive and straightforward. A Liger work-flow is comprised of nodes and links. Each node, depending on its function, has inputs and outputs. The links are used to transfer signals between nodes, so that all nodes are evaluated in the required sequence. The work-flow can be viewed as a visual description of operators, that perform a task on the data set and their execution is timed from start to end node. Work-flows are created in the "Designer" tab, which is selected in Figure 3. The most interesting elements, numbered in Figure 3, are listed in Table 1.

A single optimization is defined by the nodes enclosed between the "Start" node (1) and the "End" node (7). The number of iterations can be defined in the "End" node. This number defines the number of optimizations (experiments) to be performed. At the end of each iteration, a final population is obtained. The "Objective Function" node (3), defines our objectives. More specifically, the population of solutions, determined by the "Algorithm" node (4), is used by node 3 to evaluate the objectives, in this instance using MATLAB. Node 3 could be any of the nodes provided in the "Problems" tab of the "Node Collection Sidebar". The default "Problem" nodes are mainly some of the standard problems used to test optimization algorithms (e.g. WFG problems [9]), but the practitioner can create custom problem nodes, since the source code is provided. The "Algorithm" node is where the population of solutions is determined, based on the optimization algorithm used. The evaluation of the final population, is also an

Table 1. Nodes illustrated in the work-flow shown in Figure 3

Reference Number Figure (3)	Node Name
1	Start
2	MATLAB Import
3	Objective Function
4	Algorithm
5	Signal Split
6	MATLAB Export
7	End
8	Parallel Co-ordinates Plot
9	Matrix Scatter Plot
10	Node Collection Sidebar

iterative process and should not be confused with the iteration loop defined by the start and end nodes.

The "MATLAB Import" node (2) is used to import data with a specific format into Liger from MATLAB. The "Matlab Export" node (6) is used to save the final population of solutions. Finally the 'Parallel Co-ordinates Plot" (8) and the "Matrix Scatter Plot" (9) nodes, provide visualization tools that can assist the DM to interpret the results of the experiments.

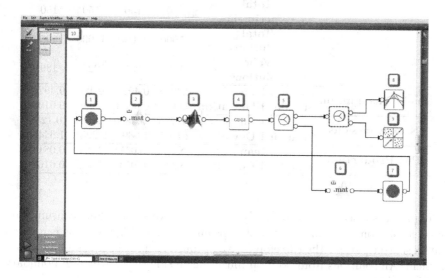

Fig. 3. An example Liger work-flow used to perform the optimizations in this case study

The "MATLAB Objective Function" node evaluates our objective functions (fuel consumption, NOx emissions, PM production), by calling a predefined MATLAB function file. Creating the function files is straightforward, since a template (.m) file is provided (in the latest versions of Liger we used). Two functions for the drive cycle were created. The first function file is used to perform the optimizations for all the operating points without constraints. The second function used constraints, as described in section 3.

5 Results

From the second part of the optimizations we obtained a total of 420 solutions, which are inside the convex hull boundary. As stated previously, solutions that are not within the convex hull boundary cannot be physically implemented on the engine. A more detailed breakdown of the results can be found in Table 2. Table 2 contains a summary of the total solutions yielded by our optimizations, for each part of the problem and for each algorithm separately. The reader can also find the mean values and standard deviations for each of the objective functions for each case.

Table 2. A summary of the results of the optimizations for each algorithm and for each part of our optimization problem

| | | GDE 3 | | NSGA - II | |
		Part 1	Part 2	Part 1	Part 2
	Total Experiments	150	150	150	150
	Total Solutions	15000	15000	15000	15000
	Valid Solutions	190	231	222	189
Fuel Consumption	Mean	0.4938	0.4922	0.489	0.4928
	Standard Deviation	0.0072	0.0083	0.0061	0.0108
NOx Emissions	Mean	7.7775	10.6889	7.8651	10.1288
	Standard Deviation	2.1947	1.6894	2.2281	1.6443
PM Production	Mean	N/A	0.077	N/A	0.0814
	Standard Deviation	N/A	0.014	N/A	0.0163

Figure 4 is a scatter plot of the NOx emissions versus the Fuel consumption for each solution of the first part of the optimizations. In Figure 5, you can find a matrix scatter plot for the three objective optimization in the second part of our problem. All solutions from both optimizations inside the convex hull boundary were included.

Figure 6 shows the parallel co-ordinates plot for the second part of the optimizations as well; it includes a small subset of the entire cohort of solutions, in order to be easy to read. All plots (Figure 4, 5 and 6) were produced with the help of the visualization plug-in of Liger.

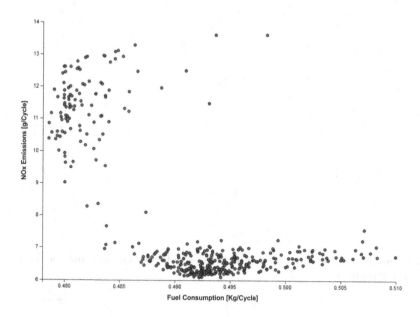

Fig. 4. Two-objective optimization with a PM drive-cycle constraint scatter plot

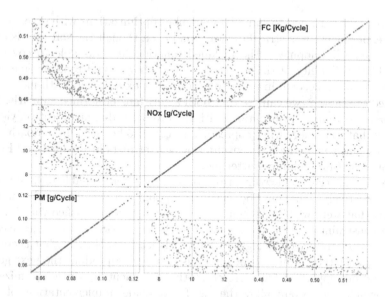

Fig. 5. A matrix scatter plot of all the solutions obtained from the cycle optimization with three objectives from both NSGA-II and GDE3. *FC* denotes the fuel consumption.

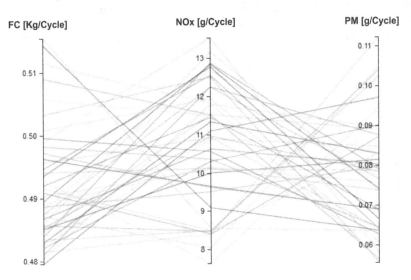

Fig. 6. Parallel co-ordinates plot of a reduced set of the valid solutions for the three-objective optimizations

6   Discussion

Using Liger we managed to run a large number optimizations with ease. Setting up our experiments was straightforward and even though the software is at an early stage of development, the included functionality was able to cover our needs. The availability of the source code also allowed us to make some small additions that we needed, in order to perform our experiments with more ease.

Based on Table 2, we can see that both algorithms performed similarly. Furthermore, performing optimizations with the PM defined as an objective, yielded solutions that were more widely spread. This can also bee seen by comparing the NOx emissions versus the fuel consumption scatter plots in Figure 4 and Figure 5 (second row third column scatter plot).

The optimization framework currently used by Liger is unable to handle constraints effectively. This could be one of the reasons that we did not obtain more solutions that were within the convex hull boundary and satisfied the PM production constraints. A penalty was used in the evaluation of the objectives, in order to indirectly drive the algorithm towards feasible solutions, but there was no control over the population of solutions that would produce the next generation. The developers of Liger are currently implementing a new optimization library, which will not only allow the use of "pre-made" implementations of algorithms, such as the ones we used, but also the construction of an algorithm itself using nodes representing algorithm operators. Additionally, this new library will be able to handle constraints natively. This will allow Liger to be used for a wider range of applications.

Lastly, our original problem is actually much larger (in terms of objectives and constraints) than it became after our reformulation. The responses for each of the operating points were combined to provide a drive-cycle response. In the case of the PM constraint, this practically means that even though the drive-cycle constraint is met, the local constraint might not be. Meeting the local constraints of a drive-cycle is very important for manufacturers, since the drive-cycle cannot represent the driving habits of all drivers. Additionally, the combustion noise is another very important constraint, which we did not consider, because this constraint can only be satisfied locally, at each operating point. In fact, the real problem originally consisted of 60 objectives and 62 constraints, for the first part, where the PM production is a constraint. For the second part it consisted of 90 objectives and 32 constraints.

7 Conclusion

In this case study, we have demonstrated the use of Liger, an open source integrated optimization environment, to find a solution for a real-world problem. Problems such as the one investigated in this paper are very common in the industry and tools like Liger, that help the practitioners tackle them, are much needed. Even though the original problem was reduced in size, the complexity of the problem and the difficulty of finding feasible solutions were still high.

What Liger offers is an open source framework that has the potential to be as efficient as other commercial alternatives. The main advantage of being open source is its inherent transparency, allowing everything implemented to be tested, verified and criticised by the research community. At the same time it offers a comprehensive interface to solve complex optimization problems and a framework that can easily be use to test new ideas, and construct workflows to support solution analysis and decision making.

Our next goal, is to tackle the original problem and meet the constraints at the local operating points, by making use of the new optimization library that the developers of Liger are currently introducing.

References

1. Deb, K., Pratap, A., Agarwal, S., Meyarivan, T.: A fast and elitist multiobjective genetic algorithm: NSGA-II. IEEE Transactions on Evolutionary Computation 6(2), 182–197 (2002)
2. Kroshko, D.: OpenOpt: Free scientific-engineering software for mathematical modeling and optimization (2007)
3. Durillo, J.J., Nebro, A.J.: jMetal: A Java framework for multi-objective optimization. Advances in Engineering Software 42(10), 760–771 (2011)
4. Fonseca, C.M., Fleming, P.J.: Multiobjective optimization and multiple constraint handling with evolutionary algorithms. I. A unified formulation. IEEE Transactions on Systems, Man and Cybernetics, Part A: Systems and Humans 28(1), 26–37 (1998)
5. Fonseca, C.M., Fleming, P.J.: Genetic Algorithms for Multiobjective Optimization: Formulation, Discussion and Generalization. ICGA 93, 416–423 (1993)

6. Giagkiozis, I., Lygoe, R.J., Fleming, P.J.: Liger: an open source integrated optimization environment. In: Proceeding of the fifteenth annual conference companion on Genetic and evolutionary computation conference companion, pp. 1089–1096. ACM (2013)

7. Giagkiozis, I., Purshouse, R.C. and Fleming, P.J.: An overview of population-based algorithms for multi-objective optimisation. International Journal of Systems Science, (ahead-of-print), 1–28 (2013)

8. Heath, C.M., Gray, J.S.: OpenMDAO: Framework for Flexible Multidisciplinary Design, Analysis and Optimization Methods. In: 8th AIAA Multidisciplinary Design Optimization Specialist Conference (MDO), pp. 1–13, Honolulu, Hawaii

9. Huband, S., Hingston, P., Barone, L., While, L.: A review of multiobjective test problems and a scalable test problem toolkit. IEEE Transactions on Evolutionary Computation 10(5), 477–506 (2006)

10. Johnson, T.: Diesel engine emissions and their control. Platinum Metals Review 52(1), 23–37 (2008)

11. Knecht, W.: Diesel engine development in view of reduced emission standards. Energy 33(2), 264–271 (2008)

12. Kukkonen, S., Lampinen, J.: GDE3: The third evolution step of generalized differential evolution. In: The 2005 IEEE Congress on Evolutionary Computation, vol. 1, pp. 443–450. IEEE (2005)

13. Lattarulo, V., Zhang, J., Parks, G.T.: Application of the MOAA to Satellite Constellation Refueling Optimization. In: Purshouse, R.C., Fleming, P.J., Fonseca, C.M., Greco, S., Shaw, J. (eds.) EMO 2013. LNCS, vol. 7811, pp. 669–684. Springer, Heidelberg (2013)

14. Minami, T., Takeuchi, K., Shimazaki, N.: Reduction of diesel engine NOx using pilot injection. Technical report, SAE Technical Paper (1995)

15. Morino, H., Obayashi, S.: Knowledge Extraction for Structural Design of Regional Jet Horizontal Tail Using Multi-Objective Design Exploration (MODE). In: Purshouse, R.C., Fleming, P.J., Fonseca, C.M., Greco, S., Shaw, J. (eds.) EMO 2013. LNCS, vol. 7811, pp. 656–668. Springer, Heidelberg (2013)

16. Munson, T., Sarich, J., Wild, S., Benson, S., McInnes, L.C.: Tao 2.0 users manual. Technical Report ANL/MCS-TM-322, Mathematics and Computer Science Division, Argonne National Laboratory (2012). http://www.mcs.anl.gov/tao

17. Tanabe, K., Kohketsu, S., Nakayama, S.: Effect of fuel injection rate control on reduction of emissions and fuel consumption in a heavy duty DI diesel engine. Technical report, SAE Technical Paper (2005)

18. Tanaka, T., Ando, A., Ishizaka, K.: Study on pilot injection of DI diesel engine using common-rail injection system. JSAE Review 23(3), 297–302 (2002)

19. Zaefferer, M., Bartz-Beielstein, T., Naujoks, B., Wagner, T., Emmerich, M.: A Case Study on Multi-Criteria Optimization of an Event Detection Software under Limited Budgets. In: Purshouse, R.C., Fleming, P.J., Fonseca, C.M., Greco, S., Shaw, J. (eds.) EMO 2013. LNCS, vol. 7811, pp. 756–770. Springer, Heidelberg (2013)

20. Zaglauer, S., Deflorian, M.: Multi-criteria Optimization for Parameter Estimation of Physical Models in Combustion Engine Calibration. In: Purshouse, R.C., Fleming, P.J., Fonseca, C.M., Greco, S., Shaw, J. (eds.) EMO 2013. LNCS, vol. 7811, pp. 628–640. Springer, Heidelberg (2013)

21. Zheng, M., Reader, G.T., Hawley, J.G.: Diesel engine exhaust gas recirculation - a review on advanced and novel concepts. Energy conversion and management 45(6), 883–900 (2004)

Re-design for Robustness: An Approach Based on Many Objective Optimization

Hemant Singh[✉], Md Asafuddoula[✉], Khairul Alam[✉],
and Tapabrata Ray[✉]

School of Engineering and Information Technology, University of New South Wales,
Canberra ACT 2600, Australia
{h.singh,md.asaf,k.alam,t.ray}@adfa.edu.au

Abstract. Re-Design for Robustness (RDR) represents a practical class of problems, where a limited set of components of an existing product are re-designed to improve the overall robustness of the product. RDR is still a common inefficient, expensive and a time consuming industry ritual, where component sensitivities are sequentially analyzed and altered with human experts in loop. In this paper, we introduce an automated approach, wherein a trade-off set of design variants (varying number of altered components) spanning the entire a range of feasibility and performance robustness are identified using a decomposition based evolutionary optimization algorithm. The benefits offered by the approach are highlighted using two re-design optimization problems from the automotive industry.

1 Introduction and Background

Virtually all products (consumer electronics, automobiles, home appliances etc.) that we are exposed on a daily basis share a number of common components with its predecessor. Usually, re-design is the underlying process resulting in improved products over time. While the objective of a re-design exercise might vary, i.e., improved performance, need to meet new industry standards/statutory requirements, improved robustness etc.; in this paper we focus on re-design for robustness, which aims to deliver a set of tradeoff robust designs with varying number of altered components. While the term *robust* refers to solutions that are less sensitive to varying loading conditions, material imperfections, inaccuracies in analyses/simulations and imprecise geometries [1], we restrict our discussion to problems involving uncertainties in the design variables only e.g., imprecise geometries. Furthermore, the uncertainties in the design variables are assumed to follow Gaussian distribution.

Robustness of solutions is typically assessed using two measures i.e., feasibility robustness (FR) and performance robustness (PR). The first measure assures feasibility, while the second provides an assurance on the performance. A comprehensive review of robustness quantification schemes, formulations and optimization strategies for robust optimization appear in [2]. This paper builds

© Springer International Publishing Switzerland 2015
A. Gaspar-Cunha et al. (Eds.): EMO 2015, Part II, LNCS 9019, pp. 343–357, 2015.
DOI: 10.1007/978-3-319-15892-1_23

upon author's previous work on robust optimization and extends the approach to deal with an important class of problems, i.e., re-design for robustness. The central question is *given an existing design, which components can be changed to improve the robustness of the product?* With this goal, enhancements have been proposed to the previously introduced decomposition based evolutionary algorithm for robust optimization (DBEA-r) algorithm. Specifically, the following research contributions are made:

- A new formulation is presented in order to deal with re-design optimization problems. Apart from feasibility and performance robustness modeled using objectives, an additional objective is introduced i.e., the number of altered components.
- A generational model of *DBEA-r* embedded with customized recombination schemes is developed to deal with re-design optimization problems.
- The efficacy of the proposed approach is illustrated using two engineering design optimization problems, namely vehicle crash worthiness optimization problem (VCOP) and car side impact problem (CSIP).

2   Problem Formulation and Robustness Quantification

Robustness has been quantified using the "sigma-levels", denoted as $sigma_g$ and $sigma_f$ for feasibility and performance robustness respectively:

- The term $sigma_g$ refers to the ratio of *expected* constraint value (μ_g) and standard deviation (σ_g) of constraint g. Since the constraints have been formulated as $g > 0$, the ratio $sigma_g = \mu_g/\sigma_g$ is equivalent to $(\mu_g - 0)/\sigma_g$, which is a measure of how many standard deviations can be fit between the constraint boundary (0) and the given solution [2]. This quantity is positive for feasible solutions and maximization of this quantity would result in solutions with high feasibility robustness.
- The term $sigma_f$ refers to the ratio of an user defined acceptable deviation σ_{f_0} and the standard deviation (σ_f) of objective f. For σ_f, the boundary is the user prescribed specification limit. Hence, the ratio $sigma_f = \sigma_{f_0}/\sigma_f$ denotes the number of standard deviations of the objective function that can be fit within the specification limit. Again, maximization of $sigma_f$ would result in solutions with high performance robustness.

If the value of $sigma_g$ is greater than a given value R_c, the value is truncated to R_c. It essentially means, the user is satisfied with the feasibility robustness level R_c and solutions having any higher robustness has the same preference as the one with R_c. A similar truncation strategy has been applied to $sigma_f$ (using R_f) to ensure performance robustness. In a problem involving multiple constraints, the *minimum sigma_g* across all constraints is considered to represent the overall $sigma_g$ of the solution. This translates to measuring the sigma-level of constraint that is most likely to be violated; which is different from traditionally used six-sigma formulation where defects caused using *all* constraints together are considered. Same strategy has been adopted for $sigma_f$ for the case of multiple objectives. This way of quantifying robustness helps comparing the robustness of solution with respect to each objective/constraint on a common scale (even though their raw values and standard deviations may be of different orders). For a six-sigma design, both R_c and R_f are set to 6.

The problem formulation introduced in this study for re-design is derived from *Form-4* discussed in [2]. For ease of reference, *Form-4* will be referred to as *Feasibility and Performance Robustness (FPR)* formulation, and the modified formulation introduced in this study will be referred to as *FPR for Re-design (FPRR)* formulation. In FPR, the objectives are the minimization of the expected values of each performance function and maximization of feasibility robustness and performance robustness measures (in terms of "sigma-level"). In order to quantify quantum of change between the base design and a candidate design, a new objective F_{nc} has been introduced. This new objective measures the number of variables that have been altered with respect to the base design. This objective is to be minimized. For example, consider a base design with design variables as $\{x_1^b, x_2^b, x_3^b\}$ and a candidate design as $\{x_1^1, x_2^1, x_3^b\}$. Since the candidate design has different means value for x_1 and x_2, therefore F_{nc} would be equal to 2. The FPRR formulation is presented in Equation 1. The objective and constraint functions involve a set of variables $\mathbf{x}$ with a given standard deviations. For deterministic variables the standard deviations will be zero.

$$\underset{(\mathbf{x})}{\text{Minimize}}\ \mu_{f_i(\mathbf{x})}, i = 1, 2, \ldots\ldots\ldots\ldots M$$

$$\underset{(\mathbf{x})}{\text{Maximize}} f_{M+1}(\mathbf{x}) = \text{Min}(sigma_g, R_c)$$

$$\underset{(\mathbf{x})}{\text{Maximize}} f_{M+2}(\mathbf{x}) = \text{Min}(sigma_f, R_f)$$

$$\underset{(\mathbf{x})}{\text{Minimize}} f_{M+3} = F_{nc}(\mathbf{x}, \mathbf{x}^b)$$

subject to

$$sigma_g \equiv \text{Min}(\mu_{g_j(\mathbf{x})}/\sigma_{g_j(\mathbf{x})}) \geq 0$$

$$\mathbf{x}^{(L)} \leq \mathbf{x} \leq \mathbf{x}^{(U)}$$

where

$$sigma_f \equiv \text{Min}(\sigma_{f_{0,i}(\mathbf{x})}/\sigma_{f_i(\mathbf{x})})$$

$$\mathbf{x}^{(L)} \leq \mathbf{x} \leq \mathbf{x}^{(U)}$$

$$(1)$$

3   Solution Strategy

Robust formulation presented in the previous section requires solution of optimization problem involving additional objectives. The total number of such objectives is four or more and hence a many objective optimization algorithm is required to solve the problem efficiently. The algorithm used in this study is a variant of the decomposition based evolutionary algorithm (DBEA-r) developed by the authors [2]. The algorithm referred to DBEA-rg relies on the use of a generational model as opposed to a steady state form reported in [2]. The method is outlined in Algorithm 1, and its components are discussed in the following subsections.

Algorithm 1.. DBEA-rg

Input: Gen_{max} (maximum number of generations), W (number of reference points), p_c (probability of crossover), p_m (probability of mutation), η_c (distribution index for crossover), η_m (distribution index for mutation)

1: $gen = 1$; Corner Set $(CS) = \emptyset$;
2: Generate the reference points using Normal Boundary Intersection (NBI) method. Each reference direction is a vector connecting origin to these reference points.
3: Initialize the population $\mathbf{P}$ consisting of W individuals. Randomly assign each individual of $\mathbf{P}$ to an unique reference direction.
4: Assign a random binary vector BV of size n to each individual, where n denotes the number of variables of the problem.
5: Repair the individuals of the population based on its BV and the base design.
6: Evaluate the initial population using prescribed robust formulation.
7: Compute the ideal point and the extreme point.
8: Normalize the individuals of the population
9: Use *corner-sort* to identify $2M$ corner solutions and assign them to Corner Set (CS).
10: **while** $(gen \leq Gen_{max})$ **do**
11:     Select $I_1 = 1 : W$ as the base parents
12:     I_2=Generate a shuffled list of individuals in the population
13:     Create offspring individuals $\mathbf{C}$ via recombination of I_1 and I_2
14:     Create offspring BV's via recombination of I_1 and I_2
15:     Repair the offspring individuals using their BV's and the base design.
16:     Evaluate the offspring individuals $\mathbf{C}$
17:     Update the corner set CS, ideal and extreme points
18:     Normalize the individuals of $\mathbf{P}$ and $\mathbf{C}$
19:     Compute the distances $(d_1$ and $d_2)$ for all members of $\mathbf{P}$ in their respective reference directions.
20:     Compute the distances $(d_1$ and $d_2)$ for all members of $\mathbf{C}$ in all reference directions.
21:     Update the parent individuals in the shuffled order of W with C_l using *single-first encounter strategy*, where C_l is the set of individuals satisfying replacement condition.
22:     $gen = gen + 1$
23: **end while**

- **Generation of reference directions:** A structured set of reference points γ is generated spanning a hyperplane with unit intercepts in each objective axis using normal boundary intersection method (NBI) [3]. The approach generates W points on the hyperplane with a uniform spacing of $\delta = 1/s$

for given number of objectives M with s unique sampling locations along each objective axis. The total number of points (W) is $W=\binom{M+s-1}{s}$. The reference directions are formed by constructing a straight line from the origin to each of these reference points.

- **Normalization of the solutions:** In DBEA-r [2], the normalization was based on intercepts calculated using M extreme points of the non-dominated set. In DBEA-rg, the extreme solution a_j has the coordinates corresponding to the maximum in each objective direction computed based on the set of non-dominated solutions and the corner solutions delivered by *corner-sort* procedure [4]. In corner sort, the top M solutions are the minimum in each objective, while the following M solutions are the minimum based on L_2 norm of all but one objectives. The ideal solution z_j has the coordinates corresponding to the minimum in each objective direction computed based on the set of non-dominated solutions.

Every solution in the population is normalized as follows:

$$f'_j(\mathbf{x}) = \frac{f_j(\mathbf{x}) - z_j}{a_j - z_j}, \quad \forall j = 1, 2, ...M \tag{2}$$

- **Computation of the distances:** For any given reference direction, the performance of a solution is judged using two measures d_1 and d_2 as shown in Equations 3 and 4. The first measure d_1 is the Euclidean distance between origin and the foot of the normal drawn from the solution to the reference direction, while the second measure d_2 is the length of the normal. Mathematically, d_1 and d_2 are computed as follows:

$$d_1 = \mathbf{w}^T \mathbf{f}'(\mathbf{x}) \tag{3}$$

$$d_2 = \|\mathbf{f}'(\mathbf{x}) - \mathbf{w}^T \mathbf{f}'(\mathbf{x})\mathbf{w}\| \tag{4}$$

where $\mathbf{w}$ is a unit vector along any given reference direction and $\mathbf{f}'(\mathbf{x})$ is a vector of normalized objective values. A value of $d_2 = 0$ ensures the solutions are perfectly aligned along with the reference directions ensuring good diversity, while a smaller value of d_1 indicates superior convergence.

- **Recombination and Repair:** In order to deal with re-design optimization problems, a modified recombination scheme and a repair method is used. The base design and the binary vector BV is used in the repair process. Let us assume that $[1.1, 4.5, 3.2]$ denotes the base design and $[\ 1.4,\ 2.6,\ 2.8]$ denotes a randomly initialized individual with a BV of $[1\ 0\ 1]$. The repair process will result in an individual with $[\ 1.4,\ 4.5,\ 2.8]$, i.e., a design with the second variable value fixed at the base design. During the process of recombination, two offsprings are generated using simulated binary crossover and polynomial mutation [5]. Thereafter, the corresponding BVs are generated in the following way. If for a given variable x_j, BV_j of the two parents are the same (either 0 or 1), then the BV_j of the offspring is set to the same value. Otherwise, if one of the parent has $BV_j = 0$ and other has $BV_j = 1$, then the corresponding BV_j of the offspring is set as 0 or 1 with equal probability.

Thereafter, the offsprings are *repaired* using BVs and one of them (selected at random) is considered as a candidate attempting to enter the population via replacement. It is to be noted that the objective F_{nc} corresponding to any given solution is simply the number of 1's in its BV.

- **Selection/replacement:**In the generational form, the non-dominated off-springs attempt to enter the population via replacement. For the entry, each offspring solution has to compete with all the solutions in the population in a random order until it makes a successful replacement. If we denote the distances as $\{d_{1_r}, d_{2_r}\}$ for a rth solution in the population and $\{d_{1_c}, d_{2_c}\}$ denotes the distances for the offspring solution along r^{th} reference direction, the offspring is considered winner if d_{2_c} is less than d_{2_r}. In the event the d_{2_c} is equal to d_{2_r}, the offspring is considered a winner if d_{1_c} is less than d_{1_r}.

- **Constraint Handling:** The constraint handling approach used in this work is based on epsilon level comparison reported earlier in [6].

4   Numerical Examples

In this section, we illustrate the behavior of FPR and FPRR strategies using two engineering design optimization problems. The first problem is a bi-objective Vehicle Crash-worthiness Optimization Problem (VCOP) while the second is a single objective Car Side Impact Problem (CSIP).

4.1   Experimental Setup

If the original problem has M objectives, FPR formulation will involve $M + 2$ objectives, while FPRR will involve $M + 3$ objectives. Therefore, for FPR formulation, the total number of objectives is four for VCOP and three for CSIP. As for FPRR, the corresponding numbers are five and four.

Population sizes of 820, 820, 969 and 1001 have been used for two, three, four and five objective (overall) problems arising out of DF, FPR and FPRR formulations. The reference directions have been created using Normal Boundary Intersection method (NBI) with the spacing parameter s is set to 819, 39, 16 and 10 for 2, 3, 4 and 5 objective formulations respectively. The probability of crossover is set to 1 and the probability of mutation is set to 0.05. The distribution index of crossover and mutation are set to $\eta_c = 30$ and $\eta_m = 20$ respectively. The population is evolved over a maximum of 164000 function evaluations and a sample size of 100 has been used to compute the *expected* value of the functions through explicit averaging. These samples have been generated using Latin-hypercube Sampling with Gaussian distribution (*LHS-Gaussian*). For comparison, we have also included the results of deterministic formulation (DF) of these problems.

4.2   Vehicle Crash Worthiness Optimization (VCOP)

The problem was first introduced by Sun *et al.* [7]. A modified bi-objective formulation of the problem is studied in this paper which seeks to maximize the post-impact energy absorption (U) of the vehicle structure and aims to minimize the structural weight (M), subject to the constraint on peak deceleration (a). Higher energy absorption lowers the risk to the occupants of the car. However, increase in energy absorption often leads to unwanted increase in the structural weight. To limit impact severity, a constraint on maximum deceleration is imposed in this formulation which is assumed to be 40g (g = 9.81 m/s^2). A full-scale finite element (FE) model of Ford Taurus was used as the baseline case obtained from the National Crash Analysis Center (NCAC) at http://www.ncac.gwu.edu/vml/models.html.

Problem Definition: Part thicknesses of three key members, inner rail, outer rail and the cradle rail of the vehicle front end structure have been chosen as design variables. Table 1 reports the bounds on the design variables together with the corresponding base thickness and mass.

Table 1. Part thicknesses (variables) and variable bounds for VCOP

Variable name	Part no.	Lower bound (mm)	Thickness (mm)	Upper bound (mm)	Mass (kg)
Inner rails (t_1)	29	1	1.50	2	7.49
Outer rails (t_2)	30	1	2.00	2	8.95
Cradle rails (t_3)	79 and 81	1	1.93	2	9.22

The multi-objective robust optimization problem is represented using Equation 5.

$$\text{Maximize } U_\mu = 72.4996 + 2.8178366t_1 - 0.0778410t_1^2 + 3.7901860t_2$$
$$+ 6.0060214t_2^2 + 52.005026t_3 - 17.599580t_3^2$$
$$+ 1.2718916t_1t_2 - 0.5211597t_1t_3 - 30.982883t_2t_3$$
$$+ 11.034587t_2t_3^2$$
$$\text{Minimize } M_\mu = 0.00392497 + 4.9603440t_1 + 4.4474721t_2 + 4.7437340t_3 \quad (5)$$
$$\text{subject to } a_\mu = 48.3807 - 8.4035115t_1 + 4.0333016t_1^2 - 17.774059t_2$$
$$+ 4.2845324t_2^2 - 11.547927t_3 + 4.3592314t_3^2$$
$$+ 4.7775756t_1t_2 + 4.5825734t_2t_3 \le 40$$

Robust Optimization: In this example, all three variables are assumed to be of uncertain nature and their uncertainties are assumed to follow a Gaussian distribution with standard deviations of $\sigma_{x_1}^2 = 2.5 \times 10^{-3}$, $\sigma_{x_2}^2 = 2.5 \times 10^{-3}$ and $\sigma_{x_3}^2 = 2.5 \times 10^{-3}$. The allowable performance variation threshold is assumed to be as follows: $\sigma_{f_{0,1}} = 2.10$ units and $\sigma_{f_{0,2}} = 2.10$ units.

Results obtained from 30 independent runs of DF and FPR formulations were accumulated. Nadir and Ideal points were identified from this set as [−100.2440, 23.3580] and [−106.1330, 14.1550] and they have been used for normalization of the objectives (to calculate hypervolumes). The hypervolume calculation is done using the method described in [8]. The hypervolume obtained using DF and FPR formulations are presented in Table 2.

Table 2. Hypervolume comparisons in (μ_f) space

Prob.		DF	FPR
VCOP	Best	0.54638	0.53840
	Median	0.54162	0.53453
	Mean	0.54191	0.53230
	Std	0.00114	0.01349

To visualize the solutions obtained, the results from the median run of FPR are color coded and shown in Figure 1(a) and Figure 1(b). One can clearly observe from Figures 1(b) that only a few solutions have six-sigma performance (in both feasibility and performance). While the performance of such six-sigma solutions are not too different from the ones identified using DF, they seem to lie on a specific region of the front as visible from Figures 1(a).

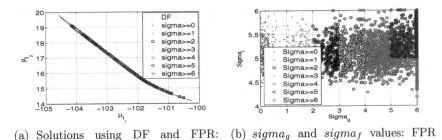

(a) Solutions using DF and FPR: Median Run

(b) *sigma$_g$* and *sigma$_f$* values: FPR Median Run

Fig. 1. Solutions of VCOP using DF and FPR formulations

Re-design for Robustness: In the previous subsection, we assumed that the designer had the flexibility to redesign/change all three component thicknesses to ensure six-sigma feasibility and performance robustness. However, as discussed before, one may often be faced with a situation where an existing design is in production and the aim is to identify the minimum set of components that need to be re-designed to deliver a six-sigma robustness. For this problem, the baseline design is $\{x_1^b, x_2^b, x_3^b\} \equiv \{1.50, 2.0, 1.93\}$ [9]. With three variables in play, there are six possibilities for re-design (i.e., changing one variable at a time and changing two at a time while keeping the other variables fixed at their base design values). The cases are listed in Table 3 and each of the six cases have been solved using FPR formulation. The hypervolumes obtained in the μ_f space are presented in Table 3, whereas the obtained fronts and their sigma-levels are shown in Figure 2. The base design under prescribed uncertainties is itself infeasible in terms of feasibility robustness i.e., it has a_{mu} of 52.412.

With one variable fixed at the base design and the other two allowed to vary (Cases 1-3), a number of solutions with various sigma levels are obtained as shown in Figure 2. The $sigma_g$ and $sigma_f$ values of the median run using FPR formulation with different number of variables allowed to change are presented in Figure 3. One can observe that, the base design can be modified to a six-sigma FPR design by changing variables x_2 and x_3, and keeping the x_1 the same as base design (Case 1).

With two variables fixed at the base design and the remaining one is allowed to vary (Cases 4-6), there is no feasible solution obtained for re-design and hence Cases 4-6 have been excluded from further discussion.

Table 3. Part thicknesses and variable bounds for each of the re-design cases

Scenarios	Variable name	Part no.	Lower bound (mm)	Thickness (mm)	Upper bound (mm)	Mass (kg)	HV (Best, Median, Mean, Std)
	Inner rails (t_1)	29	1.50	1.50	1.50	7.49	
Case-1	Outer rails (t_2)	30	1	2.00	2	8.95	(0.4174, 0.4171, 0.4170, 0.0002)
	Cradle rails (t_3)	79 and 81	1	1.93	2	9.22	
	Inner rails (t_1)	29	1	1.50	2	7.49	
Case-2	Outer rails (t_2)	30	2.00	2.00	2.00	8.95	(0.3736, 0.3729, 0.3728, 0.0005)
	Cradle rails (t_3)	79 and 81	1	1.93	2	9.22	
	Inner rails (t_1)	29	1	1.50	2	7.49	
Case-3	Outer rails (t_2)	30	1	2.00	2	8.95	(0.1738, 0.1735, 0.1734, 0.0002)
	Cradle rails (t_3)	79 and 81	1.93	1.93	1.93	9.22	
	Inner rails (t_1)	29	1.50	1.50	1.50	7.49	
Case-4	Outer rails (t_2)	30	2.00	2.00	2.00	8.95	-
	Cradle rails (t_3)	79 and 81	1	1.93	2	9.22	
	Inner rails (t_1)	29	1	1.50	2	7.49	
Case-5	Outer rails (t_2)	30	2.00	2.00	2.00	8.95	-
	Cradle rails (t_3)	79 and 81	1.93	1.93	1.93	9.22	
	Inner rails (t_1)	29	1.50	1.50	1.50	7.49	
Case-6	Outer rails (t_2)	30	1	2.00	2	8.95	-
	Cradle rails (t_3)	79 and 81	1.93	1.93	1.93	9.22	

Next, we solve the same problem using the FPRR formulation discussed in Section 2. The algorithm DBEA-rg was run for the same number of function evaluations as used in *each* of the FPR re-design cases. The results obtained are presented in Figure 4. The combined robust solutions obtained from all the cases of FPR are shown in Figure 4(a). The results obtained from a single run of

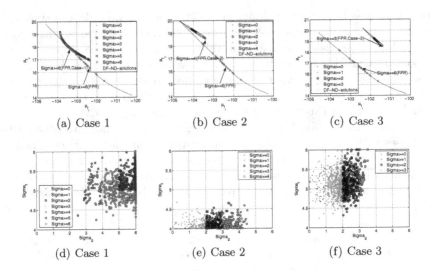

Fig. 2. Robust solutions for redesign Cases 1,2,3 and their corresponding sigma levels using FPR for a median run

FPRR are shown in Figure 4(b). It can be seen that the same level of robustness is achieved using only a fraction (roughly one-sixth) of the total evaluations using FPRR as compared to exhaustively investigating all possible re-design cases using FPR formulation.

FPRR would be the only practical alternative for problems involving large number of variables. While FPRR is meant to deliver robust re-design alternatives, it will also *include* solutions where all the variables/components might have been altered i.e., the solution to the original robust optimization problem using FPR formulations.

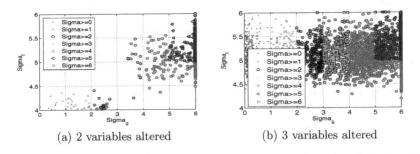

Fig. 3. *sigma$_f$* and *sigma$_g$* of the solutions obtained for VCOP from a median run of *FPR*, i.e., Case-1 to 3 (two variable altered) and the original robust optimization formulation (all three variables altered)

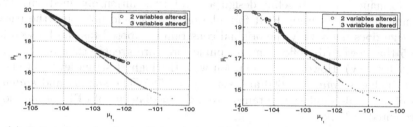

(a) Collated results from all 6 redesign (b) Results from a single run of FPRR
cases and the original FPR

Fig. 4. Comparison of robust non-dominated solutions obtained in the μ_f space for VCOP

4.3   Car Side Impact Problem (CSIP)

In this problem, the objective is to minimize the weight of the car subject to a set of constraints as presented in [10]. The design variables are described in Table 4. The baseline design listed in Table 4 is from [9]. It is important to take note that the baseline design violates the constraints listed in [10] and in particular the constraint related to lower rib deflection.

Table 4. Description of the design variables for CSIP

Design Variable	Side constraints	Base design	Deterministic optimum
Thickness of B-pillar inner	$0.5 \leq x_1 \leq 1.5$	1	0.5
Thickness of B-pillar reinforcement	$0.45 \leq x_1 \leq 1.35$	1	1.2382
Thickness of floor side inner	$0.5 \leq x_1 \leq 1.5$	1	0
Thickness of cross members	$0.5 \leq x_1 \leq 1.5$	1	1.5
Thickness of door beam	$0.875 \leq x_1 \leq 2.625$	1	0.875
Thickness of door belt line reinforcement	$0.4 \leq x_1 \leq 1.2$	1	1.1783
Thickness of roof rail	$0.4 \leq x_1 \leq 1.2$	1	0.4
Material of B-pillar inner	$0.345 \leq x_1 \leq 0.345$	0.345	0.345
Material of floor side inner	$0.192 \leq x_1 \leq 0.192$	0.192	0.192
Barrier height	$0.0 \leq x_1 \leq 0.0$	0	0
Barrier hitting position	$0.0 \leq x_1 \leq 0.0$	0	0

Robust Optimization. Seven out of eleven variables have been assumed to be of uncertain nature. The uncertainties associated with the variables have been assumed to follow a Gaussian distribution with the standard deviations of $\sigma_{x_1}^2 = 1.7 \times 10^{-3}$, $\sigma_{x_2}^2 = 1.7 \times 10^{-3}$, $\sigma_{x_3}^2 = 1.7 \times 10^{-3}$, $\sigma_{x_4}^2 = 1.7 \times 10^{-3}$, $\sigma_{x_5}^2 = 1.7 \times 10^{-3}$, $\sigma_{x_6}^2 = 1.7 \times 10^{-3}$, $\sigma_{x_7}^2 = 1.7 \times 10^{-3}$. [1]. The allowable performance function (weight) variation is set as $\sigma_{f_0} = 2.5$ units. Table 5 presents

---

[1] x_8 to x_{11} are kept fixed as in [10].

the performance of the baseline design and robust solution obtained with the maximum sigma level using FPR formulation. One should note that the baseline design does not satisfy the original constraints imposed (as can be seen in Table 5 that lower rib deflection and pubic force are greater than 32mm and 4kN respectively). However, the solution with the highest sigma level obtained using FPR has $sigma_g = 5.600$ and $sigma_f = 5.785$. The $sigma_f$ and $sigma_g$ values of the final population obtained from the median run of FPR formulation are presented in Figure 5.

Table 5. Performance of the base design and the robust design with the highest sigma level obtained for CSIP

Response		Performance of the design	Robust solution base which maximizes $(sigma_g,sigma_f)$	Constraints
Weight (kg)		29.05	26.84	–
Abdomen load		0.69	0.39	$\leq$ 1kN
Viscous Criterion	Upper	0.19	0.21	$\leq$ 0.32mm/ms
Viscous Criterion	Middle	0.19	0.19	$\leq$ 0.32mm/ms
Viscous Criterion	Lower	0.27	0.29	$\leq$ 0.32mm/ms
Rib deflection	Upper	27.19	26.92	$\leq$ 32mm
Rib deflection	Middle	25.97	25.03	$\leq$ 32mm
Rib deflection	Lower	32.0095	29.65	$\leq$ 32mm
Pubic force		4.03	3.87	$\leq$ 4kN
Velocity of V-pillar		9.23	8.98	$\leq$ 10mm/ms
Velocity of front door		15.12	15.39	$\leq$ 17.5mm/ms

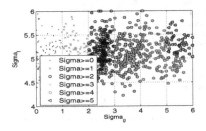

Fig. 5. Robust solutions obtained for CSIP using FPR formulation (Median run)

Re-design for Robustness–Results and Discussion: Unlike VCOP, the total number of re-design cases (if done individually) using FPR for this problem would be very large ($2^7 - 1 = 127$ cases, including the one where all variables can change from baseline). Since it is computationally prohibitive to do all cases individually via FPR, we make use of FPRR formulation. The solutions obtained using FPRR formulation are shown in Figure 6. The baseline design is itself infeasible and the effect of altering one through to seven variables is presented in Figure 6.

The effects of number of altered variables on the robustness measure and the performance measure is presented in Figure 7(a) and Figure 7(b) respectively. One can observe that by merely altering one of the variables (different combinations may be possible), the design can be improved from being infeasible to nearly 1-sigma in feasibility and performance. If two variables were allowed to be altered, a robustness of nearly 5-sigma can be achieved. It is worth noting that the infeasible baseline design has an weight of 29.05kg, while the one 5-sigma design with two or more variables altered would offer a mean weight of nearly 31.65kg. The example also illustrates that altering any more than two variables do not offer any benefit in terms of robustness.

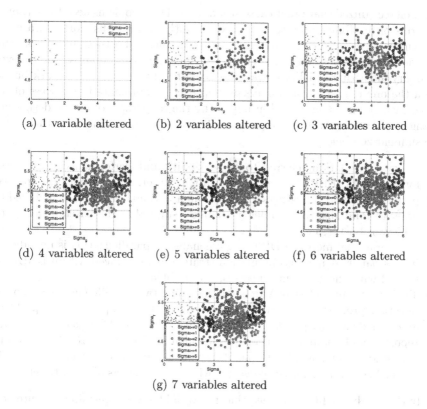

(a) 1 variable altered   (b) 2 variables altered   (c) 3 variables altered

(d) 4 variables altered   (e) 5 variables altered   (f) 6 variables altered

(g) 7 variables altered

Fig. 6. $sigma_g$ and $sigma_f$ of the solutions obtained using FPRR formulation for CSIP

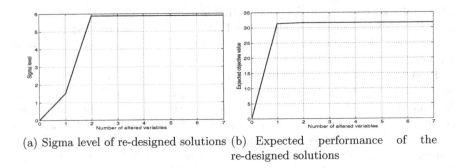

(a) Sigma level of re-designed solutions (b) Expected   performance   of   the
                                        re-designed solutions

Fig. 7. Re-designed solutions using FPRR for CSIP

5   Summary and Future Directions

It is well recognized that practical solutions of real life problems need to be robust
in terms of feasibility and performance. In authors' previous work [2], a comprehensive formulation was presented, wherein such robustness measures were
included as objectives, and an efficient decomposition based many-objective evolutionary algorithm (DBEA-r) was developed for its solution. This work builds
upon the previous study and offers an approach to automate the process of re-
design, where one is interested in improving the robustness of a base design via
changes to a selected set of components. The specific contributions of this paper
are summarized below:

- A new formulation referred to as Feasibility and Performance Robustness
  for Re-design (FPRR) is introduced in this paper. Apart from the expected
  performance and robustness measures, an additional objective is included in
  order to minimize the number of changes required in the baseline design to
  achieve robustness.
- A generational model of DBEA-rg amenable to parallelization is introduced
  in this paper. A new recombination strategy is introduced within DBEA-rg
  to deal with the redesign aspects of the problem.
- The performance of DBEA-rg is studied using two real-life engineering optimization problems, namely vehicle crash worthiness optimization problem
  and car side impact problem. Robust solutions obtained for each problem are
  reported and compared with solutions identified using FPRR formulation.
  The study clearly highlights the ability of the approach to identify valuable
  tradeoff set of alternative re-designs with various levels of robustness.

In this study, we have assumed that the variables are uncertain in nature and
follow a Gaussian distribution. The resulting performance function in the face
of uncertainties is also assumed to follow a Gaussian distribution. While explicit
averaging with 100 samples have been used in this study, alternative approaches
based on Polynomial Chaos can be used to compute the mean and variance of
the distributions using far fewer samples.

References

1. Beyer, H.G., Sendhoff, B.: Robust optimization - a comprehensive survey. Computer Methods in Applied Mechanics and Engineering **196**(33–34), 3190–3218 (2007)
2. Asafuddoula, M., Singh, H., Ray, T.: Six-sigma robust design optimization using a many-objective decomposition based evolutionary algorithm. IEEE Transactions on Evolutionary Computation (99) (2014)
3. Das, I., Dennis, J.E.: Normal-bounday intersection: A new method for generating Pareto optimal points in multicriteria optimization problems. SIAM J. Optim. **8**(3), 631–657 (1998)
4. Singh, H.K., Isaacs, A., Ray, T.: A Pareto corner search evolutionary algorithm and dimensionality reduction in many-objective optimization problems. IEEE Transactions on Evolutionary Computation **15**(4), 539–556 (2011)
5. Deb, K.: Multi-objective optimization using evolutionary algorithms. John Wiley & Sons (2001)
6. Asafuddoula, M., Ray, T., Sarker, R., Alam, K.: An adaptive constraint handling approach embedded MOEA/D. In: IEEE World Congress on Computational Intelligence, (June 10–15 2012), 1–8
7. Sun, G., Li, G., Zhou, S., Li, H., Hou, S., Li, Q.: Crashworthiness design of vehicle by using multiobjective robust optimization. Structural and Multidisciplinary Optimization **44**(1), 99–110 (2011)
8. Beume, N., Naujoks, B., Emmerich, M.: SMS-EMOA: Multiobjective selection based on dominated hypervolume. European Journal of Operational Research **181**, 1653–1669 (2006)
9. Gu, L., Yang, R.J., Tho, C.H., Makowskit, M., Faruquet, O., Li, Y.: Optimisation and robustness for crashworthiness of side impact. International Journal of Vehicle Design **26**(4), 348–360 (2001)
10. Saxena, D.K., Deb, K.: Trading on infeasibility by exploiting constraint's criticality through multi-objectivization: A system design perspective. In: IEEE Congress on Evolutionary Computation, 911–926 (2007)

A Model for a Human Decision-Maker in a Polymer Extrusion Process

Luciana Rocha Pedro[1]([✉]), Ricardo Hiroshi Caldeira Takahashi[2],
and António Gaspar-Cunha[3]

[1] COPPE, UFRJ, Rio de Janeiro, Brasil
lu.ufmg@gmail.com
[2] Department of Mathematics, UFMG, Belo Horizonte, Brasil
taka@mat.ufmg.br
[3] Department of Polymer Engineering, Universidade do Minho, Guimarães, Portugal
agc@dep.uminho.pt

Abstract. The NN-DM is a method developed to find a mathematical model that represents the Decision-Maker (DM) by employing an artificial neural network (NN) in situations in which the preferences can be represented by a utility function. This paper presents further developments to the NN-DM method to find a model in a polymer extrusion process. The form of the DM's interaction, the domain assignment, the ranking process, and the performance assessment are adapted to a real context of a multi-objective optimization problem followed by a design decision. The DM is then requested to fill a matrix expressing his preferences considering pairwise comparisons expressing ordinal relations only. Two multi-objective optimization problems are tested, each one with three estimates of different Pareto-optimal fronts. The adapted NN-DM method is able to provide a model which sorts the available solutions from the best to the worst according to the DM's preferences.

Keywords: Polymer extrusion process · Human decision-maker · Multi-objective optimization · Multi-criteria decision analysis · Utility function

1 Introduction

Most real-world optimization problems involve multiple objectives which have to be considered simultaneously. As these objectives are usually conflicting it may not be possible to find a single solution which is optimal with respect to all objectives. For obtaining only one solution a Decision-Maker (DM) has to make a choice regarding the importance of different criteria related to the optimization process [1]. Therefore, the final single solution of a Multi-Objective Optimization Problem (MOOP) results from the combined optimization and decision processes.

The importance of the decision-making process in a multi-objective environment is a characteristic of an interactive algorithm [6,10]. Two recent publications involving this kind of method are summarized here. Deb et al. proposed

© Springer International Publishing Switzerland 2015
A. Gaspar-Cunha et al. (Eds.): EMO 2015, Part II, LNCS 9019, pp. 358–372, 2015.
DOI: 10.1007/978-3-319-15892-1_24

an interactive Multi-Objective Evolutionary Algorithm (MOEA) based on progressively approximated value functions [2]. The DM's preference information is captured progressively by constructing a value function based on a preference order defined by the DM by pairwise comparison between the current solutions. In an approach considering reference points Köksalan and Karahan developed the Interactive Territory-Defining Evolutionary Algorithm (iTDEA) [5] which creates territories around the solutions with sizes reflecting the DM's preferences. The iTDEA guides the search converging to the entire Pareto-optimal front with preferable regions highlighted by its density. The quality of the described methods concerning the correspondence between the produced solutions and the DM's preferences is assessed empirically.

The NN-DM method [8], focus of this paper, is a procedure for constructing a mathematical model for the DM in situations in which the preferences are according to the Multi-Attribute Utility Theory (MAUT), that is, the preferences are represented by an underlying utility function $\mathcal{U}$. The DM is required to express his preferences by pairwise comparisons expressing ordinal relations only. Pedro and Takahashi proposed a new method derived from NN-DM and NSGA-II methods called Interactive Non-dominated Sorting algorithm with Preference Model (INSPM) [9]. INSPM is an interactive method for modeling the DM's preferences inside an adaptation of NSGA-II. In the NN-DM version discussed here, the process of providing information is made *a posteriori* to the optimization procedure and the data is employed in constructing a model, denoted NN-DM model $\hat{\mathcal{U}}$, that represents the DM's preferences in a specific domain $\mathcal{D}$. The function $\hat{\mathcal{U}}$ can be repeatedly employed whenever the available solutions are within the domain $\mathcal{D}$ without further demand to the DM.

The specific contribution of this paper is an adaptation of the NN-DM method to take into account some issues that arise in actual decision-making processes that appear in design contexts. The adaptations take place to assist the DM in the interaction process, making it simple and efficient. The original NN-DM method is divided into four steps. In the current work the domain $\mathcal{D}$ is previously provided by the DM. Thereby it is not necessary to establish the domain as the original **Step 1** has proposed. **Step 2** introduces the ranking of alternatives which is now built from a total sorting (the decision-making matrix). **Step 3** is unchanged, but additional changes are made in **Step 4** since the performance of the resulting model, assessed by the Kendall-Tau Distance (KTD) in the original method, is now evaluated by the DM himself.

The chosen application is one important polymer processing technology: the single screw extrusion [3]. The process performance depends on three different parameters: the polymer properties, the system geometry, and the operating conditions. Two MOOPs are examined and in each MOOP three sets of Pareto-optimal Front Estimates (PFE) are available, considering different sets of decision variables: the operating conditions, the screw geometry, and a combination of both. In each scenario the resulting NN-DM model $\hat{\mathcal{U}}$ provides the sorting of solutions belonging to the PFE from the best to the worst according to the DM's preferences.

This paper is organized as follows. In Section 2 the definitions of multi-objective optimization and multi-criteria decision-making problems are presented. In Section 3 the polymer extrusion process is explained and the examined data is presented. In Section 4 the adaptation of the NN-DM method is introduced and an example of the original NN-DM method is shown. In Section 5 the resulting NN-DM models are established and tests with the available data illustrate the models' behavior. Section 6 discusses the obtained results and the work under development.

2   Problem Statement

2.1   Multi-Objective Optimization

A Multi-Objective Optimization Problem (MOOP) is concerned with mathematical optimization problems involving more than one objective to be optimized simultaneously. In a MOOP, different optimal solutions usually exist such that no single solution can be considered better than all other ones with respect to all the criteria. The set of such solutions is called Pareto-optimal set and its image in the space of objectives is called Pareto-optimal front, or just Pareto-front. In the absence of any additional preference information, none of the Pareto-optimal solutions can be said to be inferior when compared to any other solution, as they are superior in at least one criterion.

Solving a MOOP is often a difficult task since it involves conflicting criteria and usually several constraints exist. Multi-Objective Evolutionary Algorithms (MOEAs) became popular in the task of solving problems of this class [11]. In the current application the Reduced Pareto Set Genetic Algorithm (RPSGA) [4] is the MOEA selected to solve the problem of parameter setting in the polymer extrusion process. RPSGA is an algorithm based on the assignment of the fitness through a ranking function obtained employing a clustering algorithm. This optimization methodology has already been applied to the optimization of the operating conditions and to the design of screws for polymer extrusion. RPSGA has shown good performance and it is able to find solutions with physical meaning in the proposed application. For further details see [3,4].

2.2   Decision-Making Methodology

The selection of a single solution from a Pareto-front resulting from an optimization process requires information that may not be present in the objective functions. This information, expressing subjective preferences, must be introduced by a Decision-Maker (DM). The insertion of the DM's preferences in the optimization procedure allows the distinction among the solutions within a non-dominated set and, as a consequence, provides a ranking of the solutions in the MOOP.

In this work the DM indicates preference relations (ordinal relations only) among simulated alternatives in the desired domain leading the NN-DM method to construct a model for the DM's preferences. The following basic elements are involved:

Set A of available alternatives This set is an estimate of the Pareto-front provided by RPSGA which works as a problem instance of the multi-criteria decision-making problem. The set A is discrete and each element $a \in A$ corresponds to a solution located on the PFE.

Decision-Maker Each alternative possesses a *value* which is assigned by a Decision-Maker (DM) that formally corresponds to a utility function $\mathcal{U}$. The best alternative $x^* \in A$ is the one that maximizes the function $\mathcal{U}$ in the set A. It is assumed here that it is not possible to directly measure the values of $\mathcal{U}(x)$, for any alternative x. Only the ordinal information, provided by a preference function U, may be extracted from yes/no queries to the DM.

Set F of simulated alternatives This set is constructed to request information from the DM about the entire domain $\mathcal{D}$ in which the utility function $\mathcal{U}$ is being approximated; the alternatives on a Pareto-optimal front usually does not fully provide this kind of information.

3 Polymer Extrusion Process

3.1 Process Description

Single screw extrusion is an important polymer processing technology allowing the production of products such as pipes, film, profiles, and fibers. The main basic functions of a single screw extruder are: to transport the solid material from the hopper to the heated barrel zone; to melt the polymer; to homogenize and mix the melted polymer with the additives usually present; and to create the necessary pressure which enables the polymer to pass through the die at the desired output. Different polymers are characterized by properties such as: thermal (heat conduction coefficient, melting temperature, heat capacity, etc.), physical (friction coefficients, density, etc.), and rheological (which is a measure of the resistance of the polymer to the flow).

In industrial practice the polymer processing technology is employed in managing a single polymer whose properties change according to pressure and temperature. Therefore, a thermo-mechanical environment is developed in which the polymer passes through different thermal and physical states. Figure 1 illustrates a simple extruder with a conventional screw with five geometrical zones:

(i) solids conveying in the hopper: the solids are fed into the hopper in which, by action of gravity, are transported inside the barrel;

(ii) solids conveying in the screw: by action of the screw rotation and due to the friction between the screw and barrel walls the solid polymer is pressurized and a solid bed is formed and, simultaneously, the polymer is transported to the heated barrel zone;

(iii) delay zone: due the the heat generated by friction and the heat conducted from the barrel a melt film is formed;

(iv) melting zone: a specific melting mechanism, characterized by the existence of a melt pool and melt films around the solid bed, is developed;

(v) conveying zone: the polymer is pressurized and it is transported to the die.

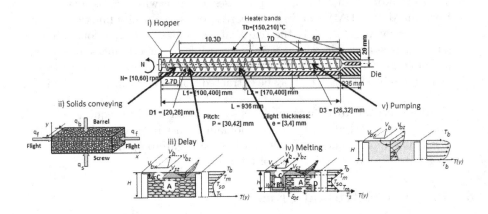

Fig. 1. Thermo-mechanical functional process developed in a single screw extruder

The modeling of the polymer processing technology involves the linkage of all those functional zones adopting the appropriate boundary conditions. The process performance depends on different type of parameters (polymer properties, system geometry, and operating conditions) which can be characterized by the mass output of the machine ($\mathbf{Q}$), the average melt temperature of the polymer at die exit ($\mathbf{T_{melt}}$), the power consumption required to rotate the screw ($\mathbf{P}$), the capacity of pressure generation ($\mathbf{P_{max}}$), the length of screw required to melt the polymer ($\mathbf{L_{melt}}$), and the degree of mixing quantified by the average of the deformation induced, denoted WATS ($\mathbf{W}$). Those are the common objectives considered in the definition of the multi-objective optimization problem related to the polymer extrusion process. Further details of the modeling routine implemented can be found elsewhere [3].

3.2 Available Data

As the single screw extrusion is a computationally expensive multi-objective optimization problem, this paper deals directly with previously estimates of different Pareto-optimal fronts obtained by the RPSGA multi-objective optimization algorithm [3]. The objectives considered in the multi-objective optimization problems are $\mathbf{Q}$, $\mathbf{P}$, and $\mathbf{W}$. Two different optimization problems are considered using only two objectives each with the aim of studying the process: $\mathbf{Q} \times \mathbf{P}$ and $\mathbf{Q} \times \mathbf{W}$. In each problem three sets of PFE are available considering different decision variables. The first set considers the operating conditions given by the screw rotation speed ($\mathbf{N}$) and the barrel temperature profile ($\mathbf{T_{b1}}$, $\mathbf{T_{b2}}$, and $\mathbf{T_{b3}}$); the second set considers the geometry, in which the variables are the screw flighted length ($\mathbf{L_1}$ and $\mathbf{L_2}$), the screw outside diameter ($\mathbf{D_1}$ and $\mathbf{D_2}$), the screw pitch ($\mathbf{Pitch}$), and the flight width ($\mathbf{e}$); and, finally, in the third set both types of decision variables are considered. Table 1 resumes the information about the available PFE and Table 2 provides the objectives, aim of optimization, range of variation, and the partitions.

Table 1. Multi-objective optimization problems in a single screw extrusion process

PFE	Objectives	Optimization Type	Decision Variables
$\mathbf{QP_1}$	Q and P	Operating conditions	$\mathbf{N}$, $\mathbf{T_{b1}}$, $\mathbf{T_{b2}}$, $\mathbf{T_{b3}}$
$\mathbf{QP_2}$	Q and P	Geometry	$\mathbf{L_1}$, $\mathbf{L_2}$, $\mathbf{D_1}$, $\mathbf{D_2}$, $\mathbf{Pitch}$, $\mathbf{e}$
$\mathbf{QP_3}$	Q and P	Both	$\mathbf{N}$, $\mathbf{T_{b1}}$, $\mathbf{T_{b2}}$, $\mathbf{T_{b3}}$, $\mathbf{L_1}$, $\mathbf{L_2}$, $\mathbf{D_1}$, $\mathbf{D_2}$, $\mathbf{Pitch}$, $\mathbf{e}$
$\mathbf{QW_1}$	Q and W	Operating conditions	$\mathbf{N}$, $\mathbf{T_{b1}}$, $\mathbf{T_{b2}}$, $\mathbf{T_{b3}}$
$\mathbf{QW_2}$	Q and W	Geometry	$\mathbf{L_1}$, $\mathbf{L_2}$, $\mathbf{D_1}$, $\mathbf{D_2}$, $\mathbf{Pitch}$, $\mathbf{e}$
$\mathbf{QW_3}$	Q and W	Both	$\mathbf{N}$, $\mathbf{T_{b1}}$, $\mathbf{T_{b2}}$, $\mathbf{T_{b3}}$, $\mathbf{L_1}$, $\mathbf{L_2}$, $\mathbf{D_1}$, $\mathbf{D_2}$, $\mathbf{Pitch}$, $\mathbf{e}$

Table 2. Objectives, aim of optimization, range of variation, and partitions

Objective	Aim of optimization	Range of variation	Partition
Mass output	Maximization	$[1, 20]$	$[f_{10}, f_{11}, f_{12}, f_{13}] = [1, 7, 14, 20]$
Power consumption	Minimization	$[0, 9200]$	$[f_{20}, f_{21}, f_{22}, f_{23}] = [0, 3067, 6134, 9200]$
WATS	Maximization	$[0, 1300]$	$[f_{30}, f_{31}, f_{32}, f_{33}] = [0, 434, 867, 1300]$

Figure 2 presents the available estimates of the Pareto-optimal fronts considering the problems $\mathbf{Q} \times \mathbf{P}$ ($\mathbf{QP}_1$, $\mathbf{QP}_2$, and $\mathbf{QP}_3$) and $\mathbf{Q} \times \mathbf{W}$ ($\mathbf{QW}_1$, $\mathbf{QW}_2$, and $\mathbf{QW}_3$). The domain is established by the range of variation presented in Table 2.

3.3   Interaction with the DM

A decision-making matrix $\mathcal{M}$ is a matrix filled by the DM to assist the NN-DM method in the construction of a model for the DM's preferences. Each element m_{ij} of $\mathcal{M}$ is defined as given in Equation 1.

$$\begin{cases} m_{i,j} = -1, & \text{if } a_i \text{ is preferable than } a_j; \\ m_{i,j} = 0, & \text{if } a_i \text{ and } a_j \text{ are equivalents;} \\ m_{i,j} = 1, & \text{if } a_j \text{ is preferable than } a_i. \end{cases} \tag{1}$$

Considering n the number of partitions in each dimension and m the number of objectives the total number of simulated alternatives is given by n^m. Therefore, the total number of pairwise comparisons is given by n^{2m} which corresponds to the number of entries of the decision-making matrix. The information required from the DM is reduced by dominance and comparisons between the same alternative (the matrix diagonal). The symmetry also develops an important role:

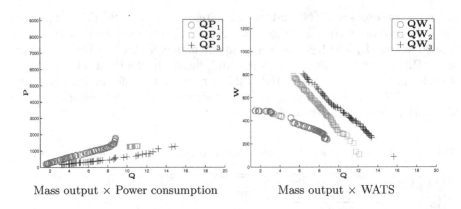

Mass output × Power consumption   Mass output × WATS

Fig. 2. Available estimates of the Pareto-optimal fronts

given a utility function $\mathcal{U}$ and two alternatives a and b, $\mathcal{U}(a, b) = \mathcal{U}(b, a) = a$ or $\mathcal{U}(a, b) = \mathcal{U}(b, a) = b$. In both scenarios only one query is required from the DM and the entries in the decision-making matrix are symmetric values ± 1.

For exemplifying the described process consider a decision-making problem with two objective functions F_1 and F_2 that should be minimized. The function F_1 is defined in the interval $[a_1, b_1]$ and the function F_2 in the interval $[a_2, b_2]$. Assuming only two partitions in each interval, the simulated decision-making problem consists of 16 queries provided by the combination of elements of the set $\{[a_1, a_2], [a_1, b_2], [b_1, a_2], [b_1, b_2]\}$. Table 3a presents the unfilled decision-making matrix $\mathcal{M}$ for this example, with variables s_i, $i = 1 \dots 16$, representing the entries. The answers to the queries s_i are then divided into four groups:

Equivalence Variables s_1, s_6, s_{11}, and s_{16} derive from queries made between the same alternative (the matrix diagonal). Therefore, their values are zero indicating that a choice is unnecessary.

Dominance Variables s_2, s_3, s_4, s_8, and s_{12} are determined considering the dominance, since $a_1 < a_2$, $b_1 < b_2$ and the aim of the optimization for both objectives is minimization.

Symmetry Variables s_5, s_9, s_{13}, s_{14}, and s_{15} result from symmetry, since if the preferred alternative between a and b is, for example, a, the preferred alternative between b and a is also a.

Decision-Maker Variables s_7 and s_{10} demand the DM's expertise. Considering that s_7 and s_{10} are provided from queries between the same alternatives, in this example only one query would be presented to the DM.

Table 3b shows the decision-making matrix partially filled by considering the equivalence, the dominance, and the symmetry among the alternatives. This matrix is then presented to the DM who needs to provide an answer to the

Table 3. Example of a decision-making matrix $\mathcal{M}$

$F_1 \times F_2$	$[a_1,a_2]$	$[a_1,b_2]$	$[b_1,a_2]$	$[b_1,b_2]$
$[a_1,a_2]$	s_1	s_2	s_3	s_4
$[a_1,b_2]$	s_5	s_6	s_7	s_8
$[b_1,a_2]$	s_9	s_{10}	s_{11}	s_{12}
$[b_1,b_2]$	s_{13}	s_{14}	s_{15}	s_{16}

$F_1 \times F_2$	$[a_1,a_2]$	$[a_1,b_2]$	$[b_1,a_2]$	$[b_1,b_2]$
$[a_1,a_2]$	0	−1	−1	−1
$[a_1,b_2]$	1	0	s_7	−1
$[b_1,a_2]$	1	s_{10}	0	−1
$[b_1,b_2]$	1	1	1	0

(a) Unfilled matrix   (b) Matrix presented to the DM

remaining queries. In this example, only one query would be required from the DM.

In the real scenario considered here the number of partitions in each dimension of the grid is established as 4. This value provides enough information for the NN-DM method for constructing suitable NN-DM models for the DM's preferences without requiring demanding information from the DM.

As each optimization problem is composed of two objective functions, there are 16 pairs of simulated alternatives which generate a total of 256 pairwise comparisons per problem. Excluding the comparison of pairs composed by the same alternatives (the matrix diagonal) and considering that $\mathcal{M}$ is anti-symmetric the resulting number of queries becomes 120. Among these 120 queries the dominance is applied considering the aim of optimization in each scenario, which solves 84 queries. Therefore the DM had to answer only 36 of the 256 queries in each optimization problem. The resulting matrix had been presented to a human DM who had to choose the best alternative of each pair of simulated alternatives whose answer was not obtained by one of those described decision criteria. The decision-making matrices employed in estimating the DM's preferences in the polymer extrusion process are presented in Figures 4 and 5. The gray cells indicate the 36 positions the DM actually filled.

Table 4. Decision-making matrix: mass output $(\mathbf{Q}) \times$ power consumption $(\mathbf{P})$

$Q \times P$	f_{10},f_{20}	f_{11},f_{20}	f_{12},f_{20}	f_{13},f_{20}	f_{10},f_{21}	f_{11},f_{21}	f_{12},f_{21}	f_{13},f_{21}	f_{10},f_{22}	f_{11},f_{22}	f_{12},f_{22}	f_{13},f_{22}	f_{10},f_{23}	f_{11},f_{23}	f_{12},f_{23}	f_{13},f_{23}
f_{10},f_{20}	0	1	1	1	−1	1	1	1	−1	1	1	1	−1	−1	−1	1
f_{11},f_{20}	−1	0	1	1	−1	−1	1	1	−1	−1	−1	−1	−1	−1	−1	−1
f_{12},f_{20}	−1	−1	0	1	−1	−1	−1	−1	−1	−1	−1	−1	−1	−1	−1	−1
f_{13},f_{20}	−1	−1	−1	0	−1	−1	−1	−1	−1	−1	−1	−1	−1	−1	−1	−1
f_{10},f_{21}	1	1	1	1	0	1	1	1	−1	1	1	1	−1	−1	−1	1
f_{11},f_{21}	−1	1	1	1	−1	0	1	1	−1	−1	1	1	−1	−1	−1	−1
f_{12},f_{21}	−1	−1	1	1	−1	−1	0	1	−1	−1	−1	1	−1	−1	−1	−1
f_{13},f_{21}	−1	−1	1	1	−1	−1	−1	0	−1	−1	−1	−1	−1	−1	−1	−1
f_{10},f_{22}	1	1	1	1	1	1	1	1	0	1	1	1	−1	−1	1	1
f_{11},f_{22}	−1	1	1	1	−1	1	1	1	−1	0	1	1	−1	−1	1	1
f_{12},f_{22}	−1	1	1	1	−1	−1	1	1	−1	−1	0	1	−1	−1	−1	1
f_{13},f_{22}	−1	−1	1	1	−1	−1	−1	1	−1	−1	−1	0	−1	−1	−1	−1
f_{10},f_{23}	1	1	1	1	1	1	1	1	1	1	1	1	0	1	1	1
f_{11},f_{23}	1	1	1	1	1	1	1	1	1	1	1	1	−1	0	1	1
f_{12},f_{23}	1	1	1	1	1	1	1	1	1	1	1	1	−1	−1	0	1
f_{13},f_{23}	−1	1	1	1	−1	1	1	1	−1	−1	−1	1	−1	−1	−1	0

Table 5. Decision-making matrix: mass output (**Q**) × WATS (**W**)

Q × W	$[f_{10},f_{30}]$	$[f_{11},f_{30}]$	$[f_{12},f_{30}]$	$[f_{13},f_{30}]$	$[f_{10},f_{31}]$	$[f_{11},f_{31}]$	$[f_{12},f_{31}]$	$[f_{13},f_{31}]$	$[f_{10},f_{32}]$	$[f_{11},f_{32}]$	$[f_{12},f_{32}]$	$[f_{13},f_{32}]$	$[f_{10},f_{33}]$	$[f_{11},f_{33}]$	$[f_{12},f_{33}]$	$[f_{13},f_{33}]$
$[f_{10},f_{30}]$	0	1	1	1	1	1	1	1	1	1	1	1	1	1	1	1
$[f_{11},f_{20}]$	-1	0	1	1	-1	-1	1	1	-1	-1	-1	1	-1	-1	-1	-1
$[f_{12},f_{20}]$	-1	-1	0	1	-1	-1	-1	-1	-1	-1	-1	-1	-1	-1	-1	-1
$[f_{13},f_{20}]$	-1	-1	-1	0	-1	-1	-1	-1	-1	-1	-1	-1	-1	-1	-1	-1
$[f_{10},f_{21}]$	1	1	1	1	0	1	1	1	-1	1	1	1	-1	-1	-1	1
$[f_{11},f_{21}]$	-1	1	1	1	-1	0	1	1	-1	-1	1	1	-1	-1	-1	-1
$[f_{12},f_{21}]$	-1	-1	1	1	-1	-1	0	1	-1	-1	-1	1	-1	-1	-1	-1
$[f_{13},f_{21}]$	-1	-1	1	1	-1	-1	-1	0	-1	-1	-1	-1	-1	-1	-1	-1
$[f_{10},f_{22}]$	1	1	1	1	1	1	1	1	0	1	1	1	-1	-1	1	1
$[f_{11},f_{22}]$	-1	1	1	1	-1	1	1	1	-1	0	1	1	-1	-1	1	1
$[f_{12},f_{22}]$	-1	1	1	1	-1	-1	1	1	-1	-1	0	1	-1	-1	-1	1
$[f_{13},f_{22}]$	-1	-1	1	1	-1	-1	-1	1	-1	-1	-1	0	-1	-1	-1	-1
$[f_{10},f_{23}]$	1	1	1	1	1	1	1	1	1	1	1	1	0	1	1	1
$[f_{11},f_{23}]$	1	1	1	1	1	1	1	1	1	1	1	1	-1	0	1	1
$[f_{12},f_{23}]$	1	1	1	1	1	1	1	1	-1	-1	1	1	-1	-1	0	1
$[f_{13},f_{23}]$	-1	1	1	1	-1	1	1	1	-1	-1	-1	1	-1	-1	-1	0

4   The Adapted NN-DM Methodology

This paper deals with a modification of the NN-DM method [8]. The NN-DM method is an algorithm developed to find a model, denoted NN-DM model, which simulates the DM's preferences in situations in which these preferences are represented by a utility function $\mathcal{U}$. First, the domain $\mathcal{D}$ of the approximation is established on the basis of the domain of the available alternatives A. Second, a partial ranking is built by answers to pairwise comparisons provided by the DM expressing ordinal relations only. Last, an artificial neural network is employed in approximating the partial ranking resulting in a model $\hat{\mathcal{U}}$ that has the same level sets of the DM's utility function $\mathcal{U}$. The NN-DM model $\hat{\mathcal{U}}$ is then able to represent the DM's preferences in alternatives belonging to the domain $\mathcal{D}$ without further queries to the DM.

The real DM considered here, a polymer engineer, is assumed to have preferences that can be represented by a utility function $\mathcal{U}$ and hence by a NN-DM model $\hat{\mathcal{U}}$. This engineer is required to fill a decision-making matrix regarding a set of unsolved queries (Section 3.3). Additionally, the existence of a real DM demands an adaptation of the NN-DM method, ultimately developed with the assistance of an underlying utility function. The original NN-DM method, divided into four steps, and the adjustments in steps one, two, and four are better described next.

4.1   Step 1: Domain Establishment

In the NN-DM method the domain $\mathcal{D}$ is established from the available alternatives A. In the original model the alternatives A are employed in constructing a box whose values vary between the minimum and maximum values of the alternatives in each dimension problem. Into this domain a set of random simulated alternatives F is created and employed in building a ranking of alternatives and therefore the NN-DM model $\hat{\mathcal{U}}$.

In this application, the DM provides the decision-making domain for the objectives of each optimization problem. Assuming that this domain represents the possible region of interest to the DM, it is then employed in establishing the domain $\mathcal{D}$ of the model $\hat{\mathcal{U}}$. Into the domain $\mathcal{D}$ a grid of simulated alternatives F is constructed to extract information about the DM's preferences. The grid is considered in an attempt to make the DM's analysis easier.

4.2   Step 2: Ranking Construction

The original NN-DM method builds a partial ranking $\mathcal{R}$ of the alternatives which assigns a scalar value to each alternative. Considering a set $\mathcal{A}$ with n alternatives, a subset with $p = \log n$ alternatives,[1] denoted *pivots*, is randomly constructed from the set $\mathcal{A}$. The pivots are sorted in ascending order of the DM's preferences and a rank is assigned to each pivot. Next the $n - p$ remaining alternatives are clustered into the classes defined by the $\log n$ pivots.

In an attempt to simplify this process to the real DM a decision-making matrix is constructed. The equivalence, the dominance, and the symmetry are first considered to take the decision in situations in which the answer is acquired without consulting the DM. The remaining queries are then presented to the DM as the decision-matrix $\mathcal{M}$ which captures the DM's preferences within the domain $\mathcal{D}$ and can be filled bu the DM in his own time.

Since the answers to all the queries are supplied by the matrix $\mathcal{M}$ a total ranking $\mathcal{R}$ is now available. Even knowing that the total sorting provides additional information the partial ranking is employed here since the DM's scale is unknown and the integer scale is inconvenient to the approximation technique. The partial ranking is built preserving the $\log n$ levels and distributing uniformly the alternatives such that the number of alternatives is the same in each level, with the possible exception of the higher level.

4.3   Step 3: RBF Approximation

A Radial Basis Function (RBF) network is an artificial neural network that uses radial basis functions as activation functions. The output of the network is a linear combination of radial basis functions of the inputs and neuron parameters. Given certain mild conditions on the shape of the activation function the RBF networks are universal approximators on a compact subset of $\mathbb{R}^n$ which means that an RBF network with enough hidden neurons can approximate any continuous function with arbitrary precision.

For training the RBF network $\hat{\mathcal{U}}$ which approximates the utility function $\mathcal{U}$ the alternatives within the domain $\mathcal{D}$ are employed as inputs and the ranking level of each alternative, as outputs. The shape of the function $\mathcal{U}$ is captured by the ranking procedure and the artificial neural network has the role of representing a function which approximates this shape and provides answers to other alternatives within the same domain.

The NN-DM model is trained in a domain standardized by scaling each dimension between zero and one. This standardization is required to make the

---

[1] The function $\log x$ is employed in this paper as representing the function $\log_2 x$.

procedure of parameters tuning easier. Once the model is constructed it is adjusted to the domain $\mathcal{D}$.

All data processing has been performed employing the commercial software package MATLAB© [7]. For the construction of the RBF network the *newrb* function has been chosen with parameters given by Table 6.

Table 6. MATLAB parameters of the *newrb* function employed in constructing the RBF network

Name	Value	Name	Value
P	Set F	SPREAD	500
T	Partial ranking	MN	200
GOAL	0	DF	25

4.4   Step 4: Performance Assessment

The original NN-DM method relies on the Kendall-Tau Distance (KTD) as an efficiency metric. The KTD is a metric that counts the number of pairwise disagreements between two ranking lists. In the NN-DM method these lists are generated by sorting the available alternatives according to the DM's underlying utility function and the resulting NN-DM model. The resulting KTD value is such that the smaller the value, the better the result. In the context of the original NN-DM method the KTD is an applicable metric because an underlying utility function is available to provide information about the quality of the resulting model.

As this paper focus in a real DM there is no underlying utility function which demands another validating process. The advantage is that here the process

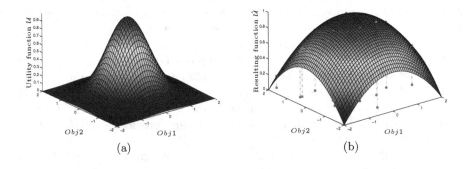

(a)   (b)

Fig. 3. (a) DM's underlying utility function $\mathcal{U}$ (b) Example of an application of the resulting function $\hat{\mathcal{U}}$

is validated by the DM himself. Once the model for the DM's preferences is constructed it is applied to sort the available data and the DM can verify the accuracy of the results.

4.5   Algorithm

Algorithm 1 presents the adapted NN-DM method to the real scenario introduced in Section 3. A grid of alternatives is constructed in the domain $\mathcal{D}$ provided by the DM. The decision-making matrix $\mathcal{M}$ is filled by the answers provided by the DM related to the alternatives belonging to the grid. A total ranking $\mathcal{R}$ is constructed and then clustered into $\log n$ levels. The RBF network converts the ranking $\mathcal{R}$ into a function $\hat{\mathcal{U}}$ that is able to provide answers to queries concerning alternatives belonging to the entire domain $\mathcal{D}$. The NN-DM model is now fit to be employed in the estimates of each Pareto-optimal front.

Algorithm 1 Adapted NN-DM method

1: Read the domain $\mathcal{D}$
2: Read the decision-making matrix $\mathcal{M}$
3: Built the total ranking of alternatives $\mathcal{R}$
4: Classify the alternatives into $\log n$ levels
5: Construct the RBF network $\hat{\mathcal{U}}$

Algorithm 2 introduces the NN-DM model applied to the polymer extrusion process. In each considered scenario the PFE and the corresponding model $\hat{\mathcal{U}}$ are loaded. The model $\hat{\mathcal{U}}$ is then employed in evaluating each solution generating a sorting of the solutions from the best to the worst.

Algorithm 2 NN-DM model applied to the polymer extrusion process

---

1: Load the Pareto-optimal front estimates
2: Load the NN-DM model $\hat{\mathcal{U}}$
3: Evaluate each available solution
4: Sort the solutions from the best to the worst

Figure 3a presents an illustrative example in which the DM is represented by an underlying utility function $\mathcal{U}$ expressed as a Gaussian. After the construction of the NN-DM model the DM is not required to provide any data related to his preferences within the domain $\mathcal{D}$. Once the the resulting model $\hat{\mathcal{U}}$ is estimated it can be employed in quantifying any alternative within its domain, as shown in Figure 3b. From this point forward the alternatives can be sorted from the best to the worst according to the DM's preferences represented by the NN-DM model.

5 Computational Experiments

The filled decision-making matrices (Tables 4 and 5) provided by the DM are taken into account to construct general NN-DM models as described in Section 4. Figures 4 and 5 present respectively the models for the two considered scenarios: $\mathbf{Q} \times \mathbf{P}$ and $\mathbf{Q} \times \mathbf{W}$. The models have been trained in the provided domain $\mathcal{D}$ (Table 2).

Figures 6 and 7 present the general NN-DM models applied to sort the estimates of the Pareto-fronts considering the objectives $\mathbf{Q}$ and $\mathbf{P}$ ($\mathbf{QP}_1$, $\mathbf{QP}_2$, and $\mathbf{QP}_3$) and $\mathbf{Q}$ and $\mathbf{W}$ ($\mathbf{QW}_1$, $\mathbf{QW}_2$, and $\mathbf{QW}_3$). The models' level sets are illustrated in the figures and the DM's preferences are represented by the external scale.

The DM's preferences, captured by the decision-making matrices, are now represented by the NN-DM models which are employed in sorting the solutions belonging to the estimates of the Pareto-fronts. Moreover, the resulting models are now available to represent the DM's preferences in any other situations without demanding further information from the DM.

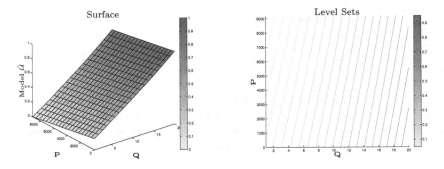

Fig. 4. General NN-DM model: $\mathbf{Q} \times \mathbf{P}$

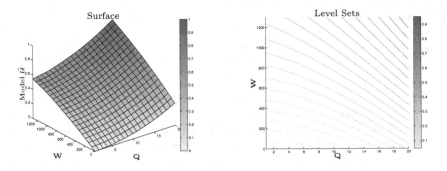

Fig. 5. General NN-DM model: $\mathbf{Q} \times \mathbf{W}$

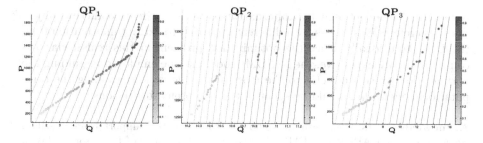

Fig. 6. NN-DM model applied to sort the estimates of the Pareto-optimal fronts in the problem $\mathbf{Q} \times \mathbf{P}$

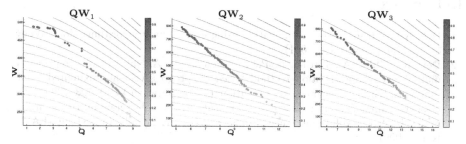

Fig. 7. NN-DM model applied to sort the estimates of the Pareto-optimal fronts in the problem $\mathbf{Q} \times \mathbf{W}$

6 Conclusions

In this paper, adaptations of the NN-DM method were executed and six estimates of different Pareto-optimal fronts derived from the polymer extrusion process were examined. The DM had to answer only 36 queries related to pairwise comparisons for each multi-objective optimization problem: $\mathbf{Q} \times \mathbf{P}$ and $\mathbf{Q} \times \mathbf{W}$. The final solutions are provided by the sorting given by the NN-DM models with a low amount of DM's intervention. The results shown that the adapted NN-DM method is able to construct models that correspond to the DM's expectations used in practice.

Once the model $\hat{\mathcal{U}}$ is trained it can be employed in quantifying any alternatives according to the DM's preferences and sort them from the best to the worst. Furthermore, the resulting models $\hat{\mathcal{U}}$ can replace the DM in recurrent decisions with alternatives within the trained domain $\mathcal{D}$.

The authors are studying improvements in the NN-DM method to consider a more complete polymer extrusion process. The average melt temperature of the polymer at die exit ($\mathbf{T_{melt}}$) and the length of screw required to melt the polymer ($\mathbf{L_{melt}}$) also characterize the process performance and could have been considered in the multi-objective optimization problem. However, in a five-objective problem the decision-making matrix is inappropriate since it is difficult for a person to decide between two alternatives considering five conflicting objectives. Therefore, it is necessary a different approach to extract information from the DM. Additionally, since the optimization problem in this real scenario is computationally expensive, the NN-DM model may be employed in guiding the optimization process directly to the most preferable region, avoiding the computational effort that would be spent in the non-preferable regions.

Acknowledgments. The authors would like to thank the support from Brazilian agencies, CNPq, CAPES, and FAPERJ, and the contributions of Professor Carlos E. Pedreira, current post-doctoral supervisor of the first author.

References

1. Deb, K.: Multi-objective optimization using evolutionary algorithms. John Wiley & Sons Inc., New York (2001)
2. Deb, K., Sinha, A., Korhonen, P.J., Wallenius, J.: An interactive evolutionary multiobjective optimization method based on progressively approximated value functions. IEEE Transactions on Evolutionary Computation **14**(5), 723–739 (2010)
3. Gaspar-Cunha, A.: Modelling and optimisation of single screw extrusion using multi-objective evolutionary algorithms. Lambert Academic Publishing (LAP), Koln (2009)
4. Gaspar-Cunha, A., Covas, J.A.: RPSGAe - Reduced Pareto Set Genetic Algorithm: application to polymer extrusion. In: Metaheuristics for Multiobjective Optimisation. Lecture Notes in Economics and Mathematical Systems, vol. 535, pp. 221–249. Springer, Heidelberg (2004)
5. Köksalan, M., Karahan, I.: An interactive territory defining evolutionary algorithm: iTDEA. IEEE Transactions on Evolutionary Computation **14**(5), 702–722 (2010)
6. Marler, R.T., Arora, J.S.: Survey of multi-objective optimization methods for engineering. Structural and Multidisciplinary Optimization **26**(6), 369–395 (2004)
7. MATLAB: Version 6.9.0 (R2009b). The MathWorks Inc., Natick, Massachusetts (2009)
8. Pedro, L.R., Takahashi, R.H.C.: Decision-maker preference modeling in interactive multiobjective optimization. In: Purshouse, R.C., Fleming, P.J., Fonseca, C.M., Greco, S., Shaw, J. (eds.) EMO 2013. LNCS, vol. 7811, pp. 811–824. Springer, Heidelberg (2013)
9. Pedro, L.R., Takahashi, R.H.C.: INSPM: An interactive evolutionary multiobjective algorithm with preference model. Information Sciences **268**, 202–219 (2014)
10. Shin, W.S., Ravindran, A.: Interactive multiple objective optimization: survey i - continuous case. Computers & Operations Research **18**(1), 97–114 (1991)
11. Veldhuizen, D.A.V., Lamont, G.B.: Multiobjective evolutionary algorithms: analyzing the state-of-the-art. Evolutionary Computation **8**(2), 125–147 (2000)

Multi-Objective Optimization of Gate Location and Processing Conditions in Injection Molding Using MOEAs: Experimental Assessment

Célio Fernandes[✉], António J. Pontes, Júlio C. Viana, and António Gaspar-Cunha

Institute for Polymer and Composites - IPC/I3N, Department of Polymer Engineering,
University of Minho, Braga, Portugal
{cbpf,pontes,jcv,agc}@dep.uminho.pt

Abstract. The definition of the gate location in injection molding is one of the most important factors in achieving dimensionally accuracy of the parts. This paper presents an optimization methodology for addressing this problem based on a Multi-objective Evolutionary Algorithm (MOEA). The algorithm adopted here is named Reduced Pareto Set Genetic Algorithm (RPSGA) and was used to create a balanced filling pattern using weld line characterization. The optimization approach proposed in this paper is an integration of evolutionary algorithms with Computer-Aided Engineering (CAE) software (Autodesk Moldflow Plastics software). The performance of the proposed optimization methodology was illustrated with an example consisting in the injection of a rectangular part with a non-symmetrical hole. The numerical results were experimentally assessed. Physical meaning was obtained which guaranteed a successful process optimization.

Keywords: Multi-Objective Evolutionary Algorithms · Gate Location · Moldflow

1 Introduction

Injection molding is a complex but efficient polymer processing technique for producing a variety of plastics parts. It is especially adequate to produce products with low dimensional tolerances and complex shapes. It consists in reproducing the required geometry previously machined in the mold by injecting molten polymer into the mold cavity. The quality of the injection moulding parts are affected by different processing parameters, machine control system (e.g., injection cycle times and injection and holding pressures), cooling system (e.g., cooling channels geometry and cooling liquid temperature), gates and runners (e.g., geometry and location) and cavities (e.g., geometry and total flow length). An important factor is the gate location, since it influences the way the polymer flows into the mold cavity, affecting the existence or not of weld lines and its eventual location, the shrinkage, mold filling pattern, dimensional tolerances, degree and direction of orientation, pressure distribution in the cavity, sink marks, gas traps and short shots, warpage and residual stress. Thus, the definition of the number, type, and location of the gate(s) is of high importance. These concepts will be explained in section 2.

© Springer International Publishing Switzerland 2015
A. Gaspar-Cunha et al. (Eds.): EMO 2015, Part II, LNCS 9019, pp. 373–387, 2015.
DOI: 10.1007/978-3-319-15892-1_25

For optimizing gate location it is necessary the integration of tools, such as, simulation software able to take into account the referred processing parameters and optimization methodologies. There are in the literature various optimization strategies using different methodologies to optimize gate location in injection molding.

Pandelidis and Zou (1990) optimized gate location based on the combination of simulated annealing with a hill-climbing method [1]. The optimization effect is restricted by the determination of some weighting factors used by the authors. Young (1994) used a genetic algorithm to optimize gate location for the case of the molding of a liquid composite based on the minimization of the mold-filling pressure, the uneven-filling pattern and the temperature difference during mold filling [2]. Lee and Kim (1996) proposed an automated selection method for gate location, in which a set of initial gate locations were proposed by a designer and, then, the optimal location of the gate was defined using the adjacent node evaluation method [3]. The scheme can be used for complicated parts, but it requires an extensive number of design evaluations to obtain the best gate location. In their work, Douglas et al. (1998) designed a mold by combining process modelling and sensitivity analysis [4]. The gate location and injection pressure profile were optimized through minimizing the filling time. The extension of the proposed methodology to more complicated geometries is not obvious. Lam and Jin (2001) proposed the optimization of gate location based on the flow path concept [5]. For complicated parts, such as ones including holes, ribs and/or boss, the appropriate boundary is not easy to select automatically by computer being the user input required. Courbebaisse and Garcia (2002) suggested a shape analysis to estimate the best gate location of injection molding [6]. This methodology can only be used for simple flat parts with uniform thickness but it is easy to use and is not time-consuming. Shen et al. (2004) optimized the gate location by minimizing a weighted sum of filling pressure, filling time difference between different flow paths, temperature difference and over-pack percentage [7]. A hill-climbing algorithm was used to search the optimal gate location. Zhai et al. (2005) developed an efficient search method based on pressure gradient (PGSS) to optimize the location of two gates for a single molding cavity [8]. The weld lines were subsequently positioned to the desired location by varying runner sizes [9]. Li et al. (2007) proposed a different objective function to evaluate the warpage of injection molded parts [10]. The quality of the warpage was defined from the "flow plus warpage" simulation outputs of Moldflow software and the optimization is made by using simulated annealing. Wu et al. (2011) developed a study where the combination of different classes of design variables are considered simultaneously, together with both the length and the position of the weld line as design constraints [11]. This study adopted an enhanced genetic algorithm, called Distributed Multi-Population Genetic Algorithm (DMPGA), combining with a commercial Moldflow software and a master–slave distributed architecture. However, only runner size, molding conditions and part geometry are taken into consideration.

The above methodologies proposed to optimize gate location have some important limitations, namely, the capacity to handle with multi-objectives simultaneously, the linkage with the simulation codes and the complexity of the part geometry.

Therefore, in the present work an automatic optimization methodology based on Multi-Objective Evolutionary Algorithms (MOEA) is used to define the processing conditions and the gate location in injection molding of a complicated part containing a hole [12]. For that purpose a MOEA is linked to an injection molding simulator code (in this case Moldflow). The proposed optimization methodology was applied to a case study where the processing conditions and the gate location are established in order to create a balanced filling pattern, achieved by weld line length minimization, to maximize part quality, guaranteed by difference between the shrinkage at the end of the flow and the pre-defined design value and to minimize the cycle time to provide low costs on part production. Finally, the optimization results were assessed experimentally. This article follows two previous papers [14, 15] with additional contribution subjected to the following tasks. First, Moldflow substituted C-Mold as the simulator program used in the other studies. Also, the way to connect Moldflow to the RPSGA algorithm is more sophisticated than with C-Mold. In this case AutoIt program was used to mimic human interface with MoldFlow because input variables and output results are not allowed to be changed/saved by command line. Second, the robustness of the optimization methodology was tested by using injection molding gate location as a case study. Finally, to our knowledge, the experimental assessment of the gate location optimization results is a relevant step in the literature.

This paper is organized as follows: first the optimization methodology used is described, specifying how the MOEA interacts with the simulation software Moldflow; second, a case study based on the use of a rectangular part to be injected with a non-symmetrical hole is presented and, finally, gate location optimization results are shown and compared with experimental measurements.

2   Optimization Methodology

2.1   Injection Molding Process

Injection molding is a process of polymer transformation involving several steps, which are performed in an order that is repeated at each cycle, such as plasticizing, packing and cooling. A typical injection molding machines (see Figure 1) have four units: power supply unit, injection unit, clamping unit and control unit. The main concern in injection molding is to produce plastic parts of the desired quality, which are related with mechanical characteristics, dimensional conformity (shrinkage, warpage) and appearance (sink marks, weld lines).

For this study it is important to clarify the concepts of shrinkage, warpage and weld lines. Shrinkage is defined as the reduction in the size of a molded component in any direction after it has been ejected from the mold do to the cooling. Warpage occurs when there are variations of internal stresses in the material caused by a variation in shrinkage. A weld line is formed when separate melt fronts travelling in opposite directions meet. Instead, a meld line occurs if two emerging melt fronts flow parallel to each other and create a bond between them. Thus, the meeting angle is used to differentiate weld lines and meld lines. If the meeting angle is smaller than 135 degrees produces a weld line. If the angle is greater than 135 degrees it will produce a meld line. In the first case, a weld line surface mark will appear in the part, but when the meeting angle reaches 120 - 150 degrees it will disappear. Weld lines are considered to be of lower quality than meld lines, since relatively less molecular diffusion occurs across a weld line after it is formed. Therefore, weld lines are the weakest areas on the part and are the potential failure locations (Moldflow reference [13]).

The major factors affecting part quality are polymer properties, mold design and operating conditions. Some of these variables will be considered later in the optimization methodology.

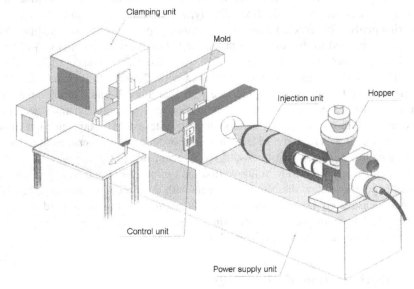

Fig. 1. Functional units of the injection molding machine

2.2   Integrated Methodology

A methodology, integrating modelling of the injection molding process and an optimization strategy based on MOEA, is proposed [14, 15]. The aim being to define the best injection gate location and injection molding operating conditions in order to minimize the cycle time, the differential shrinkage and the weld line location and length, using as example the production of an injection part with a hole (see Figure 3).

EAs are based on the principles of natural selection of survival of the fittest individual by mimicking some of the concepts of this natural process. The selection, crossover and mutation concepts are used by the EA to explore the search space in order to find an optimal solution or a set of optimal solutions. The initial population of chromosomes represents the gate location and/or the set of operative processing variables, which is generated randomly within the feasible search space. Then, these solutions are evaluated using the modelling routine (Autodesk Moldflow 2010 software). The performance of each one of the solutions (chromosomes) proposed by the MOEA is quantified using as objectives the minimization of the cycle time, the differential shrinkage and the length and location of the weld line. A MOEA was used to optimize the process [16]. The Reduced Pareto Set Genetic Algorithm (RPSGA) proposed previously was used for that purpose. First, the population is random initialized, where each individual (or chromosome) is represented by the binary value of the set of all variables. Then, each individual is evaluated by calculating the values of the relevant objectives using the modeling routine. Finally, the remaining steps of a MOEA are to be accomplished. To each individual is assigned a single value identifying its performance on the process (fitness). If the convergence objective is not satisfied (e.g., a predefined number of generations), the population is subjected to the operators of reproduction (i.e., the selection of the best individuals for crossover and/or mutation) and of crossover and mutation (i.e., the methods to obtain new individuals for the next generation). The solution must result from a compromise between the different objectives. Generally, this characteristic is taking into account using an approach based on the concept of Pareto frontiers (i.e., the set of points representing the trade-off between the objectives) together with an MOEA. This enabled the simultaneously accomplishment of the several solutions along the Pareto frontier, i.e., the set of non-dominated solutions. The performance of this algorithm was tested in a set of problems and its efficiency well demonstrated [12].

Figure 2 shows the interface for integrating Autodesk Moldflow 2010 and the GA-based optimization routine. First, coordinates of injection point are sent to Moldflow Adviser by an AutoIt script which mimics the user interface with computer. Next, geometric Moldflow Adviser file is renamed to geometric Moldflow Synergy file and an AutoIt script is executed to remesh the part and define processing conditions to be used in the simulation. A Fill+Pack+Warp analysis is done through command files provided by Moldflow Synergy software. When the analysis is finished, an AutoIt script is executed in order to obtain differential shrinkage in two different locations and weld line/meld line results files are saved. The optimization routine will use these results to calculate the cycle time, the dispersion of differential shrinkage and the length of the weld plus meld line, as described in next section.

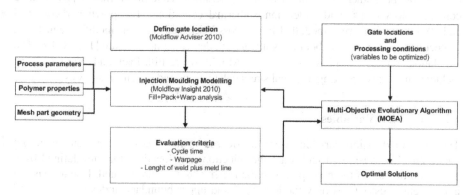

Fig. 2. Interfacing between optimization routine and Moldflow software

3   Case Study

3.1   Problem Description

To demonstrate the validity of the proposed optimization methodology, the study of the gate location of a rectangular molding with a hole, as represented in Figure 3, was used. The molding has a thickness of 1.5mm, a width of 60 mm and a length of 140 mm. The finite element mesh used has 940 triangular elements and 525 nodes (see Figure 3). The part is molded in a polypropylene, PPH 5060, from TOTAL Petrochemicals. The polymer properties used in the simulations were obtained from the Moldflow database (Moldflow 2010): melt density of 0.73406 g/cm^3, solid density of 0.90032 g/cm^3, melting temperature of 230 °C, ejection temperature of 60 °C, maximum shear stress of 0.25 MPa, maximum shear rate of 100000 1/s, specific heat of 2700 J/kg °C, thermal conductivity of 0.17 W/m °C, elastic module of 1400 MPa and Poisson ratio of 0.42. The material selected for the mold was a P20 steel. Concerning the processing conditions only the mold open time was maintained constant and equal to 5 seconds.

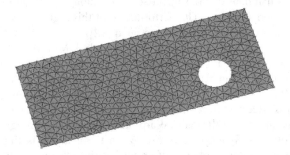

Fig. 3. Injection molding part to be used

The simulations in Moldflow are based on a hybrid finite-element/finite-difference/control-volume numerical solution of the generalized Hele-Shaw flow equation of a compressible viscous fluid under non-isothermal conditions. The polymer rheological and PVT behaviors were modelled by a Cross-WLF and the Tait modified equations, respectively. More details about the software are described in literature [17, 18, 19]. The simulations phases considered when using Moldflow are Fill, Pack and Warp analysis, including melt flow, packing, residual stress calculations and structural analysis.

3.2   Decision Variables

Two type of decision variables were considered in this study, design and operating conditions. The design variables are the injection gate location, that are defined by x and y coordinates of the node points in the mesh were they are located. Due to existing experimental restrictions only the left, right and upper boundary nodes were admitted as injection gate location. Six operating conditions were considered, filling time, melt and mold temperatures, holding time, holding pressure and cooling time. Table 1 summarizes the design variables selected and the corresponding range of variation.

Table 1. Range of variation of the decision variables

Decision variables	Range of variation
X coordinate (mm) – x	[0,140] subject to constraint
Y coordinate (mm) – y	[0, 60] subject to constraint
Fill time (s) - t_f	[1, 5]
Melt temperature (°C) - T_{inj}	[190, 270]
Mold temperature (°C) - T_w	[10, 50]
Holding pressure (MPa) - P_h	[30, 60]
Packing time (s) - t_p	[1, 20]
Cooling time (s) - t_c	[5, 20]

The RPSGA uses a real representation of the variables, a simulated binary crossover, a polynomial mutation and a roulette wheel selection strategy [12]. The following RPSGA parameters were selected: 10 generations, crossover rate of 0.8, mutation rate of 0.05, internal and external populations with 30 individuals, limits of the clustering algorithm set at 0.2 and NRanks at 30. These values resulted from a carefully analysis made in a previous work [12]. The computation time required by the MoldFlow software to evaluate a single candidate solution is approximately 5 minutes. Thus, the time necessary for a complete optimization is circa of 25 hours. This multi-objective problem used a 'budget' of 300 evaluations because of the expensive nature of evaluating candidate solutions, namely, the time taken to perform one evaluation, only one evaluation can be performed at one time and also the dimensionality of the search space is low-to-medium [20]. The proposed optimization methodology will be used for setting the injection location and to define the selected processing conditions that satisfy the objectives defined.

3.3   Objective Functions

The optimization problem consists in defining the values of the decision variables that allow the production of a part with the minimum cycle time, to minimize the production costs, the minimum of warpage due to the anti-symmetric shrinkage and the minimum of weld plus meld line length, so that weakest areas are minimized.

These objectives are defined as follows:

- Minimize cycle time, CT:

$$\min CT = t_f + t_p + t_c + t_o \tag{1}$$

where t_f is the filling time, t_p is the packing time, t_c is the cooling time and t_o is the mold open time.

- Minimize warpage, *WARP*:

$$\min WARP = \sqrt{\frac{ds_1^2 + ds_2^2}{2}} \tag{2}$$

where ds_1 and ds_2 are the differential shrinkage values measured in longitudinal and transversal directions, respectively.

- Minimize length of weld plus meld line, *LWML*:

$$\min LWML = \Sigma_{i,j} \sqrt{(x_i - x_j)^2 + (y_i - y_j)^2 + (z_i - z_j)^2} \tag{3}$$

where (x_*, y_*, z_*) represents the coordinates of the nodes in the finite element mesh were the weld and meld lines are located as calculated by Moldflow.

4   Results and Discussion

4.1   Optimization Results

Figure 4 shows the results obtained for an optimization run considering simultaneously the three objectives defined before (minimization of cycle time, length of weld plus meld line and warpage), the aim being to define the best values for the decision variables presented in Table 1. The figure represents all solutions of the initial population and the non-dominated solutions of the final population (10th generation). The operating conditions and objectives of the optimal solutions found are presented in Table 2.

Table 2. Solutions for gate location optimization

Solutions	Variables						Objectives		
	T_w (°C)	T_{inj} (°C)	t_f (s)	P_h (%)	t_p (s)	t_o (s)	CT (s)	$WARP$ (%)	$LWML$ (m)
P1	46	268	2.20	42	1.16	5.00	13.36	0.815	0.0331
P2	28	268	4.30	56	12.7	10.1	32.06	0.800	0.0216
P3	46	269	2.84	44	6.57	6.39	20.80	0.800	0.0219
P4	38	265	2.61	47	1.49	5.30	14.20	0.820	0.0103
P5	40	265	2.44	46	1.00	5.35	13.79	0.820	0.0216
P6	46	268	2.27	42	1.21	5.40	13.88	0.815	0.0216
P7	44	264	1.37	44	3.02	6.23	15.62	0.815	0.0106

There is a clear improvement from the initial population to the 10th generation, since the optimal solutions found have better values for the objectives considered. Also, there is not a solution that, simultaneously, provides the better (minimum) values for all three objectives. Therefore, three cases will be analyzed considering each one of the objectives as the most important.

If cycle time is considered the most important objective, the solution with lower cycle time is P1 in Figure 4. In this case *CT* is equal to 13.36 s, *WARP* is equal to 0.815 % and *LWML* is equal to 0.0331 m that is the highest value found for the length of weld plus meld line. Therefore, this solution is unsatisfactory when the length of weld plus meld line is considered. The injection molding machine must operate with a fill time of 2.2 s, melt and mold temperatures of 268 °C and 46 °C, respectively, holding pressure of 42 MPa, packing time of 1.16 s and cooling time of 5 s.

Figure 5 (A) shows the gate location and the filling pattern, while Figure 5 (B) shows weld line location for solution P1 as calculated by Moldflow. Since gate location is positioned in the top left corner of the part the two melt fronts will meet in the bottom right corner of the hole (were the weld line is plotted in Figure 5 (B)). In this case, the flow pattern design does not reach the top, bottom and left cavity boundaries uniformly, as shown by the filling pattern in Figure 5 (A). However, no weld lines were formed, since the meeting angles of the flow fronts are kept higher than 135 °. The line shown in Figure 5 (B) represents both a weld and a meld line, but that does not represent a problem for the molded part since the weld line does not reach the external boundary of the part. Figure 5 (C) shows the warpage distribution for solution P1. The value of WARP represents the standard deviation of differential shrinkage values measured on the boundary midpoints in horizontal and vertical directions. For the present case means that the distances between these points after part production only differs of 0.815 % when compared with the distances in the mold cavity.

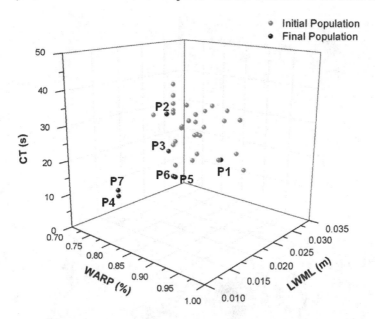

Fig. 4. Optimization results for three objectives in the objectives domain. Black symbols: Pareto frontier at 10th generation; grey symbols: initial population (CT – cycle time, WARP – warpage, LWML – length of weld plus meld line).

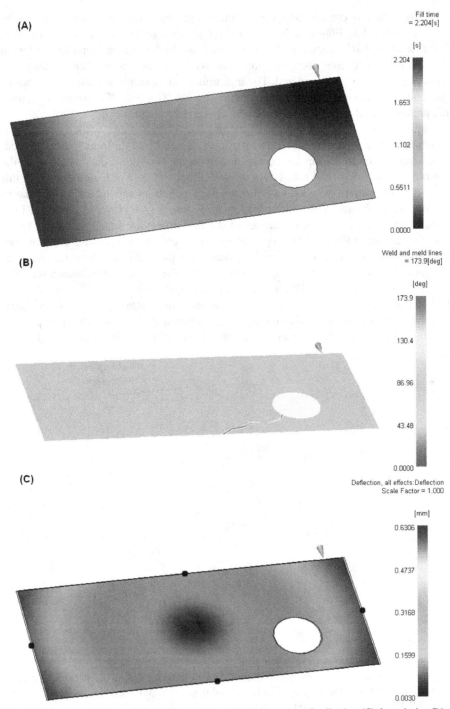

Fig. 5. Filling pattern (A), weld/meld line position (B) and warpage distribution (C) for solution P1

Weld and meld lines
= 169.7[deg]

[deg]

169.7

127.3

84.87

42.44

0.0000

Fig. 6. Weld/meld line position for solution P2

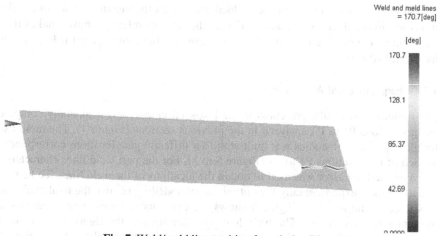

Weld and meld lines
= 170.7[deg]

[deg]

170.7

128.1

85.37

42.69

0.0000

Fig. 7. Weld/meld line position for solution P3

When warpage is considered the most important objective to satisfy, two solutions were obtained with lower warpage values, i.e., solutions P2 and P3. The main difference between these two solutions is related with the cycle time, which is 32.06 s for P2 and 20.80 s for P3. Since the values of WARP and LWML are very similar, it is clear that P3 is a much better solution than P2. Concerning solutions P2, P3 and P4, only the location of the weld plus meld lines are represented, since these are the results used in the experimental assessment. Figures 6 and 7 represents the location of weld plus meld line, respectively for P2 and P3. In both cases, the gate is located in the top right corner of the part and thus the weld plus meld line appear in the bottom left corner of the hole. Also, only meld lines are formed for these two solutions in the external boundary of the part. Figure 8 presents the modelling results for solution P4, i.e., the solution with lower length of weld plus meld line. In this case only a weld line

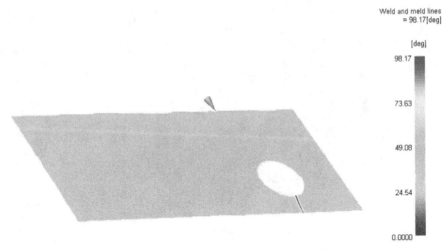

Weld and meld lines
= 98.17[deg]

[deg]

98.17

73.63

49.08

24.54

0.0000

Fig. 8. Weld/meld line position for solution P4

forms on the bottom surface of the molded part (as all the meeting angles of the flow fronts are lower than 135 °), which, due to the reasons referred above, makes this a bad solution concerning this aspect. Simultaneously, this solution has the higher value for WARP (0.820 %).

4.2   Experimental Assessment

The optimization results presented in the last section were experimentally compared using solutions P1 to P4 analyzed in the previous sections (Figure 4). The mold used for the experimental studies was built with four different gate locations corresponding to each of the solutions P1 to P4 (Figure 5 to 8). For the part weld lines characterization a crossed polarizer was used to obtain the locations of the weld lines. To be possible to perform experimentally the optimization/modelling results, the total cycle time was fixed in the machine. Figure 9 shows the experimental weld lines locations for the four solutions chosen. The weld lines are identified in the figure with arrows to better understand their location. As can be seen the experimental results confirm the location of the weld lines predicted by the optimization methodology (Figure 5 to 8).

Figure 10 shows the simulation vs. experimental warpage measurements. The results are graphically represented in two different plots to better distinguish the shapes of the curves due to differences in warpage scale. As can be seen the experimental results follow the same tendency of the simulated ones. Solution P1 is the only one that has a different pattern comparing to the simulated one, due to the use of a lower packing pressure in few packing time (see Table 2). In agreement with the simulation results the experimental solutions P2 and P3 have similar values. Moreover, those values are the lowest ones for warpage criteria. This can be explained by the use of a packing pressure during more time than in solutions P1 and P4. Quantitative differences between experimental and numerical results are explained by the fact that in the numerical warpage calculation the effect of differential shrinkage, differential cooling and orientation effects are taken into consideration. Simultaneously, the experimental measurements were made considering only the differential shrinkage effect.

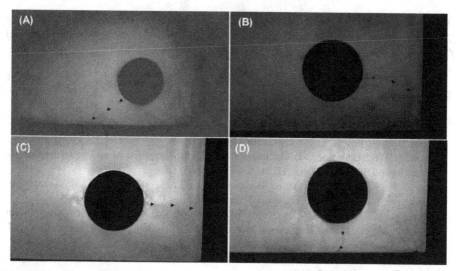

Fig. 9. Experimental weld lines location for solution P1 (A), solution P2 (B), solution P3 (C) and solution P4 (D)

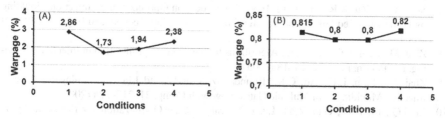

Fig. 10. (A) Experimental and (B) optimization warpage results for solutions P1 to P4

5   Conclusions

In this work, a multi-objective optimization methodology based on Evolutionary Algorithms (MOEA) was applied to the optimization of processing conditions and gate location of a rectangular molding with a hole in order to minimize the cycle time, the warpage and the length of weld plus meld line.

The methodology proposed was able to produce results with physical meaning. The optimization algorithm is able to minimize simultaneously the three objectives defined through the generation of optimal Pareto frontiers showing the trade-off between the solutions found. This allows the user the comparison between these solutions to select the best that corresponds to their design purposes.

Finally, the optimization results were assessed experimentally. The experiments obtained in an injection machine available shows identical behavior when compared with the computational ones.

Acknowledgments. This work was supported by the Portuguese Fundação para a Ciência e Tecnologia under grant SFRH/BD/28479/2006 and IPC/I3N – Institute for Polymers and Composites, University of Minho.

References

1. Pandelidis, I., Zou, Q.: Optimization of Injection Moulding Design. Part I: Gate Location Optimization. Polym. Eng. and Sci. **30**, 873–882 (1990)
2. Young, W.B.: Gate Location Optimization in Liquid Composite Moulding Using Genetic Algorithm. J. Compos. Mater. **12**, 1098–1113 (1994)
3. Lee, B.H., Kim, B.H.: Automated Selection of Gate Location Based on Desired Quality of Injection Moulded Part. Polym.-Plast Techno. and Eng. **35**, 253–269 (1996)
4. Douglas, E.S., Daniel, A.T., Charles III, L.T.: Analysis and Sensitivity Analysis for Polymer Injection and Compression Moulding. Comput. Methods Appl. Mech. Eng. **167**, 325–344 (1998)
5. Lam, Y.C., Jin, S.: Optimization of Gate Location for Plastic Injection Moulding. J. Inject. Mold. Tech. **5**, 180–192 (2001)
6. Courbebaisse, G., Garcia, D.: Shape Analysis and Injection Moulding Optimization. Comput. Mat. Sci. **25**, 547–553 (2002)
7. Shen, C.Y., Yu, X.R., Li, Q., Li, H.M.: Gate Location Optimization in Injection Moulding by Using Modified Hill-Climbing Algorithm. Polym.-Plast Techno. and Eng. **43**, 649–659 (2004)
8. Zhai, M., Lam, L.C., Au, C.K.: Algorithms for Two Gate Optimization in Injection Moulding. Int. Polym. Proc. **20**, 14–18 (2005)
9. Zhai, M., Lam, L.C., Au, C.K.: Runner Sizing and Weld Line Positioning for Plastics Injection Moulding with Multiple Gates. Eng. with Comp. **21**, 218–224 (2006)
10. Li, J.Q., Li, D.Q., Guo, Z.Y., Lv, H.Y.: Single Gate Optimization for Plastic Injection Mould. J. of Zhejiang University Science A **8**, 1077–1083 (2007)
11. Wu, C.Y., Ku, C.C., Pai, H.Y.: Injection Moulding Optimization with Weld Line Design Constraint Using Distributed Multi-Population Genetic Algorithm. Int. J. Adv. Manuf. Technol. **52**, 131–141 (2011)
12. Gaspar-Cunha, A., Covas, J.A.: RPSGAe – reduced pareto set genetic algorithm: application to polymer extrusion. In: Gandibleux, X., et al. (eds.) Lecture Notes in Economics and Mathematical Systems, pp. 221–255. Springer, Berlin (2004)
13. Moldflow plastic insight 2010. Moldflow Corporation, Wayland, MA (2010)
14. Fernandes, C., Pontes, A.J., Viana, J.C., Gaspar-Cunha, A.: Using Multi-Objective Evolutionary Algorithms in the Optimization of Operating Conditions of Polymer Injection Moulding. Polym. Eng. and Sci. **50**, 1667–1678 (2010)
15. Fernandes, C., Pontes, A.J., Viana, J.C., Gaspar-Cunha, A.: Using Multi-Objective Evolutionary Algorithms for Optimization of the Cooling System in Polymer Injection Moulding. Int. Polym. Proc. **2**, 213–223 (2012)
16. Deb, K.: Multi-Objective Optimization using Evolutionary Algorithms. John Wiley & Sons Publishers, New York (2001)
17. Hieber, C.A., Shen, S.F.: A Finite-Element/Finite-Difference Simulation of the Injection-Moulding Filling Process. J. of Non-Newtonian Fluid Mech. **7**, 1–32 (1980)

18. Chiang, H.H., Hieber, C.A., Wang, K.K.: A Unified Simulation of the Filling and PostFilling Stages in Injection Moulding. Part I: Formulation. Polym. Eng. Sci. **31**, 116–124 (1991a)
19. Chiang, H.H., Hieber, C.A., Wang, K.K.: A Unified Simulation of the Filling and PostFilling Stages in Injection Moulding. Part II: Experimental Verification. Polym. Eng. Sci. **31**, 125–139 (1991b)
20. Knowles, J., Hughes, E.J.: Multiobjective optimization on a budget of 250 evaluations. In: Coello Coello, C.A., Hernández Aguirre, A., Zitzler, E. (eds.) EMO 2005. LNCS, vol. 3410, pp. 176–190. Springer, Heidelberg (2005)

A Multi-Criteria Decision Support System for a Routing Problem in Waste Collection

João A. Ferreira[(⊠)], Miguel Costa, Anabela Tereso, and José A. Oliveira

University of Minho, Department of Production and Systems, Campus de Azurém,
4800-058 Guimarães, Portugal
{joao.aoferreira,miguelpintodacosta}@gmail.com,
{anabelat,zan}@dps.uminho.pt

Abstract. This work presents a decision support system for route planning of vehicles performing waste collection for recycling. We propose a prototype system that includes three modules: route optimization, waste generation prediction, and multiple-criteria decision analysis (MCDA). In this work we focus on the application of MCDA in route optimization. The structure and functioning of the DSS is also presented.

We modelled the waste collection procedure as a routing problem, more specifically as a team orienteering problem with capacity constraints and time windows. To solve the route optimization problem we developed a cellular genetic algorithm. For the MCDA module, we employed three methods: SMART, ValueFn and Analytic Hierarchy Process (AHP).

The decision support system was tested with real-world data from a waste management company that collects recyclables, and the capabilities of the system are discussed.

Keywords: Waste collection · Vehicle routing · Team Orienteering Problem · Decision Support System · Multiple-Criteria Decision Analysis · Cellular Genetic Algorithm · AHP · SMART · ValueFn

1 Introduction

Over the years, decision sciences have been applied in resource management to achieve success in the business world. In order to avoid losses and increase income while depending on the use of resources, companies tend to rely on the decisions carried out by their managers. The process of decision-making often happens over the analysis of several criteria that sometimes have conflicting objectives or induce divergent results while aiming for the same goal. Those different outcomes can either complement each other or create problems while deciding the best course of action to complete a certain task or project. Common conflicting criteria are cost (or price) and quality measures (or performance indicators). Trying to achieve a good balance between important criteria implies weighing their contributions and (adverse) consequences to the final objective, and opting

© Springer International Publishing Switzerland 2015
A. Gaspar-Cunha et al. (Eds.): EMO 2015, Part II, LNCS 9019, pp. 388–402, 2015.
DOI: 10.1007/978-3-319-15892-1_26

for the best alternative for that specific situation. This process requires structuring the problem in a proper way to reduce its complexity and making better decisions.

MCDA is the acronym for multiple-criteria decision analysis, which involves the application of models and methodologies to provide the decision maker (DM) with tools that enable solving decision and planning problems, although, when there are several criteria to consider, it is very difficult, if not impossible, to determine an unique best solution. To find an overall best solution, the DMs preferences need to be included in the decision process, since there are trade-offs that need to be considered, and the importance given to each criteria depends on the DM.

The dawn of computer age enabled the development of software implementing various models and methods to solve multiple-criteria decision-making problems, and there are plenty of commercially available general-purpose products, as well as more specific ones. This kind of software became extremely useful in many different management scenarios. Often, MCDA software is included in larger (software) systems called Decision Support Systems (DSS). A DSS is an information system that supports business decision-making activities, and helps making decisions at different levels: 1) management, 2) operations and 3) planning. A DSS is usually an interactive software tool that combines data from various sources and formats and presents the user (i.e. the DM) with useful information, displayed on graphical interfaces, that helps solving decision-making problems.

In this study, we focus on a real-world problem, where management issues are faced by a waste management company (WMC) that needs to pick-up recyclable materials stored along a network of collection points. In general, successful waste management highly depends on good performing logistic systems that keep track of all needed requirements and the goals/objectives to be met by the companies. Resource management (i.e. vehicles, drivers, assistants) and designing cost-efficient waste collection routes are some of the major issues a DM in a WMC has to deal with on a daily basis, and it often involves weighting the importance of several criteria. This creates different operation scenarios that need to be assessed for the DM to determine the best solution for a certain situation. Therefore, a DSS is an important tool to assist in this decision-making process. The aim of this study was to develop a DSS that includes route optimization models and employs MCDA methods to compare different operational procedures carried out by a WMC when performing waste collections.

2   Problem Description

The recycling of waste materials has earned great importance over the years, and today is a vital process to our survival in a clean and healthy environment, and also to move towards a more sustainable future. Regarding waste composition, in Portugal, more than 50% of municipal solid waste is composed of recyclable materials, where paper accounts for 20.3%, plastics 18%, glass 6%, metals 5%, and textiles 3.8% [19]. The collection of recyclable materials have lately become a fertile subject for the development of new ideas to improve resource management and global efficiency. There is also a constant need for improvements in waste management, especially in terms of resource management.

The process of collecting recyclables usually involves three main resources that need management: 1) workers, 2) vehicles and 3) time. In addition, there are elements that impose constraints to resource management such as cost limitations in fuel expenditure, avoidance of high vehicle wearing, minimum (daily) quotas of waste to collect to be met, and also certain collection performance standards that should be attained. Many improvements often occur through successful fleet management, which greatly relies on optimization procedures applied to the design of collection routes. Designing more efficient routes implies balancing the use of resources while respecting all constraints that may be imposed.

Focusing on route design, the objective is to visit a set of waste collection points using a vehicle fleet, while respecting constraints such as vehicle capacity and maximum route duration and/or length (time spent and/or travelled distance). This kind of situation can be addressed as a vehicle routing problem (VRP) [13]. Although this description fits the mentioned waste collection problem, more flexibility is needed when designing the routes. Each collection point is assigned a certain priority level to be visited and emptied, based on their waste generation rates and current filling status. While scheduling collection routes, there is a need to select which points to visit during those routes, and only a part of those points may be collected. The VRP models can be too restrictive when employed in this situation, since the premise is to visit all points in the network, regardless of its filling status, and using as many vehicles as needed to do so. So, instead of targeting all collection points, a more fitting model named team orienteering problem (TOP) [8,11], can be applied. In this context, the TOP can be described as the problem of designing routes and assign them to a limited fleet of vehicles performing collection of recyclable waste stored along a network of collection points; each collection point has a priority level; the collection routes have maximum durations and/or distances; the vehicles have capacity limits; the selection of collection points to be visited by the vehicles is made by balancing their priorities and their contributions for route duration, route distance and quantity of waste collected per vehicle. The objective is to maximize the total amount of waste collected by all routes while respecting the time and/or distance constraints, and also capacity constraints or even time windows.

In Portugal, a major source of potentially recyclable materials is household packaging waste (HPW), which is usually composed of materials made of glass, paper/cardboard, plastic or metal. HPW is separated by citizens at the local recycling site (collection point), named ecopoint (ecological point). Given the goals Portugal has to fulfil for the recycling and recovery of HPW, there is a permanent need for increased efficiency in waste collection performed by waste management companies. Therefore, the main goal of this study is to explore new solutions and management options for a real problem faced by a Portuguese inter-municipal waste management company (WMC) that takes action across six municipalities and currently operates a network of more than 1,200 ecopoints. This WMC's area of operations is a mix of urban and rural areas, which prompts the demand of different strategies for waste management.

There is no requirement for the mentioned WMC to visit all their ecopoint sites every workday, as it would be unprofitable and inefficient, and of course impossible, since there is a limited vehicle fleet available. Therefore, it is necessary to select a subset of ecopoints to visit each day. Furthermore, given a planning horizon of, for example, a week, or a month, the WMC must decide, based on the priority levels of ecopoints, which ones must be visited, which ones can be visited, and which ones can be skipped during the collection routes, and then design effective routes to perform the collection of HPW. Therefore, modelling this collection procedure as a TOP is suitable. In fact, this routing problem can be modelled as a capacitated team orienteering problem with time windows (CTOPTW), due to capacity constraints on the vehicles, and because there is a time interval specified for each ecopoint, during which the waste containers must be emptied.

It is not uncommon for performance requirements in waste collection to change while aiming to accomplish different goals. There are times when a DM might find himself in difficulty to choose the best strategy for a given situation, and when performance indicators (PIs) are presented, conflicting objectives can arise. For example, the company aims to collect as much quantities of waste per route as possible, but also needs to minimize the total distance travelled by the collection vehicles. Handling these situations and deciding on what the most suitable solution is in order to attend acceptable values for the PIs, can be a time-consuming task, and so, a valuable tool to use is a DSS with MCDA capabilities, which can process a great amount of information and present solutions to the DM at a faster pace. Nonetheless, the DM assumes a central role during decision-making processes in a DSS.

The work presented in this paper is integrated in a R&D project named Genetic Algorithm for Team Orienteering Problem (GATOP). The main goal of GATOP is the development of more complete and efficient solutions for several real-life multi-level vehicle routing problems, with an emphasis on waste collection management. In this work, we intend to present a prototype DSS for management of HPW collection, and it shall be composed of different modules or elements. Our purpose is to design the proper functioning of the DSS by determining how the modules interact, how information flows between them, and how the information is processed and solutions are presented to the users. In addition, we needed to do some improvements and changes to the route optimization module for it to be able to handle different objective functions, which were formulated based on the study of real case scenarios faced by a WMC. Briefly, the tasks for this study were:

– Define alternative objectives for route design in waste collection for recycling;
– Develop a solution method to solve different objective functions for the routing problem;
– Define the set of criteria that will be used in the MCDA module;

- Develop and implement the MCDA module;
- Define the structure and way of operation of the DSS prototype;
- Test the DSS for real case scenarios and evaluate its functionality.

3 Literature Review

3.1 Solution Methods for the TOP and Other Variants

In the context of fleet management there is a high demand for more efficient procedures and techniques to perform route planning and design. There are plenty of problems involving transportation of people and/or items/commodities that need route optimization for various purposes. This kind of problems can be addressed as a vehicle routing problem (VRP). As we stated previously, the VRP model is too restrictive for the collection of recyclable waste, and more flexibility is required to enable the selection of locations to visit. Such flexibility is achievable by modelling the collection process as a team orienteering problem (TOP). A wide variety of algorithms have been developed to solve the TOP, and some successful ones are tabu search [4,35], branch and price [7], guided local search [43], path-relinking [33], ant colony optimization [22], memetic algorithm [6], particle swarm optimization [12,32]. For the TOP with time windows (TOPTW), efficient algorithms are iterated local search [42], variable local search, ant colony system [25], variable neighbourhood search [24,41], and hybrid iterated local search [34]. Other variant of the TOP includes capacity constraints (CTOP) has been receiving attention lately, with tabu search being a good solution option [2,3]. Other algorithms followed, as well as new variants of the TOP and CTOP. Promising algorithms based on other methodologies are also found in the literature of routing problems, such as genetic algorithms [15,16,27,28,36,37] and cellular genetic algorithms [1].

3.2 Application of Multi-Criteria Decision Analysis to Routing Problems

After a wide range review of literature on the application of multi-criteria decision analysis in decision support systems for routing problems over the last decade, the authors came across some interesting works in this research field.

In 2002, the author Jacek Zak [45] brought attention to several problems faced by many transportation companies. One problem was about the acceptance or rejection of incoming orders based on the definition of minimum price for the orders and the assignment of vehicles to orders. This multi-objective problem was solved with the ELECTRE III method.

Cavar et al. [10], in 2005 applied AHP to the selection of the best VRP algorithm to be used for a particular case, considering factors like the number of vehicles to use, the time necessary to calculate the routes and the overall travelled distance of the vehicle fleet.

Later, in 2008, an interesting application of MCDA to a routing problem was carried out by Tavana et al. [39], within the subject of Joint Air Operations. Their goal was to model a problem that comprises the assignment of aerial vehicles to mission packages. The MCDA model focus on four competing objectives considered for the assessment of vehicle-target allocation, and to solve the problem, the Analytic Hierarchy Process (AHP) method is also used in the model.

In [20] the authors intended to optimize the design of a supply centre for public service, and they employ Fuzzy-AHP to decide on the service facility types and then use VRP solution methods to find the number and location of the facilities. Criterion like delivery level, service level, supply cost, customers response (satisfaction), transportation and service information were used.

Still in 2008, an application of goal programming (GP) methodology to model a single vehicle routing problem with multiple routes was proposed in [21]. The developed model is solved using a heuristic method based on an elementary Shortest Path Algorithm with Resource Constraints.

Later, in 2010, Tavana and Bourgeois [38] focused on the problem of operational planning and navigation of autonomous underwater vehicles. A dynamic multiple criteria support system was developed and the authors employed MCDA methods, along with other methodologies, to assist in mission planning carried out by the United States Navy.

A combination of GP and genetic algorithm was employed in [18] to model and solve a multi-objective VRP with time windows. The considered objectives were the minimization of total required fleet size and minimization of total travelling distance, while constraints such as capacity and time windows are fulfilled.

A more recent work was presented in 2012 by Ries and Ishizaka [29]. The authors developed a DSS to solve a routing problem of Unmanned Aerial Vehicles within the scope of maritime surveillance. They applied MCDA methods such as AHP and PROMETHEE to evaluate the operational scenarios produced by the routing problem algorithm.

Also in 2012, a combination of AHP and TOPSIS was presented in [44] as a solution method for assessing alternative routes for a VRP.

4   Methodology

Previously in this paper we stated that our intent was to present a prototype Decision Support System (DSS) to help managers deal with multi-criteria decision-making applied to route planning in the context of waste collection. In our concept, the DSS to be developed shall include three different modules: 1) a route optimization module, 2) a MCDA module, and 3) a waste generation prediction module to predict waste generation rates and determine priority levels of the collection points. The route optimization module has its origin on improvements made to previously presented works within the scope of the GATOP project [15,16,26,27,28]. The same is applicable to the waste generation prediction module [17]. The base for the MCDA module development was the beSmart software v1.1 [5]. Improvements were made to the software so it could better meet our purpose with the DSS. In this section we will present the modifications and improvements made to the route optimization module and the MCDA module, as well as the adopted methodologies.

4.1 Route Optimization for Waste Collection

Defining Different Alternatives for Waste Collection Routes. Waste management companies usually have special concern about their waste collection system and how efficient their route planning procedures currently are and how they can be improved. Many companies employ fixed collection routines and schedules, and surely that simplifies resource management. However, performing route optimization according to specific objectives may represent a relevant source of extra revenue, and that is a great motivator for the employment of better management practices, the application of better route planning algorithms and/or decision aid software. From now on in this document, the term solution refers to a set of optimized routes that can be assigned to a vehicle fleet.

The definition of objectives when performing route planning is crucial in order to meet certain goals or quotas for waste collection. For example, one common objective is to minimize the distance travelled by the vehicles while visiting a set collection points, and less mileage means less vehicle maintenance costs and fuel expenses and also less greenhouse gas emissions. For this study, we used a list of important objectives that were agreed with the WMC, that should be taken into consideration while aiming to optimize routes:

- Minimize Total Distance Travelled - MinD
- Maximize Total Collected Quantity - MaxQ
- Maximize Performance - MaxP (quantity collected per kilometre travelled)
- Maximize Number of Ecopoints Collected - MaxE
- Minimize Number of Vehicles Used - MinV
- Maximize Number of Priority Points Collected - MaxPP

Performance Indicators in Waste Collection. The route planning procedure often relies on specific algorithms and/or software tools. The quality of the solutions obtained with those tools and techniques needs to be assessed using some performance indicators. A performance indicator is a type of performance measurement, and in this case it quantifies a certain factor related to waste collection routes. All the values for the performance indicators are quantitative, which makes it easier for comparisons when using some MCDA techniques. These performance indicators are in fact a set of criteria that have influence in the context of route planning for waste collection. The performance indicators considered for this study are the following:

- Total Distance Travelled (per solution)
- Average Distance Travelled (per route)
- Total Collected Quantity (per solution)
- Average Collected Quantity (per route)
- Total Collection Performance (per solution)
- Average Collection Performance (per route)
- Number of Collected Ecopoints (per solution)
- Average Number of Collected Ecopoints (per route)

- Number of Vehicles used (per solution)
- Number of Priority Points Collected (per solution)
- Average Number of Priority Points Collected (per route)
- Number of High Priority Ecopoints Collected (per solution)

A Cellular Genetic Algorithm to solve the CTOPTW. The final model we used for route planning was the capacitated team orienteering problem with time windows (CTOPTW). As mentioned in a previous section, there are several methodologies capable of dealing with (solving) the TOP and its variants. Nonetheless, we opted to follow a research line we have been pursuing, which is the application of genetic algorithms (GAs) to solve optimization problems, specially routing problems [15,16,26,27,28]. More recently, one of our focus has been the employment of cellular genetic algorithms (cGAs) since the experimental results achieved with this method performed better than our previously developed GAs. So, in order to solve the CTOPTW, we developed an algorithm based on the cGA methodology. We also made adjustments to the algorithm so it could deal with different objective functions and output optimized routes accordingly.

4.2   beSmart A Multi-Criteria Decision Aid Software Application

Software Description and Improvements Made. There are many different MCDA methods, and there are also many tools developed to implement the MCDA methods. In 2010, Seixedo and Tereso [31] have assembled a list of the available MCDA tools and developed a Multi-Criteria Decision Aid Software Application for selecting MCDA software using the Analytic Hierarchy Process (AHP) method. Later, in 2011, improved MCDA software for the same purpose was developed by Tereso et al. [40]. The software tool was named beSmart [5], and was designed to be a general-purpose application, which can load and process data to help solving any MCDA problem. The beSmart became an interesting tool for us to explore and integrate in the DSS prototype we intend to develop.

Although the beSmart software was at a good state of development, some improvements were made in order to offer a better user experience. This was achieved by enhancing the Graphical User Interface (GUI), using better displaying of options and commands in the menu bar, and by improving the contents of tips and instructions displayed to the user throughout the decision-making process. Changes to the solution explorer module were also made, in order to present more useful information on the solutions produced. In addition, the option to perform sensitivity analysis was included in the solution explorer.

MCDA Methods Available in beSmart. There are three MCDA methods embedded within beSmart: 1) SMART, 2) AHP, and 3) ValueFn. The SMART method [14] consists of assigning a score to each alternative, and the higher the scores are, the more importance the alternative represents. In the AHP method [30], a structuring process of the problem occurs so that it is decomposed into a hierarchy of sub-problems. Then, the DM evaluates the relative importance of these sub-problems (criteria) by pairwise comparison, where a degree of importance is given to each criterion in relation to another. The AHP method converts these evaluations to numerical values (weights or priorities), which are used to calculate a score for each alternative. A consistency rate measures the extent to which the DM has been consistent in the comparison done. This rate should be lower than 0.10 [30]. In the ValueFn method [9,23], the evaluation of alternatives is directly fitted into a function. It can be a maximization or a minimization function, depending on the DMs intention to maximize or minimize a given attribute. For further information and more detailed explanations on how these MCDA methods were implemented into beSmart, one should check the work in [31] and [40].

4.3   A Prototype Decision Support System to Improve Waste Collection

Since each module of the DSS prototype is also a stand-alone software tool, a proper design is needed in order to establish how each module interacts with each other, how information flows between them and how solutions are presented to the end-user. A schematic in figure 1 shows how the DSS's operates. Initially, the waste generation prediction module forecasts the current state of each collection point in terms of quantity of waste stored in its containers. These informations are fed to the route optimization module, which in turn proceeds with the computation of collection routes for all objectives previously presented, although not in a multi-objective approach, with a solution for each objective function being output separately. Once the routes are calculated, their values for each performance indicator (PI) are saved in a text file, which is later loaded in beSmart.

The beSmart software can help to select the best set of routes for specific scenarios of waste collection where the DM needs to weigh the importance of each PI. The decision process occurs during five steps (see figure 2). The first one is the selection of alternatives for comparison, which are sets of routes called solutions. Each solution results from the output of the route optimization module according to a certain objective function. In the next decision step, the DM chooses the PIs he finds relevant for solution assessment. The third step is the definition of weights, and the DM expresses the relative importance of each previously selected PI using either the SMART or AHP method. The DM assigns values to each PI to denote its importance relative to the others. The fourth step is the definition of priorities, and using the AHP or ValueFn methods, one can determine the priority level each PI represents to each solution. Finally, in the fifth step, the comparison results are displayed using charts and ranking lists. It is possibile to perform real-time sensitivity analysis for the selected PIs. This feature enables the DM to examine the trade-offs between a PI and the rest, and also the impact of changing the priority level of each PI individually. Once the decision process in beSmart ends, the DM is presented with a final ranking of alternatives, sorted from the most to the least suitable solution. Using the DSS makes the DM more aware of solution possibilities and their outcomes.

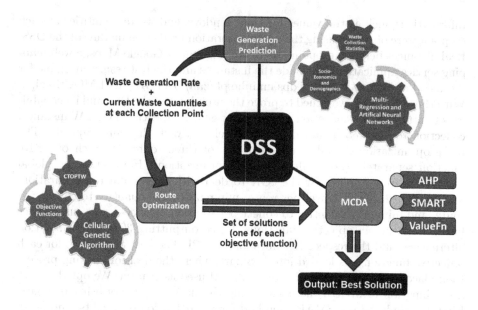

Fig. 1. DSS's Operational Schematic

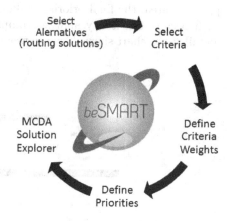

Fig. 2. Steps for Decision-Making in beSmart

5   Experiments and Results

The assessment of the proposed DSS was done by simulation of real situations of HPW collection, using the WMC's data to design 10 CTOPTW instances. The data sources used to assemble the instances consisted of a list of 94 ecopoints and their GPS coordinates. These 94 ecopoints represent a whole municipality.

Informations such as the values for time windows and waste quantities at each ecopoint were obtained using the waste generation prediction module of the DSS. Real distances between ecopoints were obtained using Google Maps, a web mapping service application. We made the instances and the test results available for download at "http://pessoais.dps.uminho.pt/zan/GATOP /instEMO2015.zip". With those instances we aimed to prove the usefulness of the DSS and its capabilities to attend waste collection management goals. For each CTOPTW instance, collection routes were planned for three vehicles with the same capacity. The route optimization module assembled sets of three routes for each objective function separately (not multi-objective optimization). However, some global constraints were applied for the cGA to deal with: maximum route duration, maximum distance to travel per route, minimum waste quantity to collect per route, and minimum quantity to collect per solution. The cGA was run 5 times for each of the six objective functions, hence outputting 30 different solution alternatives and the respective values for each PI. The best alternative for each objective function was loaded into beSmart, where the decision-making process took place, with the 12 PIs previously stated used as criteria. We opted to test a combination of MCDA methods during the decision process using beSmart. First we employed the SMART method for the definition of weights, and later, for the definition of priorities, the choice relied on a mixed application of AHP and ValueFn, choosing the method that seemed most appropriate for each criteria. In figure 3, the solution achieved using beSmart for one tested instance is presented. On the upper-left area, the final priority ranking of alternatives is shown. On the bottom-left area, a sensitivity analysis component is available. On the bottom-right a detailed bar chart shows how the weight of each criterion

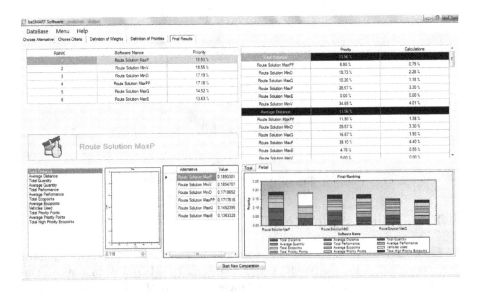

Fig. 3. Results with beSmart for the simulation scenario

contributed for the ranking of each alternative. Although the best alternative found was "Route Solution MaxP", if the DM expresses different preferences, alternative solutions can be found. Throughout several similar tests, we observed that the DSS has proven its usefulness, as it allows the user to solve waste collection problems, and is able to help analyse the influence of several criteria, weighing their contribution while aiming to attend certain performance levels.

6   Conclusions and Future Work

In this work we presented a decision support system (DSS) that provides solutions for problems that arise in waste collection management. One main issue is to perform route planning for selective collection of recyclable waste. We modelled this routing problem as a capacitated team orienteering problem with time windows (CTOPTW), and we developed a cellular genetic algorithm (cGA) to produce solutions. The cGA is able to deal with different objective functions. We presented twelve criteria based on waste collection performance indicators (PIs), and we employed multi-criteria decision analysis (MCDA) methods to assist in decision-making when weighing the importance of each PI in route planning.

The developed DSS includes three modules: waste generation prediction, route optimization, and MCDA. We presented a design for the DSS and assembled a prototype to run experiments at this stage of development. We performed simulations of real problems faced by a waste management company (WMC) responsible for collecting waste for recycling. The DSS was tested with real instances of the problem. In our best knowledge, there are no similar approaches to the one we presented in this paper for dealing in an unified way with several problems related to waste collection management, and so, direct comparisons with alternative systems were not possible. Nevertheless, we assessed the DSS's capabilities in terms of route planning and optimization, with a focus on MCDA. The DSS provides different optimization options and can present alternative solutions for the same instance of HPW collection, depending on the preferences of decision-makers (DMs). These features are advantageous, since the DM's preferences play a central role in selecting the best routes for a certain planning period, according to the WMC's logistic strategy and the importance given to each PI. We obtained positive feedback from the WMC and more improvements are foreseen. We are positive that our DSS design can be of great assistance in the context of waste collection, and WMCs would be able to improve their performances by exploring better collection routines and by adapting to challenges that may arise before them. Overall, the interpretation of computational results output by the DSS can provide meaningful information to waste collection management practitioners. However, it is not advisable to apply the DSS on a daily basis, as it would be an impractical and time-consuming task to go through the decision process every day. Instead, the DSS should be used to outline the WMC's logistic strategy for a longer period such as a week, month or trimester.

For future work, we intend to simplify the user's interaction with the DSS to enhance overall user experience and accelerate work flow. More experiments with the DSS shall be conducted in more real-case situations to fully validate our approach.

Acknowledgments. This work has been supported by FCT Fundação para a Ciência e Tecnologia within the Project Scope: PEst-OE/EEI/UI0319/2014.

References

1. Alba, E., Dorronsoro, B.: Solving the vehicle routing problem by using cellular genetic algorithms. In: Gottlieb, J., Raidl, G.R. (eds.) EvoCOP 2004. LNCS, vol. 3004, pp. 11–20. Springer, Heidelberg (2004)
2. Archetti, C., Bianchessi, N., Speranza, M.G.: The capacitated team orienteering problem with incomplete service. Optimization Letters 7(7), 1405–1417 (2013)
3. Archetti, C., Feillet, D., Hertz, A., Speranza, M.G.: The capacitated team orienteering and profitable tour problems. Journal of the Operational Research Society 60, 831–842 (2009)
4. Archetti, C., Hertz, A., Speranza, M.G.: Metaheuristics for the team orienteering problem. Journal of Heuristics 13, 49–76 (2007)
5. BeSmart software development webpage. https://code.google.com/p/besmart/
6. Bouly, H., Dang, D.-C., Moukrim, A.: A memetic algorithm for the team orienteering problem. In: Giacobini, M., Brabazon, A., Cagnoni, S., Di Caro, G.A., Drechsler, R., Ekárt, A., Esparcia-Alcázar, A.I., Farooq, M., Fink, A., McCormack, J., O'Neill, M., Romero, J., Rothlauf, F., Squillero, G., Uyar, A.Ş., Yang, S. (eds.) EvoWorkshops 2008. LNCS, vol. 4974, pp. 649–658. Springer, Heidelberg (2008)
7. Boussier, S., Feillet, D., Gendreau, M.: An exact algorithm for team orienteering problems. 4OR 5, 211–230 (2007)
8. Butt, S.E., Cavalier, T.M.: A heuristic for the multiple tour maximum collection problem. Computers and Operations Research 21, 101–111 (1994)
9. Canada, J.R., Sullivan, W.G.: Economic and Multiattribute Evaluation of Advanced Manufacturing Systems. In: Multiattribute Decision Analysis: utility models. ch. 9. Prentice Hall College Div. (1989)
10. Cavar, I., Gold, H., Caric, T.: Assessment of heuristic algorithms for solving real capacitated vehicle routing problems by analytic hierarchy process. In: 12th World Congress in Intelligent Transport Systems, San Francisco (2005)
11. Chao, I., Golden, B.L., Wasil, E.A.: The team orienteering problem. European Journal of Operational Research 88, 464–474 (1996)
12. Dang, D.-C., Guibadj, R.N., Moukrim, A.: A PSO-based memetic algorithm for the team orienteering problem. In: Di Chio, C., Brabazon, A., Di Caro, G.A., Drechsler, R., Farooq, M., Grahl, J., Greenfield, G., Prins, C., Romero, J., Squillero, G., Tarantino, E., Tettamanzi, A.G.B., Urquhart, N., Uyar, A.Ş. (eds.) EvoApplications 2011, Part II. LNCS, vol. 6625, pp. 471–480. Springer, Heidelberg (2011)
13. Dantzig, G.B., Ramser, J.H.: The Truck Dispatching Problem. Management Science 6(1), 80–91 (1959)
14. Doran, G.T.: There's a S.M.A.R.T. way to write management's goals and objectives. Management Review 70(11), 35–36 (1981)

15. Ferreira J., Oliveira J., Pereira G., Dias L., Vieira F., Macedo J., Carção T., Leite T., Murta D.: Developing tools for the team orienteering problem - a simple genetic algorithm. In: Proceedings of the 2nd International Conference on Operations Research and Enterprise Systems, Barcelona, pp. 134–140 (2013)
16. Ferreira, J., Quintas, A., Oliveira, J.A., Pereira, G., Dias, L.: Solving the team orienteering problem – developing a solution tool using a genetic algorithm approach. In: Snášel, V., Krömer, P., Köppen, M., Schaefer, G. (eds.) Soft Computing in Industrial Applications. AISC, vol. 223, pp. 365–375. Springer, Heidelberg (2014)
17. Ferreira, J.A., Figueiredo, M.C., Oliveira, J.A.: Forecasting household packaging waste generation: a case study. In: Murgante, B., Misra, S., Rocha, A.M.A.C., Torre, C., Rocha, J.G., Falcão, M.I., Taniar, D., Apduhan, B.O., Gervasi, O. (eds.) ICCSA 2014, Part III. LNCS, vol. 8581, pp. 523–538. Springer, Heidelberg (2014)
18. Ghoseiri, K., Ghannadpour, S.F.: Multi-objective vehicle routing problem with time windows using goal programming and genetic algorithm. Applied Soft Computing 10(4), 1096–1107 (2010)
19. Gomes, A., Matos, M., Carvalho, I.: Separate collection of the biodegradable fraction of MSW: An economic assessment. Waste Manage 28, 1711–1719 (2008)
20. Hwang, H., Choi, B., Lee, K., Cho, G.: Supply Center Planning Model Using Fuzzy-AHP and VRP. In: 3rd International Conference on Innovative Computing Information and Control - ICICIC 2008, p. 109 (2008)
21. Jolai, F., Aghdaghi, M.: A Goal Programming Model for Single Vehicle Routing Problem with Multiple Routes. Journal of Industrial and Systems Engineering 2(2), 154–163 (2008)
22. Ke, L., Archetti, C., Feng, Z.: Ants can solve the team orienteering problem. Computers and Industrial Engineering 54, 648–665 (2008)
23. Keeney, R., Raiffa, H.: Decision with Multiple Objectives: preferences and value tradeoffs. Cambridge University Press (1993)
24. Labadi, N., Mansini, R., Melechovský, J., Calvo, R.W.: The Team Orienteering Problem with Time Windows: An LP-based Granular Variable Neighborhood Search. European Journal of Operational Research 220(1), 15–27 (2012)
25. Montemanni, R., Weyland, D., Gambardella, L.M.: An enhanced ant colony system for the team orienteering problem with time windows. In: International Symposium on Computer Science and Society (ISCCS), pp. 381–384 (2011)
26. Mota, G., Abreu, M., Quintas, A., Ferreira, J., Dias, L.S., Pereira, G.A.B., Oliveira, J.A.: A genetic algorithm for the TOPdTW at operating rooms. In: Murgante, B., Misra, S., Carlini, M., Torre, C.M., Nguyen, H.-Q., Taniar, D., Apduhan, B.O., Gervasi, O. (eds.) ICCSA 2013, Part I. LNCS, vol. 7971, pp. 304–317. Springer, Heidelberg (2013)
27. Oliveira, J.A., Mota, G., Ferreira, J., Figueiredo, M., Dias, L., Pereira, G.: A decision support system for waste collection modeled as TOPTW variant. In: WASTES: Solutions, Treatments and Opportunities - 2nd International Conference (accepted, 2013)
28. Oliveira, J.A., Ferreira, J., Figueiredo, M., Dias, L., Pereira, G.: Comparação de dois algoritmos genéticos aplicados ao TOP. XI Congreso Galego de Estatística e Investigación de Operacións (accepted, 2013)
29. Ries, J., Ishizaka, A.: A multi-criteria support system for dynamic aerial vehicle routing problems. In: Proceedings of IEEE 2nd International Conference on Communications, Computing and Control Applications (CCCA), pp. 1–4 (2012)
30. Saaty, T.L.: The Analytic Hierarchy Process. McGraw-Hill, New York (1980)

31. Seixedo, C., Tereso, A.: A Multicriteria Decision Aid Software Application for selecting MCDA Software using AHP. In: 2nd International Conference on Engineering Optimization, Lisbon, Portugal (2010)

32. Sevkli, A.Z., Sevilgen, F.E.: Discrete particle swarm optimization for the team orienteering problem. Turk. Journal Elec. Eng. & Comp. Sci. 20(2) (2012)

33. Souffriau, W., Vansteenwegen, P., Van Oudheusden, D.: A Path Relinking Approach for the Team Orienteering Problem. Computers & Operations Research, Metaheuristics for Logistics and Vehicle Routing 37(11), 1853–1859 (2010)

34. Souffriau, W., Vansteenwegen, P., Berghe, G.V., Oudheusden, D.V.: The Multiconstraint Team Orienteering Problem with Multiple Time Windows. Transportation Science 47(1), 53–63 (2013)

35. Tang, H., Miller-Hooks, E.: A TABU search heuristic for the team orienteering problem. Computers & Operations Research 32, 1379–1407 (2005)

36. Tasgetiren, M.F.: A Genetic Algorithm with an Adaptive Penalty Function for the Orienteering Problem. Journal of Economic and Social Research 4(2), 1–26 (2002)

37. Tasgetiren, M.F., Smith, A.E.: A genetic algorithm for the orienteering problem. In: Proceedings of the 2000 Congress on Evolutionary Computation (CEC2000), vol. 2, pp. 910–915 (2000)

38. Tavana, M., Bourgeois, B.S.: A multiple criteria decision support system for autonomous underwater vehicle mission planning and control. International Journal of Operational Research 7(2), 216–239 (2010)

39. Tavana, M., Bailey, M.D., Busch, T.E.: A multi-criteria vehicle-target allocation assessment model for network-centric Joint Air Operations. International Journal of Operational Research 3(3) (2008)

40. Tereso, A., Sampaio, A., Frade, H., Costa, M., Abreu, T.: beSMART: a software tool to support the selection of decision software. In: International Conference on Engineering UBI2011 (ICEUBI2011), Covilhã, Portugal (2011)

41. Tricoire, F., Romauch, M., Doerner, K.F., Hartl, R.F.: Heuristics for the multi-period orienteering problem with multiple time windows. Computers & Operations Research 37, 351–367 (2010)

42. Vansteenwegen, P., Souffriau, W., Van Oudheusden, D.: Iterated local search for the team orienteering problem with time windows. Computers & Operations Research 36, 3281–3290 (2009)

43. Vansteewegen, P., Souffriau, W., Van Oudheusden, D.: A guided local search metaheuristic for the team orienteering problem. European Journal of Operational Research 196(1), 118–127 (2009)

44. Yilmaz Z., Aplak, H.S., Vehicle routing by revaluing the alternative routes by using AHP-TOPSIS Combination. In: Proceedings of X International Logistics & Supply Chain Congress, İstanbul, Turkey, pp. 304–310 (2012)

45. Zak, J.: The MCDA methodology applied to solve complex transportation decision problems. In: Proceedings of the 13th Mini Euro Conference, Bari, Italy (2002)

Application of Evolutionary Multiobjective Algorithms for Solving the Problem of Energy Dispatch in Hydroelectric Power Plants

Carolina G. Marcelino[1]([✉]), Leonel M. Carvalho[3], Paulo E.M. Almeida[1], Elizabeth F. Wanner[2], and Vladimiro Miranda[3]

[1] Centro Federal de Educação Tecnológica de Minas Gerais, Laboratório de Sistemas Inteligentes, Belo Horizonte, MG, Brazil
{carolina,pema}@lsi.cefetmg.br
[2] Centro Federal de Educação Tecnológica de Minas Gerais, Departamento de Computação, Belo Horizonte, MG, Brazil
efwanner@decom.cefetmg.br
[3] Institute for Systems and Computer Engineering of Porto, Centre for Power and Energy Systems, Porto, Portugal
{lcarvalho,vmiranda}@inesctec.pt

Abstract. The Brazilian population increase and the purchase power growth have resulted in a widespread use of electric home appliances. Consequently, the demand for electricity has been growing steadily in an average of 5 % a year. In this country, electric demand is supplied predominantly by hydro power. Many of the power plants installed do not operate efficiently from water consumption point of view. Energy Dispatch is defined as the allocation of operational values to each turbine inside a power plant to meet some criteria defined by the power plant owner. In this context, an optimal scheduling criterion could be the provision of the greatest amount of electricity with the lowest possible water consumption, i.e. maximization of water use efficiency. Some power plant operators rely on "Normal Mode of Operation" (NMO) as Energy Dispatch criterion. This criterion consists in equally dividing power demand between available turbines regardless whether the allocation represents an efficient good operation point for each turbine. This work proposes a multiobjective approach to solve electric dispatch problem in which the objective functions considered are maximization of hydroelectric productivity function and minimization of the distance between NMO and "Optimized Control Mode" (OCM). Two well-known Multiobjective Evolutionary Algorithms are used to solve this problem. Practical results have shown water savings in the order of million m^3/s. In addition, statistical inference has revealed that SPEA2 algorithm is more robust than NSGA-II algorithm to solve this problem.

Keywords: Multiobjective optimization · NSGA-II · SPEA2 · Energy efficiency

© Springer International Publishing Switzerland 2015
A. Gaspar-Cunha et al. (Eds.): EMO 2015, Part II, LNCS 9019, pp. 403–417, 2015.
DOI: 10.1007/978-3-319-15892-1_27

1   Introduction

The demand for electricity is one of the key emerging issues in current energy management in Brazil. According to the annual report of Brazilian Energy Planning Company ("Empresa de Planejamento Energético" - EPE, in Portuguese), the share of renewable energy in electricity mix has dropped to 79.3% in 2013 due to unfavourable hydrological conditions despite owning the most diverse and extensive river networks from around the world. Consequently, thermal generation has taken over which has resulted in increased CO_2 emissions and a less environmentally friendly generation mix. Fig. 1 shows a comparison between generation of electric power in Brazil in relation to the World and OECD member countries.

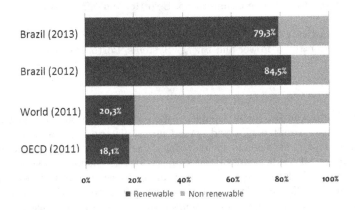

Fig. 1. Comparing energy generation matrix - 2013 (Source: EPE, Brazil)

This comparison reveals that Brazil has an advantage regarding the use of renewable energy over the rest of the World. Also notable is the reduction in renewable energy use between years of 2012 and 2013. This was caused by a widespread drought in 2013 causing thermal power plants to be called for production in order to maintain domestic demand for electricity. Hence, proper use of water resources is now an emerging topic in Brazil to guarantee that power system remains sustainable in future years.

The optimal scheduling of hydroelectric power plants (HPP), which are composed by several turbines, pipes turbines and connected electric generators or, in simple terms, generation units, is known as the hydro unit commitment problem. The objective consists in determining which generation units need to be on and their respective electric power set-point (in MW) so that overall hydroelectric power plant operation cost is minimized while meeting with the power required to be produced by the whole plant and satisfying the constraint set.

Electric power set-points are defined to each available generation unit at the hydroelectric plant, given some criteria to be met, such as operating limits, etc.

This last problem, which is named as energy dispatch optimization problem, can only be solved if the production model for the whole hydroelectric plant is available. The majority of hydroelectric plants operators in Brazil equally distribute total power required to be produced by the plant among the available units. In this paper this will be referred as the "Normal Mode of Operation" (NMO). However, one cannot say that this simple dispatch criterion presents a good operation point for each unit since it does not take into account whether each unit will be operating close to its optimal operational point or not. The problem of finding optimal distribution of power demand among units of a power plant is complex, due to non-linearities of the productivity function and the high number of continuous and discrete constraints involved.

1.1 State of the Art

Several optimization techniques to improve energy production efficiency in power systems were discussed in [1]. That study was motivated by the signing of Kyoto Protocol by European Union in 1997 which has led to the definition of 2020 climate and energy package commonly known as "20-20-20" targets. Accordingly, researchers have sought to find new methods and to use new optimization techniques to improve EU's energy efficiency in 20% by 2020, which is one of the goals of that agreement. Some of the techniques described in [1] are: Search Algorithms, Evolutionary Algorithms, Simulated Annealing, Tabu Search, Ant Colony Optimization, Particle Swarm Optimization (PSO), Genetic Algorithms (GA) and Evolutionary Programming. Among them, GA is recommended to minimize energy losses and to maximize efficiency.

Baños [2] conducted a review of metaheuristic-based optimization techniques that have been applied so far to solve renewable energy optimization problems. The main conclusion of his survey is that the number of scientific papers that used metaheuristics to solve these problems has dramatically increased over the last few years. However, he has also reported that, in many cases, computational cost of using these methods is high even when using parallel processing techniques.

Finardi [3] proposed a new mathematical model for long-term planning of hydroelectric power plants. Linear programming was used to solve the problem of energy dispatch. This approach was shown to have a high computational cost which makes this model infeasible to be used for real time Energy Dispatch. Despite of the interesting results obtained, the author has not clearly discussed how important variables of the production function were discarded making the model very difficult to be understood and validated.

Abrao [4] proposed to use an artificial neural network to model the production function of a single generation unit. The author solved short-term planning problem, which consists on defining the respective dispatch of each generation unit for a specific period of time, using a version of Differential Evolution (DE) algorithm and a version of a PSO algorithm. However, computational cost of this solution is relatively high.

Marcelino [5] proposed a new mathematical model to solve HPP energy dispatch problem using DE. He showed that evolutionary strategy DE/best/1/bin is the most efficient for solving the mono-objective version of this problem. This model proved to be efficient and provided very promising results.

The research work reported in [6] shows the application and multiobjective algorithms to solve the classic electrical dispatch problem. The test case used, which is based on IEEE 30-bus system, comprises thermal and hydro units. Results indicate that SPEA algorithm achieved the best results compared to algorithms NSGA and NPGA, where the goals are to minimize carbon emissions and to minimize production cost.

Zhou [7] proposes a new multiobjective algorithm, which is named Multiple Group Search Optimizer (MGSO), to solve the classic electrical dispatch problem for IEEE-30 bus and IEEE 118-bus systems. The objectives are to minimize carbon emissions and production cost of power plants. Practical results of MSGO proved to be competitive when compared to results of NSGA-II and SPEA2.

This work proposes a multiobjective approach to solve energy dispatch problem in HPP using a mathematical model very similar to the one proposed in [5]. For this purpose, two objective functions are defined: maximization of hydroelectric production function of whole HPP and minimization of distance between "Normal Mode of Operation" (NMO) and "Optimized Control Mode" (OCM). Note that OCM is the outcome of former objective. The latter objective is very interesting from practical point of view. Usually, HPP operators are not used to employ OCM and sometimes they can be restrictive. Moreover, from experience, the maintenance cost of the turbine set will be increased since some components can be subject to more wear reducing the turbine lifetime. As a case study, the proposed approach is applied to a large HPP operating in Brazil.

The paper is organized as follows: Section 2 describes the problem of Energy Dispatch in HPP; Section 3 presents the multiobjective problem proposed and algorithms used to solve it; Section 4 shows the case study, outcome of experiments and a simple comparative statistical analysis between different algorithms used; Section 5 presents final conclusions.

2   Multiobjective Problem

A multiobjective problem is characterized by having two or more objective functions, which are generally self conflicting. This type of problem does not have a single solution but a set of optimal solutions. A multi-objective optimization problem can be formulated as:

$$\mathbf{x}^* = \min_x \mathbf{f}(\mathbf{x})$$
$$\text{subject to:} \begin{cases} \mathbf{g_i}(\mathbf{x}) \le 0; \ i = 1, 2, \cdots, r \\ \mathbf{h_j}(\mathbf{x}) = 0; \ j = 1, 2, \cdots, p \end{cases} \tag{1}$$

in which $\mathbf{x} \in \mathbb{R}^n$, $\mathbf{f}(\cdot) : \mathbb{R}^n \to \mathbb{R}^m$, $\mathbf{g}(\cdot) : \mathbb{R}^n \to \mathbb{R}^r$, and $\mathbf{h}(\cdot) : \mathbb{R}^n \to \mathbb{R}^p$. Functions $\mathbf{g_i}(\mathbf{x})$ and $\mathbf{h_j}(\mathbf{x})$ are, respectively, inequality and equality constraints. Vectors $\mathbf{x} \in \mathbb{R}^n$ are called *parameters* of the multiobjective problem and belong to a *parameter space*. Vector function $\mathbf{f}(\mathbf{x}) \in \mathbb{R}^m$ belongs to a vectorial space called *objective space*.

In multobjective problems, there is not a single solution which is best or global optimum with respect to all objectives. Presence of multiple objectives in a problem usually gives rise to a family of non dominated solutions, called Pareto-optimal set, where each objective component of any solution along Pareto front can only be improved by degrading at least one of its other objective components.

Given two solutions, $\mathbf{x}$ and $\mathbf{y}$, it is said that $\mathbf{x}$ dominates $\mathbf{y}$ (denoted $\mathbf{x} \succeq \mathbf{y}$) if following conditions are met:

1. The solution $\mathbf{x}$ is at least equal to $\mathbf{y}$ for all objective functions;
2. The solution $\mathbf{x}$ is better than $\mathbf{y}$ for at least one objective.

2.1   Evolutionary Multiobjective Algorithms

Studies related to evolutionary multiobjective algorithms date back to 1980s. The first algorithm of this class based on Pareto Front was proposed in early 1990s and is named as Multiobjective Genetic Algorithm (MOGA) [8]. After this, some other algorithms have emerged: Niched Pareto Genetic Algorithm (NPGA) [9], Nondominated Sorting Genetic Algorithm (NSGA) [10] (and its evolution NSGA-II [11]) and Strength Pareto Evolutionary Algorithm (SPEA) [12] (and its evolution SPEA2 [13]). Since then, several other evolutionary multiobjective algorithms were proposed and published. A general evolutionary multiobjective algorithm can be run by the following pseudocode:

Step 1. Initialize population;
Step 2. $Q(t=0) = q^1 \ldots q^\mu$;
Step 3. Initialize population of archive $A(t=0) = 0$;
Step 5. While ($\to$ stop criteria) do
    1. $P(t) \leftarrow Q(t)$;
    2. $S(t) \leftarrow$ selection $(P(t))$;
    3. $R(t) \leftarrow$ crossover $(S(t))$;
    4. $Q(t) \leftarrow$ mutation $(R(t))$;
    5. $A(t) \leftarrow$ file update $(Q(t), A(t))$;
    6. $t \leftarrow t + 1$.

This paper uses well-known evolutionary multiobjective algorithms, NSGA-II and SPEA2, to solve the proposed problem.

3   The Problem of Electric Dispatch

The electricity production of HPP is result of a process of potential and kinetic energy transformations. The potential energy stored in reservoir is transformed into mechanical energy by the turbine through its shaft, which, in turn, is transmitted to a electric generator unit. The electrical generator transforms mechanical energy into electrical energy. The power produced goes through collector electrical substation and is injected in transmission system to be delivered to consumption centres. A turbine-generator set has a specific hydraulic curve which characterizes its efficiency according to specific water flow and reservoir net head. This curve is called Hill Curve or Efficiency Curve, see Fig. 2.

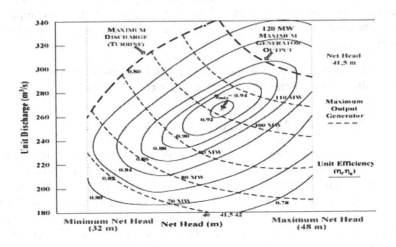

Fig. 2. Hill Curve example [5]

The efficiency curve contains important information to be considered when planning a HPP operation. Given this, it is possible to extract operating limits of turbine-generator set, allowable range for net head in the dam, and also minimum and maximum points of efficiency, where the point of maximum efficiency is in the center of its contours.

From this, one can easily understand that the efficiency curve must be taken into account to ensure power generation with minimal use of water resources while considering operational constraints of a hydroelectric plant. This optimization problem can be characterized as the maximization of electricity production efficiency of whole HPP. In other words, the solution of this problem aims to generate more power with minimal water discharge needed.

3.1 Mathematical Modeling of Power Productivity

This section presents brief summary of the mathematical model that describes the energy dispatch problem HPP, which has been discussed in Marcelino's [5] work. Table 1 describes model parameters.

The equation which defines production of energy, in general, can be described according to Eq. (2),

$$ph_{jt} = g \cdot \eta_{jt} \cdot hl_{jt} \cdot q_{jt}. \tag{2}$$

Table 1. Parameters used in model

Parameter	Description
ph_{jt}	is power generated by unit j at time t
g	is acceleration of gravity
η_{jt}	is global efficiency of unit j at time t
hl_{jt}	is net water head of unit j at time t
q_{jt}	is water discharge of unit j at time t
Hb_t	is hydraulic head of the reservoir
Δ_{Hjt}	is sum of pen-stock losses
$\rho_{0j}...\rho_{5j}$	are coefficients obtained from the *Hill Curve*
Dm	is requested demand (MW)
$q_{jt}min$	is minimum water discharge
ph_{jk}^{min}	is minimum power
$q_{jt}max$	is maximum water discharge
ph_{jk}^{max}	is maximum power
Z_{jk}	is operative zone of a generator unit
q_{cc}	is total water discharge in normal mode of operation

Having that in mind and assuming that the model presented in this work is the best representation for a power plant which uses Kaplan generators, as in the case study, power production performed by an hydroelectric unit, in MW, can be calculated by Eq. (3),

$$ph_{jt} = g \cdot [\rho_{0j} + \rho_{1j}hl_{jt} + \rho_{2j}q_{jt} + \rho_{3j}hl_{jt}q_{jt} + \tag{3}$$

$$\rho_{4j}hl_{jt}^2 + \rho_{5j}q_{jt}^2] \cdot [Hb_t - \Delta_{Hjt}] \cdot q_{jt}.$$

Table 2 presents coefficients obtained by a multi-variable regression process, representing 99% of accuracy, see [5].

Table 2. Efficiency Coefficients

Coefficient	Value
ρ_{0j}	1.4630e-01
ρ_{1j}	1.8076e-02
ρ_{2j}	5.0502e-03
ρ_{3j}	-3.5254e-05
ρ_{4j}	-1.1234e-03
ρ_{5j}	-1.4507e-05

3.2 Multiobjective Optimization Model

According to the mathematical model presented so far, the problem multiobjective goals are to maximize the hydroelectric productivity function (4), which is derived from the electric power function (3), and to minimize distance between

NMO and OCM in function (5). The optimization variables are water flow rate of each generation unit in the following vector,

$$x = [q_{1t}, q_{2t}...q_{jt}].$$

and the bi-objective problem can be described as

$$Maximize\ F_1(x) = \frac{\sum_{j=1}^{J(r)} ph_{jt}}{\sum_{j=1}^{J(r)} q_{jt}}, \tag{4}$$

$$Minimize\ F_2(x) = \sqrt{\sum_{j=1}^{J(r)} (q_{jt} - q_{cc})^2} \tag{5}$$

subject to:

$$\sum_{j=1}^{J(r)} ph_{jt} \cong Dm,\ q_{jt}min \le q_{jt} \le q_{jt}max,$$

$$ph_{jk}^{min} \sum_{k=1}^{\emptyset_j} Z_{jk} \le ph_{jt} \le ph_{jk}^{max} \sum_{k=1}^{\emptyset_j} Z_{jk},$$

$$Z_{jk} \in \{0,1\}, \sum_{k=1}^{\emptyset_j} Z_{jk} \le 1.$$

The first objective function determines how much power the plant is able to produce with a given volume of water. Maximizing this function means to produce more power using less water. The numerator of F_1 is the production function: as this number increases, objective function value also grows. When the denominator of F_1 is decreased, productivity ratio is also reduced. This function is subject to operational constraints, i.e., the sum of all generation units production must be equal to total power demand required to be produced by the HPP. Power production must also comply with generation units operational limits, represented by inequality constraints of objective function.

The second objective function, F_2, measures distance between water discharge used in NMO and water discharge used in OCM. This function was proposed because, in practice, the plant technical operation staff has the predilection for using NMO. This function shows that there are operation points in OCM which are closer to NMO but still ensure the maximization of energy production efficiency. This contributes to a new culture development by the HPP operation staff, increasing their confidence on OCM.

The first constraint indicates that power to be delivered should be equal to power requested to be produced by the HPP. The second constraint states that calculated flow rate must comply with the minimum and maximum flow capacity of each generation unit. The third constraint requires that the corresponding generated power must comply with the minimum and maximum power capacity of each generation unit. At last, the fourth constraint ensures that each generation unit maintains its operating status, i.e. it stays ON or OFF during the whole production period.

4 Experiments

As a computational simulation test, this paper proposes as test scenario a HPP in Brazil with nominal installed capacity of $396MW$. The HPP has in its powerhouse 6 power generators. Water discharge varies between $[70,140]$ m^3/s and power generators are operating in $[35,66]$ MW range. The plant value of Hb ranges between $[32,56]$ m. All generation units are considered identical, so *Hill Curve* coefficients are the same for each unit. Two experiments are presented to validate the adopted multiobjective approach.

The first experiment aims to compare and to analyse the value found in mono-objective approach proposed by Marcelino [5] (the solution was found via a Differential Evolution algorithm [14]) and results found in proposed multiobjective approach. The second experiment aims to assess Pareto front quality obtained by NSGA-II and SPEA2 algorithms. Initialization parameters of both multiobjective algorithms are: population size (50 individuals), crossover probability (80%), mutation probability (2%) and iterations (50 generations).

4.1 Experiment 1

The main goal of this experiment is to verify if the mono-objective solution presented in [5] is a reasonable solution for the multiobjective approach solved via NSGA-II and SPEA2. For that, a demand of $320MW$ is established, since this is a typical demand of the HPP. The reservoir hydraulic head, Hb is set to $54m$.

Fig. 3 shows that there is a point, belonging to Pareto Front obtained by NSGA-II and SPEA2, which is very close to the result reported by DE/best/1/bin [5].

A simulation report for DE/best/1/bin, NSGA-II and SPEA2 is shown in Table 3. This table presents results for the best individual obtained by total water flow (q_{jt}) and, using these values, other parameters are calculated from the mathematical model.

As a comparison to multiobjective optimization results, resulting power production per unit would be $53.33MW$ per unit if NMO is used, which corresponds to a global water flow rate of $655.05m^3/s$. In this context, the productivity of NMO for this experiment is 0.48. The value found by DE algorithm after maximizing productivity is 0.4906 (DE/best/1/bin). Thus, DE/best/1/bin configuration achieves higher productivity rate than NMO and consequently higher economy of water discharge, corresponding to a water flow rate of 2.54 m^3/s.

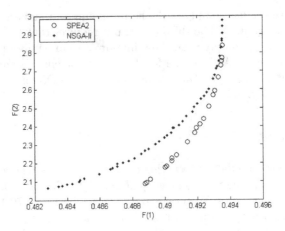

Fig. 3. Pareto Front to 320 MW

For the solution point obtained using SPEA2, water flow economy is even better with 6.14 m^3/s whereas water flow rate is 5.66 m^3/s for case of NSGA-II. Expanding to one month, this is equivalent to saving approximately 6.5 *million* m^3 of water using mono-objective approach, 14.4 *million* m^3 of water using the solution of NSGA-II and 15.7 *million* m^3 of water using the solution of SPEA2 algorithm.

According to [15], monthly water consumption for a city of 300,000 inhabitants is, on average, 1.1 *million* m^3. Belo Horizonte, which is the 6th biggest Brazilian city, has a population of 2.4 million habitants. In a simple analogy, 15.7 *million* m^3 is sufficient to supply the city of Belo Horizonte for almost 2 months, on average. It is also easy to check that, in OCM operation, all units reached maximum efficiency of 93% by using determined water flow rate.

4.2   Experiment 2

In this experiment, NSGA-II and SPEA2 are executed for 30 times each, and a combined Pareto Front is generated for each algorithm. After that, a dominance routine is applied to generate the final Pareto Front for power demand of $320MW$. Fig. 4 shows the final Pareto Front for NSGA-II and SPEA2.

Note that the multiobjective approach presents several solutions for the problem. The Pareto optimal set indicates different solutions among NMO (characterized as the lowest point of the Pareto Front) and OCM (other points on Pareto front). The solution set has an important role in operational scope since the HPP operation team can realize the OCM is not far, in terms of water discharge, from NMO. In this way, it is shown that OCM is a type of control that can be used without harming departing from current operational practises. This leads to optimization technique usage acceptance in industrial environment.

Table 3. General Simulation Report for a Power Demand of 320 MW

Mono-objective algorithm: DE/best/1/bin — $(Hb) = 54m$ — [5]				
UN	$ph_{jt}\ (MW)$	$q_{jt}\ (m^3/s)$	$\eta_{jt}\ (\%)$	$hl_{jt}\ (m)$	$\Delta_{Hjt}\ (m)$
1	48,763	99,441	**0,93**	53,804	0,19595
2	54,589	111,26	**0,93**	53,836	0,16354
3	55,81	113,74	**0,93**	53,839	0,16103
4	56,429	115,00	**0,93**	53,841	0,15925
5	53,122	108,26	**0,93**	53,841	0,15925
6	51,438	104,83	**0,93**	53,839	0,16103
SUM	**320,15**	**652,51**	Flow in SCM: **655,05** (m^3/s)		
DIF	**+0,15**	**2,54**	Productivity index: 0.4906		
Multiobjective algorithm: SPEA2 — $(Hb) = 54m$				
UN	$ph_{jt}(MW)$	$q_{jt}(m^3/s)$	$\eta_{jt}(\%)$	$hl_{jt}(m)$	$\Delta_{Hjt}(m)$
1	53,475	108,274	**0,93**	53,759	0,24149
2	54,229	109,805	**0,93**	53,798	0,20155
3	53,363	108,047	**0,93**	53,802	0,19845
4	53,325	107,971	**0,93**	53,804	0,19625
5	52,891	106,094	**0,93**	53,804	0,19625
6	53,182	108,110	**0,93**	53,802	0,19845
SUM	**320,467**	**648,903**	Flow in SCM: **655,05** (m^3/s)		
DIF	**+0,46**	**6,14**	Productivity index: 0.4936		
Multiobjective algorithm: NSGA-II — $(Hb) = 54m$				
UN	$ph_{jt}(MW)$	$q_{jt}(m^3/s)$	$\eta_{jt}(\%)$	$hl_{jt}(m)$	$\Delta_{Hjt}(m)$
1	52,221	105,747	**0,93**	53,796	0,20377
2	51,751	104,804	**0,93**	53,83	0,17007
3	51,976	105,255	**0,93**	53,833	0,16745
4	55,687	112,787	**0,93**	53,834	0,1656
5	54,457	110,269	**0,93**	53,834	0,1656
6	54,358	110,526	**0,93**	53,833	0,16745
SUM	**320,452**	**649,390**	Flow in SCM: **655,05** (m^3/s)		
DIF	**+0,45**	**5,66**	Productivity index: 0.4935		

5   Statistical Analysis

5.1   ANOVA

Analysis of variance (ANOVA) is a statistical technique that evaluates hypotheses about several populations means and variances. This analysis evaluates primarily if there is a significant difference between the mean and the influence factors on some dependent variable. In this way, ANOVA is used when one wants to decide if sample differences are real (i.e., caused by significant differences in observed populations) or casual (resulting from mere sampling variability) [16]. Therefore, this analysis assumes that chance only produces small deviations, the major differences being generated by real causes. The null and alternative hypotheses to be tested by ANOVA here are:

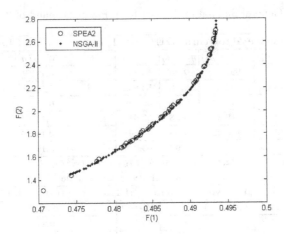

Fig. 4. Comparison between combined Pareto Fronts

- Null hypothesis H_0: populations means are equal;
- Alternative hypothesis H_1: populations means are different, i.e. , at least one of the means is different from the others.

To perform ANOVA hypothesis test, previously tested power demand (320 MW) is used in 30 runs of NSGA-II and SPEA2 algorithms. For each run, the obtained Pareto Front quality is assessed using S-Metric, and the mean value of S-Metric is obtained for each result. S-Metric is a commonly accepted quality measure for comparing approximations of Pareto fronts generated by multiobjective optimizers [12]. This metric calculates hypervolume of a multi-dimensional region enclosed by β and a reference point, thus calculating the region extent that β dominates. Table 4 shows results of ANOVA. Since the obtained P-value is 0.0008, the hypothesis of equality between NSGA-II and SPEA2 S-Metric means is rejected with statistical significance of 95%.

Table 4. Reports by Analysis of variance

ANOVA					
Source	SS	dF	MS	F	Prob>F
Columns	8.1496	1	8.14963	12.48	**0.0008**
Error	37.8698	58	0.65293		
Total	46.0194	59			

Despite the indication of ANOVA that there is a significant difference between the S-Metric values for the algorithms, it is not possible to say which algorithm is the best one when the S-Metric values are compared. Since the statistician often disagree over the efficiency of pairwise comparison method, two tests, Tukey and Permutation tests, are used to determine which algorithm has a higher S-Metric value.

5.2 Tukey and Permutation Tests

Given its ability to analyse multiple data sets, this study used ANOVA with
Tukey test to find some information that differentiates the algorithms mentioned above. This statistical method can be interpreted as a comparison of
means between different groups, with variance between all individuals within
those groups. Tukey's strategy is to define the least significant difference between
means. The hypothesis to be considered in this test is the equality of data sets
series results and adopted a confidence interval of 95% [17]. Permutation tests
are non-parametric statistical methods which estimate a reference distribution
by calculating all possible values (or at least a considerably large set) of a test
statistic under rearrangements of labels on a set of observed data points [17].
The mean difference between S-Metric values, NSGA-II - SPEA2, is used in this
test. Fig. 5 and Fig. 6 shows results of the performed Tukey and Permutation
tests.

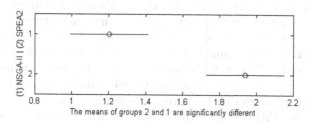

Fig. 5. Left: Tukey test results

Tukey test shows that there is a difference between sets of data tested indicating that, on the mean, the SPEA2 algorithm has higher value of S-Metric
with 95% confidence.

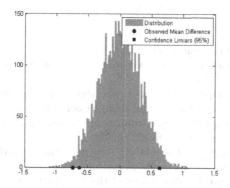

Fig. 6. Permutation test results

Permutation test confirms this information by the fact that the average value represented by the observed mean difference ("black ball") is outside the confidence interval tested, indicating that the data sets are different. This proves that SPEA2 solves the multiobjective energy dispatch of HPP problem better when compared to NSGA-II. It is worthwhile to notice that both SPEA2 and NSGA-II showed superior results to those found in mono-objective approach using DE algorithm, as shown in Table 3.

6   Conclusion

This paper presented a multiobjective approach to solve the energy dispatch problem of Hydroelectric Power Plants using NSGA-II and SPEA2 algorithms. Results of practical experiments indicate that it is possible to identify operating points near NMO that present high productive efficiency. In one experiment, a selected point in Pareto Front with power demand of 320 MW showed a productivity index equal to 0.4936. This point is very similar, in both objective functions, to results found in previous mono-objective approaches, granting reliability to the results and indicating a saving in energy production of 15.7 $million\ m^3$ of water using SPEA2. This amount of water is able to supply a city of 2.4 million people for 2 months. The water savings are relevant with economic, environmental and social implications. Through statistical inference, it was possible to see that SPEA2 algorithm is shown to have greater robustness than NSGA-II algorithm to solve this problem. To conclude, it is important to mention that OCM approach can be easily adapted to run inside other kinds of plants, similar to HPP case study discussed here, as it is a generalist approach.

Acknowledgements. The authors would like to thank CEMIG and ORTENG, for the cession of operational data and support, CEMIG and ANEEL for the grant GT333/2011, CEFET-MG and INESC TEC for the infrastructure used for this project, and CAPES, CNPq and FAPEMIG for financial support.

References

1. Pezzini, P., Gomis-Bellmunta, O., Sudri-Andreua, A.: Optimization techniques to improve energy efficiency in power systems. Renewable and Sustainable Energy Reviews 15, 2028–2041 (2011)
2. Baños, R., et al.: Optimization methods applied to renewable and sustainable energy: a review. Renewable and Sustainable Energy Reviews **15**, 1753–1766 (2011)
3. Finardi, E., da Silva, E.L.: Unit commitment of single hydroelectric plant. Electric Power System Research **75**, 116–123 (2005)
4. Abrao, P. L., Wanner, E., Almeida, P.: A novel movable partitions approach with neural networks and evolutionary algorithms for solving the hydroelectric unit commitment problem. In: Proceeding of the 15h Annual Conference on GECCO, pp. 1205–1212 (2013)

5. Marcelino, C., Wanner, E., Almeida, P.: A novel mathematical modeling approach to the electric dispatch problem: case study using differential evolution algorithms. In: Proceedings of Conference: IEEE Congress on Evolutionary Computation (CEC), pp. 400–407 (2013)
6. Abido, M.A.: Multiobjective evolutionary algorithms for electric power dispatch problem. IEEE Transactions on Evolutionary Computation 10(3), 315–319 (2006)
7. Zhou, B., Chan, K.W., Yu, T., Chung, C.Y.: Equilibrium-inspired multiple group search optimization with synergistic learning for multiobjective electric power dispatch. IEEE Transactions on Power Systems 28(4), 3534–3545 (2013)
8. Fonseca, C. M., Fleming, P.J.: Genetic algorithms for multiobjective optimization: Formulation, discussion and generalization. In: Forrest, S., (ed.) Proceedings of the Fifth International Conference on Genetic Algorithms, San Mateo, California, pp. 416–423. Morgan Kaufmann (1993)
9. Horn, J., Nafpliotis, N., Goldberg, D.E.: A niched pareto genetic algorithm for multiobjective optimization. In: Proceedings of the First IEEE Conference on Evolutionary Computation, IEEE World Congress on Computational Computation, vol. 1, pp. 82–87. IEEE Press, Piscataway (1994)
10. Srinivas, N., Deb, K.: Multiobjective optimization using nondominated sorting in genetic algorithms. Evolutionary Computation 2(3), 221–248 (1994)
11. Deb, K., Agrawal, S., Pratap, A., Meyarivan, T.: A fast elitist nondominated sorting genetic algorithm for multi-objective optimization: Nsga-ii. In: Schoenauer, M., Deb, K., Rudolph, G., Yao, X., Lutton, E., Merelo, J.J., Schwefel, H.-P. (eds.) PPSN VI. LNCS, vol. 1917, pp. 849–858. Springer, Heidelberg (2000)
12. Zitzler, E.: Evolutionary algorithms for multiobjective optimization: Methods and applications. PhD thesis, Swiss Federal Institute of Technology (ETH), Zurich (1999)
13. Zitzler, E., Laumanns, L., Thiele, M.: Spea 2: Improving the strength pareto evolutionary algorithm. TIK-Report 103 (May 2001)
14. Storn, R., Price, K.: Differential evolution: a simple and efficient adaptative scheme for global optimization over continuous spaces. Techinical report TR-95-012, ICSI, Berkley (1995)
15. Carneiro, G., Chaves, J.: Pilot study to establish the flow of comfort for residential water consumption in the city of ponta grossa. In: 4th Meeting of the General Engineering and Technology Fields, Brazil (2008)
16. Montgomery, D., Runger, G.: Applied statistics and probability for engineers. 4th edn., Rio de Janeiro (2009)
17. Carrano, E.G., Wanner, E.F., Takarashi, R.: A multicriteria statistical based comparison methodology for evaluating evolutionary algorithms. IEEE Transactions on Evolutionary Computation 15, 848–870 (2010)

Solutions in Under 10 Seconds for Vehicle Routing Problems with Time Windows Using Commodity Computers

Pedro J.S. Cardoso[1]([⊠]), Gabriela Schütz[2], Andriy Mazayev[3],
and Emanuel Ey[4]

[1] Instituto Superior de Engenharia, LARSys, University of Algarve, Faro, Portugal
pcardoso@ualg.pt
[2] CEOT, Instituto Superior de Engenharia, University of Algarve, Faro, Portugal
gschutz@ualg.pt
[3] Departamento de Engenharia Eletrónica e Informática, University of Algarve,
Faro, Portugal
amazayev@ualg.pt
[4] Instituto Superior de Engenharia, University of Algarve, Faro, Portugal
eevieira@ualg.pt

Abstract. I3FR (Intelligent Fresh Food Fleet Router) is a project in development by the University of the Algarve and X4DEV, Business Solutions. One of the I3FR's main goals is to build a system for the optimization of the distribution of fresh goods using a fleet of vehicles. The most similar problem in literature is the well established and experimented Vehicle Routing Problem with Time Windows, which is by nature a multiple objective problems where the number of vehicles and the total traveled distance must be minimized.

In this paper we propose a hybrid variant of the Push Forward Insertion Heuristic with post optimizers, with the intention of achieving good solutions under 10 seconds using a commodity computer. We show that it is possible to obtain solution with "fair errors" when compared with the best solutions of a well known benchmark.

Keywords: VRPTW · Hybrid PFIH · Commodity computers

1 Introduction

I3FR (Intelligent Fresh Food Fleet Router) is a project in development by the University of the Algarve and X4DEV, Business Solutions. One of the I3FR's main goals is to build a system for the optimization of the distribution of fresh goods. The system will be integrated with an existing ERP, namely the SAGE ERP X3 [1], and will compute routes between the delivery points using cartography informations.

The most similar problem to the one to be solved is the Vehicle Routing Problem with Time Windows (VRPTW), common to the majority of the distribution fleets. In its simplified form, the VRPTW can be stated as the problem

A. Gaspar-Cunha et al. (Eds.): EMO 2015, Part II, LNCS 9019, pp. 418–432, 2015.
DOI: 10.1007/978-3-319-15892-1_28

of designing an optimum set of routes that deliver a set of goods, to a set of costumers, within predefined time windows.

In more detail, the objective is to compute routes from one or more depots that visit each costumer once, within given time intervals, without violating the vehicles capacities. Several approaches were made in the past regarding the VRPTW. While there are methods which can solve instances of the Travelling Salesman Problem (a subproblem of the VRP) with a few thousand nodes, the VRP shows to be much harder to be solved exactly.

The problem is intrinsically a multi-objective problem. As expectable the multi-objective optimization instances of the VRP are even harder than the single objective. This combinatorial problems, as well as the majority of the multi-objective optimization problems (see for example the Multiple Objective Minimum Spanning Trees problem case in [3,4,13]) are classified as NP-complete and even, in some cases, NP-# [9,12]. Academic papers and the majority of the know libraries of problems, consider firstly the minimization of the necessary number of routes, followed by the total travel distance. In [6] a study is conducted taking into consideration conflicting relationships between 5 objectives: number of vehicles, total travel distance, makespan, total waiting time, and total delay time. The same work addresses real scenarios, where consideration like the traveled distance vs. traveled time are made, unlike the common datasets which consider that an unit of distance (generally computed as Euclidean) always corresponds to a time unit (e.g., [23]). Other objectives and constraints arise in real problems. For instance, in the distribution of frozen, refrigerated and fresh goods it may be important to minimize the distribution time (correlated with the traveled distance) since maintaining the temperature inside the refrigerated compartments ads significant costs, associated with the consumption of fuel to keep the dedicated freezing engines working. In this sense, a method that minimizes not only the fixed costs for dispatching vehicles, but also the transportation, inventory, energy and penalty costs for violating time-windows is presented in [14]. The same work discusses the time-dependent travel and time-varying temperatures, during the day, which led to the modification of the objective functions as well as the constraints. In [7] a mathematical model is presented which combines production scheduling and vehicle routing with time windows for perishable food products. The objective of the model is to maximize the expected total profit of the supplier by optimizing the optimal production quantities, the time to start producing and the vehicle routes. In [10] the authors develop the MIXALG method as a way of solving routing problems with moderated size. The good behavior of the method was verified by its application to a real logistic problem.

The use of heuristics and meta-heuristics is therefore a common solution to solve problems of the VRP class (namely in the fresh goods deliveries). The problem in study has an extra constraint: the solution should be achieved in near real time since costumers tend to send their orders very near to the loading time of the vehicles. In some extreme cases the orders are even made after the beginning of the vehicle's loading, which on those cases will not be unloaded. These last minute orders are then engaged in an existing route, or into a new one if they are not engageable without violating feasibility (e.g., time windows or vehicles capacity).

Recent solutions to solve the problem include the use of meta-heuristics such as Genetic Algorithms, Ant Colony Optimization algorithm, among others [2, 8, 11, 16, 18, 19, 24]. However, the time required by these methods is a major problem when we take our premises into consideration, namely that good solutions should be achieved fast. Besides, many of these meta-heuristics do not scale well.

In this paper we present a solution which integrates an ERP with an optimization module communicating through a web services. Data comes from a local database. To solve the problem in under 10 seconds, we propose an adapted Push Foward Insertion Heuristic with extra operators. In those operators we include a seed procedure to start building the routes, the well known 2-Opt and cross routes operators, and we adapt the radial ejection operator to a band operator. The results show that for the majority of cases, good solutions (in some cases, the best known in literature) are achievable in less than 10 seconds using a commodity computer.

The paper is structured as follows. Section 2 presents the formulation of the problem. The proposed variations to the PFIH method are presented in Section 3. Results, conclusions and future work are presented in the last two sections.

2   Problem Formulation

A solution of an instance of the VRPTW is a set of routes which serve a set of costumers, using a set of vehicles, within costumers and depots time windows, and without exceeding the vehicles capacities.

Mathematically, we have considered an instance of the VRPTW composed by a set of locations L (partitionable in two sets: costumers, C, and depots, D), and distances $d(i, j) \in \mathbb{R}$ and times $t(i, j) \in \mathbb{N}$ between each pair of locations $i, j \in L$. Furthermore, each vehicle has a capacity q, and each customer, i, has associated a time window $[a_i, b_i]$ (time interval in which the costumers are available or wish to receive their orders), a service time s_i and a demand d_i (volume), for $i \in C$.

A solution of the VRPTW is composed by set of m routes, $T_i = (l_{i,1}, l_{i,2}, l_{i,3}, \ldots, l_{i,m_i})$ with $i \in \{1, 2, \ldots, m\}$, such that $l_{i,1} = l_{i,m_i} \in D$ is a depot (i.e., a route starts and ends at the same depot), $l_{i,2}, l_{i,3}, \ldots, l_{i,m_i-1} \in C$ are costumers, and a costumer is served by a single route ($\{l_{i,2}, l_{i,3}, \ldots, l_{i,m_i-1}\} \cap \{l_{j,2}, l_{j,3}, \ldots, l_{j,m_j-1}\} = \emptyset, i \neq j$).

The problem also has a set of restrictions. The feasible routes take into consideration the travel time between costumers and the associated service time, such that (1) each vehicle leaves and arrives at the depots in their time windows, and (2) reaches the customers also within their time windows (or before the time window opening, forced in this case to wait until the opening moment). In this formulation it is assumed that the vehicles are allowed to wait at the delivery site and the wait has no cost, except for time. Another restriction states that

(3) the capacity of the vehicle should not be exceeded, i.e.,

$$\sum_{l \in T_i - \{l_{i,1}, l_{i,m_i}\}} d_l \leq q, i \in \{1, 2, \ldots, m\}.$$

Taking the previous formulation into consideration several criteria can be optimized, e.g.: (1) Minimize the total number of vehicles required to achieve the service within the restrictions; (2) Minimize the total traveled distances (i.e., the sum of the distances made by each vehicle); (3) minimize the difference between the longest and shortest route; or (4) maximize the minimum load of the vehicles. The first two objectives are common to the majority of the datasets, as already mentioned. The third and fourth objective try to produce balanced routes, in terms of the workers effort and vehicles occupancies.

Particularly in this work, we will consider the first three objectives.

In the next section we will explore a set of heuristics keeping in mind the achievement of solutions in limited time.

3   Algorithmic Approaches

Is this section we will describe the methods used in this work. Starting with the definitions of methods already established in literature, we will propose some variations and improvements that were tested. The established methods give us an upper bound for our results.

3.1   Adapted Push Forward Insertion Heuristic

The Push Forward Insertion Heuristic (PFIH) is a greedy constructive heuristic [23,27] proposed by M. Solomon. The method has been implemented and tested by several authors [20,25,26,28]. In general, the PFIH tour-building procedure sequentially inserts costumers into the solution. Our option of implementation consists in the simultaneous construction of the routes. The overall steps are described in Algorithm 1.

The Algorithm's Step 4 requires the computation of the PFIH cost to set the order in which the costumers are inserted in the solution. The i-costumer's cost, $PFIHCost_i$, was defined as

$$PFIHCost_i = -\alpha d(i, o) + \beta b_i + \gamma \theta_i \frac{d(i, o)}{360} + \lambda (b_i - a_i)^{-1}, i \in C \qquad (1)$$

where $d(i, o)$ is the distance between costumer i and the depot, a_i, b_i are respectively the lower and upper limit of the i-costumer's time window, θ_i is the polar angle for costumer i (considering the origin at the depot), and α, β, γ and λ are parameters such that $\alpha + \beta + \gamma + \lambda = 1$. Different α, β, γ and λ values allow to give more or less importance to formula parcels, resulting in distinct orderings of the costumers. For instance, large values of α will prioritize the insertion of costumers near the depot. Larger values of β will make costumers with earlier

---

Algorithm 1. Adapted Push Forward Insertion Heuristic Algorithm

---

Step 1. Instantiate an empty set of routes, S, which will contain the final solution.

Step 2. Optionally, seed S by starting a predefined number of routes (see Section 3.2)

Step 3. If all costumers were placed in a route go to Step 8;

Step 4. For all non inserted customers compute their PFIH cost and choose the one with smallest value.

Step 5. If S is empty, start a new route (depot-customer-depot) and add it to S. Return to Step 3.

Step 6. Try to insert the costumer into an existing route in S, minimizing the traveled distance and taking into consideration the constraints (time windows and vehicle capacities). If the insertion of the costumers is impossible without violating the constraints, start a new route (depot-customer-depot) and add it to S.

Step 7. Update the distances, delivery times and vehicle capacities. Return to Step 3.

Step 8. Stop the procedure and return the built solution.

---

closing window preferable. Larger values of γ will make preferable the insertion of the costumers in a circular spanning mechanism (like a "radar"). Finally, larger values of λ will prefer costumers with smaller time windows to be inserted first. Using Eq. (1) it was possible to set the order in which the costumers were inserted in the building of the solutions.

Some considerations include the fact that this might be a good approach to insert late costumers demands, in a dynamic situation, and that it can be used to produce a set of relatively good solutions to be used in a meta-heuristic (e.g., initial population of a Genetic Algorithm).

3.2   Seeding the Solutions

Step 2 of Algorithm 1, optionally, starts a predefined number of routes. Since the PFIH is capable of inserting nodes in any position of the routes, provided feasibility, it is possible to create a partial solution *a priori*, i.e., before the execution of the algorithm. However, the creation of the partial solution shouldn't be randomly generated.

Given the customers' demands, $\sum_{i \in C} d_i$, and the vehicles capacity, q, it is possible to estimate the minimum required number of routes to solve the problem by making a simple math division. This estimation of the number of routes is a lower bound since time restrictions and possible temporal conflicts are not taken into account. The PFIH algorithm (Step 6 of Algorithm 1) will latter take those conflicts into account and solve them by starting new routes if necessary.

Once the number of initial routes is computed it is necessary to choose a set of costumers and assign them to the new routes, of the type (*depot, customer, depot*). Our proposal consists in consecutively choosing the farthest from all the routed customers to start a new route. On other words, the first customer to be given a new route is the one farthest away from the depot, $c_1 = \arg\max_{i \in C} d(depot, i)$. The second route will be started with the customer farther away from the depot and the customer in the first route, $c_2 = \arg\max_{i \in C - \{c_1\}} d(depot, i) + d(c_1, i)$.

Algorithm 2. Seeded partial solution

Require: Set of costumers (C), depot, distance function ($d(\cdot,\cdot)$), vehicle capacity (q) and costumers demands (d_i)
Ensure: Partial Solution, S
1: $S \leftarrow \emptyset$ ▷ Set of routes
2: $C' \leftarrow \{depot\}$ ▷ Set of "served" costumers
3: $numberOfRoutes \leftarrow \left\lceil \frac{\sum_{i \in C} d_i}{q} \right\rceil$
4: **for** $i \in \{1, 2, \ldots, numberOfRoutes\}$ **do**
5:      $farthestNode \leftarrow \arg\max_{i \notin C-C'} \sum_{j \in C'} d(j, i)$
6:      $C' \leftarrow C' \cup \{farthestNode\}$
7:      $S \leftarrow S \cup \{(depot, farthestNode, depot)\}$
8: **end for**

The $n - th$ customer to be given a route will be the one farthest away from all the previous "served" customers,

$$c_n = \arg\max_{i \notin C-C'} \left(d(depot, i) + \sum_{j \in C'} d(j, i) \right),$$

where $C' = \{c_1, c_2, \ldots, c_{n-1}\}$. The process is summarized in Algorithm 2.

Once the partial solution is built, the adapted PFIH will complete the solution by inserting the remaining customers, as each client is evaluated and inserted in a route where the insertion causes the minimum increment of the total distance and feasibility is satisfied. Figure 1 sketches on the left an instance with 101 nodes of the Solomon's benchmark [22,23] ("clustered" type), while on the right is shown the partial solution obtained with the previously described process.

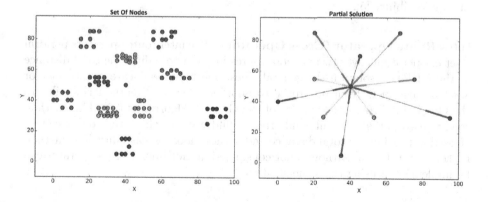

Fig. 1. Partial solution obtained using the seed method

A simple test set was developed to study the performance of the PFIH algorithm with and without a partial initial solution . The obtained results have

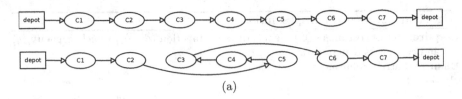

Fig. 2. 2-Opt post optimization operator

shown that the creation of a partial solution, especially in clustered cases as shown in Fig. 1, by the Algorithm 2 and the subsequent use of it in PFIH algorithm considerably improves the quality of the solution (see Section 4).

3.3   Post-PFIH Improvement Methods

Once the extended PFIH algorithm is executed, a feasible solution is produced which, in most cases, can be optimized by post-optimization methods. Due to the multi-objective nature of the problem the post-optimization operators (can) have different goals, namely to reduce the total distance, the number of routes, or the maximum time difference between routes. The remainder of this section will expose some of the tested operators.

Intra Route Operator (2-Opt). An intra route operator, as the name suggests, performs operations on a single route. One of the most commonly used operators from this set of methods is the 2-Opt operator (see Fig. 2). This operator iterates through all routes, one by one, and tries to rearrange the sequence by which the customers are visited in order to reduce the route distance, maintaining feasibility [5].

Inter Route Operator (Cross Operator). The inter route operators perform a set of operations over two or more routes in order to reduce the total distance of the VRPTW solution. In general, these methods reallocate a customer or a set of customers from one route to another. The used method is similar to the One Point Crossover operator of the Genetic Algorithms [17]. This method receives two paths as input, and tries to find a cross where the routes can be crossed improving the total distance and without loosing feasibility. The method is sketched in Fig. 3. Although not considered, a multi-point cross operator can be implemented using the same strategy.

Reduce Number of Routes Operators (Costumers Ejection). The previous operators are capable of diminishing the total distance, i.e., doing route optimization. However, they are not capable of reducing the number of routes present in the original solution. Adapted, they could also be used to reduce the maximum time difference between routes.

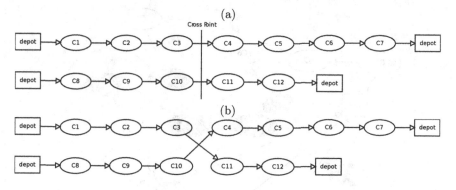

Fig. 3. Cross operator: (a) original solution and (b) returned solution

In order to reduce the number of routes, another set of methods, based on the ejection of customers, was implemented. The ejection of customers consists in their removal and possible reinsertion in other routes, without violating the problems' constraints. Several hypotheses arise on how to select the ejected customers. A naive approach would randomly select one or more routes, eject all customers from those routes and reinsert them using the previously described PFIH. Tests showed that the optimization rapidly reaches a local optimum.

The proposed method is a generalization of the radial ejection presented in [21]. The method selects a route and for each customer located in it, ejects a certain number of geographical neighbors. The ejection is based on the proximity and similarity of the nodes.

In more details, the first step is to choose a route to start the ejection. The route is chosen by a roulette method, inspired by the rank selection operator of the Genetic Algorithms [15]. To build our roulette it is necessary to sort the routes in an ascending manner, from the shortest to the longest in terms of the number of costumers. The next step is to give each route its own "slice" in the roulette. We consider that shorter routes (less costumers) are more "defective", so they will have a bigger probability of being selected by the roulette procedure. This means that costumers in those routes will be most likely to be chosen for ejection. The probabilities of ejection for the i-worst route, previously sorted by the number of costumers, is calculated according to expression

$$\frac{2i}{n(n+1)}$$

where n is the number of routes.

Once the route is chosen and the first set of costumers to be ejected is known, it becomes necessary to find the neighbors from other routes that will also be ejected. The second phase ejects the neighbor of the first ejected customers which are located in their proximity and have similar time windows. This phase takes an ejection rate which gives the number of customers to be ejected.

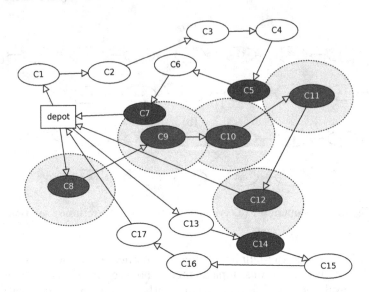

Fig. 4. Band ejection operator. In red the customers of the selected route, in light blue the red customers' "influence" area and in blue the customers ejected from other routes.

Figure 4 exemplifies the procedure. In this case, the route in dark red, $\pi = (depot, C8, C9, C10, C11, C12, depot)$, was the one chosen by the rank procedure. Next, costumer $C5$, $C7$ and $C14$ will be also ejected since they are in the geographical neighborhood of the customers of π and, for the sake of the example, we have considered that their time windows are "compatible" with the ones in π.

3.4   Overall Procedure

Algorithm 3 describes the overall procedure proposed in this document. The process starts with a seeded partial solution, S_{seed}. Next, for all the PFIH parameters (such that their sum equals 1) the seeded partial solution is completed and stored in a set of solutions, $PFIHSet$. The next steps are computed until a stopping criteria is met, in our case a maximum computation time is reached. The cycle start by getting (and removing) the most promising solution from $PFIHSet$ and setting the ejection rate value. A tabu list, T, is started which will contain all computed solutions before applying post-optimization, for each ejection rate. Then try at most $MaxTries$ times to improve the solution by applying the 2-Opt, the cross route and the band ejection operators to $S_{current}$. Reset the tries counter if the solution is improved or it is worst within a threshold. If the number of tries ($MaxTries$) is reached or the obtained solution was already post-optimized in a previous cycle then change the ejection rate, reset the tabu list and repeat the improvement procedure. When all ejection rate values where tested, get a new solution for *PFIHSet* repeat the above post-optimization process.

Algorithm 3. Overall PFIH procedure with seeded solution and band ejection

---

Require: Instance of the VRPTW, PFIH parameters (ejection rate, α, β, γ and λ) ranges, *MaxTries*

Ensure: Set of solutions

1: Generate a seeded partial solution, S_{seed} ▷ Section 3.2
2: $PFIHSet \leftarrow \emptyset$
3: **for all** Combination of the PFIH parameters such that $\alpha + \beta + \gamma + \lambda = 1$ **do**
4:     Set the PFIH parameters
5:     Run the PFIH starting from S_{seed} to obtain a solution, $S_{current}$ ▷ Section 3.1
6:     $PFIHSet \leftarrow PFIHSet \cup \{S_{current}\}$
7: **end for**
8: Sort *PFIHSet* by the number of routes followed by distance and maximum time difference between routes
9: **while** stopping criteria not met **do**
10:     $S_{current} \leftarrow pop(PFIHSet)$ ▷ Get the most promising solution from *PFIHSet* and remove it
11:     Set the ejection rate   ▷ *With its minimum value*
12:     **repeat**
13:         $NoSuccess \leftarrow 0$
14:         Set a tabu list, $T = \emptyset$ ▷ *will contain all computed solutions before applying post-optimization, for each ejection rate*
15:         **while** $NoSuccess < MaxTries$ and $S_{current} \notin T$ **do**
16:             $S_{backup} \leftarrow S_{current}$
17:             Update the tabu list, $T = T \cup \{S_{current}\}$
18:             Apply 2-Opt, cross route and band ejection operators to $S_{current}$ ▷ Section 3.3
19:             **if** $S_{current}$ improves S_{backup} or is not worst within a threshold **then**
20:                 $NoSucces \leftarrow 0$
21:             **else**
22:                 $NoSuccess \leftarrow NoSucces + 1$
23:             **end if**
24:         **end while**
25:         Change ejection rate   ▷ *Increase its value*
26:     **until** all ejection rates values where tested
27: **end while**

---

4 Experimental Results

This section presents a compilation of the achieved results when experimenting the above methods. All tests were run on a PC containing an Intel i7-4770 processor, with 16Gb of RAM and Kubuntu 14.04. As a base algorithm we have considered the PFIH with $\alpha, \beta, \gamma, \lambda \in \{0, 0.1, 0.2, \ldots, 1.0\}$. The set of instances of the VRPTW was taken from Solomon's benchmark [22,23], namely, instances with 100 nodes in the form of random (R), clustered (C) and random-clustered (RC) geometric distributions. Instances with numbering of the form 2xx represent the cases with vehicles with large capacity.

Table 2 summarizes the obtained results where the first two columns are the best known values for the number of vehicles (routes) and total distance, and the remaining columns are the results for seven variants of the algorithm, according to Table 1. The stopping criteria was a single run of the operators for the non random methods ($PFIH_0$, $PFIH_1$ and $PFIH_2$) and a maximum of 10 seconds for the remaining ones, which correspond to the ones with ejection. In this sense, as a final observation, the values in Table 2 are the mean values for 25 runs of the methods ($PFIH_3$, $PFIH_4$, $PFIH_5$ and $PFIH_6$). For this set of results, the problem was considered as a multi-objective problem but with preference for the number of routes, as is normal in majority of the VRPTW literature.

As we can see, there is a significant improvement in the majority of the cases when comparing the basic PFIH ($PFIH_0$) with the one with seeds ($PFIH_1$). Comparing the PFIH with seeds ($PFIH_1$) with PFIH with seeds and 2Opt/cross operators also shows general improvements. Finally, the use of the injection methods allowed us in some cases to reduce the number of routes and the total distance as desired, providing "fair" solutions given the computational time.

Table 1. Variants of the algorithm

	PFIH	Seeds	2-Opt and cross operators	Radial ejection on single node	Radial ejection on multiple nodes	Threshold
$PFIH_0$	✓					
$PFIH_1$	✓	✓				
$PFIH_2$	✓	✓	✓			
$PFIH_3$	✓	✓	✓	✓		0
$PFIH_4$	✓	✓	✓	✓		0.01
$PFIH_5$	✓	✓	✓		✓	0
$PFIH_6$	✓	✓	✓		✓	0.01

Some tests were also made considering the three mentioned objectives, namely: number of vehicles, total traveled distances and the length difference between the longest and shortest route. Figure 5 shows three clouds of solutions (in the objective space) obtained with $PFIH_6$, for the (a) C102, (b) R102, and (c) RC102 instances of the Solomon's benchmark. In red the non-dominated solutions and in dark the dominated ones.

Table 2. Results obtained with the proposed methods for instances (with 100 nodes) of the Solomon's benchmark

	Best known		$PFIH_0$		$PFIH_1$		$PFIH_2$		$PFIH_3$		$PFIH_4$		$PFIH_5$		$PFIH_6$	
Instance	V	D	$\Delta v\%$	$\Delta_D\%$	$\Delta v\%$	$\Delta_D\%$	$\Delta v\%$	$\Delta_D\%$	$\Delta v\%$	$\Delta_D\%$	$\Delta v\%$	$\Delta_D\%$	$\Delta v\%$	$\Delta_D\%$	$\Delta v\%$	$\Delta_D\%$
C101	10	828,9	0,0	0,0	0,0	0,0	0,0	0,0	0,0	0,0	0,0	0,0	0,0	0,0	0,0	0,0
C102	10	828,9	0,0	28,9	0,0	2,0	0,0	0,3	0,0	0,0	0,0	0,0	0,0	0,3	0,0	0,4
C103	10	828,1	0,0	57,4	0,0	4,0	0,0	1,8	0,0	0,0	0,0	0,0	0,0	2,8	0,0	2,9
C104	10	824,8	0,0	78,3	0,0	1,0	0,0	2,7	0,0	0,1	0,0	0,4	0,0	1,0	0,0	1,0
C105	10	828,9	10,0	11,3	0,0	0,0	0,0	0,0	0,0	0,0	0,0	0,0	0,0	0,0	0,0	0,0
C106	10	828,9	10,0	35,2	0,0	0,0	0,0	0,0	0,0	0,0	0,0	0,0	0,0	0,0	0,0	0,0
C107	10	828,9	0,0	9,3	0,0	0,0	0,0	0,0	0,0	0,0	0,0	0,0	0,0	0,0	0,0	0,0
C108	10	828,9	10,0	59,3	0,0	4,1	0,0	0,0	0,0	0,0	0,0	0,2	0,0	4,1	0,0	4,1
C109	10	828,9	0,0	42,9	0,0	0,4	0,0	0,4	0,0	0,0	0,0	0,0	0,0	0,4	0,0	0,4
C201	3	591,6	0,0	0,5	0,0	0,0	0,0	0,0	0,0	0,0	0,0	0,0	0,0	0,0	0,0	0,0
C202	3	591,6	0,0	0,0	0,0	0,0	0,0	14,3	0,0	0,0	0,0	0,0	0,0	0,0	0,0	0,0
C203	3	591,2	0,0	29,8	0,0	4,1	0,0	1,5	0,0	0,0	0,0	0,0	0,0	3,9	0,0	4,0
C204	3	590,6	33,3	91,8	0,0	7,9	0,0	5,0	0,0	1,1	0,0	0,8	0,0	7,9	0,0	7,8
C205	3	588,9	0,0	7,8	33,3	11,7	0,0	3,5	0,0	0,0	0,0	0,0	0,0	0,0	0,0	0,0
C206	3	588,5	0,0	45,6	0,0	26,0	0,0	0,4	0,0	0,0	0,0	0,7	0,0	0,3	0,0	0,4
C207	3	588,3	0,0	13,4	0,0	2,1	0,0	0,0	0,0	0,0	0,0	0,0	0,0	0,7	0,0	0,6
C208	3	588,3	0,0	17,6	0,0	2,6	0,0	1,1	0,0	0,0	0,0	0,0	0,0	0,8	0,0	0,8
R101	19	1650,8	5,3	5,6	5,3	11,2	5,3	4,4	0,0	1,0	0,0	1,8	0,0	1,0	0,0	1,1
R102	17	1486,1	5,9	1,7	5,9	3,3	5,9	2,7	4,7	-0,2	4,9	0,6	5,9	-0,7	5,2	-0,5
R103	13	1292,7	7,7	-1,2	7,7	-0,9	0,0	4,6	7,7	-5,5	7,7	-5,3	7,7	-5,4	7,7	-5,0
R104	9	1007,3	22,2	27,8	11,1	19,0	11,1	9,7	11,1	3,1	11,1	-0,7	11,1	4,2	11,1	2,1
R105	14	1377,1	7,1	15,1	7,1	10,3	7,1	8,4	7,1	0,6	5,4	1,8	7,1	1,6	5,4	1,4
R106	12	1252,0	8,3	15,9	8,3	9,8	8,3	3,8	8,3	2,3	6,7	1,7	8,3	3,6	6,0	3,0
R107	10	1104,7	10,0	16,5	10,0	7,9	10,0	3,7	10,0	-1,9	10,0	-2,0	10,0	2,0	10,0	-0,2
R108	9	960,9	22,2	45,8	11,1	16,4	11,1	13,6	11,1	2,0	11,1	3,3	11,1	6,6	11,1	5,1
R109	11	1194,7	27,3	24,4	18,2	31,2	9,1	20,3	16,4	0,1	9,1	-1,3	11,3	2,6	10,2	1,9
R110	10	1118,8	30,0	49,5	20,0	18,6	20,0	7,4	20,0	1,0	14,4	-0,5	18,8	2,4	14,8	2,5
R111	10	1096,7	20,0	32,3	10,0	12,7	10,0	7,0	10,0	0,6	10,0	-1,4	10,0	2,5	10,0	0,4
R112	9	982,1	44,4	67,2	22,2	21,5	11,1	15,9	15,1	3,2	11,1	0,6	12,9	6,9	11,1	6,5
R201	4	1252,4	0,0	26,5	0,0	31,1	0,0	15,4	0,0	5,6	0,0	3,5	0,0	8,5	0,0	7,6
R202	3	1191,7	33,3	10,7	33,3	9,1	33,3	2,7	33,3	-7,9	33,3	-7,7	33,3	-0,6	33,3	-1,0
R203	3	939,5	0,0	15,7	0,0	14,7	0,0	6,0	0,0	3,5	0,0	0,8	0,0	11,5	0,0	11,8
R204	2	825,5	50,0	7,7	50,0	3,8	0,0	7,9	50,0	-4,7	50,0	-7,7	50,0	0,6	50,0	2,8
R205	3	994,4	0,0	45,3	0,0	34,9	0,0	21,1	0,0	13,2	0,0	5,7	0,0	22,7	0,0	21,3
R206	3	906,1	0,0	25,6	0,0	30,9	0,0	22,6	0,0	6,8	0,0	3,4	0,0	17,8	0,0	19,5
R207	2	890,6	50,0	34,9	50,0	23,9	50,0	13,4	50,0	-0,8	50,0	-2,4	50,0	8,4	50,0	8,6
R208	2	726,8	0,0	22,3	50,0	26,3	0,0	11,8	50,0	0,5	46,0	-0,5	50,0	11,3	44,0	12,7
R209	3	909,2	0,0	56,0	0,0	45,8	0,0	26,6	0,0	7,1	0,0	1,8	0,0	23,7	0,0	21,1
R210	3	939,4	0,0	25,4	0,0	27,1	0,0	20,9	0,0	4,1	0,0	5,0	0,0	18,9	0,0	17,3
R211	2	885,7	50,0	26,9	50,0	15,4	50,0	10,6	50,0	-3,1	50,0	-9,1	50,0	5,6	50,0	7,9
RC101	14	1696,9	21,4	9,8	14,3	7,9	14,3	2,6	14,3	-1,4	8,3	-1,2	14,3	-0,3	9,7	0,1
RC102	12	1554,8	16,7	14,2	16,7	11,9	16,7	2,6	16,7	-3,2	14,3	-3,8	16,7	-1,9	16,0	-2,7
RC103	11	1261,7	18,2	25,6	0,0	9,6	0,0	6,1	0,0	2,1	0,0	2,3	0,0	4,0	0,0	3,1
RC104	10	1135,5	20,0	35,0	0,0	9,2	0,0	7,0	0,0	5,7	0,0	5,3	0,0	6,6	0,0	6,0
RC105	13	1629,4	15,4	7,0	7,7	5,5	7,7	-1,2	7,7	-0,9	7,7	-0,2	7,7	-0,6	7,7	-1,4
RC106	11	1424,7	27,3	28,2	18,2	14,9	18,2	9,7	18,2	0,1	13,5	-0,9	17,8	1,2	18,2	0,6
RC107	11	1230,5	27,3	47,9	9,1	12,4	9,1	8,5	9,1	1,6	8,7	1,1	9,1	2,4	8,0	2,8
RC108	10	1139,8	40,0	55,2	10,0	5,6	10,0	7,9	10,0	0,3	10,0	0,9	10,0	4,2	10,0	4,6
RC201	4	1406,9	0,0	28,3	25,0	21,7	0,0	19,6	25,0	-2,2	25,0	-3,3	5,0	9,9	7,0	10,0
RC202	3	1365,7	33,3	18,4	33,3	14,4	33,3	6,4	33,3	-8,4	33,3	-9,6	33,3	-1,9	33,3	-2,9
RC203	3	1049,6	0,0	23,7	0,0	31,8	0,0	21,0	0,0	6,4	0,0	8,1	0,0	15,3	0,0	15,6
RC204	3	798,5	0,0	35,9	0,0	29,9	0,0	19,6	0,0	5,7	0,0	4,4	0,0	11,7	0,0	12,6
RC205	4	1297,7	0,0	29,1	0,0	19,6	0,0	17,5	0,0	4,9	0,0	4,3	0,0	9,6	0,0	9,1
RC206	3	1146,3	33,3	41,9	33,3	38,8	33,3	32,0	33,3	-2,7	33,3	-3,7	30,7	11,5	26,7	11,1
RC207	3	1061,1	33,3	51,3	33,3	40,6	0,0	39,9	22,7	1,5	13,3	1,0	22,7	14,7	21,3	14,4
RC208	3	828,1	0,0	84,3	0,0	43,3	0,0	40,5	0,0	8,4	0,0	7,0	0,0	19,7	0,0	21,5

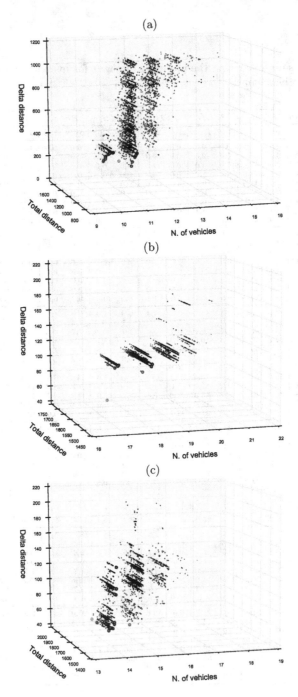

Fig. 5. Examples of the clouds of solutions (in the objective space) obtained with $PFIH_6$ for instances (a) C102, (b) R102, and (c) RC102 of the Solomon's benchmark. In red the non-dominated solutions and in dark the dominated ones.

5   Conclusions

This work, part of the I3FR project, proposes a set of algorithms to be integrated as an optimization module for an ERP system. The present ERP system is used to plan the delivery of fresh goods to customers within time windows and other constraints. The problem is dynamic as last minute order occur in a common basis. To guarantee responses within the strict time constraints, we adopted a Push Forward Insertion Heuristic combined with pre and post-optimizers, namely the a seed partial solution generator, and the 2-Opt, the cross and the band ejection operators. The last operator is a proposal of adaption of the radial ejection operator. The results shown that the proposed algorithms achieved acceptable solutions in less than 10 seconds while using a commodity computer, when compared with the ones in a known benchmark.

Acknowledgments. This work was partly supported by project i3FR: Intelligent Fresh Food Fleet Router – QREN I&DT, n. 34130, POPH, FEDER, the Portuguese Foundation for Science and Technology (FCT), project LARSyS PEstOE/EEI/LA0009-/2013. We also thanks to project leader X4DEV, Business Solutions

References

1. Sage, ERP X3. http://www.sageerpx3.com/ (accessed september 28, 2014)
2. Bräysy, O., Gendreau, M.: Tabu search heuristics for the vehicle routing problem with time windows. Top **10**(2), 211–237 (2002)
3. Camerini, P., Galbiati, G., Maffioli, F.: Complexity of spanning tree problems: Part I. European Journal of Operational Research **5**, 346–352 (1980)
4. Camerini, P., Galbiati, G., Maffioli, F.: On the complexity of finding multi-constrained spanning trees. Discrete Applied Mathematics **5**, 39–50 (1983)
5. Carić, T., Galić, A., Fosin, J., Gold, H., Reinholz, A.: A modelling and optimization framework for real-world vehicle routing problems. Vehicle Routing Problem **15** (2008)
6. Castro-Gutierrez, J., Landa-Silva, D., Moreno Perez, J.: Nature of real-world multi-objective vehicle routing with evolutionary algorithms. In: 2011 IEEE International Conference on Systems, Man, and Cybernetics (SMC), pp. 257–264. IEEE (2011)
7. Chen, H.K., Hsueh, C.F., Chang, M.S.: Production scheduling and vehicle routing with time windows for perishable food products. Computers & Operations Research **36**(7), 2311–2319 (2009)
8. Chiang, W.C., Russell, R.A.: Simulated annealing metaheuristics for the vehicle routing problem with time windows. Annals of Operations Research **63**(1), 3–27 (1996)
9. Ehrgott, M., Gandibleux, X.: Approximative solution methods for multiobjective combinatorial optimization. TOP **12**(1), 1–63 (2004)
10. Faulin, J.: Applying MIXALG procedure in a routing problem to optimize food product delivery. Omega **31**(5), 387–395 (2003)
11. Gambardella, L.M., Taillard, É., Agazzi, G.: Macs-vrptw: A multiple colony system for vehicle routing problems with time windows. In: New Ideas in Optimization, Citeseer (1999)

12. Garey, M., Johnson, D.: Computers and Intractability. A Guide to the Theory of NP-Completeness. W. Freeman & Co., New York (1990)

13. Hamacher, H., Ruhe, G.: On spanning tree problems with multiple objective. Annals of operation research **52**, 209–230 (1994)

14. Hsu, C.I., Hung, S.F., Li, H.C.: Vehicle routing problem with time-windows for perishable food delivery. Journal of Food Engineering **80**(2), 465–475 (2007)

15. Jebari, K., Madiafi, M.: Selection methods for genetic algorithms. International Journal of Emerging Sciences 3(4) (2013)

16. Lin, S.W., Yu, V.F., Lu, C.C.: A simulated annealing heuristic for the truck and trailer routing problem with time windows. Expert Systems with Applications **38**(12), 15244–15252 (2011)

17. Magalhães-Mendes, J.: A comparative study of crossover operators for genetic algorithms to solve the job shop scheduling problem. WSEAS Transactions on Computers 12(4), 164–173 (2013)

18. Moura, A.: A multi-objective genetic algorithm for the vehicle routing with time windows and loading problem. In: Intelligent Decision Support, pp. 187–201. Springer (2008)

19. Ombuki, B., Ross, B.J., Hanshar, F.: Multi-objective genetic algorithms for vehicle routing problem with time windows. Applied Intelligence **24**(1), 4–5 (2006)

20. Russell, R.A.: Hybrid heuristics for the vehicle routing problem with time windows. Transportation science **29**(2), 156–166 (1995)

21. Schrimpf, G., Schneider, J., Stamm-Wilbrandt, H., Dueck, G.: Record breaking optimization results using the ruin and recreate principle. Journal of Computational Physics **159**(2), 139–171 (2000)

22. Solomon, M.M.: Solomon's VRPTW benchmark. http://w.cba.neu.edu/~msolo~mon/problems.htm (accessed september 28, 2014)

23. Solomon, M.M.: Algorithms for the vehicle routing and scheduling problems with time window constraints. Operations Research **35**(2), 254–265 (1987)

24. Tam, V., Ma, K.: An effective search framework combining meta-heuristics to solve the vehicle routing problems with time windows. Vehicle Routing Problem 35 (2008)

25. Tan, K., Lee, L., Ou, K.: Artificial intelligence heuristics in solving vehicle routing problems with time window constraints. Engineering Applications of Artificial Intelligence **14**(6), 825–837 (2001)

26. Thangiah, S.R.: A hybrid genetic algorithms, simulated annealing and tabu search heuristic for vehicle routing problems with time windows. Practical Handbook of Genetic Algorithms **3**, 347–381 (1999)

27. Thangiah, S.R., Osman, I.H., Sun, T.: Hybrid genetic algorithm, simulated annealing and tabu search methods for vehicle routing problems with time windows. Computer Science Department, Slippery Rock University, Technical Report SRU CpSc-TR-94-27 69 (1994)

28. Thangiah, S.R., Osman, I.H., Vinayagamoorthy, R., Sun, T.: Algorithms for the vehicle routing problems with time deadlines. American Journal of Mathematical and Management Sciences **13**(3–4), 323–355 (1993)

A Comparative Study of Algorithms for Solving the Multiobjective Open-Pit Mining Operational Planning Problems

Rafael Frederico Alexandre[1,2,3]([⊠]), Felipe Campelo[1],
Carlos M. Fonseca[4],
and João Antônio de Vasconcelos[1,2]

[1] Graduate Program in Electrical Engineering, Federal University of Minas Gerais,
Av. Antônio Carlos, 6627, Pampulha, Belo Horizonte, MG 31270-901, Brazil
rfalexandre@decea.ufop.br
[2] Evolutionary Computation Laboratory, Federal University of Minas Gerais,
Av. Antônio Carlos, 6627, Pampulha, Belo Horizonte, MG 31270-901, Brazil
[3] Department of Computer and Systems, Federal University of Ouro Preto,
Rua 37, No 115, Loanda, João Monlevade, MG 35931-008, Brazil
[4] Department of Informatics Engineering,
University of Coimbra Poló II, Pinhal de Marrocos, 3030-290 Coimbra, Portugal

Abstract. This work presents a comparison of results obtained by different methods for the Multiobjective Open-Pit Mining Operational Planning Problem, which consists of dynamically and efficiently allocating a fleet of trucks with the goal of maximizing the production while reducing the number of trucks in operation, subject to a set of constraints defined by a mathematical model. Three algorithms were used to tackle instances of this problem: NSGA-II, SPEA2 and an ILS-based multiobjective optimizer called MILS. An expert system for computational simulation of open pit mines was employed for evaluating solutions generated by the algorithms. These methods were compared in terms of the quality of the solution sets returned, measured in terms of hypervolume and empirical attainment function (EAF). The results are presented and discussed.

Keywords: Open pit mines · Dispatch · Multiobjective optimization · Performance comparison

1 Introduction

The efficient use of available resources by companies is a requirement in any highly competitive market. For mining companies, using the fleet of trucks and shovels in the best possible way can enable a significant reduction in operational costs and a considerable improvement in productivity. According to Nel et al. [21] the cost of operating trucks and shovels in a open pit mine corresponds to between 50 to 60 percent of the total cost of operation. Moreover, trucks ranging from 100 to 240 tonnes of transport capacity usually cost from

© Springer International Publishing Switzerland 2015
A. Gaspar-Cunha et al. (Eds.): EMO 2015, Part II, LNCS 9019, pp. 433–447, 2015.
DOI: 10.1007/978-3-319-15892-1_29

$1.8 to $4.7 million dollars, respectively [8]. Therefore, investment in efficient usage of available equipments can result in significant reductions in the total costs of a mining operation.

The solution to the problem of truck dispatching in open pit mines consists basically of answering the question: where each truck should go after leaving each place? Any answer must be provided with the aim of satisfying the needs of the mine using the available resources in the best possible way. Thus, the answer to this question must consider issues such as what should be produced, what is the expected quality, travel time to the next location, and even possible queues that may occur on the way to a given destination. When requesting a new dispatch, a truck moves to a pit, which must have a shovel compatible with that particular truck. The material removed from each pit has a certain quality that is associated with the proportion of chemical elements such as Iron, Silicon, Manganese, among others. If there is no queue at the place of loading, the truck is loaded and moves up to a crusher. Each crusher has quality requirements that the material produced must meet. Material that has no commercial value (that is, waste) is conducted by the trucks to mine sites reserved for storage of this type of material (rock piles).

The objectives of this work are twofold: first, to present a multiobjective model that defines, for a given fleet of trucks, a sequence of dispatches for the efficient use of equipment, minimizing the occurrence of queues and idle shovels. The proposed multiobjective model for the open-pit mining operational planning problem (OPMOPP) additionally includes the modeling of possible queues for truck loading operations as well as different speeds for loaded and empty trucks. The second objective is to propose and compare the performance of three approaches for the solution of the proposed model: two multiobjective evolutionary algorithms (MOEAs) and a metaheuristic based on the Pareto Iterated Local Search (PILS). A specific solution encoding and operators for generating candidate solutions are proposed for the evolutionary approaches, in order to generate feasible solutions given the operational constraints of the problem, therefore enabling a more effective search for the solution of this class of problems. The algorithms are compared using standard quality indicators: hypervolume and empirical attainment function (EAF).

2   Previous Works

The work of Doig and Kizil [8] studied the impact of the truck cycle time differences in mine productivity. The authors conclude in their work that the cycle time and the subutilization of the truck fleet impacts significantly on productivity in a mine, thus justifying the efficient use of available equipment. Additionally, roads in good condition for transportation were also found to be relevant. The work of Topal [26] asserts that proper planning of maintenance of trucks is essential to minimize its costs. That is, assuming availability of the entire fleet of trucks when looking for a solution may lead to oversensitive solutions, as units may be unavailable due to the preventive maintenance schedule. A case study of a large-scale gold mine showed a significant reduction (10%) of annual maintenance costs and more than 16% of overall reduction in maintenance costs over 10 years of operation, in comparison with the baseline spreadsheet used in operation [25].

Tan *et al.* [24] presented a procedure for obtaining the optimal number of trucks in operation at the mine and also to estimate the capacity of the fleet. For the simulation of mine the software *Arena* [17] was employed. The data used for the simulations were collected using a GPS system and used weekly average values as the reference. Souza *et al.* [22] proposed a solution to an open-pit mining planning problem with dynamic truck allocation. The objective considered in their work was the minimization of the number of trucks used in the mine, and determination of the extraction rate at each pit to fulfill production and quality goals. They developed a heuristic called GGVNS, which combines ideas from both the Greedy Randomized Adaptive Search Procedure (GRASP) [9] and General Variable Neighborhood Search (GVNS) [19]. The GGVNS was successfully applied to solve the 8 distinct testing scenarios, with results validated using the commercial optimization software CPLEX [16]. More recent work presented three heuristics to solve the same problem, considering a multiobjective approach. Moreover, the work does not consider a possible queue to load and unload the trucks and also does not define the order of dispatches [3].

Subtil *et al.* [23] proposed a multi-stage approach for dynamic allocation of trucks in real environments for open pit mines. The proposed approach was validated through a simulation model based on discrete events. The authors report significant results using the algorithm, yielding increased production and also reduced operational delays of equipments. The work also states that, although the model is able to predict ore quality, this ability was not studied due to lack of relevant data for analysis. He *et al.* [14] sought to reduce the number of vehicles used in a mine by minimizing transportation costs and maintenance using GAs. Although satisfactory results were achieved, the model employed does not consider multiple constraints (compatibility between vehicles, production equipment and shovels, among others) found in dispatching problems in mines.

Given the many works in the literature, one realizes that they each have a different mathematical model and treat different objectives using techniques such as weighted sum of funcions or goal programming. None of these works directly address the multiobjective nature of the problem by using multicriteria optimization techniques. Moreover, a large portion of these works aims at optimizing functions related to production, but fail to consider the quality of material produced or even operational constraints such as compatibility between shovels and trucks. In the next section we propose a multiobjective model to address these issues.

3 The Multiobjective Open-Pit Mining Operational Planning Problem

This section presents a new multiobjective mathematical model that includes two objectives: the first one is to maximize production at the mine, be it ore or waste rock. The second one is to minimize the number of trucks in operation. However, as there are trucks with different capacities it is necessary to take their size into consideration. To facilitate understanding of the model, the parameters and variables are first presented. The parameters are defined by the test instances discussed in section 6.1. Let the parameters be:

- C is the set of crushers;
- O is the set of active ore pits;
- P is the set of pits formed by $O \bigcup W$;
- Q is the set of chemical elements of the ore;
- S is the set of shovels;
- T represents the set of trucks available;
- W is the set of active waste pits;
- Cap_t is the payload (in tonnes) of the truck t;
- f_{ts} is a flag variable. 1, if truck t is compatible with shovel s and 0, otherwise.
- $Limc_p$ is the number of shovels that can be allocated to pit p;
- Ql_{qc} is the lower limit of the amount of concentration (in percent) of the qth chemical element to the crusher c;
- Qu_{qc} is the upper limit of the amount of concentration (in percent) of the qth chemical element to the crusher c;
- q_{qo} is the content of chemical concentration (in percent) of the element q in the oth pits of the ore;
- $y_{sp} \in \{0,1\}$ is a flag variable. 1, if shovel s operates in pit p and 0, otherwise;
- y_{tp} is a flag variable. 1, if truck t can operate in pit p and 0, otherwise;

Let the variables be:

- $\bar{v} \in \{0,1\}^{|T|}$ is the vector of optimization variables responsible for representing the availability of the trucks, with the tth position of the vector (v_t) indicating whether the truck is in operation ($v_t = 1$) or not ($v_t = 0$);
- $\widetilde{M}$ defines the sequence of dispatches received for each truck in operation inside the mine;
- x_o is the production (in tonnes) of the ore pit o;
- x_w is the production (in tonnes) of the waste pit w;
- x_{oc} is the production of the oth ore pit, crusher c (in tonnes).

Next, the Eqs. (1)-(9) present the mathematical model for the problem under consideration. It is important to highlight at this point that x_o, x_w, and x_{oc} are calculated as a function of optimization variables $\bar{v}$ and $\widetilde{M}$. The objective functions are given as:

$$\text{Maximize:} \quad \sum_{\forall o \in O} x_o(\bar{v}, \widetilde{M}) + \sum_{\forall w \in W} x_w(\bar{v}, \widetilde{M}) \tag{1}$$

$$\text{Minimize:} \quad \sum_{\forall t \in T} \bar{v}_t \times Cap_t \tag{2}$$

subject to a number of operational constraints, that define key aspects of the operating environment of a mine:

$$\frac{\sum\limits_{\forall o \in O} q_{qo} x_{oc}(\overline{v}, \widetilde{M})}{\sum\limits_{\forall o \in O} x_{oc}(\overline{v}, \widetilde{M})} \geq Ql_{qc}, \qquad \forall q \in Q; c \in C \qquad (3)$$

$$\frac{\sum\limits_{\forall o \in O} q_{qo} x_{oc}(\overline{v}, \widetilde{M})}{\sum\limits_{\forall o \in O} x_{oc}(\overline{v}, \widetilde{M})} \leq Qu_{qc}, \qquad \forall q \in Q; c \in C \qquad (4)$$

$$\sum\limits_{\forall s \in S} y_{sp} \leq Limc_p, \qquad \forall p \in P \qquad (5)$$

$$\sum\limits_{\forall p \in P} y_{sp} \leq 1, \qquad \forall c \in C \qquad (6)$$

$$y_{sp} + y_{tp} - 2f_{ts} = 0 \qquad (7)$$

$$|C|, |S|, |P|, |Q|, |T| > 0 \qquad (8)$$

$$Ql_{qc}, q_{qo}, h, u_t > 0, \qquad \forall q \in Q; c \in C; t \in T; o \in O \qquad (9)$$

The optimization variables $\overline{v}$ and $\widetilde{M}$ are discussed in detail in section 4.1. The constraints of the model represent the *limits of chemical quality deviation* (3)–(4); the *shovel allocation constraints* (5)–(6); the *compatibility between shovel and trucks constraint* (7); and the*ensures that the variables are greater than zero* (8)–(9).

4   Multiobjective Evolutionary Algorithms

The optimization problem presented in the previous section can be solved using evolutionary algorithms. Evolutionary algorithms (EAs) [5] represent a family of metaheuristics that perform an adaptive iterative sampling of the design space by means of a population of candidate solutions. EAs generally work by iteratively updating the current population to create a new population by means of four main operators: selection, crossover, mutation and elite-preservation. Evolutionary methods can be easily designed or adapted to solve multiobjective problems, with or without constraints [7]. Moreover, these algorithms are easily adjusted to handle a diversity of problem domains, which allows for their straightforward adaptation to the multiobjective OPMOPP.

In this work two algorithms were adapted to solve the multiobjective OPMOPP: the NSGA-II [6] and the SPEA2 [27]. A detailed description of these two algorithms can be found in the references, and will not be provided here. In common with other EAs, successful multiobjective implementations require well-designed representation systems for individual problems and also genetic operators that are appropriate for the task. Recombination (crossover) operators can be particularly problematic. In the following sections a new representation to allow the dispatch fleet of trucks in a open pit mine is presented, together with operators to perform the crossover and mutation of candidate solutions coded according to this representation. Moreover, it is important to note that in this work the initial populations of all algorithms were randomly initialized. Additionally, binary tournament selection [5] was used in all cases.

4.1   Representation

The proposed codification initially builds a matrix $\widetilde{P}$ wherein each column j represents a location of the mine. For each location, a subset of the possible places to where a truck can be dispatched is defined. Therefore, each cell p_{ij} of the matrix indicates a possible target location for a truck that is in location p.

From the initial matrix $\widetilde{P}$, the candidate solutions can be created without the need for additional information from the mine, ensuring that the constraint (7) is satisfied. For the generation of individuals it is necessary to inform the value of k which aims to define the number of rows (i) of the matrix $\widetilde{M}$ of the solution $s = [\bar{v}|\widetilde{M}]$. The number of columns (j) is the same as in matrix $\widetilde{P}$. For each cell of column j a random place p in column j of the matrix $\widetilde{P}$ is chosen. The vector $\bar{v} \in \{0,1\}^{|T|}$ is randomly constructed, indicating whether the truck is in operation ($v_t = 1$) or not ($v_t = 0$). With this structure, for each request for a new order by a truck in operation the candidate solution informs the next destination for that truck, considering the location of the truck at the time of the request.

4.2   Crossover Operator

The crossover operators proposed for this representation are based on cutting operators, as discussed in several studies of the literature [12] [4]. Cutoff crossing (1PX) considers two candidate solutions x' and x'' represented by matrices $\widetilde{M}$ of dimension $I \times J$. An integer cutoff value $c \in [1, J]$ is randomly drawn from a discrete uniform random variable, and a new candidate solution y' is generated by combining the first c columns from x' and the final $J - c$ columns from x''. A second candidate solution y'' is also generated with the c first columns of x'' and the last $J - c$ columns from x', as is the case of the usual 1-point vector crossover employed in the EAs. Vector $\bar{v}$ uses binary crossover [5].

4.3   Mutation Operator

The mutation proposed for this representation is known as flip mutation [2]. In this case, each cell of the $\widetilde{M}$ of the solution s selected for mutation receives a new value obtained from the random matrix $\widetilde{P}$. This operator is applied, with a certain probability of occurrence p_m, to the candidate solutions generated by the crossover operator. For the vector $\overline{v}$, the bits are changed by turning individual trucks on or off.

5 Multiobjective Iterated Local Search

To provide a comparison baseline for the evolutionary approaches NSGA-II [6] and SPEA2 [27] using the operators defined in the previous section, and to evaluate the potential of the specific operators proposed for the multiobjective OPMOPP, we employ a method based on the Pareto Iterated Local Search (PILS) [11], which is an adaptation of the Iterated Local Search (ILS) [18] for multiobjective problems.

Algorithm 1. MULTIOBJECTIVE ITERATED LOCAL SEARCH (MILS)

    Input: $maxIter$
    Input: $maxCount$
    Output: $Front$
1 $Front \leftarrow makeInitialSolutions()$
2 $iter \leftarrow 1$
3 **while** $iter \leq maxIter$ **do**
4     $s' \leftarrow selection(Front)$
5     $labeled(s')$
6     $count \leftarrow 1$
7   **while** $count \leq maxCount$ **do**
8        $s'' \leftarrow perturbation(s')$
9        $s'' \leftarrow localSearch(s'')$
10       $inserted \leftarrow refresh(Front, s'')$
11   **if** $inserted$ **then**
12          $count \leftarrow 1$
13          $s' \leftarrow s''$
14   **else**
15          $count \leftarrow cont + 1$
16   **end**
17   **end**
18     $iter \leftarrow iter + 1$
19 **end**
20 **return** $Front$

The operation of the MILS is illustrated in Algorithm 1. It starts by generating an initial population and extracting the nondominated set, which gets stored in the $Front$ set (line 1). After this initial step, the iterative cycle is started. For $maxIter$ iterations, a solution from $Front$ is selected and the iteration of the main algorithm (lines 7-17) is executed. In this step, the procedures of perturbation (line 8) and local search (line 9), similar to those existing in

PILS, are executed. The front set is updated (line 10) with the refined solution obtained after local search. If the solution generated after the procedures of perturbation and local search is nondominated, it is inserted into the set $Front$, and the $count$ variable is reset (line 12). Otherwise, count is incremented by one (line 15). The $maxCount$ variable indicating the maximum number of times the solution is operated without inserting a non-dominated solution in the set front.

The procedure defined as perturbation (line 8) is responsible for generating the solutions known as neighbors. For the problem addressed in this work, the neighboring solutions are constructed as follows: two random integers p_1 and p_2 are generated such that $0 \leq p_1 \leq (J - j_d)$ and $p_2 = p_1 + j_d$, where J represents the number of columns of the matrix $\widetilde{M}$ and j_d is the number of columns to be changed. All values in the interval $[p_1, p_2]$ of the matrix $\widetilde{M}$ are changed, creating a new solution.

The other procedure used by Algorithm 1 is responsible for performing a local search (line 9) with the objective of exploring neighboring regions of the search space. To accomplish this task we use an algorithm known as reduced VNS (RVNS) [13]. The RVNS is a simplified version of the Variable Neighbourhood Search (VNS), where the deterministic local search procedure (the most time-consuming part of VNS) is removed in order to reduce the computational cost. This algorithm receives the solution to be perturbed and uses the mutation operator (line 3) defined in this work. If the solution changed (s'') dominates the current solution (s'), it is replaced (line 4-5) and the variable $iter$ is reset (line 6). The procedure for generating neighboring solutions is performed N times, where N is an input of the algorithm.

Algorithm 2. REDUCED VARIABLE NEIGHBOURHOOD SEARCH (RVNS)

    Input: s'
    Output: s'
1 $iter \leftarrow 1$
2 **while** $iter \leq N$ **do**
3      $s'' \leftarrow MakeNeighborhood(s')$
4   **if** $s'' \prec s'$ **then**
5         $s' \leftarrow s''$
6         $iter \leftarrow 1$
7   **else**
8         $iter \leftarrow iter + 1$
9   **end**
10 **end**
11 **return** s'

---

6 Experimental Setup

In this section we define the test problems and the experimental design of the computational experiments employed to verify the ability of the NSGA-II and

SPEA2 heuristics to obtain a good set of tradeoff solutions for the multiobjective OPMOPP. This experiment has essentially two goals: to evaluate whether any of the algorithms will be able to find feasible, interesting solution tradeoffs for the multiobjective OPMOPP instances considered, and to check whether the algorithms will yield significantly different performances.

First we describe the test scenarios employed and the configurations of the algorithms. Afterwards the performance metrics and experimental design are provided.

6.1 Test Problems

In this study we considered benchmark instances of problems based on those proposed by Souza et al. [22] [1]. Table 1 describes the main characteristics of the test instances. Columns **# Pits**, **# Shovels**, **# Trucks** and **# Par** indicate the number of pits, shovels, trucks and control parameters (chemical), respectively. The column **Details** provides the number and capacity (in case of trucks), or the productivity (in case of shovels). For example, the pair (15;50t) means there are 15 shovels (or trucks) of 50 tonnes of capacity (or maximum productivity). The difference between Mines 1 and 2 are the levels of quality of chemical elements.

Table 1. Test Instances

Instance	# Pits	Details		# Par
		(# Shovels,capacity)	(# Trucks,capacity)	
Mine1	8	(4,900t) (2,1000t) (2,1100t)	(15,56t) (15,90t)	10
Mine2	8	(4,900t) (2,1000t) (2,1100t)	(15,56t) (15,90t)	10
Mine3	7	(2,500t) (2,400t) (1,600t) (1,800t) (1,900t)	(30,56t)	5
Mine4	10	(2,400t) (2,500t) (1,600t) (1,800t) (1,900t) (3,1000t) (3,2600t)	(22,56t) (7,90t)	5

6.2 Evaluation of the Solutions

An expert simulation system, based on discrete events, was built to evaluate the solutions generated by the optimization algorithms. This system has an interface with these algorithms, in which candidate solutions are processed and returned by the simulator to the algorithms, including the values of the objectives and constraints. Dispatches for mining fronts consider the distance and the average speed of trucks to calculate the time required for the trucks reaching their destination. In addition, the simulator considers the possbilidade queue occur when loading trucks. The load time of each truck depends on the productivity of the shovels and truck capacity. The trucks are then dispatched to the crusher or

---

[1] The definitions of the test instances used can be retrieved online [1].

waste piles, according to the quality of material produced. The stopping criterion of the simulation is the operation time of the mine.. This simulator was built using a programming language Java JDK 1.7.

6.3 Algorithm Setup

All the experiments considered the following (arbitrarily set) parameters: Population size = 200; Maximum number of evaluations = 20,000; Crossover rate = 0.9; and Mutation rate = 0.4. The dispatch matrices $(\widetilde{M})$ have $J=20$, i.e., twenty columns. The selection operator employed was the Binary Tournament [5]. Initial populations were generated randomly, and all trucks were considered as starting their operation in the crusher. The MILS used $maxIter = 100$ and $maxCount = 20$, and $N = 10$ for the RVND. All runs consider one hour of operation of the mine. All algorithms were coded in Java and compiled with JDK 1.7, and were tested in a PC Intel(R) Core(TM) i7-3632, 2.2 GHz, with 8 GB of RAM, running Windows 8.1.

6.4 Quality Indicators

Evolutionary multiobjective optimization techniques usually need to consider complementary goals, namely the acquisition of a set of tradeoff solutions that are at the same time near the true (oftentimes unknown) Pareto-optimal front, and to have this set evenly covering the whole extension of the Pareto-optimal front - dual objectives usually referred to as *convergence* and *diversity*. To consider this multi-criterion nature in the evaluation of multiobjective algorithms, regarding the convergence and diversity of the solutions, the following quality indicator is used in this work.

Hypervolume or S-Metric. Proposed by Zitzler and Thiele [28], returns the hypervolume of the region covered between the points present in the frontier and a P_{ref} point. This point (P_{ref}) is used as a reference and is dominated by all solutions presented on this frontier. For each solution $i \in \mathcal{PF}$ is constructed a hyperrectangle (c_i) with reference to P_{ref}. The result of this metric can be calculated as:

$$HV(\mathcal{PF}) = \sum_{i \in \mathcal{PF}}^{|\mathcal{PF}|} v_i \qquad (10)$$

where v_i provided by c_i. The higher the value of HV better the quality of the solution indicating that there was a better spread and also a better convergence although the metric is more sensitive to convergence of solutions in relation to the real Pareto frontier. For all the test problems we considered a reference point 10% higher than the upper limits of the Pareto optimal frontier.

Empirical Attainment Function. In the face of random Pareto-set approximations, unary quality indicators provide a convenient transformation from random sets to random variables. To prevent the transformation of sets of solutions in a unary indicator and allow at the same time, a statistical analysis of the set of solutions obtained by multiobjective algorithm was proposed calls Empirical Attainment Function (EAF) [10]. Furthermore, an analysis using EAF allows one to identify in which regions of the objective space one algorithm is better than another, and to visualize this difference. The attainment function gives the probability of a particular point in the objective space vector being attained by (dominated by or equal to) the outcome of a single run of an algorithm. This probability can be estimated from several runs of an algorithm, in order to calculate the EAF of an algorithm. The EAF from to is defined as:

$$\alpha_n(z) = \frac{1}{N} \cdot \sum_{i=1}^{n} b_i(z) \tag{11}$$

where $b_1(z), ..., b_n(z)$ be n realizations of the attainment indicator $b_x(z)$, $z \in \mathbb{R}^d$. Then, the function defined as $\alpha_n : \mathbb{R}^d \rightarrow [0, 1]$.

In the case of bi-objective optimization problems, the empirical attainment function (EAF) is fast to compute, and its graphical representation provides more intuitive information about the distribution of the output of an algorithm than unary (or binary) quality indicators. A tool for graphical analysis of the EAF is proposed on the work of Ibáñez et al. [15].

6.5   Experimental Design

The algorithms NSGA-II, SPEA2, and MILS were applied for the solution of the four test instance on 33 independent runs, after which each quality metric described in the previous section was calculated. The experimental model used was a 2-way factorial design, with both the algorithms and instances as factors [20]. Since our main interest is on the effects of the algorithms, only their effects were analyzed.

We first assessed the convergence of the three algorithms used considering the hypervolume for the four scenarios considered. Figure 1 considers the average of these metrics. The estimated Pareto frontier of the problem was constructed assessing 10^6 solutions that aim to cover the search space of the problem.

The results presented by Figure 1 suggest that NSGAII and SPEA2 algorithms have similar behavior except for instance 2, wherein the NSGAII has a relatively better performance. Additionally, it is important to note that MILS has worse performance for all test instances.

Tables 2 to 5 shows the results obtained by comparing the algorithm used in the experiments for the four scenarios mine. The tests considered as null hypothesis ($\mathcal{H}_0$) that the two proposed algorithms have the same performance. Otherwise, there is a statistical difference between the algorithms. We consider first-and second tests similar to those proposed in order Fonseca et al. [10].

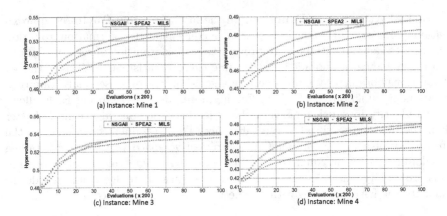

Fig. 1. Average hypervolume for the algorithms on each test case considered

Table 2. Hypothesis test results for Mine 1 ($\alpha = .05$)

Optimiser	Hypothesis test	Test statistic	Critical value	p-value	Decision
MILS – **NSGAII**	1st-order EAF	0.696	0.454	0	Reject H_0
MILS – **NSGAII**	2nd-order EAF	0.848	0.575	0	Reject H_0
MILS – **SPEA2**	1st-order EAF	0.727	0.454	0	Reject H_0
MILS – **SPEA2**	2nd-order EAF	0.878	0.575	0	Reject H_0
NSGAII – SPEA2	1st-order EAF	0.424	0.454	> 0.05	Do not Reject H_0
NSGAII – SPEA2	2nd-order EAF	0.606	0.575	0.044	Reject H_0

Table 3. Hypothesis test results for Mine 2 ($\alpha = .05$)

Optimiser	Hypothesis test	Test statistic	Critical value	p-value	Decision
MILS – **NSGAII**	1st-order EAF	0.636	0.454	0	Reject H_0
MILS – **NSGAII**	2nd-order EAF	0.787	0.575	0	Reject H_0
MILS – **SPEA2**	1st-order EAF	0.727	0.454	0	Reject H_0
MILS – **SPEA2**	2nd-order EAF	0.606	0.575	0.044	Reject H_0
NSGAII – SPEA2	1st-order EAF	0.333	0.454	> 0.05	Do not Reject H_0
NSGAII – SPEA2	2nd-order EAF	0.454	0.575	> 0.05	Do not Reject H_0

Tables 2-5 show the comparisons between pairs of algorithms on each scenario, regarding the EAF indicator. The Optimiser column of the tables highlight the algorithms performed better when $\mathcal{H}_0$ was rejected. Overall, these results suggest that NSGAII and SPEA2 algorithms perform better when compared with MILS algorithm. The comparison between the NSGAII and SPEA2 algorithms does not allows to identify statistical differences between them except for Mine 1 (Table 2).

Table 4. Hypothesis test results for Mine 3 ($\alpha = .05$)

Optimiser	Hypothesis test	Test statistic	Critical value	p-value	Decision
MILS – NSGAII	1st-order EAF	0.393	0.393	> 0.05	Do not Reject H_0
MILS – NSGAII	2nd-order EAF	0.575	0.484	0.004	Reject H_0
MILS – SPEA2	1st-order EAF	0.393	0.393	> 0.05	Do not Reject H_0
MILS – SPEA2	2nd-order EAF	0.636	0.484	0	Reject H_0
NSGAII – SPEA2	1st-order EAF	0.242	0.393	> 0.05	Do not Reject H_0
NSGAII – SPEA2	2nd-order EAF	0.333	0.484	> 0.05	Do not Reject H_0

Table 5. Hypothesis test results for Mine 4 ($\alpha = .05$)

Optimiser	Hypothesis test	Test statistic	Critical value	p-value	Decision
MILS – NSGAII	1st-order EAF	0.757	0.454	0	Reject H_0
MILS – NSGAII	2nd-order EAF	0.909	0.575	0	Reject H_0
MILS – SPEA2	1st-order EAF	0.727	0.454	0	Reject H_0
MILS – SPEA2	2nd-order EAF	0.909	0.575	0	Reject H_0
NSGAII – SPEA2	1st-order EAF	0.454	0.454	> 0.05	Do not Reject H_0
NSGAII – SPEA2	2nd-order EAF	0.575	0.575	> 0.05	Do not Reject H_0

7   Conclusions

This work presented the definition of a multiobjective formulation for the open-pit mining operational planning problem. This model considers as objectives the maximization of production (ore and waste) and the minimization of the number of trucks in operation. An innovative representation of candidate solutions was proposed and employed by three multiobjective optimization methods: SPEA2, NSGA-II, and MILS. The proposed encoding enables the use of algorithms for heterogeneous fleets and also ensures that the solutions created are operationally feasible.

An experiment to compare the algorithms in terms of hypervolume and empirical attainment function values was performed. The results suggest that NSGA-II and SPEA2 algorithms have a better performance when compared with MILS for the problems considered, with the NSGA-II being marginally better than the SPEA2. As future work, we intend to evaluate the idleness of trucks and shovels. Moreover, the mathematical model can be extended to consider other variables, such as, operating conditions of the mine.

Acknowledgments. This work has been supported by the Brazilian agencies National Council for Research and Development (CNPq, grants 475763/2012-2 and 306022/2013-3), Research Foundation of the State of Minas Gerais (FAPEMIG, grant CEX APQ-04611-10), and Coordination for the Improvement of Higher Education Personnel (CAPES, grant 012322/2013-00).

References

1. Alexandre, R., Vasconcelos, J., Campelo, F.: Additional electronic files. http://cpdee.ufmg.br/~fcampelo/files/MOPMOPP/ (2014)
2. Chicano, F., Alba, E.: Exact computation of the expectation curves of the bit-flip mutation using landscapes theory. In: Proceedings of the Genetic and Evolutionary Computation Conference, GECCO 2011, Dublin, Ireland, pp. 2027–2034 (July 2011)
3. Coelho, V., Souza, M., Coelho, I., Guimarães, F., Lust, T., Cruz, R.C.: Multi-objective approaches for the open-pit mining operational planning problem. Electronic Notes in Discrete Mathematics **39**, 233–240 (2012)
4. Coello, C., Lamont, G., Veldhuizen, D.: Evolutionary multi-objective optimization: A historical view of the field. IEEE Computational Intelligence Magazine **1**(1), 28–36 (2006)
5. Coello, C., Lamont, G., Veldhuizen, D.: Evolutionary Algorithms for Solving Multi-Objective Problem, 2nd edn. Springer (2007)
6. Deb, K., Pratap, A., Agarwal, S., Meyarivan, T.: A fast and elitist multiobjective genetic algorithm: NSGA-II. IEEE Evolutionary Computation **6**(2), 182–187 (2002)
7. Dias, A., Vasconcelos, J.: Multiobjective genetic algorithms applied to solve optimization problems. IEEE Transactions on Magnetics **38**(2), 1133–1136 (2001)
8. Doig, P., Kizil, M.: Improvements in truck requirement estimations using detailed haulage analysis. In: 3th Coal Operators Conference, The Australasian Institute of Mining and Metallurgy and Mine Managers Association of Australia, pp. 368–375 (February 2013)
9. Feo, T., Resende, M.: Greedy randomized adaptive search procedures. Journal of Global Optimization **6**(2), 109–133 (1995)
10. Fonseca, C.M., da Fonseca, V.G., Paquete, L.: Exploring the Performance of Stochastic Multiobjective Optimisers with the Second-Order Attainment Function. In: Coello Coello, C.A., Hernández Aguirre, A., Zitzler, E. (eds.) EMO 2005. LNCS, vol. 3410, pp. 250–264. Springer, Heidelberg (2005)
11. Geiger, M.: The PILS metaheuristic and its application to multi-objective machine scheduling. In: Kfer, K.H., Rommelfanger, H., Tammer, C., Winkler, K. (eds.) Multicriteria Decision Making and Fuzzy Systems Theory, Methods and Applications. pp. 43–58. Shaker Verlag, Industriemathematik und Angewandte Mathematik (2006)
12. Goldberg, D.: Genetic Algorithms in Search, Optimization and Machine Learning, 1st edn. Addison-Wesley (1989)
13. Hansen, P., Mladenovic, N., Pérez, J.M.: Variable neighbourhood search: methods and applications. 4OR **6**(4), 319–360 (2008)
14. He, M., Wei, J., Lu, X., Huang, B.: The genetic algorithm for truck dispatching problems in surface mine. Information Technology Journal **9**, 710–714 (2010)
15. Ibáñez, M., Stützle, T., Paquete, L.: Graphical tools for the analysis of bi-objective optimization algorithms. In: Proceedings of the 12th Annual Conference Companion on Genetic and Evolutionary Computation, GECCO 2010, pp. 1959–1962. ACM, New York (2010)
16. ILOG: Users Manual. IBM (2008)
17. Kelton, W., Sadowski, R., Sturrock, D.: Simulation with Arena. McGraw-Hill series in industrial engineering and management science, 4. ed. internat. ed. McGraw-Hill Higher Education, Boston (2007)

18. Loureno, H., Martin, O., Stützle, T.: Iterated local search. ArXiv Mathematics e-prints. (Feburary 2001), arXiv:math/0102188

19. Mladenovic, N., Hansen, P.: Variable neighborhood search. Computers & Operations Research **24**(11), 1097–1100 (1997)

20. Montgomery, D.: Design and Analysis of Experiments, 7th edn. Wiley (2008)

21. Nel, S., Kizil, M., Knights, P.: Improving truck-shovel matching. In: 35TH APCOM Symposium, The Australasian Institute of Mining and Metallurgy, Wollongong, NSW, pp. 381–391 (September 2011)

22. Souza, M., Coelho, I., Ribas, S., Santos, H., Merschmann, L.: A hybrid heuristic algorithm for the open-pit-mining operational planning problem. European Journal of Operational Research **207**(2), 1041–1051 (2010)

23. Subtil, R., Silva, D., Alves, J.: A practical approach to truck dispatch for open pit. In: 35th International Symposium on Application of Computers in the Minerals Industry (35th APCOM) pp. 765–777 (2011)

24. Tan, Y., Chinbat, U., Miwa, K., Takakuwa, S.: Operation modeling and analysis of open pit copper mining using GPS tracking data. In: Proceedings of the 2012 Winter Simulation Conference. pp. 1–12. IEEE, Berlin (2012)

25. Topal, E., Ramazan, S.: A new MIP model for mine equipment scheduling by minimizing maintenance cost. European Journal of Operational Research **207**(2), 1065–1071 (2010)

26. Topal, E., Ramazan, S.: Mining truck scheduling with stochastic maintenance cost. Journal of Coal Science and Engineering (China) **18**(3), 313–319 (2012)

27. Zitzler, E., Laumanns, M., Thiele, L.: SPEA2: Improving the Strength Pareto Evolutionary Algorithm for Multiobjective Optimization. In: Giannakoglou, K., et al. (eds.) Evolutionary Methods for Design, Optimisation and Control with Application to Industrial Problems (EUROGEN 2001), pp. 95–100. International Center for Numerical Methods in Engineering (CIMNE) (2002)

28. Zitzler, E., Thiele, L.: Multiobjective evolutionary algorithms: A comparative case study and the strength Pareto approach. IEEE Transactions on Evolutionary Computation **3**(4), 257–271 (1999)

A Model to Select a Portfolio of Multiple Spare Parts for a Public Bus Transport Service Using NSGA II

Rodrigo José Pires Ferreira[1(✉)], Eduarda Asfora Frej[1],
and Roberto Klecius Mendonça Fernandes[2]

[1] Universidade Federal de Pernambuco, Recife, Brazil
{rodrigo,eduarda.asfora}@ufpe.br
[2] Instituto Federal de Educação, Ciência e Tecnologia do Rio Grande do Norte, Natal, Brazil
robertokmf@hotmail.com

Abstract. This study proposes a model for managing spare parts in urban passenger bus transport companies so as to support maintenance planning decisions. Spare parts play a significant role in the assets of these companies because inappropriate management of these inventories can cause significant losses to the business. A multi-objective model based on NSGA-II is developed to aid the management of spare parts in corrective maintenance. A multiple item portfolio approach is defined instead of a traditional single item approach. As a typical portfolio problem, a portfolio of multiple spare parts combines "n" items while competing for the same resources. Two criteria were considered: the level of service and the total acquisition cost. An adaptation of non-dominated sorting genetic algorithm II (NSGA II) was used to solve the problem. The model was tested in an urban passenger bus transport company in the city of Natal, Brazil.

Keywords: Inventory management · Spare parts · Urban bus · Genetic algorithm

1 Introduction

Spare parts have low (or very low) consumption and forecasting demand is difficult and erratic. They have high unit costs, long lead-times, and are of high criticality for the operation (missing cost). It is common for enterprises to relegate these items to the background, but in some companies – such as steel, mining, petrochemical and automotive ones, where in the latter alone, the annual costs of opportunity, storage, depreciation, insurance and handling of spare parts range from 25% to 35% of the book value of all stocks in any company – spare parts are a significant part of all product inventories, and therefore, need to be better controlled. [1] confirmed this when they showed that managing MRO (Maintenance, Repair and Operations) inventories represents 36% of the overall costs while the procurement process represents 25% of them.

This is what happens especially in companies offering an urban passenger bus transport service: for them, spare parts are critical and have relevant value for the business. In Brazil, among the modalities of urban passenger transport, travelling by bus is the primary means of transportation for people within their cities and metropolitan areas. This is evidenced by [2] which demonstrates that urban transportation

A. Gaspar-Cunha et al. (Eds.): EMO 2015, Part II, LNCS 9019, pp. 448–457, 2015.
DOI: 10.1007/978-3-319-15892-1_30

by bus was 11.4 billion passengers in 2009, while railroad transportation, the second placed, carried 2.1 billion passengers in same period. For this, buses ran 6.9 billion kilometers in 2009. Given that the average cost of a ticket was R$ 2.50, this sector of the economy produced revenue of about R$ 28.5 billion, only from this source of revenue. So, faulty parts and/or lack of spare parts possibly needed for replacement purposes, as well as vehicles being laid up for these reasons, can result in serious losses to any transportation company. Thus, good inventory management of spare parts certainly has a positive influence on maintenance management, since this leads to the higher reliability and greater availability of equipment and therefore has a direct impact on business profitability.

In this context, [3] dealt with production jointly with spare parts inventory control strategy driven by condition-based maintenance (CBM) for a part of equipment, where the objectives to be minimized were the stock of spare parts and the total expected operating cost. [4] established a systematic method for the storage of aeronautic spare parts, by analyzing the distribution of demand probability of the spare parts, and solved the model using dynamic programming. In the present work, the spare parts will be separated into critical and non-critical items and treated separately, as did [5], who prioritized those critical items, which are expensive, highly reliable, with higher lead times, and are not available in store.

The application made in this work it is a typical problem of portfolio assets. [6] reported that the three points to be investigated in a portfolio problem are the measure of the profitability of a portfolio of investments, the selection and planning of an optimal set of investments and measure the risk of a portfolio of investments.

Multi-objective approaches are widely used to deal with problems related to maintainability of systems, where usually there are conflicts between relevant criteria for decision making. [7] proposed a multi-objective approach to find out an optimal periodic maintenance policy for a repairable and stochastically deteriorating multi-component system over a finite time horizon, in which the objectives to be minimized are the total cost of maintenance and total time of system unavailability. [8] dealt with the problem of scheduling bus maintenance activities using heuristics, looking to minimize the interruptions in the daily bus operating schedule, and maximize the utilization of the maintenance facilities.

Multi-objective genetic algorithms (MOGA) are often chosen to solve different problems with conflicting criteria, mainly due their high computational complexity. [9] used multiobjective genetic algorithm coupled with discrete event simulation to analyze the trade-off between reliability and cost in a system reliability when handling redundancy allocation problems, considering series-parallel systems comprised of components subjected to corrective maintenance actions with failure-repair cycles modeled by renewal processes. [10] applied genetic algorithm to a probabilistic-based levelized cost of energy problem to assess investment risks, taking into account four main factors: wind speed, system availability, maintenance policy and spare parts stock level. [11] proposed an efficient algorithm to find the Pareto optimal frontier to the problem of multi-objective optimization which aims to minimize both the global maintenance cost and the total maintenance time, while [12] presented a methodology for multi-objective optimization using an evolutionary algorithm to find out the best distribution network reliability while minimizing the system expansion costs.

In this context, there are many papers that address issues of spare parts policy using multi-objective genetic algorithms, as [13], who explored the possibility of using genetic algorithms to optimize the number of spare parts into a multicomponent system in the optics of various goals, such as, for example, the maximization of system revenues and the minimization of the total spares volume, where the modeling of the system failure, repair and replacement stochastic processes is done by means of Monte Carlo simulation. [14] proposed an approach using genetic algorithm to optimize preventive maintenance and spares policies of a manufacturing system operating in the automotive sector, as [15] developed a framework that integrates multiobjective evolutionary algorithm (MOEA) with multiobjective computing budget allocation (MOCBA) method for the multi-objective simulation optimization problem of allocation of spare parts for aircraft.

In this paper, to solve the multi–objective problem of spare parts policy, we used an adaptation of the multi-objective genetic algorithm proposed by [16], NSGA II, which consists in an elitist version and less complex computing than NSGA, formulated by [17]. Several situations where NSGA II is applied to solve problems with more than one criteria are found in the literature, like [18], that presented an application of multi-objective optimization techniques, NSGA and SPEA, to a project of power distribution systems, and concluded that, despite the differences between the two models, they have similar efficiency to solve the problem, while [19] showed an application of NSGA II for the multi-objective generation expansion planning (GEP) problem, where the objectives to optimize are the total minimum investment cost maximum reliability.

This study covers spare parts, with a failure rate and purchase cost, classified into critical and non-critical items which compete for the same funds of a budget that may or may not have constraints. The model proposed identifies for purchase the spare part that has the best cost-benefit ratio, i.e., the spare part that offers the minimum cost and the maximum service level. It is important to emphasize that only spare parts used in corrective maintenance, the demand for which is random, are dealt with in this paper rather than parts used in preventive maintenance for which consumption can be defined by a periodic replacement strategy.

The content of this paper is organized as follows. Section 2 describes the mathematical model with its structure and algorithm. Section 3 presents a case study and Section 4 provides the main results and a conclusion.

2   The Mathematical Model

The mathematical model proposed to the spare parts inventory problem was a multiobjective optimization model, where the objectives to optimize are the total level of service, that should be maximized, and the total cost of the spare parts purchased, which should be minimized.

Data analysis reveals a typical problem of a portfolio of assets, where each of these (in this case, the spare parts) vies for the resources available such that preference is given to purchasing the one that is based on the maximum return (or benefit) and at the least risk that it could cause for the "investor", in this case, the manager of the stock. [20] showed that the basic elements of his portfolio theory were based on these two criteria - the expected return and the risk - with which the investor seeks to choose the optimal point (the "efficient frontier", in the words of that author) at which to apply his resources.

It is worth noting that the model was designed to make use of consumable parts, used only in corrective maintenance, which if broken, will be replaced immediately if available in stock.

The model aims to answer the main question inherent in any process of inventory management: what is the ideal inventory level for a spare part that guarantees the minimum cost and maximum availability by means of the Poisson distribution.

As shown by [21], the Poisson distribution is the most widely-used mathematical-statistical model in the literature for optimizing inventories of spare parts, and is premised on modelling the behaviour of demand for the item by a probability distribution, which is widely used to describe rare random events, such as, for example, the unforeseen failure of certain types of equipment, and hence is adhered to when representing demand for some cases of spare parts replacement. Among the main properties of the Poisson distribution, it can be stressed that it is discrete and assumes independence between events, and is represented by Expression 1.

$$P_x(t) = \frac{(\lambda t)^x e^{-\lambda t}}{x!} \tag{1}$$

where:
x = consumption of replacement parts by time interval for which the wish is to estimate the probability;
t = time interval considered;
λ = historical consumption rate of the replacement parts by unit of time;
$P_x(t)$ = probability of there being "x" requests for replacement parts during time interval t.

The model can be represented as follows:

SP_i = each bus spare part (critical or non-critical);
x_i = amount in stock of each bus spare part;
λ_i = monthly rate of consumption of the i-th spare part;
LS_i = level of service of the i-th spare part associated with the quantity x_i;
C_i = unit cost of the i-th spare part;

Such that:

$$LS_i = P(x \leq x_i) = \sum_{i=0}^{i} \frac{(\lambda t)^i e^{-\lambda t}}{i!} \tag{2}$$

Thus the problem consists of:

$$Max \sum_{i=1}^{n} LS_i \qquad (3)$$

$$Min \sum_{i=1}^{n} C_i \qquad (4)$$

3   Multi-Objective Genetic Algorithm

In this paper, the genetic algorithm used to solve the spare parts inventory problem was an adaptation of the elitist multi-objective genetic algorithm proposed by [16] NSGA II. The algorithm is based on sorting the chromosomes based on non-dominance to find the Pareto front of multi-objective problems, as well as it maintains the good solutions during the evolutionary process, since it's an elitist algorithm.

NSGA II differs from NSGA - proposed by [17] - mainly due to its lower computational complexity, elitism and mechanism of diversity preservation. The genetic algorithm proposed in this paper keeps the main characteristics of NSGA II, remaining with non-dominated sorting mechanism and preservation of diversity based on *crowding distance* ([16]). What differentiates, essentially, the proposed algorithm from the original one are the genetic operators of crossover and mutation, which are presented below, and also how the chromosomes are represented. In the proposed algorithm, the length of the chromosome is equal to the number of different spare parts, where each gene represents the amount to be purchased.

3.1   Genetic Operators: Crossover and Mutation

The selection of parents who will undergo crossover operation to generate offspring is randomly made, and, for each set of parents chosen, two descendants are made using the genetic operators. Crossover and mutation are mutually exclusive, ie, if one occurs, the other will not occur. The probability that the crossover happens is 90%, and the remaining 10% are the probability of mutation.

The single point crossover is used, which is done by defining arbitrarily a position of the chromosome that will be the cutoff point, ie, offspring 1 inherits the genes of parent 1 until the position of the cutoff and, from this on, inherits the genes of parent 2; offspring 2, then, inherits the genes of parent 2 until the cutoff point, and, from that point on, inherits the genes of parent 1.

Mutation is incorporated in the evolutionary process in order to preserve diversity, and ensure that solutions are not concentrated in a local optimum. In this algorithm, a point corresponding to the position where the individual suffers mutation is randomly defined, ie, the gene in that point will come out of the chromosome, while another gene, also randomly chosen, will become part of that chromosome, in the place of the excluded one. The mutation may improve or worsen the quality of the solution, but the aim is that it will ensure the population diversity.

4   Case Study

For the case study, first of all, the procedure followed was: (1) to define the replacement spare parts components of the buses to be studied, such that 33 items used were defined in corrective maintenance actions alone, and (2) to define the parameters to be quantified, which in this case were the consumption rate λ, the unit price, the criticality, and so forth.

The data were collected from an urban collective public transport company that has been operating buses in Natal for more than 25 years and it has been regarded as anonymous in this study. This company has a fleet of 83 buses the average age of which is 5.69 years, which run 600,000 km per month. The initial data collected are in Table 1.

Table 1. Initial data on the 33 replacement spare parts of a bus

Spare Part	Monthly Consumption (λ)	Unit Price	Criticality	Initial Stock
P1	0.636	168	Y	0
P2	0.364	660	Y	0
P3	0.727	2,700.00	Y	0
P4	1	1,843.00	Y	0
P5	1	229	N	0
P6	0.364	23	Y	0
P7	2.273	882	Y	0
P8	2.727	1,176.00	Y	1
P9	1.364	136	Y	0
P10	0.273	168	N	0
P11	1.273	197	N	0
P12	1.818	129	N	0
P13	1.273	70	N	0
P14	0.909	200	Y	0
P15	27.636	12.97	N	0
P16	18	1,180.00	Y	13
P17	6.364	1.25	N	0
P18	7.727	8.9	N	0
P19	5.364	1.25	N	0
P20	6.273	0.77	N	0
P21	3.091	134	N	0

Table 1. (*Continued*)

P22	7.636	380	Y	4
P23	85.455	14.49	Y	74
P24	68.545	14.1	N	0
P25	63.364	16.65	N	0
P26	97.455	18.21	N	0
P27	1.273	268	Y	0
P28	1.636	279	N	0
P29	0.909	265	N	0
P30	1.636	76.31	N	0
P31	2,642.36	0.45	N	0
P32	1,279.36	0.54	N	0
P33	3	30	Y	1

With these initial parameters, the critical items (which cause the bus to stop running when they fail) and non-critical ones (which do not immobilize the vehicle) were determined.

Initially, the algorithm was run for the critical items. 99.9% was established as the upper limit of the level of service, which ensures a high quality of service obtained by the purchase of critical items. As mentioned in the previous section, the initial solution of the genetic algorithm was generated through the result of applying the model based on cost/benefit (CB), which is obtained through the ratio of the cost by varying the level of service caused by the purchase of spares. The algorithm based on cost-benefit presented a total of 142 solutions in the Pareto frontier. The population size chosen to use in NSGA II was twice the number of solutions obtained by CB model, ie, 284. The first 142 chromosomes of the initial solution are the same chromosomes obtained by the CB model, and the other half of the chromosomes is generated randomly, so as the diversity in the solutions is preserved.

After 250 iterations of the genetic algorithm, we obtained a total of 276 solutions on the Pareto frontier. Of this number, only 21 coincide with the solutions generated by the CB model, which shows that the genetic operators have diversified a lot the initial solutions. If analyzed together, the two models generated a total of 397 different solutions, of which 363 are non-dominated.

A comparative graph of the solutions of the model based on cost-benefit and NSGA II for critical items is shown in Figure 1.

On analysing Figure 1, it is easy to see that, near the origin of the graph, a lot is gained in the level of service with little investment. From an LS of 90% ahead, it is soon seen that there is a "saturation" in the curve, thus reversing the prevailing logic, i.e. there are then high investments for little return (low increments in the level of service), which clearly it is not worth the company's spending resources on in this situation.

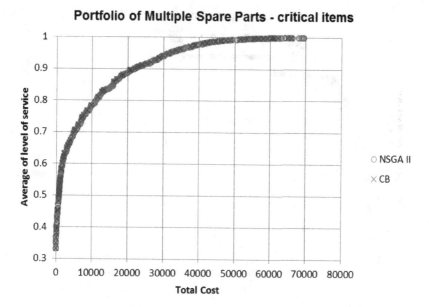

Fig. 1. Budget versus Level of Service (critical items)

For non-critical items, an upper limit of 80% service level was established. The algorithm based on the cost-benefit presented 371 solutions in Pareto frontier. The population size chosen to use in NSGA II was equal to 400. The first 371 chromosomes of the initial solution were the same chromosomes obtained by the first model, and the other 29 were randomly generated, so that the diversity in the solutions is guaranteed.

After 250 iterations of the genetic algorithm, we obtained a total of 313 solutions on the Pareto frontier. Of this number, only 6 coincide with the solutions generated based on the cost-benefit model, showing, again, a considerable diversification of the initial solutions.

A comparative graph of the solutions of the model based on cost-benefit and NSGA II for non-critical items is shown below, where you can observe that the genetic algorithm greatly improved the initial solution, since from a level of service of approximately 50%, most of the solutions obtained by the cost-benefit model were dominated by the solutions presented by NSGA II.

The option to deal separately with the critical and non-critical items allows the manager to have greater flexibility in managing the contingency element of his/her budget, and certainly yields a better result for inventory management as it allows the logic of the program, based on the typical problem of a portfolio of assets, in which several items (within its group of criticality) compete for resources simultaneously, thus gaining the one that presents the lowest cost-benefit index, which brings a gain to the operation as a whole.

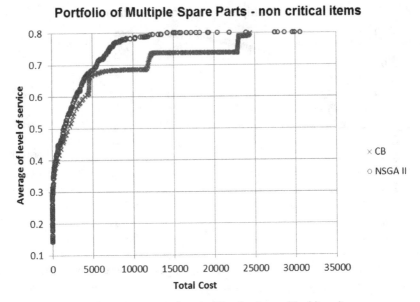

Fig. 2. Budget versus Level of Service (non critical items)

5   Conclusions

It can be concluded that the model developed and applied in a real situation reached its objective, as it allowed important parameters in controlling the inventory of replacement spare parts to be monitored efficiently, thus contributing to the management of an urban bus company in the city of Natal. It is further understood that this model can be replicated in any other company which has replacement spare parts in its inventory and consumes them when carrying out corrective maintenance.

Although a cost benefit analysis is able to find a subset of the Pareto front set, the NSGA-II approach proposed assumed that a cost benefit analysis can support the initial solutions generations satisfactorily and a more complete and disperse set of solutions of the problem can be reached.

A study of the impact that the crossover and mutation operators have on the final solutions can be recommended to investigate in a future research. Besides, a performance analysis of the algorithm taking into account recognized metrics is a relevant step to check the quality of the solutions proposed in the algorithm.

References

1. Cheng, C.-Y., Prabhu, V.: Evaluation models for service oriented process in spare parts management. Journal of Intelligent Manufacturing (2010). doi:10.1007/s10845-010-0486-0
2. IDET/FIPE. Índice de Desempenho Econômico do Transporte (online). http://fipe.org.br/web/index.asp (accessed June 5, 2010)

3. Rausch, M., Liao, H.: Joint production and spare part inventory control strategy driven by condition based maintenance. IEEE Transactions on Reliability **59**(3), 507–516 (2010)
4. Xingfang, F., Juheng, F.: Study on the inventory optimization model of aeronautic spare parts under the condition of uncertain demand. In: 2011 International Conference on Business Management and Electronic Information (BMEI). IEEE (2011)
5. Godoy, D.R., Pascual, R., Knights, P.: Critical spare parts ordering decisions using conditional reliability and stochastic lead time. Reliability Engineering & System Safety **119**, 199–206 (2013)
6. Lonchampt, J., Fessart, K.: Investments Portfolio Optimal Planning for industrial assets management: Method and Tool. In: IAEA 3rd International Conference on NPP Life Management (PLIM) for Long Term Operations (LTO), Salt Lake City, UT, USA (2012)
7. Certa, A., et al.: A multi-objective approach to optimize a periodic maintenance policy. International Journal of Reliability, Quality and Safety Engineering 19(6) (2012)
8. Haghani, A., Shafahi, Y.: Bus maintenance systems and maintenance scheduling: model formulations and solutions. Transportation Research Part A: Policy and Practice **36**(5), 453–482 (2002)
9. Lins, I.D., Droguett, E.L.: Multiobjective optimization of availability and cost in repairable systems design via genetic algorithms and discrete event simulation. Pesquisa Operacional **29**(1), 43–66 (2009)
10. Jin, T., et al.: Coordinating maintenance with spares logistics to minimize levelized cost of wind energy. In: 2012 International Conference on Quality, Reliability, Risk, Maintenance, and Safety Engineering (ICQR2MSE). IEEE (2012)
11. Certa, A., et al.: Determination of Pareto frontier in multi-objective maintenance optimization. Reliability Engineering & System Safety **96**(7), 861–867 (2011)
12. Ramírez-Rosado, I.J., Bernal-Agustín, J.L.: Reliability and costs optimization for distribution networks expansion using an evolutionary algorithm. IEEE Transactions on Power Systems **16**(1), 111–118 (2001)
13. Marseguerra, M., Zio, E., Podofillini, L.: Multiobjective spare part allocation by means of genetic algorithms and Monte Carlo simulation. Reliability Engineering & System Safety **87**(3), 325–335 (2005)
14. Ilgin, M.A., Tunali, S.: Joint optimization of spare parts inventory and maintenance policies using genetic algorithms. The International Journal of Advanced Manufacturing Technology **34**(5–6), 594–604 (2007)
15. Lee, L.H., et al.: Multi-objective simulation-based evolutionary algorithm for an aircraft spare parts allocation problem. European Journal of Operational Research **189**(2), 476–491 (2008)
16. Deb, K., et al.: A fast and elitist multiobjective genetic algorithm: NSGA-II. IEEE Transactions on Evolutionary Computation **6**(2), 182–197 (2002)
17. Srinivas, N., Deb, K.: Muiltiobjective optimization using nondominated sorting in genetic algorithms. Evolutionary Computation **2**(3), 221–248 (1994)
18. Mendoza, F., Bernal-Agustin, J.L., Dominguez-Navarro, J.A.: NSGA and SPEA applied to multiobjective design of power distribution systems. IEEE Transactions on Power Systems **21**(4), 1938–1945 (2006)
19. Murugan, P., Kannan, S., Baskar, S.: NSGA-II algorithm for multi-objective generation expansion planning problem. Electric Power Systems Research **79**(4), 622–628 (2009)
20. Markowitz, H.M.: Foundations of portfolio theory. The Journal of Finance **46**(2), 469–477 (1991)
21. Bevilacqua, M.; Ciarapica, F. E.; Giacchetta, G.: Spare parts inventory control for the maintenance of productive plants. In: 2008 IEEE International Conference on Industrial Engineering and Engineering Management, IEEM 2008, art. no. 4738096, pp. 1380–1384 (2008)

A Multi-objective Optimization Approach Associated to Climate Change Analysis to Improve Systematic Conservation Planning

Shana Schlottfeldt[1,2]([⊠]), Jon Timmis[2], Maria Emilia Walter[1],
André Carvalho[3], Lorena Simon[4], Rafael Loyola[4],
and José Alexandre Diniz-Filho[4]

[1] Department of Computer Science, University of Brasilia, Brasilia, Brazil
shanass@unb.br
[2] Department of Electronics, University of York, York, UK
[3] Department of Computer Science, SCC-ICMC-USP, São Paulo, Brazil
[4] Institute of Biological Sciences, Federal University of Goiás, Goiânia, Brazil

Abstract. Biodiversity conservation has been since long an academic community concern, leading scientists to propose strategies to effectively meet conservation goals. In particular, Systematic Conservation Planning (SCP) aims to determine the most cost effective way of investing in conservation actions. SCP can be formalized by the Set-Covering Problem, which is NP-hard. SCP is inherently multi-objective, although it has been usually treated with a monobjective and static approach. Here, we propose a multi-objective solution for SCP, increasing its flexibility and complexity, and, at the same time, augmenting the quality of provided information, which reinforces decision-making. We used ensemble forecasting, considering future climate simulations to estimate species occurrence projected to 2080. Our method identifies sites: 1) of high priority for conservation; 2) with significant risk of investment; and, 3) that may become attractive in the future. To the best of our knowledge, this application to a real-world problem in ecology is the first attempt to apply multi-objective optimization to SCP associated to climate forecasting, in a dynamic spatial prioritization analysis for biodiversity conservation.

Keywords: Multi-objective optimization · Systematic conservation planning · Spatial conservation prioritization · Biodiversity conservation · Climate change · Uncertainty in simulations · Parameter tuning

1 Introduction

Effective conservation of biodiversity is essential for continued human well-being, and has been since long a concern of the academic community, but only in recent years it has been faced as a political, economic and social affair [20]. In this context, the growing interest and concern regarding biodiversity is leading scientists

A. Gaspar-Cunha et al. (Eds.): EMO 2015, Part II, LNCS 9019, pp. 458–472, 2015.
DOI: 10.1007/978-3-319-15892-1_31

to develop effective strategies to meet conservation goals. The underlying principle of these strategies lies on the Systematic Conservation Planning (SCP), which determines the most cost effective way of investing in conservation actions.

SCP can be formalized by the Set-Covering Problem [5], which is NP-hard [9]. SCP can be enunciated as the problem of finding a minimum set of sites (among several available ones), simultaneously maximizing the other features under study. Thus, there are at least two conflicting objectives to be optimized, making SCP a natural candidate for Multi-Objective Optimization (MOO).

Several parameters, e.g., vegetation remnants, annual actual evapotranspiration (AET), and human occupation, among other environmental, social, and political objectives, can be incorporated to SCP, adding more dimensions to the problem, therefore increasing its complexity.

Albeit inherently multi-objective, SCP has been usually dealt with a monobjective approach through the assignment of weights to dimensions of the problem aiming to obtain a unique objective function [2,3,7,8,19,22]. Moreover, the most known techniques for SCP are static, implicitly adopting the hypothesis that conserved biodiversity does not change throughout time [17]. However, this is not really accurate, and climate change analyses should be incorporated into conservation plans to more properly reflect the biodiversity dynamics [15].

Quite a few reasons justify the use of the multi-objective approach to deal with SCP. First, a set of solutions can be found, instead of just one, and this can be of great interest to decision makers. In addition, flexibility of data type is increased and constraints can be integrated, at the same time that the problem is kept tractable [9].

In this paper, we propose a MOO approach for SCP, which significantly augments the amount and quality of information provided to users, reinforcing decision-making. We employ the well known NSGA-II, given the wide success of the algorithm, this seemed a logical place to start before developing more sophisticated multi-objective approaches.

To the best of our knowledge, this application to a real-world problem in ecology is the first attempt to apply MOO to SCP associated to climate forecasting, in a dynamic spatial prioritization analysis for biodiversity conservation. In particular, our analysis considered future climate simulations to estimate species occurrence projected to 2080. Our method suggests sites of high priority for conservation, regions with significant risks of investment and those ones that may become attractive in the future.

The remainder of the paper is structured as follows. In Section 2 we discuss the approaches previously used to deal with SCP. Section 3 describes the materials and methods adopted in this study. In Section 4, we discuss the results obtained so far. Conclusions and possible future work are presented in Section 5.

2   Previous Approaches to the SCP Problem

The SCP problem aims to minimize the number of sites, total area or cost and at the same time guarantee the representation of natural features (objects of conservation) [25]. In order to achieve this, the problem can been formulated as follows [5]:

Let $A_{m \times n}$ be a matrix where $m = sites$ and $n = natural\ features$, whose element $a_{ij} \in \{0,1\}$, and $a_{ij} = \begin{cases} 1, & \text{if the natural feature } j \text{ occurs in the site } i; \\ 0, & \text{otherwise.} \end{cases}$

Let each site i have a cost c_i, and each feature j a desired representation level r_j. Let $x_i \in \{0,1\}$, where $x_i = \begin{cases} 1, & \text{if the site } i \text{ is included in the solution;} \\ 0, & \text{otherwise.} \end{cases}$

The SCP problem consists in minimizing Eq. 1:

$$\sum_{i=1}^{m} c_i x_i \qquad (1)$$

Subject to Eq. 2 (for all j, each feature should be represented at least r_j times):

$$\forall j \in \{1, 2, ..., n\}, \sum_{i=1}^{m} a_{ij} x_i \geq r_j \qquad (2)$$

The development of algorithms and tools for SCP began in the 1980s [24]. Since then, several approaches have been suggested, ranging from a simple scoring system to more complex optimization techniques. Commonly, in these approaches, algorithms select complementary sites in a sequential order, until they reach the goal of representing all the species (in effect, a greedy algorithm). Alternatively, the adoption of an exact approach (which ensures the production of optimal solutions, e.g., integer linear programming) was initially discussed by Cocks and Baird, in 1989 (mentioned in [29]). However, as SCP is a NP-hard problem, even the available software packages computing exact algorithms are not able to solve some large data sets [26]. Due to these characteristics, meta-heuristics are used as an alternative approach to SCP. The most widely used metaheuristics for SCP are Simulated Annealing (SPEXAN [3], SITES [22], and Marxan [2]), and the Tabu Search (ConsNet [7]). Nonetheless, as previously mentioned, these approaches have treated SCP in a monobjective way by combining the different problem objectives in one single objective function.

On many occasions, it is difficult to work exclusively with agregated values in a monobjective function. Often the subjectivity associated to such an approach can drive to distinct results for the same data set [3,19]. Furthermore, when two criteria represent distinctive value systems it can be impossible to combine and/or compare such criteria in a meaningful manner. To insist in a single objective function can lead to disparate values, conducing to inaccurate results and/or requiring assumptions that some decision makers would find inappropriate [8].

3   Materials and Methods

3.1   Data

Plant Species. We used data of occurrence of 96 plants with economic importance in Cerrado, a large biome in Central Brazil, occupying around $1,500,000$ km^2. Satellite-based estimates of habitat transformations in Cerrado show rates that are still very high and far from diminishing, which will likely put many endemic and rare species under high threat levels or extinction [14]. Besides the importance of the biome conservation, plant species used in this research have historical and cultural relevance, being widely used as part of the culture and development of regional communities [10].

Information of the 96 plant species under study were obtained from Centro de Referência em Informação Ambiental (CRIA; www.cria.org.br), from Flora Integrada da Região Centro-Oeste (Florescer; www.florescer.unb.br), from the scientific literature index in ISI (apps.isiknowledge.com) and from Scielo (www.scielo.org). A total of 8,896 points were compiled and used for modelling the 96 species. The Cerrado region was overlapped by a 181-cell grid, in which cells were 1° of latitude by 1° of longitude. The occurrences were modelled as a function of several environmental variables using different methods (for details see [31]), and results were combined to generate the distribution data, which was later converted into a matrix of presence-absence of species.

Climate Forecast. To evaluate the effects of future climate changes on the species geographical distribution, we used an ensemble forecast approach, a conjunction of different climate models, modelling methods and carbon emission scenarios [13], obtaining what would be the distribution of the species in the considered region by 2080 (for details see [31]).

Additional Objectives. Three additional objectives were used in this study: annual actual evapotranspiration, human occupancy and vegetation remnants.

Annual Actual Evapotranspiration (AET). A measure of the joint availability of energy and water in the environment. Information came from many databases, and our dataset was obtained according to Rodriguez et al. [28].

Human Occupancy (H_O). Human population density (H) has been often used as a criterion to be minimized [18] or as an evidence of conflicts between economic/social interests and biological conservation. Although, Rangel et al. [27] showed that, in Brazilian Cerrado, species richness was positively correlated with patterns of modern agriculture and cattle ranching, but not with human population density. Consequently, other socio-economic variables should be considered to minimize costs when establishing regional programs for conservation planning in Brazilian Cerrado. Therefore, this study considered the human occupancy (H_O), a measure obtained compiling data on social and economic variables indicating conservation conflicts [14,27]. Data was obtained from the Brazilian Institute of Geography and Statistics (IBGE; www.ibge.gov.br).

Vegetation Remnants (VR). These refer to the proportion of each 1° grid cell covered by natural vegetation, based on remote sense information (Moderate-Resolution Imaging Spectroradiometer (MODIS)). Data used in this article are detailed described in Carvalho et al [6].

Conservation Scenarios. For present and future data, we have a presence-absence matrix $A_{m \times n}$, where $m = 181$ *sites* and $n = 96$ *plant species*. In addition, for each site over time, we have information about AET, H_O, and VR. Hence, we have five different objectives to be optimized: 1) minimize the number of sites (among the 181 grid cells); 2) maximize the number of 96 represented plant species; 3) maximize AET; 4) minimize H_O; 5) maximize VR.

Our fitness functions were developed by having as many representations of Eq. 1 as objectives to be optimized, and varying c_i according to the objective under consideration. This allowed to simultaneously optimize distinct objectives instead of aggregating them into one single function.

We worked with minimization, so, based on the duality principle, w.l.o.g., we converted all objectives to their equivalent minimization representation (e.g., for the objective mentioned in item 2, optimization consisted in minimizing the number of missing species – which is the same as maximizing the number of represented species).

The experts defined that all the species should be represented at least once, i.e. in Eq. 2, $r_j = 1$, $j \in \{1, ..., 96\}$.

We defined three conservation scenarios:

- *Scenario 1*: to represent all species in current time, applying optimization in 2 dimensions (we optimized objectives 1 and 2, respectively, the number of sites and the number of plant species);
- *Scenario 2*: to represent all species in current time, using optimization in 5 dimensions (i.e., optimizing simultaneously objectives 1 to 5); and,
- *Scenario 3*: to represent all species in 2080 (since it happens to be a forecast, objectives 3 to 5 are not available, and optimization was performed in two dimensions, considering only objectives 1 and 2).

3.2   Experimental Setup

Algorithm. We used the Non-Dominated Sorting Genetic Algorithm-II (NSGA-II) [11]. For each run, a population of initial solutions was randomly generated. These solutions were then evolved using NSGA-II, which was implemented in Matlab®.

Aleatory Uncertainties. In order to determine the number of runs required to mitigate aleatory uncertainty in the stochastic algorithm employed, we used Spartan (Simulation Parameter Analysis R Toolkit Application) [1], a package of statistical techniques designed to support the identification of which simulation results can be attributed to the dynamics of the modelled system, rather than artefacts of uncertainty or parametrisation, or simulation stochasticity. More specifically, we applied the Spartan's Technique 1 (Aleatory Uncertainty

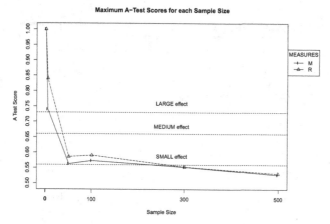

Fig. 1. Spartan's Technique 1 applied to Scenario 1. At 300 runs, stochasticity over the measures M (missing species) and R (number of selected sites) attains a small effect.

Analysis). In order to do so, we analysed 20 subsets sample sizes of 1, 5, 50, 100, 300 and 500 runs each, requiring, therefore, 19,120 individual runs for each previously described optimization scenario (a total of 57,360 individual runs). It was found that, for all the scenarios, 300 runs were sufficient to reduce the effect magnitude of aleatory uncertainty on results to less than "small" (the desired level) (Fig.1).

Parameter Settings. Almost all of the heuristic procedures involve some parameter tuning. The task of setting parameter values is notably challenging because we do not know, in advance, the impact of parameter values on the performance of the algorithm, specially when the algorithm to be tuned is stochastic in nature [32] . We used Spartan's Technique 2 (Robustness Analysis) [1] to investigate the impact of different parameter settings on the quality of the solutions, and to estimate the most suitable values for the following parameters: population size, crossover probability, mutation probability, and mutation rate. The sampling method begins at the parameters lower value and increases the value by a set increment until the upper limit is reached. Each parameter is addressed in turn, and simulation results for value assigned to that parameter analysed (Fig.2). We analysed 26 subsets sample (resulting from the combinations of the different parameter values) sizes of 300 runs each (in accordance with results obtained from Spartan's Technique 1), requiring, therefore, 7,800 individual runs for each optimization scenario (a total of 23,400 individual runs). Based on the obtained results, parameter values were set to: population size = 500; crossover probability = 0.90; mutation probability = 1/L (where L is the number of regions); mutation rate = 0.5. Besides we used: crossover operator = single point crossover (SPX); selection by binary tournament; number of objective functions evaluation = 250.000.

Computer Infrastructure. The experiments were performed on two servers running Ubuntu Linux 12.04 LTS, a HP ProLiant DL585 G7, 4xAMD Opteron

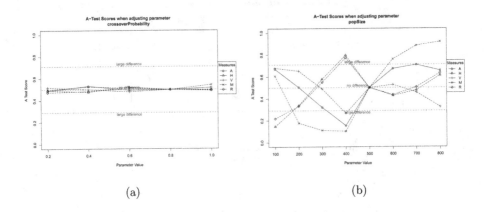

(a)   (b)

Fig. 2. Spartan's Technique 2 applied to Scenario 2. The x-axis shows the range of values explored and the y-axis displays the scores obtained by contrasting response values for perturbed parameter values to calibrated values. Solutions are considered over the measures A (AET), H (human occupancy), V (vegetation remnants), M (missing species), and R (number of seleted sites). (a) Scores for different values of crossover probability, which when perturbed has no significant effect on solutions. (b) Scores for simulations varying population size, this parameter has a strong effect on the obtained solutions, and its most suitable value is 500. Results suggest that a change in the population size has a statistically significant effect on solutions, and it is more critical than the crossover probability, which has no statistically significant impact.

6386 SE 2.8Ghz 16-cores (64 physical CPU cores), 512GB RAM, and a HP ProLiant DL385p Gen8, 2xAMD Opteron 6386 SE 2.8Ghz 16-cores (32 physical CPU cores), 256GB RAM.

Evaluation Metric. Due to the stochasticity of the algorithm, we used the *selection frequency* metric (SF) [17] to compare the outcomes of our analysis. This measure represents the number of times each site is selected in the solutions to the overall problem. Once the SF to all cells was calculated, grid cells were ranked based on the result. Grid cells with the highest SF were assigned the first rank and those having SF value zero received the last rank. Next, cell relative importance in both axes was rescaled to 0–100 (zero being not important, and 100 being highly important). Then, grid cells with value zero were excluded and all the remaining grid cells ordered in a bi-dimensional plot showing the relative importance of each cell related to current time and to 2080. This graph epitomizes the scheme for dynamic spatial prioritization analyses for biodiversity conservation.

Cells with rank higher than 90 for both axes were considered high-priority. Cells ranking higher than 90 in the present, but not in 2080, are important now, but will become climatically unsuitable in the future. Cells ranking higher than 90 in 2080, but not now, will become suitable in the future. High-priority cells are those ranking higher than 90, now and in the future.

It is worth noting that we settle the lower limit rank to 90 following the literature [17], but this value is arbitrarily defined and, depending on the context, can be relaxed assuming other lower reference values (e.g., considering cells ranking higher than 50 as important, instead of higher than 90).

4   Results and Discussion

4.1   Scenario 1

The objective of the optimization in this context was to select the smallest set of sites, among the 181 available ones, capable of representing all the 96 species (the species diversity) in current time. This also allowed to establish a lower bound for Scenario 2.

We found that the minimum number of sites required to represent all of the species was 2. We found 35 distinct solutions with these characteristics, reflecting diversity in solution, which is important since it provides more options to decision makers. It is important to note that we have no hierarchy amongst results, which means that all the solutions are equal in the considered context.

A relative frequency map of the multiple solutions indicates the relative importance of a cell in order to fullfil the objectives of optimization (Fig.3). This frequency can be taken as an estimator of *irreplaceability*[1] of the cell [21], e.g., the rarest plant appears in only 10 regions in current time, these sites tend to be irreplaceables, so that if at least one of them is not selected, the conservation goal may not be achieved. One of such a site is #105, the most frequent site in solutions (associated to the presence of the rarest specie, it has the greatest diversity of species).

4.2   Scenario 2

The objective of the optimization was to select the smallest set of sites capable of representing all the 96 species in current time, but at the same time optimizing the additional objectives AET, H_O, and VR.

Although there is some empirical inferences, and correlational data in the literature, experts did not know, in principle, what to expect from the optimization in 5 dimensions, since a behaviour was not determined with respect to optimizing AET, H_O and VR simultaneously. The initial expectation was that this additional information would bring some advantage selecting sites to compose solutions, improving, therefore, the overall quality of results.

NSGA-II was not able to find (at least in the number of evaluations performed) the lower bound of 2 sites established in Scenario 1, being 3 the smallest set of sites found. This can be due to the use of a multi-objective algorithm in this scenario, when maybe the most appropriated would be to apply a many-objective

---

[1] A measure that indicates the proportion a cell contributes to the overall solution, e.g., cells with this measure converging to 1 often tends to be irreplaceable, in the sense that if they are lost, the conservation goal is not accomplished.

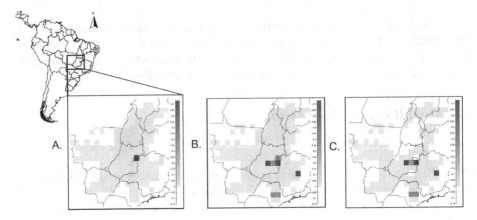

Fig. 3. Irreplaceability for: A. Scenario 1. B. Scenario 3. C. Synthesis of Scenarios 1 and 3. Irreplaceability scales from 0 to 1 (since it express the frequency a given site appears in the solutions). Cells shown in the darkest colour tend to be irreplaceable, which means that if they are lost, the conservation target (to represent all existing species) may not be achieved. Sites presenting rare species tend to have higher irreplaceability scores. Regarding the synthesis of two scenarios, irreplaceability can assume a negative value (which is plotted as value zero; white cells), this indicates that the site has lost importance (its capability to fullfil the requirements to achieve the objective).

approach [16]. Although, it can indicate that, in the context of using additional objectives, better results are obtained not through the minimum absolute possible representation, but through a trade-off between minimum representation and the other considered objectives.

The portfolio of solutions increased significantly, which was expected, since it is known in the literature that as the number of objectives increases, the number of solutions enlarge exponentially [9,16]. Thus, almost all new combination of sites will give a different result with all species being represented, so it is included in the portfolio.

Results (using Pearson correlation) confirmed the empirical conflict between H_O and VR ($r = -.84, p < .0001$), as well as between H_O and AET ($r = -.93, p < .0001$) (Fig. 4.a and Fig. 4.b), which corroborates with the evidence of *conservation conflicts* [4]. This means that H_O reflects properly the antropical effect over biodiversity by the conversion of natural habitats in antropical ones [14].

This is a strong evidence that in addition to the standard biological data used to guide planning decisions, some kind of human settlement patterns (here H_O) have to be explicitly considered from the very beginning of planning processes [18]. This is essential to reduce the conflict between population density and biodiversity and to minimize the cost of conservation (since land prices inexorably rise as human population density increases).

In addition, a positive relationship between H_O and species richness may be expected because both increase with AET [27]. This was confirmed by results obtained with a steady number of sites, where for higher values of H_O, higher

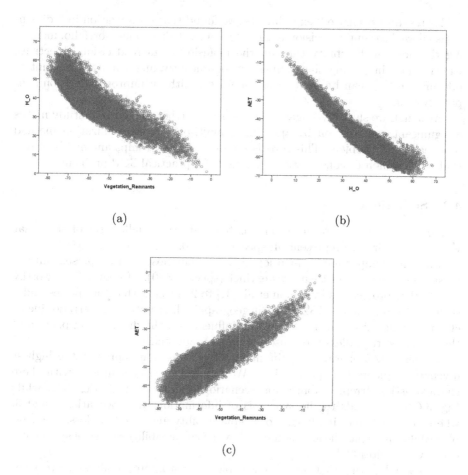

Fig. 4. Scenario 2, optimization of five objectives simultaneously. Scatterplot of additional objectives in pairs. A. VR vs H_O ($r = -.84, p < .0001$). B. H_O vs AET ($r = -.93, p < .0001$). C. VR vs AET, showing a positive correlation between them ($r = .84, p < .0001$), this is a poor pair for optimization. A and B show a negative correlation, revealing a conflicting behaviour which means that they are proper candidates for optimization.

values of AET were observed (human settlement follows better conditions patterns), even though the relationship between H_O and AET is inversely proportional (H_O has a deleterious effect on AET).

The scatterplot of AET and VR (Fig. 4.c) shows that these objectives have a positive correlation ($r = .84, p < .0001$), i.e., there is no conflict between them, revealing that this pair would be a poor candidate for multi-objective optimization. But since by definition, in optimal solutions, improving the value in one dimension of the objective function vector leads to a degradation in at least one other dimension of it, the other objectives (H_O and AET) hold optimization conditions.

As result of Scenario 2 experiments, we found that optimization in 5 dimensions allowed to supply decision makers with a more diversified portfolio, increasing the problem flexibility through the inclusion of more decision objectives, which whilst increasing the complexity, significantly augments the amount of information that can be used to provide users with an improved decision support system.

Although we shall investigate these aspects further, the current study makes a significant contribution by applying a multi-objective optimization method to a real-world problem. This reveals important relationships among objectives that are common to conservation scenarios of a practical SCP problem.

4.3   Scenario 3

The objective of this optimization was to locate the smallest set of sites that would be required to represent all species in 2080.

First, it is important to mention that it is not possible to represent all the 96 species, since one of them was extinct (species #80). Moreover, it is worthy of note that projections by Simon et al. [31] to 2080 show that the species under study will reduce about 78% of their geographic distribution in Cerrado due to climate change that will have a strong influence on the distribution pattern of these species, regardless the conservation plan adopted.

In this new scenario, the mimimum set of sites that represent the highest diversity of plants (95 species) is 5. We found 4 distinct solutions with these characteristics. Irreplaceability for Scenario 3 can be seen in Fig. 3.B, while Fig. 3.C corresponds to the synthesis of information from Scenarios 1 and 3, where positive values imply gain of irreplaceability and negative, loss. The irreplaceability map has the advantage of showing the flexibility degree of systematic conservation sites [23].

It is worth noting that despite ensemble forecast approach allows more accurate predictions on changes in the species bioclimatic envelope, it is not possible to remove all uncertainty associated with projections of future climates [31].

4.4   Dynamic Spatial Prioritization

Having a picture of how future scenarios will look like can be extremely useful for decision makers. To assess the relative importance of sites in achieving conservation targets both for current time (Scenarios 1 and 2) and for future (Scenario 3), we compared the variation of site *selection frequency* scores under dynamic conditions, using a bi-dimensional graph (Fig. 5).

Grid cells populating the *upper right corner* of the graph (framed by a square) are important both for current time and for future scenarios of climate change, therefore it would be a good choice to invest in them. However, grid cells located in the *lower right corner* (framed by a rectangle) represent a risk of conservation investment given their low relative importance in 2080, so, based in this information, the decision maker might opt not to invest in these regions, redirecting

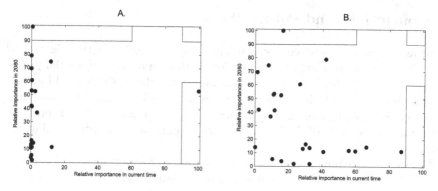

Fig. 5. Graphs for establishing a dynamic spatial conservation prioritization analysis. A. Scenario 1 x Scenario 3. B. Scenario 2 x Scenario 3. The relative importance of grid cells is given by a rank ranging from 0 to 100, based on their selection frequency. Grid cells placed in the *upper right corner* are important in the present time and in a future scenario. Grid cells in the *lower right corner* represent a risk of conservation investment as they seems to be very important in current time, but of low relative importance in future. Grid cells in the *upper left corner* gain attention since they are not critical for current time, but might became important in the future.

funds to another more promising area. Nevertheless, grid cells in the *upper left corner* (framed by a rectangle) deserve attention as they might become very important in the future even though they are not critical at present time. In this case, careful land-use planning is imperative because these regions can represent good cost-benefit in the long-term.

This information, associated to data displayed in Fig. 3.C, provides important knowledge support to decision makers. Supported by scientific data, they can scrutinize the options available in current time and decide how to define their spatial conservation priorities, reviewing them if necessary.

For optimization in two dimensions (Scenario 1) (Fig. 5.A), results show data concentrated along the vertical axis (that represents importance in 2080). We were able to identify a region in the *upper left corner*, representing a location that probably will become very important in the future, although not being critical at present time. We also found a region in the *lower right corner* that can represent a risk to conservation investment given its low relative importance in 2080.

Our results show that with optimization using additional objectives (in 5 dimensions; Scenario 2) (Fig. 5.B), we were able to find data more smoothly spread along both relative importance axes, and specially leading to the *upper right corner* and closer to the *upper left corner* that could be the most attractive locations to invest.

Although it would be interesting to find data in the *upper right corner* of Fig. 5, the results reflected the available data, and it strongly indicates that these solutions simply do not exist. However, our method is able to identify (if they exist) sites of high priority for conservation, regions with high risk of investment and sites that may become attractive options in the future. And these data can be used in order to help decision makers to select their schemes of conservation.

5   Conclusions and Future Work

As far as we know, this is the first attempt to apply multi-objective algorithms to a SCP problem associated to climate forecasting, in a dynamic spatial prioritization analysis for biodiversity. Our work improves the methods used by most of the tools for SCP, which in general apply a static and monobjective approach.

We applied the proposed new approach to a real and important SCP problem that is the conservation of the Brazilian Cerrado obtaining consistent and useful results.

The use of more dimensions allows to incorporate relevant information in the context of SCP, increasing the complexity of the process but in a more intuitive and simpler way (without the assistance of an expert).

We suggest that priorities for conservation could be integrated into a strategy that considers different additional objectives helping to select areas, which results in a conservation plan that is likely to be more effective taking into account the impact of climate change.

The dynamic analysis is an improvement compared to the static approach since it reflects a significant opportunity to adjust priorities into biodiversity conservation plan, by comparing the relative importance of conservation targets in current time and in the future.

Although bioclimatic models are effective and widely used to evaluate the consequences of climate changes for biodiversity, there are still many uncertainties associated to projections to the future.

Our results show that, despite the encouraging achievements, efforts to address the loss of biodiversity need to be strengthened by complementary policies, since changes in climate are inevitable and tend to strongly affect conservation projects as result of the direct influence on the persistence of species.

This was an exploratory study that showed the advantages of the new approach with respect to previous solutions. Having established that the approach is viable with a standard MOO algorithm, our future work will focus on the development of a multi-objective algorithm more specialized to the SCP problem. Given the success of a variety of work in the Artificial Immune System area, e.g. [30], who showed better solutions (closer to the origin axes an more regularly spread throughout the known Pareto Front), we will build on that work to improve the work presented in this paper.

We also plan to perform further comparative studies addressing SCP problem scenarios that deal with optimization of more than three objectives (e.g. Scenario 2), applying approaches as many-objective optimization [16] and bilevel optimization [12].

Acknowledgments. SS wishes to thank the University of York and Prof. Jon Timmis for the PhD stay, and the support from CNPq throughout a Science without Borders scholarship. Jon Timmis is part funded by The Royal Society. GENPAC has been supported by CNPq/MCT/CAPES (projects #564717/2010-0 and #563624/2010-8) and by the GECER (PRONEX/FAPEG/CNPq CP 07-2009). Work by MEW, AC, RL and JADF have been continuously supported by productivity fellowships from CNPq.

References

1. Alden, K., Read, M., Timmis, J., Andrews, P.S., Veiga-Fernandes, H., Coles, M.: Spartan: A Comprehensive Tool for Understanding Uncertainty in Simulations of Biological Systems. PLoS Comput. Biol. 9(2), e1002916+ (2013)
2. Ardron, J., Possingham, H.P., Klein, C. (eds.): Marxan Good Practices Handbook. Pacific Marine Analysis and Research Association (PacMARA), Victoria, BC, Canada (July 2010)
3. Ball, I.R.: Mathematical Applications for Conservation Ecology: The Dynamics of Tree Hollows and the Design of Nature Reserves. PhD thesis, University of Adelaide, Dept. Applied Mathematics, Env. Science and Management (2000)
4. Balmford, A., Moore, J., Brooks, T., Burgess, N., Hansen, A., Williams, P., Rahbek, C.: Conservation Conflicts Across Africa. Science 291, 2616–2619 (2001)
5. Cabeza, M., Moilanen, A.: Design of Reserve Networks and the Persistence of Biodiversity. Trends Ecol. Evol. 16(5), 242–248 (May 2001)
6. Carvalho, F., Ferreira, L., Lobo, F., Diniz-Filho, J., Bini, L.: Spatial Autocorrelation Patterns of The Modis Vegetation Indices for the Cerrado Biome. Revista Árvore. 32(4), 279–290 (2008)
7. Ciarleglio, M.: Modular Abstract Self-Learning Tabu Search (MASTS) Metaheuristic Search Theory and Practice. PhD thesis, Univ. Texas at Austin, Texas (2008)
8. Ciarleglio, M., Barnes, J., Sarkar, S.: ConsNet - A Tabu Search Approach to the Spatially Coherent Conservation Area Network Design Problem. J. Heuristics 16, 537–557 (2010)
9. Coello Coello, C.A., Lamont, G.B., Van Veldhuizen, D.A.: Evolutionary Algorithms for Solving Multi-Objective Problems, 2nd edn. Springer, New York (2007) ISBN 978-0-387-33254-3
10. de Almeida, S.P.: Cerrado: Aproveitamento Alimentar (Cerrado: Food Utilization). Embrapa - CPAC, Planaltina (1998) (in Portuguese)
11. Deb, K., Rudolph, G., Lutton, E., Merelo, J.J., Schoenauer, M., Schwefel, H.-P., Yao, X. (eds.): PPSN 2000. LNCS, vol. 1917, pp. 849–858. Springer, Heidelberg (2000)
12. Deb, K., Sinha, A.: Evolutionary Bilevel Optimization (EBO). In: Proceedings of the 2014 Conference Companion on Genetic and Evolutionary Computation Companion, GECCO Comp 2014, pp. 857–876. ACM, New York (2014)
13. Diniz-Filho, J., Bini, L., Rangel, T., Loyola, R., Hof, C., Nogués-Bravo, D., Araújo, M.: Partitioning and Mapping Uncertainties in Ensembles of Forecasts of Species Turnover Under Climate Change. Ecography 32(6), 897–906 (2009)
14. Diniz-Filho, J., Bini, L., Vieira, C., Blamires, D., Terribile, L., Bastos, R., Oliveira, G., Souza, B.: Spatial Patterns of Terrestrial Vertebrate Species Richness in the Brazilian Cerrado. Zool. Stud. 47(2), 146–157 (2008)
15. Groves, C., Game, E., Anderson, M., Cross, M., Enquist, C., Ferdaña, Z., Girvetz, E., Gondor, A., Hall, K., Higgins, J., Marshall, R., Popper, K., Schill, S., Shafer, S.: Incorporating Climate Change into Systematic Conservation Planning. Biodivers. Conserv. 21, 1651–1671 (2012)
16. Jain, H., Deb, K.: An Evolutionary Many-Objective Optimization Algorithm Using Reference-Point-Based Nondominated Sorting Approach, Part I: Solving Problems With Box Constraints. IEEE Journal 18(4), 577–601 (2013)
17. Loyola, R., Lemes, P., Nabout, J., Trindade-Filho, J., Sagnori, M., Dobrovolski, R., Diniz-Filho, J.: A Straightforward Conceptual Approach For Evaluating Spatial Conservation Priorities Under Climate Change. Biodivers. Conserv. 22, 483–495 (2013)

18. Luck, G., Ricketts, T., Daily, G., Imhoff, M.: Alleviating Spatial Conflict Between People and Biodiversity. Proc. Natl. Acad. Sci. USA **101**(1), 182–186 (Jan. 2004)

19. Margules, C.R., Pressey, R.L., Nicholls, A.O.: Nature Conservation: Cost Effective Biological Surveys and Data Analysis, chapter Selecting Nature Reserves. Commonwealth Scientific & Industrial Research (CSIRO), Dickson, Australia (1991)

20. McCarthy, D.P., Donald, P.F., Scharlemann, J.P.W., et al.: Financial Costs of Meeting Global Biodiversity Conservation Targets: Current Spending and Unmet Needs. Science **338**(6109), 946–949 (2012)

21. Meir, E., Andelman, S., Possingham, H.P.: Does Conservation Planning Matter in a Dynamic and Uncertain World? Ecol. Lett. **7**(8), 615–622 (2004)

22. Possingham, H.P., Ball, I., Andelman, S.: Mathematical Methods for Identifying Representative Reserve Networks, ch. 17, pp. 291–305. Springer, New York (2000)

23. Pressey, R.L.: Ad Hoc Reservations: Forward or Backward Steps in Developing Representative Reserve Systems? Conserv. Biol. **8**, 662–668 (1994)

24. Pressey, R.L.: The First Reserve Selection Algorithm: a Retrospective on Jamie Kirkpatrick's 1983 Paper. Prog. Phys. Geog. **26**(3), 434–441 (2002)

25. Pressey, R.L., Possingham, H.P., Day, J.R.: Effectiveness of Alternative Heuristic Algorithms for Identifying Indicative Minimum Requirements for Conservation Reserves. Biol. Conserv. **80**(2), 207–219 (1997)

26. Pressey, R.L., Possingham, H.P., Margules, C.R.: Optimality in Reserve Selection Algorithms: When Does it Matter and How Much? Biol. Conserv. **76**(3), 259–267 (1996)

27. Rangel, T., Bini, L., Diniz-Filho, J., Pinto, M., Carvalho, P., Bastos, R.: Human Development and Biodiversity Conservation in Brazilian Cerrado. Appl. Geogr. **27**(1), 14–27 (2007)

28. Rodríguez, M., Belmontes, J.A., Hawkins, B.: Energy, Water and Large-scale Patterns of Reptile and Amphibian Species Richness in Europe. Acta. Oecol. **28**, 65–70 (2005)

29. Sarkar, S.: Complementarity and the Selection of Nature Reserves: Algorithms and the Origins of Conservation Planning, 1980–1995. Arch. Hist. Exact. Sci. **66**, 397–426 (2012)

30. Schlottfeldt, S., Saéz, Y., Isasim P.: Sistemas Inmunológicos Artificiales aplicados al Problema de Optimización Multiobjetivo Radio Network Design (Artificial Immune Systems applied to the Radio Network Design Problem). Technical Report UC3M-TR-CS-2009-01, Universidad Carlos III de Madrid (2009) (in Spanish)

31. Simon, L., Oliveira, G., Barreto, B., Nabout, J., Rangel, T., Diniz-Filho, J.: Effects of Global Climate Changes on Geographical Distribution Patterns of Economically Important Plant Species in Cerrado. Rev. Árvore. **37**(2), 267–274 (2013)

32. Sinha, A., Malo, P., Xu, P., Deb, K.: A Bilevel Optimization Approach to Automated Parameter Tuning. In: Proceedings of the 2014 Conference Companion on Genetic and Evolutionary Computation Companion, GECCO Comp 2014, pp. 847–854, ACM, New York (2014)

Marginalization in Mexico: An Application of the ELECTRE III–MOEA Methodology

Jesús Jaime Solano Noriega[1], Juan Carlos Leyva López[2,3] (✉)
and Diego Alonso Gastélum Chavira[2,3]

[1] Universidad Autónoma de Ciudad Juárez, Ave. Del Charro 450 Norte, Ciudad
Juárez, Chihuahua, Mexico
al127821@alumnos.uacj.mx
[2] Universidad de Occidente, Blvd. Lola Beltrán y Blvd. Rolando Arjona,
Culiacán, Sinaloa, México
{juan.leyva,diego.gastelum}@udo.mx
[3] Universidad Autónoma de Sinaloa,
Ciudad Universitaria, Culiacán, Sinaloa, México

Abstract. In this paper, a multi-criteria approach for ranking the Municipalities of the States of Mexico by their levels of marginalization is proposed, and the case for the State of Jalisco is presented. The approach uses the ELECTRE III method to construct a medium-sized valued outranking relation and then employs a new multi-objective evolutionary algorithm (MOEA) based on non-dominated sorting genetic algorithm II (NSGA-II) to exploit the relation to obtain a recommendation. The results of this application can be useful for policy-makers, planners, academics, investors, and business leaders. This study also contributes to an important, yet relatively new, body of application-based literature that investigates multi-criteria approaches to decision making that use fuzzy theory and evolutionary multi-objective optimization methods. A comparison of the ranking obtained with the proposed methodology and the stratifications created by the National Population Council of Mexico shows that the methodology presented consistent and reliable results for this problem.

Keywords: Multi-criteria decision analysis · Municipal marginalization · Ranking problem · ELECTRE III · Multi-objective evolutionary algorithms

1 Introduction

Marginality can be defined as *an involuntary position and condition of an individual or group at the margins of social, political, economic, ecological, and biophysical systems that prevents them from accessing resources, assets, and services, and that restrains their freedom of choice, preventing the development of capabilities, and eventually causing extreme poverty* [7]. Such phenomena, which have caused the social inequality that has characterized Mexico, have persisted despite important advances. Related social, economic and demographics indicators have forced the Mexican government to endorse the commitment to continue fighting conditions that disadvantage certain population groups and certain regions of the country.

© Springer International Publishing Switzerland 2015
A. Gaspar-Cunha et al. (Eds.): EMO 2015, Part II, LNCS 9019, pp. 473–486, 2015.
DOI: 10.1007/978-3-319-15892-1_32

Reference [6] emphasizes the need for coherent policies and strategies to address multiple factors that constitute marginalization in socio-ecological systems. As part of demographic planning in Mexico, the National Population Council's (CONAPO) mission is to involve people in economic and social development programs that are formulated within the government sector and to link their goals to the needs posed by this socio-demographic phenomenon. To clarify how this phenomenon occurs in different regions of the country, the CONAPO has built nine socio-demographic indicators that, using the statistical technique of principal component analysis, generates a marginalization index (MI). The MI is a summary measure to differentiate entities and municipalities of Mexico according to the overall impact of shortages faced by the people because of lack of access to education, inadequate housing, insufficient funds and residency in rural areas.

The relative marginalization of a region within a given country can be assessed using different types of traditional methods. However, such an evaluation method should be multi-criteria in nature because of the multidimensional nature of social and economic marginalization. This case study utilizes a Multi-criteria Decision Aiding (MCDA) method to construct an aggregation model of preferences and then employs a new MOEA to exploit the model to rank the municipalities of the State of Jalisco, Mexico, according to their marginalization level using the same socio-demographic indicators constructed by CONAPO. While such an application has practical implications, the methodology has not yet been sufficiently developed. The proposed methodology is based on a previous work [8]. It is a MCDA method with an outranking approach which makes use of a MOEA to construct a partial order of classes of alternatives from a medium-sized set of decision alternatives.

This paper is organized as follows: Section 2 describes the ELECTRE III method and the MOEA based on NSGA-II. Section 3 describes the case study, focusing on the procedure and method used. The final section presents conclusions, comments and future research.

2   The (ELECTRE III – MOEA) Methodology

2.1   The ELECTRE III Method

As part of a philosophy of decision aid, ELECTRE (in its various forms) was conceived by [10] in response to deficiencies of existing decision-making solution methods. Roy's philosophy of decision aid is well exposed in [11]; moreover, of the different versions of ELECTRE methods (I, II, III, IV, IS and TRI), this paper only uses the method referred to as ELECTRE III, which is used when it is possible and desirable to build valued outranking relationships and quantifying the relative importance of criteria.

As part of the principle of the outranking approach, the ELECTRE III method comprise two phases: the construction of a so called outranking relation followed by an exploitation procedure to deliver a ranking. In the first step, basic information for the ELECTRE III method is composed by a set of n pseudo-criteria $\left\{\left(g_j, q_j, p_j\right), j = 1, 2, ..., n\right\}$ on a set of alternatives A and for each criterion is given: a weight w_j expressing the relative importance of the g_j criterion, and an veto threshold $v_j(g_j) > 0$. For each ordered pair $(a,b) \in A$ are defined a concordance index $C(a,b)$ and a discordance index $d_j(a,b)$ as follows:

$$C(a,b) = \frac{1}{W} \sum_{j=1}^{n} c_j(a,b)$$ (1)

where

$$W = \sum_{j=1}^{n} w_j$$ (2)

and

$$c_j(a,b) = \begin{cases} 1 \ \text{if} \ g_j(a) + q_j(g_j(a)) \geq g_j(b) \\ 0 \ \text{if} \ g_j(a) + p_j(g_j(a)) \leq g_j(b) \\ \text{linearly increasing with } g_j(a) \text{ in the intermediate region} \end{cases}$$ (3)

and

$$d_j(a,b) = \begin{cases} 0 \ \text{if} \ g_j(a) + p_j(g_j(a)) \geq g_j(b) \\ 1 \ \text{if} \ g_j(a) + v_j(g_j(a)) \leq g_j(b) \\ \text{linearly decreasing with } g_j(a) \text{ in the intermediate region} \end{cases}$$ (4)

The concordance index is considered as a measure indicating whether "action a is at least as good as action b" (usually called "a outranks b", denoted aSb) on criterion g_j while the discordance index of a criterion g_j aims to take into account the fact that this criterion is more or less discordant with such assertion. Once these components are known, an outranking relation for each pair of alternatives $S_A^\sigma(a,b)$ is constructed as follows:

$$S_A^\sigma(a,b) = \begin{cases} C(a,b) \ \text{if} \ d_j(a,b) \leq C(a,b), \forall j \\ C(a,b) \cdot \prod_{j \in J(a,b)} \dfrac{1 - d_j(a,b)}{1 - C(a,b)} \ \text{otherwise} \end{cases}$$ (5)

where $J(a,b)$ is the set of criteria j such $d_j(a,b) > C(a,b)$. In this way, $\sigma(a,b)$ can be interpreted as a credibility index expressing comprehensively in what measure "a outranks b" using both the comprehensive concordance index and the discordance indices for each criterion g_j.

For a detail explanation of ELECTRE III method, readers can review [12].

The second step in the outranking approach is to exploit the aggregation model of preferences represented by a valued outranking relation S_A^σ, and produce a ranking of alternatives from such valued outranking relation. Our proposed approach for the exploitation step, is to use a MOEA-based heuristic method, briefly explained in the next section.

2.2   A Multi-Objective Evolutionary Algorithm for Deriving Final Ranking

In this subsection, we present a MOEA based on a posterior articulation of preferences, that is able to exploit a known valued outranking relation with the purpose of constructing a recommendation for the multi-criteria ranking problem with a medium-sized set of alternatives. The algorithm borrows fundamental elements from NSGA-II [3]. In the following subsections, we present in further detail the fundamental aspects of the algorithm.

Comparing Alternatives and Set of Alternatives

First, we present a general structure in which alternatives and classes of alternatives can be compared to each other. We then use this structure to highlight and model the objective functions.

Each potential solution in the population is associated with a number λ, the cut level, where $0 \leq \lambda \leq 1$. Each cut level λ is associated with the given valued outranking relation S_A^σ. We then can induce a crisp outranking relation S_A^λ. From S_A^λ, we can deduce the followings preference relations:

Indifference
$$a_i I_A a_j \quad \leftrightarrow \quad \sigma(a_i, a_j) \geq \lambda \wedge \sigma(a_j, a_i) \geq \lambda$$
Preference
$$a_i P_A a_j \quad \leftrightarrow \quad \sigma(a_i, a_j) \geq \lambda \wedge \sigma(a_j, a_i) \leq \lambda - \beta$$
Incomparability
$$a_i R_A a_j \quad \leftrightarrow \quad \sigma(a_i, a_j) \leq \lambda - \beta \wedge \sigma(a_j, a_i) \leq \lambda - \beta$$
where λ is a constant cutting level and β is threshold level indicating the minimum values a decision-maker (DM) may accept that "a_i outranks a_j" ($a_i S_A^\lambda a_j$).

Let $P_k(A) = \{C_1, C_2, ..., C_k\}$ be a partition of A.

S_A^λ induces an antisymmetric crisp outranking relation $S^*_{P_k(A)}$ between the determined classes in the following form:

For each pair of classes (C_r, C_q), $r, q = 1, 2, ..., k$ we compute

$$l^*_{rq} = \begin{cases} \arg \max_{l_{rq} \in \{2,3,4\}} (\sum_{a_i \in C_r} \sum_{a_j \in C_q} \chi_{l_{rq}}(a_i, a_j) & if \quad r \neq q \\ 1 & if \qquad\qquad r = q \end{cases} \tag{6}$$

where

$$\chi_{l_{rq}}(a_i, a_j) = \begin{cases} 1 & if \quad (a_i, a_j) \in O \\ 0 & otherwise \end{cases} \quad \forall a_i, a_j \in A, O \in \{I_A, P_A^+, P_A^-, R_A\} \tag{7}$$

and l_{rq}^* represent the crisp preference relation between classes C_r and C_q $(2 \to P_{P_k(A)}^+, 3 \to P_{P_k(A)}^-, 4 \to R_{P_k(A)})$ where $P_{P_k(A)}^+, P_{P_k(A)}^-, R_{P_k(A)}$ are the preference relations between classes "strictly preferred to", "strictly preferred by" and "incomparable to" respectively. Reference [2] proofs that this procedure leads to the optimal crisp outranking relation $S_{P_k(A)}^*$ on such partition $P_k(A) = \{C_1, C_2, \ldots, C_k\}$ of A. We construct $S_{P_k(A)}^*$ in such a form that it fulfills the reflexive and antisymmetric properties.

Representation of a Potential Solution in the Ranking Problem

The locus-based adjacency representation proposed in [9] is used. This is a graph-based representation, as illustrated in Fig. 1, where each individual p consists of m genes $p_1, p_2, \ldots, p_m$, where $m = |A|$, and each gene p_i can take allele values j between 1 and m. A value of j assigned to gene p_i is interpreted as a link from alternative i to j. The set of all linked alternatives forms a graph that can have 1 to m connected components. Then, each connected component in the resulting graph is considered a class of alternatives formed with all of the alternatives that belong to each of them. The decoding of this representation requires the identification of all connected components.

The locus-based adjacency encoding scheme has several major advantages; most importantly, there is no need to fix the number of classes in advance, as it is automatically determined in the decoding step. Hence, it is possible to evolve and compare solutions with a different number of classes in just one run of the evolutionary algorithm.

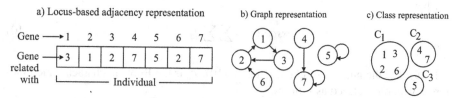

Fig. 1. Locus-based adjacency representation. A set of seven alternatives is partitioned. Figure (a) shows one possible genotypes of an individual of the population. It is transformed into the graph structure shown in Figure (b). Every connected component with this graph is interpreted as an individual class, as visualized by the circles in Figure (c).

Objective Functions

Maximizing the Cutting Level λ

From the valued outranking relation S_A^σ, it is possible to define a family of nested crisp outranking relations S_A^λ ($S_A^\lambda = \{(a,b) \in A x A : \sigma(a,b) \geq \lambda\}$, $\lambda \in [\lambda_0, 1]$). These crisp relations correspond to λ-cuts of S_A^σ, where cutting level λ represents the minimum value for S_A^σ so that $a S_A^\lambda b$ is true (see [5] for details).

Each potential solution is associated with a λ-cut, that is connected with the credibility level of a crisp outranking relation S_A^λ defined on the set of alternatives.

We want potential solutions for which credibility level λ is near 1. This indicates to us that the ranking obtained from a decoded potential solution with credibility level λ is more trustworthy. We call this objective the *maximum cut level objective*. In practice, we use an additional condition for credibility level λ— the function f, which does not permit λ values to approach one— because in this case, we could have many incomparable genes. The function f is defined as follows:

$$f(\tilde{p}) = \left| \begin{cases} (a_{k_i}, a_{k_j}): a_{k_i} \sim S a_{k_j} \ and \ a_{k_j} \sim S a_{k_i} \ ; \\ i = 1, 2, ..., m-1, \ j = 2, 3, ..., m, \ i < j \end{cases} \right| \tag{8}$$

$f(\tilde{p})$ is the number of incomparabilities between pairs of actions (a_{k_i}, a_{k_j}) in the individual $\tilde{p} = a_{k_1} a_{k_2} ... a_{k_m}$ in the sense of the crisp relation S_A^{λ}. Note that the quality of a solution increases with decreasing f score. In this case, we are interested in individuals whose f values are equal (or close) to zero. This condition improves the comparability of S on A.

The Min Cut Objective
The min cut shown in (9) aims to maximize the indifference within each class. We proceed by penalizing the pairs of alternatives inside the class that are not indifferent.

$$MinCut = \sum_{m=1}^{k} assoc(C_m) \tag{9}$$

where

$$assoc(G) = \sum_{i \in G} \sum_{j \in G} \eta_{ij} \tag{10}$$

In (9), k is the number of classes, and in (10), η_{ij} is the Boolean characteristic function η, which is defined as follows:

$$\eta_{ij} = \eta(a_i, a_j) = \eta(a_i I a_j) = \begin{cases} 0 & if \quad (a_i, a_j) \in I \\ 1 & otherwise \end{cases} \quad \forall a_i, a_j \in A \tag{11}$$

This objective function is minimized in the corresponding multi-objective optimization problem. We call this objective the *min cut objective*.

The Minimum Pair-wise Preference Disagreement Objective
Let $P_k(A) = \{C_1, C_2, ..., C_k\}$ be a partition of A. Suppose that $C_i O C_j$, where $O \in \{P_A^+, P_A^-, R_A\}$. Supposing that $C_i P_{P_k(A)}^+ C_j$, it is natural that in the beginning of the procedure, some pair of alternatives (a_r, a_s), $a_r \in C_i$, $a_s \in C_j$ is not in concordance with (C_i, C_j), i.e., [while $C_i P_{P_k(A)}^+ C_j$ in $S_{P_k(A)}^*$], $[a_r I_A a_s$, or $a_r P_A^- a_s \ a_r P^- a_s$, or $a_r R_A a_s$ in S_A^{λ}]. In these conditions, we have an inconsistency between the aggregation model of preferences S_A^{λ} and the crisp outranking relation of classes. The quality of the final crisp

outranking relation $S^*_{P_k(A)}$ should also be judged according to the number of its discrepancies and concordances with S^σ_A and the crisp outranking relation S^λ_A. Let V be the set of strong discrepancies defined as:

$$V = \{(a_r, a_s) \in A x A : a_r \in C_i, a_s \in C_j, a_r O_1 a_s, C_i O_2 C_j, O_1 \in \{I_A, P^+_A, P^-_A, R_A\},$$
$$O_2 \in \{P^+_{P_k(A)}, P^-_{P_k(A)}, R_{P_k(A)}\}, O_1 \neq O_2, i, j = 1, 2, ..., k\}$$
(12)

and n_V = cardinality of $V = |V|$.

O_1 and O_2 are preference relations in different sets of alternatives and $O_1 \neq O_2$ means that there is not concordance between the preference relations $a_r O_1 a_s$ in S^λ_A and $C_i O_2 C_j$ in $S^*_{P_k(A)}$. n_V is a function that counts the number of the pair-wise preference disagreements.

We quantify the number of preferences between alternatives into the crisp outranking relation S^λ_A that are in disagreement in the sense of $S^*_{P_k(A)}$. We call this objective the *minimum pair-wise preference disagreement objective*.

Based on these defined objectives, the multi-objective optimization problem that the MOEA aims to solve is as follows:

$$Min(MinCut(\tilde{p})), \quad Min(n_V(\tilde{p})), \quad Max(\lambda(\tilde{p}))$$
(13)

subject to:

$$\tilde{p} \in \Omega$$
(14)

$$\lambda \in [0,1], \quad \lambda \geq \lambda_0$$
(15)

where Ω is the set of antisymmetric crisp outranking relations of classes of alternatives of A, $\tilde{p}$ is an antisymmetric crisp outranking relation of classes of alternatives of a given set of data A, and λ_0 is a minimum level of credibility. Usually, no single best solution for this optimization task exists, but, instead, the framework of Pareto optimality is embraced.

Preference Incorporation in NSGA-II

Most approaches in the evolutionary multi-objective optimization literature concentrate mainly on adapting an evolutionary algorithm to generate an approximation of the Pareto frontier. However, this does not solve the problem. We present an idea: incorporate into NSGA-II the DM's preferences, expressed in a set of solutions assigned to ordered categories. We modified the NSGA-II to make selective pressure toward non-dominated solutions that belong to the *Region of Interest* (ROI) of the DM.

Along with convergence to the Pareto optimal set, it is also desired that an evolutionary algorithm maintains a good spread of solutions in the obtained set of solutions. The original NSGA-II used the well known crowded comparison approach, which has been found to maintain sustainable diversity in a population by controlling crowding of solutions in a deterministic and prespecified number of equal sized cells in the search space.

To solve the multi-criteria ranking problem using the NSGA-II, it is not necessary to seek the entire *Pareto optimal set* P_{true} or the associated Pareto front PF_{true} because many of the non-dominated solutions are not of interest to the DM. We will use the strategy of attempting to find in each NSGA-II generation the most promising and attractive solutions for the DM, which in our case are those individuals $\tilde{p}(MinCut, n_V)$ whose *MinCut* and n_V scores are close to a value of zero and have a acceptable high value of λ. It is sufficient to seek a *restricted Pareto optimal set*, which for our purpose is defined as follows:

$$P_{true}^{restricted} = \begin{cases} \tilde{p} \in P_{true} : \left\| (MinCut(\tilde{p}), n_V(\tilde{p})) \right\|_\infty \leq \varepsilon, \\ where \ \varepsilon \ is \ a \ small \\ non-negative \ number, \quad \lambda > 0.5 \end{cases} \tag{16}$$

Based on this strategy, the proposed method attempts to evolve a population toward the *true restricted Pareto frontier* $(PF_{true}^{restricted})$ by means of a succession of the restricted non-dominated solutions subset $PF_{current}^{restricted}(t) = \{P_1(t), P_2(t), ..., P_n(t)\}$. At each generation the method computes the non-dominated solutions for the ranking problem that are closest to the fixed aspiration level $(MinCut, n_V)$, with $MinCut(\tilde{p}) = 0$ and $n_V(\tilde{p}) = 0$ according to the Tchebycheff metric.

Note that the *true restricted Pareto frontier* $(PF_{true}^{restricted})$ is the ROI of the Pareto front for the DM, the privileged zone of the Pareto frontier that best matches the DM's preferences.

In the modified NSGA-II, we use a modified crowded comparison approach to identify a small, privileged subset of the Pareto front $(PF_{true}^{restricted})$. The new approach does not require *any* user defined parameter to identify the subset of the Pareto front. To describe this approach, we first define a Fixed Aspiration Point (FAP) metric and then present the FAP comparison operator.

Fixed Aspiration Point distance: To identify the solutions surrounding the fixed aspiration level $(MinCut, n_V)$, with $MinCut(\tilde{p}) = 0$ and $n_V(\tilde{p}) = 0$ according to the Tchebycheff metric, we calculate the center of mass $P^{CM(r)}$ of the set $P^{(r)} = \{P_1^r, P_2^r, ..., P_{\mu(r)}^r\}$ of solutions in rank r. The infinity norm of this point $\sigma^r = \left\| P^{CM(r)} \right\|_\infty$ serves as threshold value.

The *Center of Mass* of a group of points is defined as the weighted mean of the points' positions. The weight applied to each point is the point's mass. $\|\bullet\|_\infty$ is the maximum holder metric. Note that $P^{(1)} = PF_{current}^{restricted}$.

For each solution i in rank r, calculate the distance count d_fal_i using the following equation:

$$d\_fal(P_i^r) = d\_fal_i = \begin{cases} \dfrac{\|P_i^r\|_\infty}{\sigma^r}, & if \quad \|P_i^r\|_\infty > \sigma^r \\ 1 & , \quad otherwise \end{cases} \tag{17}$$

This quantity serves to measure the proximity of the solution P_i^r to the *fixed aspiration level* (FAL) (call this the *distance to the fixed aspiration level (d_fal_i)*).

The d_fal_i distance computation requires sorting the population according to each objective function value in ascending order of magnitude.

After all population members in the set are assigned a distance, we can compare two solutions by their extent of proximity with the FAL. A solution with a smaller value of this distance measure is, in some sense, closest to the *fixed aspiration point* (FAP). This is exactly what we compare in the proposed *Fixed_Aspiration_Point-Comparison Operator* described below.

Fixed_Aspiration_Point (FAP)-Comparison Operator: The FAP-comparison operator ($\prec_n$) guides the selection process at the various stages of the algorithm toward the ROI of the Pareto optimal front. Assume that every individual P_i in the population has two attributes:

1. Non domination rank (i_{rank});
2. FAL_to distance ($i_{distance}$).

We now define a partial order $\prec_n$ as $P_i \prec_n P_j$ if (i_{rank} *is less to* j_{rank}) or (($i_{rank=}j_{rank}$) and ($i_{distance}$ *is less to* $j_{distance}$)), where n is the number of non-domination ranks.

That is, between two solutions with different non-domination ranks, we prefer the solution with the lower (better) rank. Otherwise, if both solutions belong to the same front, then we prefer the solution that is closest to the FAP.

The ROI of the Pareto front for the DM is reached by using the FAP-comparison procedure, which is used in tournament selection and during the population reduction phase.

Because of space limitations, we omitted the presentation of the Neighbourhood-biased mutation operator, the initialization procedure, and the final step for obtaining a recommendation.

3   Case Study

3.1   Research Framework

In this study, we embrace the framework of multi-criteria decision aid to achieve the goal of ranking the municipalities of the State of Jalisco, Mexico by their marginalization level. Due to the complexity that represents working with a medium-sized set of municipalities (alternatives) we use the methodology presented in Section 2, drawing on the logic of outranking models (the ELECTRE III procedure [12]) complemented with a MOEA to "solve" the ranking problem.

3.2   Data Source

The data used in this study are part of the socio-demographic indicators constructed by the CONAPO based on data obtained from the 2010 Census of Population and Housing for generating the 2010 marginalization index for the 125 Municipalities of the State of Jalisco. The data was provided by the CONAPO in www.conapo.gob.mx. We omit a complete list of the municipalities of Jalisco due to lack of space.

3.3   Criteria

The criteria used to rank the municipalities are the same socio-demographics indicators constructed by the CONAPO to calculate the marginalization index. They are presented as follows:

- Percentage of population aged 15 or more who are illiterate.
- Percentage of population aged 15 or more who did not completed primary school.
- Percentage of occupants in private homes without sewage.
- Percentage of occupants in private homes without electric power.
- Percentage of occupants in private homes without running water.
- Percentage of private homes with some level of overcrowding.
- Percentage of occupants in private homes with dirt floors.
- Percentage of population living in towns fewer than 5000 inhabitants.
- Percentage of working population with incomes up to twice the minimum wage.

We have not reported the values of the criteria for each municipality (the performance matrix) due to lack of space, but the reader can refer to [1] for further detail.

3.4   Computations with the ELECTRE III-MOEA Methodology

A number of factors influenced the specific selection of the ELECTRE III-MOEA methodology for the problem of ranking the municipalities from the states of Mexico by their level of marginalization. First, in this paper, we presented a MOEA to exploit a valued outranking relation, but it is desirable to demonstrate the functionality of the combination of ELECTRE III and MOEA with a real-world application. Second, there exist a set of municipalities and a set of socio-demographic dimensions that can be easily converted into a set of alternatives and a set of criteria. Additionally, the problem type addressed in this study can be modeled as a multi-criteria ranking problem. Based on the literature, the ELECTRE family of methods is considered appropriate for working with problem types such as the one presented in this study (see [13]). This is especially true for the ELECTRE III method. Third, ELECTRE was originally developed by Roy to incorporate the fuzzy (imprecise and uncertain) nature of decision-making, by using thresholds of indifference and preference. This feature is appropriate for solving this problem. Finally, the choice of ELECTRE III was also influenced by successful applications of the approach (see [4] for a list of successful application of ELECTRE).

Following the MCDA methodology presented in Section 2, we first applied ELECTRE III to construct a valued outranking relation. Then, the obtained valued outranking relation was processed with the proposed MOEA to derive a final partial order of classes of alternatives.

Due to the lack of space in this paper, the steps of the construction of the valued outranking relation are not shown; rather, we highlight the whole process followed by the proposed MOEA.

To find the most promising solutions, we performed the MOEA 10 times with the following parameters: number of generations = 500, population size = 40, crossover probability = 0.9, lambda's value range = [60, 75]. The mutation probability is automatically deduced from the mutation operator.

The top ten solutions, with lower numbers of inconsistencies of the restricted Pareto front, $PF_{known}^{restricted}$ returned by the MOEA at termination, are presented in Table 1.

Table 1. Objective values, overall inconsistencies ($MinCut+n_V$), and number of classes (# Classes) of the top ten solutions with lower numbers of inconsistencies

Solution	λ	MinCut	n_V	$MinCut+n_V$	# Classes
1	0.65	483	782	1265	10
2	0.65	504	767	1271	9
3	0.65	510	764	1274	9
4	0.65	510	766	1276	12
5	0.65	516	761	1277	11
6	0.65	489	788	1277	11
7	0.65	545	736	1281	10
8	0.65	534	748	1282	9
9	0.65	508	775	1283	11
10	0.65	551	733	1284	10

From this set of solutions, we were inclined toward solution #2 because it is the one presented with fewer classes and inconsistencies. In Fig. 2 is shown the decoded representation as a partial order of classes of alternatives for the individual associated to the solution #2 and the stratifications according to the CONAPO study.

These results indicate that the municipalities belonging to the first class C1 of the MOEA are the best evaluated according to their socio-demographic information. Additionally, all of these municipalities are in agreement with the first CONAPO's stratification labeled *Very Low*. The second class C2 of the MOEA presents some differences regarding the CONAPOS's stratification labeled *Low*; municipalities a3, a9, a15, a36, a37, a51, a82, a83, a93 and a121 are better according to the CONAPOS's stratification than the MOEA's ranking. In addition municipalities a4, a52 and a89 are worse according to the CONAPOS's stratification than in the MOEA's ranking. For the third class C3, all but three of the municipalities are in agreement with the third CONAPO stratification labeled *Medium*. The three exceptions are a79, a91 and a125, which are better according to the CONAPOS's stratification than in the MOEA's ranking. The municipalities in the

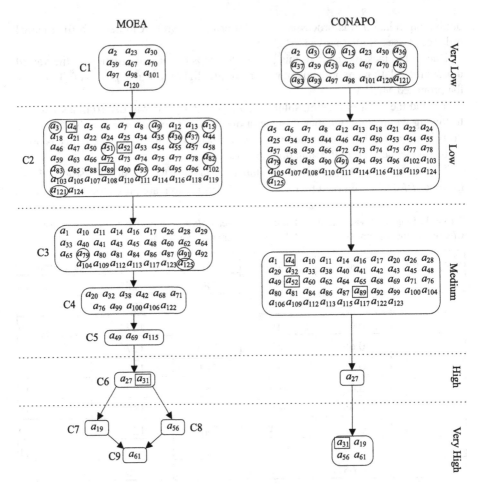

Fig. 2. Left: Decoded representation as a partial order of classes of alternatives of the associated individual of solution #2. Right: ranking of classes (stratifications) according to the CONAPO study. At the left of each class of the MOEA ranking is the identification of each class; at the right of each class of the CONAPO ranking is a label indicating the marginalization level of each class. Inside each class there are the municipalities that belong to them. Municipalities are identified by "*a* + municipality's number" (*a* stands for *alternative*). Municipalities in circles are better ranked in the CONAPO ranking that the MOEA ranking; municipalities in squares are worst ranked in the CONAPO ranking that the MOEA ranking.

fourth and fifth classes C4 and C5 could be considered at the same level of the third CONAPO stratification and are in agreement with it. We can see here the MOEA's ability to appreciate some indifferent patterns and group them apart. Municipalities in the *Medium* stratification are grouped in three different classes in the MOEA results. The sixth class, C6, is formed by just two municipalities and can be compared with the *High*-labeled CONAPO stratification, which is formed only by the *a*27 municipality. The other municipality *a*31 from class C6 is worse according to the CONAPOS's stratification than in the MOEA's ranking. The next two classes, C7 and C8, each have just one municipality. They are considered mutually incomparable; this means that there are no strong arguments to compare them. These two municipalities could be considered in agreement with the *Very high* label of CONAPO stratification. Finally, the last class, C9, has just the $a61$ municipality; it is fully in agreement with the *Very High* CONAPO stratification.

Based on this analysis, we determine that in the MOEA's result 108 of the 125 municipalities were well placed with respect to the CONAPO's result, representing it at a level of consistency of 86.4%.

We omitted the result of a sensibility analysis on the intercriteria parameters of ELECTRE III due to lack of space.

4   Conclusions and Future Research

The aim of this paper was to offer a novel procedure for integrated assessment and comparison of marginalization of municipalities from the States of Mexico – considering the State of Jalisco as a case study – using a multi-criteria decision aiding approach. The proposed procedure for multi-criteria ranking of municipalities uses the ELECTRE III method to construct a medium-sized valued outranking relation, and then employs a MOEA based on the NSGA-II to exploit such relation to obtain a ranking of the municipalities in increasing order of marginalization. The achieved results of the stratification are at least as good as the results obtained by multivariate analysis' traditional methods such as principal component analysis.

The case study presented underpins the use of multi-objective evolutionary algorithms to real life problems in a multi-criteria decision problem context. Thus, contributes to a growing body of application-based knowledge, that was until recently the exclusive domain of engineering and the natural sciences.

In this paper, we addressed the problem of multi-criteria ranking with a medium-sized set of alternatives. The main methodological contribution is a multi-objective evolutionary approach that can be applied on a medium-sized valued outranking relation to solve this problem.

In the future, we intend to use an empirical approach to test our method on a medium set of benchmarks with a wide variety in their structure to highlight the efficiency of the proposed method. Validation tests will be conducted on both artificial and real data sets. We expect to show how close the results from our method come to the optimal solutions. It will also be important to explore the limits of this approach by finding the top size within instances that can be solved with acceptable performance.

References

1. CONAPO: Índice de marginación por entidad federativa y municipio (2010) (retrieved from) http://www.conapo.gob.mx/work/models/CONAPO/indices_margina/mf2010/ CapitulosPDF/ 1_4.pdf (2011)
2. De Smet, Y., Eppe, S.: Multicriteria Relational Clustering: The Case of Binary Outranking Matrices. In: Ehrgott, M., Fonseca, C.M., Gandibleux, X., Hao, J.-K., Sevaux, M. (eds.) EMO 2009. LNCS, vol. 5467, pp. 380–392. Springer, Heidelberg (2009)

3. Deb, K., Pratap, A., Agarwal, S., Meyarivan, T.: A fast and elitist multiobjective genetic algorithm: NSGA-II. IEEE Transactions on Evolutionary Computation **6**(2), 182–197 (2002)

4. Figueira, J., Greco, S., Roy, B., Slowinski, R.: An Overview of ELECTRE Methods and their Recent Extensions. Journal of Multi-Criteria Decision Analysis **20**(1–2), 61–85 (2013)

5. Fodor, J., Roubens, M.: Fuzzy Preference Modeling and Multicriteria Decision Support. Kluwer Academic Publishers, Dordrecht (1994)

6. Gatzweiler, F., Baumüller, H., Ladenburger, C., Joachim, V.B.: Marginality: Addressing the Nexus of Poverty, Exclusion and Ecology. Springer (2014)

7. Gatzweiler, F.W., Baumüller, H., Ladenburger, C., von Braun, J.: Marginality: addressing the root causes of extreme poverty. ZEF working paper 77, Center for Development Research, University of Bonn, Bonn (2011)

8. Leyva, J.C., Solano, J., Gastelum, D., Sanchez, M.: A multiobjective evolutionary approach to a medium-sized multicriteria ranking problem. In: Leyva López, J.C., Espin Andrade, R., Bello Pérez, R., Álvarez Carrillo, P.A. (eds) Studies on Knowledge Discovery, Knowledge Management, and Decision Making, pp. 188-197. Atlantis Press, Eureka, Mazatlán, México (2013)

9. Park, Y.J., Song, M.S.: A genetic algorithm for clustering problems. In: Proceedings of the Third Annual Conference on Genetic Programming, pp. 568–575, July 22–25. University of Wisconsin, M. Kaufmann Publishers Madison (1998)

10. Roy, B.: The outranking approach and the foundations of ELECTRE methods. In: Bana e Costa, C.A. (ed.) Readings in Multiple Criteria Decision Aid, pp. 155–183. Springer, Berlin (1990).

11. Roy, B., Bouyssou, D.: Aide multicritère à la décision: Méthodes et cas. Economica, mai, Paris (1993)

12. Roy, B.: Multi-criteria Methodology for Decision Aiding. Kluwer Academic Publishers, Dordrecht (1996)

13. Roy, B.: The outranking approach and the foundations of ELECTRE methods. Theory and Decision **31**, 49–73 (1991)

Integrating Hierarchical Clustering and Pareto-Efficiency to Preventive Controls Selection in Voltage Stability Assessment

Moussa R. Mansour[1]([✉]), Alexandre C.B. Delbem[1], Luis F.C. Alberto[2], and Rodrigo A. Ramos[2]

[1] Institute of Mathematics and Computer Sciences (ICMC/USP), São Carlos, Brazil
mrmansour@ieee.org
[2] São Carlos School of Engineering (EESC/USP), São Carlos, Brazil

Abstract. Many methods to estimate the cut-off value in order to determine the actual groups from a dendrogram given via hierarchical clustering methods have been proposed in the litetarure. However, in most of the cases, the determination of this value is critical and based on heuristics. In this context, a new method based on Pareto-optimality and on the hierarchical clustering method called Data Mine of Code Repositories (DAMICORE) to determine the most promising groups in a given dendrogram is proposed. This method is called Pareto-Efficient Set Algorithm (PESA). In order to validate the proposed method, PESA was applied find the most promising groups for the preventive control selection problem in the context of voltage stability assessment in electrical power systems. PESA was able to design a set of controllers to eliminate all critical contingencies and was successfully tested in a reduced south-southeast Brazilian system composed of 107 buses.

1 Introduction

The occurrence of recent blackouts, with large impact in the system, associated with voltage stability problems justifies the necessity of developing Voltage Stability Analysis (VSA) tools to assess the security of Electrical Power Systems (EPS), specially in large power systems, on real time. The main aim of a VSA tool is the screening and ranking of a large number of contingencies and the selection of preventive and/or corrective controls. Contingencies are ranked according to severity, which is measured in terms of Voltage Stability Margin (VSM). In case of existence of critical contingencies, the system is considered insecure and preventive control actions have to be designed and implemented to turn them into non-critical ones.

Many methods were developed for preventive control selection in the context of VSA. A natural choice for the design of preventive control actions are the techniques based on optimization methods. In this approach, the VSM is treated as a constrain in the optimization problem [1,2]. One problem of optimization approaches is that a large number of control variables are usually taken into

A. Gaspar-Cunha et al. (Eds.): EMO 2015, Part II, LNCS 9019, pp. 487–497, 2015.
DOI: 10.1007/978-3-319-15892-1_33

account and many control actions have to be performed to achieve the optimal control. In order to avoid this problem, techniques to select the most effective control actions have been desired such that a small number of control variables, the most effective ones, are taken into account in the optimization phase.

Most of the techniques for selection of preventive control actions are based on a sensitivity analysis of the Maximum Loadability Point (MLP) with respect to control variables [3,4]. In [3], for example, the sensitivity of the MLP with respect to control variables is computed using the information of eigenvectors associated with the null eigenvalue of the Jacobian of the power flow equations in the MLP. In [5], a fast method for sensitivity calculation, which does not require an accurate computation of the MLP, but relies on the estimation of the MLP via solution of two power flows [6], was developed.

All the aforementioned methods are capable of ranking the most effective controls for each individual contingency of the list, however, they are not suitable for providing coordination of these controls when a large number of critical contingencies coexist. In [7], a new method to group and coordinate the most effective preventive controls for a set of critical contingencies is proposed. For this purpose, the method employs the sensitivity analysis proposed in [5] evaluating the sensitivity of the VSM with respect to the variation of a control parameter.In addition, a hierarchical clustering method [8] to group the preventive controls that are efficient and sufficient to eliminate the criticality of all contingencies simultaneously.

In [7], the selection of the groups is given by cutting the dendrogram, obtained by the clustering method, in a certain level. However, the computation of this level is very critical, since inadequate cutting levels can result in groups with either large numbers of control devices or too small numbers in the sense they are enougth to eliminate the criticality of all contingencies.

Another strategy to solve a problem of selection of groups is through the Multi-Criteria Decision Analysis (MCDA) [9]. In [10] the selection and screening of hard coating material was studied. MCDA were applied to rank and select these hard coating materials. Pareto-optimality was used to select materials that satisfies simultaneous optimization parameters. Moreover, the hierarchical clustering was employed to group the materials with respect to their physical behavior. Another method based on the multi-attribute utility theory that simulates preferences of a decision marker like the electrical systems operator was proposed in [11]. An artificial neural network is constructed to approximate the decision-maker preferences, reproducing an level sets of the underlying utility function. Pareto-optimal was used to select the most appropriate solutions to balance the training data error and the weight vector norm in order to avoid underfitting.

The afformentioned methods use the Pareto-optimality technique to make the multi-criteria decisions. In this work, the Pareto-optimal was defined as a set of all non-dominated solutions from a given solution space. In the typical problem of MCDA, the solution space is defined as a region consisting of all possible solutions [10]. In the context of selection of control actions, the

Pareto-optimal technique can be applied to determine the most promising groups of actions.

The idea of clustering solutions in a Pareto-optimal set is not new (e.g., [13,14]). An interesting study to understand the clusters of the optimal solutions in multi-objective decision problems was proposed in [15]. The main goal of the authors is the determination of groups with strongly related solutions in an efficient set aiding the practitioner maker a better decision. In summary, this is done in three steps: (1) a dendogram is obtained via hierarchical clustering technique; (2) a cut-off value is estimated by heuristic method to generate the actual clusters; and (3) the selected clusters are analyzed via Pareto-optimal to inform possible trade-offs between conflicting criteria.

This paper proposes a multi-criteria analysis to determine the most promising groups control actions for an electrical system. The proposed method takes into account the effectiveness and the cost of each group in order to eliminate the criticality of all selected contingencies, and avoids heuristic choices of thresholds and cut sets. Moreover, the proposed approaches uses the DAMICORE (DAta MIning COde Repositories) as a hierarchical clustering method, since it has shown to be able to work with increased numberand size os samples. In this sens, the DAMICORE is an interesintg technique since several electrical system contingencies an controllers can be analyzed in order to make an decision.

In summary, the main contributions of the proposed method are: (1) it does not require a cut-off value to determine the promising clusters; (2) the DAMI-CORE benefits analyses with large number of contingencies and controllers; and (3) the Pareto-optimality technique is applied to analyze all the clusters by the DAMICORE.

2   Preliminary Formulation

After screening a large number of credible contingencies, a VSA tool offers a list of critical contingencies. In the context of voltage stability, a contingency is considered critical if its voltage stability margin, measured as the difference of load powers of the MLP and the current operating condition, is lower than a certain threshold. System operators usually define acceptable voltage stability margins for planning studies and operation. The National System Operator in Brazil, for example, establishes a difference of 7% bellow the MLP as a excepatable margind when studies of expansions, reinforcements and planning of operation [16].

Given a list of critical contingencies, the problem consists of designing preventive control to eliminate these criticalities. The design of preventive control is commonly divided in two phases. In the first phase, a list of the most effective controls is determined and, in the second phase, these selected controls are adjusted to bring the VSM to acceptable levels.

In the process of selecting the most effective controls, three main aspects be considered: (i) the effectiveness of each control element in improving the stability margin, (ii) the availability of each control and (iii) the cost of choosing each control action.

The effectiveness of each control can be estimated via sensitivity analysis of the VSM with respect to the control variable. Let λ be a real variable that parametrizes the load and generation increasing and define λ_{max} as the maximum loadability of the system. The sensitivity of the MLP with respect to a control variable u_c is given by the derivative $d\lambda_{max}/du_c$. Let $u_c \in \mathbb{R}$ represent the c-th control. In order to consider the availability of the control element, a parametrization is chosen such that $u_c = 0$ represents the actual value of the control, while $u_c = 1$ corresponds to its maximum value.

In [7], a methodology inspired in the Look-Ahead method to compute the sensitivity of VSM with respect to control variables was proposed. The control actions are ranked according to this sensitivity level and those that are most effective are employed to restore the stability margin for a determined set of critical contingencies.

3   Clustering Method to Group the Most Promising Controls

In order to determine groups of promising preventive controls for a set of critical contingencies, a new approach based on a clustering method is proposed in [7]. The Hierarchical Clustering Method proposed in this paper is called DAMICORE (Data Mine of Code Repositories) [8]. Basically, this method is a propoer combination of successful algorithms from other science fields: Normalized Compression Distance (NCD) from Information Theory [17], Neighbor Joining (NJ) by Phylogenetic Reconstruction [18], and Newman's Fast Algorithm (FA) from Complex Networks [19].

It is known that some clustering techniques show reduced performance with increased number and size of samples. Fortunately, the DAMICORE does not present such restrictions, this method performs better when there are a large number of samples and size. Thus, it is believed that the DAMICORE method can present relevant performance with increasing number of clusters of controllers and contingencies.

Let $\mathcal{C}$ be the group of critical contingencies; λ_{max_i} be the MLP of the i-th contingency, and $d\lambda_{max_i}/du_c$ be the sensitivity of control λ_{max_i} with respect to u_c. The DAMICORE works as follows: the NCD calculates a matrix of distances between the sensitivities $d\lambda_{max_i}/du_c$; the NJ constructs a phylogenetic tree describing the probable relationships among the sensitivities for $\mathcal{C}$; and, finally, the FA performs a hierarchical cluster a tree of preventive controls (leaves of the tree) by grouping the most similar leaves according to the constructed (phylogenetic) tree.

The DAMICORE generates a hierarchical cluster and proposes an initial partitioning of the data in groups. Finally, such groups can be re-evaluated according to some knowledge of the problem domain.

4   Selecting the Most Promising Groups

The clustering method discussed in Section 3 provides in the data structure of phylogenetic tree with the preventive controls grouped in a hierarchical form. However, the cost of each group is not included in the aforementioned procedure. Besides the effectiveness of each control action, the cost is a key attribute to determine the implementation of some control action, in this case, a low cost is desirable.

Let C_c be the cost of the c-th control action (where $c = 1, 2, ..., NC$), the cost of the g-th group is given as follows:

$$C_g = \sum_{c=1}^{NC_g} C_c \tag{1}$$

where NC_g is the number of controls and C_g the cost of the g-th group. In order to determine the effectiveness of each group the harmonic mean ($\mathcal{H}_g$) is applied. In [7] the authors show that the harmonic mean gives an adequate measure for this kind of problem. In this context, the groups with the most effective controls actions and the lowest cost can be determined by Pareto-Efficiency Set Algorithm (PESA).

Algorithm 1. Pareto-Efficient Set Algorithm (PESA)

```
 1: Let G given by DAMICORE
 2: P ← ∅
 3: for Gi ∈ G do
 4:     dGi ← 0
 5:     for Gj ∈ G AND i ≠ j do
 6:         if (CGj ≤ CGi AND HGj > HGj) OR
 7:         (CGj < CGi AND HGj ≥ HGj) then
 8:             dGi ← dGi + 1
 9:         end if
10:     end for
11:     if dGi = 0 then
12:         P ← P ∪ Gi
13:     end if
14: end for
15: return P
```

Determining the groups considering the most effective control actions with the lowest cost involves conflicting objectives. To overcome this hurdle a multi-criteria analysis is proposed in order to take into account the cost implementation of each group as well as the controls actions. The goal of this approach is offering to operator the best groups considering a combination of effectiveness and low cost.

Algorithm 1 describes the main procedure to determine the best groups according to both criteria. Basically, given a set $\mathcal{G}$ composed of all the possible groups in the dendrogram obtained via DAMICORE, Algorithm 1 evaluates the dominance of each group $G_i \in \mathcal{G}$ with respect to the group $G_j \in \mathcal{G}$, when $G_i \neq G_j$. If G_i is dominated by G_j, i.e., the cost and harmonic mean of G_j is better than G_i, the dominance variable nd is increased by 1. The group G_i is called optimal for our problem if the nd is equal to 0, in this case, G_i is added to the Pareto set $\mathcal{P}$.

5   Simulations and Results

The proposed algorithm has been tested in a reduced south-southeast Brazilian system (test-system), Which is composed of 107 buses and 171 lines [20](See Fig. 1). For voltage control, this system offers 20 shunt reactors, 13 shunt capacitors, 1 synchronous compensator and 1 static compensator. Only shunt capacitors were considered available for voltage control. It is noteworthy to mention that the shunt capacitor 959 is being used in its maximum capacity, i.e., $u_{959} = 1$ (100%). As a consequence, this capacitor will not be available for control. Table 1 presents the localization, the capacity and the cost of each shunt capacitor.

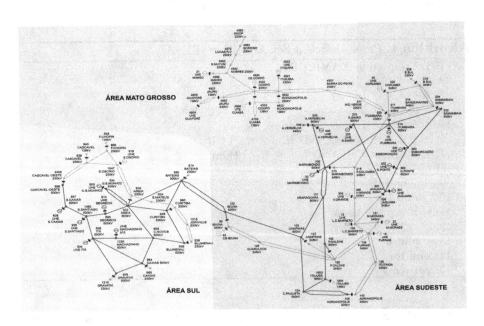

Fig. 1. Reduced south-southeast Brazilian system with 107 buses

The contingencies were classified using Look-Ahead method. The critical ones were selected according to the guidelines for operation and planning provided by

Table 1. Shunt capacitors available for control

#Control	Name	MVAr	Cost
u_{1210}	Gravata	400	1230
u_{939}	Blumenau	250	900
u_{959}	Curitiba	100	700
u_{104}	Cachoeira Paulista	200	850
u_{122}	Ibina	200	850
u_{1504}	Itajub	200	850
u_{123}	Campinas	200	850
u_{120}	Poos de Caldas	200	850
u_{234}	Samambaia	150	700
u_{4522}	Rondonpolis	30	500
u_{4533}	Coxip	86	650
u_{4582}	Sinop	30	500
u_{231}	Rio Verde	30	530

Brazilian National System Operator [16], thus, critical contingencies are those whose VSM is lower than 7% 2, see Table 2.

Table 2. Set $\mathcal{C}$ of critical contingencies

i	Outage Line	λ_{max}	i	Outage Line	λ_{max}
1	(100-101)	1.0374	6	(136-120)	1.0564
2	(101-102)	1.022	7	(136-138)A	1.0359
3	(106-104)	1.0526	8	(136-138)B	1.0395
4	(122-103)	1.0326	9	(140-138)A	1.0346
5	(136-120)	1.0564	10	(140-138)B	1.0383

*i is the i−th contingency in $\mathcal{C}$.
*A and B means parallel transmission lines.

The algorithm proposed in [5] was employed to estimate the sensitivity of λ_{max_i} with respect to changes in the control parameter u_c for each contingency in Table 2. Next, these sensitivities were used to compute the clusters via DAM-ICORE, thus obtaining the tree represented by the dendrogram of Fig. 2. Note that control u_{959} does not appear in the dendrogram since its sensitivity is equal to zero, thus, this control is being used to its fullest.

The most promising groups in the dendrogram to eliminate the criticality of all the contingencies in set $\mathcal{C}$ can be obtained by Algorithm 1. The proposed method selects 9 promising groups from a total of 23 groups found by the DAM-ICORE. Figure 3 shows the classification obtained by the proposed algorithm.

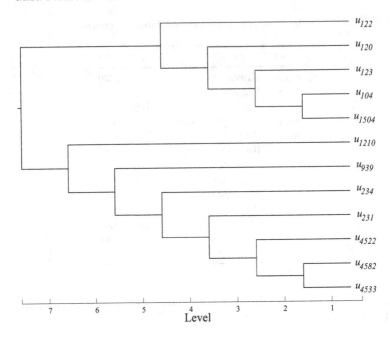

Fig. 2. The dendrogram of preventive controls of Set $\mathcal{C}$ of critical contingencies

The Pareto set ($\mathcal{P}$) is formed by the most promising groups represented by balls in the Figure3 and the worst choices are represented by crosses.

Table 3 describes the capacitors in each group of the Pareto set, the corresponding harmonic mean ($\mathcal{H}$) and the cost associated with each of these groups. The selection of the best group to be applied to eliminate the criticality of set $\mathcal{C}$ can be determined by the operator of the electrical system.

For example, group G_{20} thus it is a relevant choice of control. Table 3 presents high effectiveness and lower cost. Using the controls in this group, the MLP (λ_{max}) was evaluated via Continuation Power Flow (CPFLOW)[1] method adopting $u_c = 0.75$ (75%)[2] and the results are presented in Table 4.

Analyzing Table 4, one observes that group G_{20} eliminates all the critical contingencies of set $\mathcal{C}$ by increasing the VSM to values large than 7%. Other groups can be chosen according to the experience of the operator, for example G_{20} is the cheapeast group does not violate the VSM.

---

[1] CPFLOW is a comprehensive software tool for tracing power system steady-state behaviors due to large or small variations in loads, generation, transactions, interchanges, and imports and exports. CPFLOW is designed for the analysis of large-scale power systems and can trace a solution curve through the nose point (λ_{max}) without the numerical difficulties of repeated power flow solvers [21].

[2] The values of u_c were obtained by our experience simulations, optimal values of the controls actions can be obtained by an optimization method.

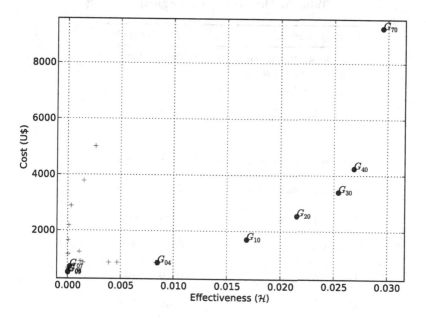

Fig. 3. Evaluation of cost and effectiveness of all the obtained groups

6   Conclusions

In this paper, a new method was proposed to group the most efficient preventive controls for a set of critical contingencies and to select the promising ones according to their cost and effectiveness. This method is based on a sensitivity analysis presented in [5], based on a hierarchical clustering method proposed in [8].

Tests were conducted in the reduced version of the south-southeast Brazilian system with 107 buses and 171 lines. The control actions were well-grouped in terms of their efficiency to increase the VSM for all the critical contingencies from a particular set control actions. the implementation of selected actions has shown that the criticality of that set can be eliminated. It is noteworthy to mention that although the simulations were performed considering only the shunt capacitors, other control elements can be easily incorporated in the proposed methodology.

Future directions of this research include the generalization of the PESA formulation to other kind of problems, the use of different hierarchical clustering methods by DAMICORE in order to better sample the space of possible groups which may increase the number of solutions in the Pareto set benefitting decision of a system operator.

Table 3. The most promising groups

Group	#Controls	$\mathcal{H}$	Total Cost
G_{70}	u_{104}, u_{1504}, u_{123}, u_{120}, u_{122}, u_{4582}, u_{4588}, u_{4522}, u_{231}, u_{234}, u_{939}, u_{1210}	0.02952	9260
G_{40}	u_{104}, u_{1504}, u_{123}, u_{120}, u_{122}	0.02688	4250
G_{30}	u_{104}, u_{1504}, u_{123}, u_{120}	0.02541	3400
G_{20}	u_{104}, u_{1504}, u_{123}	0.02153	2550
G_{10}	u_{104}, u_{1504}	0.01685	1700
G_{04}	u_{1504}	0.00851	850
G_{07}	u_{234}	0.00020	700
G_{08}	u_{231}	5.9e-05	530
G_{09}	u_{4522}	2.8e-05	500

Table 4. Set $\mathcal{C}$ of contingencies with best preventive controls the Pareto Efficient set

i	Outage Line	λ_{max}	i	Outage Line	λ_{max}
1	(100-101)	1.0717	6	(136-120)	1.0741
2	(101-102)	1.0719	7	(136-138)	1.0738
3	(106-104)	1.0737	8	(136-138)	1.0739
4	(122-103)	1.0745	9	(140-138)	1.0735
5	(136-120)	1.0741	10	(140-138)	1.0737

*i is the i−th contingency in $\mathcal{C}$.

Acknowledgement. The authors thank the financial support provided by FAPESP, under the grants 2009/05167-5 and 2012/14194-9.

References

1. Granville, S., Mello, J.C.O., Melo, A.C.G.: Application of interior point methods to power flow unsolvability. IEEE Transactions on Power Systems **11**, 1096–1103 (1996)
2. Bedrinana, M., Castro, C., Bedoya, D.: Maximization of voltage stability margin by optimal reactive compensation. In: 2008 IEEE Power and Energy Society General Meeting - Conversion and Delivery of Electrical Energy in the 21st Century, pp. 1–7, July 2008

3. Greene, S., Dobson, I., Alvarado, F.: Sensitivity of the loading margin to voltage collapse with respect to arbitrary parameters. IEEE Transactions on Power Systems **12**(1), 262–272 (1997)
4. Zhao, J., Chiang, H.-D., Li, H., Zhang, B.: A novel preventive control approach for mitigating voltage collapse. In: Power Engineering Society General Meeting, p. 6. IEEE (2006)
5. Mansour, M., Alberto, L., Ramos, R.: Look-ahead based method for selection of preventive control for voltage stability analysis, pp. 469–473, March 2012
6. Chiang, H.-D., Wang, C.-S., Flueck, A.: Look-ahead voltage and load margin contingency selection functions for large-scale power systems. IEEE Transactions on Power Systems **12**(1), 173–180 (1997)
7. Mansour, M.R., Alberto, L.F.C., Ramos, R.A., Delbem, A.C.B.: Identifying groups of preventive controls for a set of critical contingencies in the context of voltage stability. In: ISCAS, pp. 453–456 (2013)
8. Sanches, A., Cardoso, J., Delbem, A.: Identifying merge-beneficial software kernels for hardware implementation. In: International Conference on Reconfigurable Computing and FPGAs (ReConFig), November 30 to December 2, 2011, pp. 74–79 (2011)
9. Figueira, J., Greco, S., Ehrgott, M.: Multiple criteria decision analysis: state of the art surveys. International Series in Operations Research & Management Science. Springer (2005). http://books.google.co.in/books?id=YqmvlTiMNqYC
10. Chauhan, A., Vaish, R.: Hard coating material selection using multi-criteria decision making. Materials & Design **44**, 240–245 (2013)
11. Pedro, L.R., Takahashi, R.H.C.: Modeling decision-maker preferences through utility function level sets. In: Takahashi, R.H.C., Deb, K., Wanner, E.F., Greco, S. (eds.) EMO 2011. LNCS, vol. 6576, pp. 550–563. Springer, Heidelberg (2011)
12. Pareto, V.: Manual of political economy (manuale di economia politica). Kelley, New York. Translated by Ann S. Schwier and Alfred N. Page
13. Morse, J.N.: Reducing the size of the nondominated set: Pruning by clustering. Computers & OR **7**(1–2), 55–66 (1980)
14. Mattson, C.A., Mullur, A.A., Messac, A.: Smart Pareto filter: obtaining a minimal representation of multiobjective design space. Engineering Optimization **36**(6), 721–740 (2004)
15. Veerappa, V., Letier, E.: Understanding clusters of optimal solutions in multi-objective decision problems. In: Proceedings of the 2011 IEEE 19th International Requirements Engineering Conference, ser. RE 2011, pp. 89–98. IEEE Computer Society, Washington, DC (2011)
16. ONS. Submódulo 23.3: Diretrizes e critérios para estudos elétricos, November 2011. http://www.ons.org.br/procedimentos/index.aspx
17. Cilibrasi, R., Vitányi, P.M.B.: Clustering by compression. IEEE Transactions on Information Theory **51**, 1523–1545 (2005)
18. Felsenstein, J.: Inferring Phylogenies, 2nd edn. Sinauer Associates, September 2003
19. Newman, M.: Networks: An Introduction. Oxford University Press Inc., New York (2010)
20. Alves, W.F.: Proposition of test-systems to power systems analysis, Ph.D. dissertation, Universidade Federal de Fluminense, Niteroi, RJ (2007). http://www.sistemas-teste.com.br/ (in Portuguese)
21. Chiang, H.-D., Flueck, A., Shah, K., Balu, N.: Cpflow: a practical tool for tracing power system steady-state stationary behavior due to load and generation variations. IEEE Transactions on Power Systems **10**(2), 623–634 (1995)

Multi-objective Evolutionary Algorithm with Discrete Differential Mutation Operator for Service Restoration in Large-Scale Distribution Systems

Danilo Sipoli Sanches[1]([✉]), Telma Worle de Lima[2],
João Bosco A. London Junior [3], Alexandre Cláudio Botazzo Delbem[4],
Ricardo S. Prado[5], and Frederico G. Guimarães[6]

[1] Federal Technological University of Paraná, Cornélio Procópio, Brazil
danilosanches@utfpr.edu.br
[2] Institute of Informatics, Federal University of Goias, UFG, Goiânia, Brazil
telma@inf.ufg.br
[3] São Carlos Engineering School of University of São Paulo, São Carlos, SP, Brazil
jbalj@usp.br
[4] Institute of Mathematics and Computer Science,University of São Paulo,
São Carlos, SP, Brazil
acbd@icmc.usp.br
[5] Federal Institute of Minas Gerais, Ouro Preto, Brazil
ricardo.prado@ifmg.edu.br
[6] Department of Electrical Engineering, Universidade Federal de Minas Gerais,
UFMG, Belo Horizonte, Brazil
fredericoguimaraes@ufmg.br

Abstract. The network reconfiguration for service restoration in distribution systems is a combinatorial complex optimization problem that usually involves multiple non-linear constraints and objectives functions. For large networks, no exact algorithm has found adequate restoration plans in real-time, on the other hand, Multi-objective Evolutionary Algorithms (MOEA) using the Node-depth enconding (MEAN) is able to efficiently generate adequate restorations plans for relatively large distribution systems. This paper proposes a new approach that results from the combination of MEAN with characteristics from the mutation operator of the Differential Evolution (DE) algorithm. Simulation results have shown that the proposed approach, called MEAN-DE, properly designed to restore a feeder fault in networks with significant different bus sizes: 3,860 and 15,440. In addition, a MOEA using subproblem Decomposition and NDE (MOEA/D-NDE) was investigated. MEAN-DE has shown the best average results in relation to MEAN and MOEA/D-NDE. The metrics R_2, R_3, Hypervolume and ϵ-indicators were used to measure the quality of the obtained fronts.

© Springer International Publishing Switzerland 2015
A. Gaspar-Cunha et al. (Eds.): EMO 2015, Part II, LNCS 9019, pp. 498–513, 2015.
DOI: 10.1007/978-3-319-15892-1_34

1   Introduction

There are many Multi-objective optimization problems (MOP) in real world such as: vehicle routing, phylogenetic reconstruction and service restoration (SR) in distribution systems (DS). MOP are characterized by the presence of multiple objective functions to be optimized simultaneously, since such objectives can be conflicting and there is no single optimal solution that satisfies all objectives equally.

In order to find feasible solutions for MOP, Multi-objective Evolutionary Algorithms (MOEAs), such as Nondominated Sorting Genetic Algorithm II (NSGA-II) [1], Strength-Pareto Evolutionaty Algorithm 2 (SPEA2) and Multi-objective Evolutionary Algorithm Based on Decomposition (MOEA/D) [2] were proposed in the literature.

The performance obtained by MOEAs for service restoration in distribution systems is dramatically affected by the data structure used to represent computationally the electrical topology of the distribution systems. Inadequate data structure may reduce drastically the MOEA performance [3, 4]. Other critical aspects of MOEAs are the genetic operators that are used. Generally these operators do not generate radial configurations [3].

In this context, a new dynamic data structures (encodings) that exclusively generate feasible solutions have been investigated. Those encodings allow a suitable exploration of the search space, increasing the quality of solutions provided by MOEAs. Among the encondigs from the literature, the Node-depth encoding (NDE) [5] better scales, enabling its use for optimization methods applied to large networks. In this sense, some MOEAs using NDE have been investigated: MoEA with Node-depth encoding (MEAN) [4,6], NSGA-II with NDE (NSGAN) [7] and the integration of MEAN with both NSGA-II and SPEA-2 [8].

In this sense, the main contribution of this paper is to propose a differential mutation operator based on the NDE. The new operator can extract the essential difference between two DS feasible configurations and use it in order to compose new feasible configurations. Moreover, the average time complexity of the proposed operator is $O(\sqrt{n})$, enabling efficient manipulation of large-scale networks. In addition, the differential mutation operator based on the NDE is combined with MEAN, producing a new powerful MOEA (called MEAN-DE) to solve SR problems for large-scale DSs.

Although the majority of MOEA has successfully worked with combinatorial Multi-Objective Problems (MOPs) with at most two objectives, the MEAN have solved the SR problem formulated with more than two objectives [4]. Other MOEA that has obtained interesting results for MOPs with more than two objectives is the MOEA/D [2]. As a consequence, we also proposed an extension of MOEA/D using NDE, called MOEA/D-NDE, adapted for the SR problem. However, the MEAN-DE also presented better results in relation to MOEA/D-NDE for the SR problem.

2   Service Restoration Problem

The Network reconfiguration for the Service Restoration Problem is the process of opening and closing of some switches to modify the topology of a distribution network modeled by a forest. Fig. 1 (a) illustrates an example of SR in a DS with three feeders that are represented by nodes 1, 2 and 3. Each feeder supplies a subset of consumer load points (sectors) represented by other nodes . The sectors are interconnected by edges that indicate the switches (feeder lines). The switches can be Normally Closed (NC) (solid lines) and Normally Opened (NO) (dotted lines). Each tree of the forest corresponds to a feeder with its sectors and Normally Closed switches.

Assuming that a fault occurred in sector 10 (Fig. 1 (a)), all the switches connected to sector 10 (switches 10-11, 10-7 and 10-9) must be opened in order to isolate the sector in fault, thus, Sectors 11, 9 and 28 are in an out-of-service area. One way to restore energy for those sectors is by closing the switches 24-28 and 28-11 (Fig. 1 (b)).

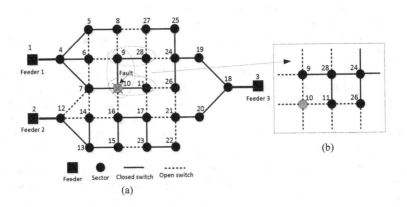

Fig. 1. DS modeled by a graph, a fault is simulated in the sector 10 (a) and then is restored energy for Sectors 11, 9 and 28 by closing the switches 24-28 and 28-11 (b)

The SR problem emerges after the faulted areas has been identified and isolated. Its solutions is the minimal number of switching operations that results in a configuration with minimal number of out-of-service loads, without violating the DS operational and radialily constraints. The minimization of the number of switching operations is important since the time required by the restoration process depends on the number of switching operations. The SR problem can be formalized as follows:

$$Min. \quad \phi(G), \psi(G, G^0) \ and \ \gamma(G)$$
$$subject \ to$$
$$Ax = b$$
$$X(G) \leq 1 \tag{1}$$
$$B(G) \leq 1$$
$$V(G) \leq 1$$
$$G \ is \ a \ forest,$$

where G is a spanning forest of the graph representing a system configuration [9] (each tree of the forest [10] corresponds to a feeder or to an out-of-service area, nodes correspond to sectors and edges to switches); $\phi(G)$ is the number of consumers that are out-of-service in a configuration G (considering only the reconnectable system); $\psi(G, G^0)$ is the number of switching operations to reach a given configuration G from the configuration just after the isolation of the faulted areas G^0; $\gamma(G)$ are the power losses, in p.u. (per-unit system), of configuration G; A is the incidence matrix of G [10]; x is a vector of line current flow; b is a vector containing the load complex currents (constant) at buses with $b_i \leq 0$ or the injected complex currents at the buses with $b_i > 0$ (substation); $X(G)$ is called network loading of configuration G, that is, $X(G)$ is the highest ratio $x_j/\overline{x}_j$, where $\overline{x}_j$ is the upper bound of current magnitude for each line current magnitude x_j on line j; $B(G)$ is called substation loading of configuration G, that is, $B(G)$ is the highest ratio $b_s/\overline{b}_s$, where $\overline{b}_s$ is the upper bound of current injection magnitude provided by a substation (s means a bus in a substation); $V(G)$ is called the maximal relative voltage drop of configuration G, that is, $V(G)$ is the highest value of $|v_s - v_k|/\delta$, where v_s is the node voltage magnitude at a substation bus s in p.u. and v_k the node voltage magnitude at network bus k in p.u. (obtained from a Forward-Backward Sweep Load Flow Algorithm (SLFA) for DSs) and δ is the maximum acceptable voltage drop (in this paper $\delta = 0.1$, i.e. the voltage drop is limited to 10%). The formulation of Equation 1 can be synthesized by considering:

- Penalties for violated constraints $X(G)$, $B(G)$ and $V(G)$;
- The use of the NDE [4], i.e. an abstract data type for graphs that can efficiently manipulate a network configuration (spanning forest) and guarantee that the performed modifications always produce a new configuration G that is also a spanning forest (a feasible configuration);
- The nodes are arranged in the Terminal-Substation Order (TSO) for each produced configuration G in order to solve $Ax = b$ using an efficient SLFA for DSs. The NDE stores nodes in the TSO. Through x obtained from a backward sweep, the complex node voltages are calculated from a forward sweep;
- $\phi(G) = 0$. The NDE always generates forests that correspond to networks without out-of-service consumers in the re-connectable system.

Equation 1 can be rewritten as follows:

$$Min. \quad \psi(G, G^0), \; \gamma(G) \; and$$
$$\omega_x X(G) + \omega_b B(G) + \omega_v V(G)$$

subject to $\qquad\qquad\qquad\qquad\qquad\qquad\qquad$ (2)
$$G \; is \; a \; forest \; generated \; by \; the \; NDE,$$
$$Load \; flow \; calculated \; using \; the \; NDE$$

where ω_x, ω_b and ω_v are weights balancing among the network operational constraints. In this paper, these weights are set as follows:

$$\omega_x = \begin{cases} 1, & \text{if, } X(G) > 1 \\ 0, & \text{otherwise;} \end{cases}$$

$$\omega_b = \begin{cases} 1, & \text{if, } B(G) > 1 \\ 0, & \text{otherwise;} \end{cases}$$

$$\omega_v = \begin{cases} 1, & \text{if, } V(G) > 1 \\ 0, & \text{otherwise.} \end{cases}$$

3   Evolutionary Algorithms with NDE

The SR problem, as formulated in the previous section, is based on NDE, thus, the efficiency in solving it depends on such encoding. Two operators were developed to efficiently manipulate a forest stored in NDEs producing a new one: the Preserve Ancestor Operator (PAO) and the Change Ancestor Operator (CAO). Each operator modifies the forest encoded by NDE arrays, which is equivalent to pruning and grafting a sub-tree of a forest generating a new forest. The NDE operators generate only feasible configurations (radial configurations able to supply energy for the whole re-connectable system). As a consequence, such abstract data type does not require a specific routine to verify and to correct unfeasible configurations. Those aspects enable the construction of new configurations in a fast way for large-scale DSs (average-time complexity $O(\sqrt{n})$, where n is the number of sectors in DS). In addition, a SLFA [11] based on the TSO provided by the NDE fast evaluates each new produced configuration for large-scale DSs (average-time complexity $O(\sqrt{n_b})$, where n_b is the number of load buses of the DS).

Moreover, the formulations of Equations 1 and 2 correspond to a Multi-Objective Problem (MOP). MOEAs are among the most relevant methods to deal with MOPs [12,13]. However, these and other methods have shown success to work with combinatorial MOPs with at most two objectives. In fact, problems with more objectives have been called many-objective problems and relatively few approaches were developed for them. Fortunately, MOEA combined with the NDE proposed in [4] has properly solved DSR problems formulated with more than two objectives. Additional information about the NDE and its operators applied to DSR problems are described in [4].

3.1 Multi-objective EA with Subpopulation Tables (MEAN)

MEAN was proposed in [4] and uses a simple and computationally efficient strategy to deal with several objectives and constraints. The basic idea is to subdivide a population into subpopulation tables related to different objectives and constraints. The MEAN is different from VEGA (Vector Evaluated Genetic Algorithm [14]), since it adds a fundamental subpopulation table that stores individuals assessed by at least one aggregation function (see Equation 3), moreover, any individual can be simultaneously evaluated using weighted (by table(s) of aggregation function(s)) and non-weighted scores (through the remaining tables) from objectives and no additional heuristic is required to induce middling values as proposed in [14]. The ability of simultaneously searching for the extreme points of the Pareto-front and the best values of the aggregation function makes MEAN more similar to the MOEA/D [2].

The whole set of tables is organized as follows:

1. Tables associated with each objective and constraint:
   (a) T_1 - solutions with low $\gamma(G)$; T_2 - solutions with low $V(G)$; T_3 - solutions with low $X(G)$; T_4 - solutions with low $B(G)$ and T_5 - solutions with low values of an aggregation function, defined as follows:

$$f_{agg}(G) = \psi(G, G^0) + \gamma(G) + \\ \omega_x X(G) + \omega_b B(G) + \omega_v V(G), \tag{3}$$

   where $\psi(G, G^0)$, $\gamma(G)$, $X(G)$, $B(G)$, $V(G)$, ω_x, ω_b and ω_v were defined in Section 2;
2. Tables denoted T_{5+p} that are related to the required pair of switching operations after fault isolation:
   (a) Each Table T_{5+p}, with $p = 1, ..., 5$, stores the best solutions found with more than $p - 1$ and at most p pairs of switching operations. In these tables the solutions are ranked (in increasing order) according to the value of $V(G) + X(G)$. Solutions with similar value, considering precision 10^{-2}, are randomly ranked.

The reproduction operators used to generate new individuals are the NDE operators PAO and CAO and more informations are described in [4].

3.2 Multi-Objective EA Based on Decomposition (MOEA/D)

MOEA/D is a multi-objective EA that uses a technique of decomposition [2] of a problem into subproblems. This algorithm simultaneously optimizes V single objective subproblems, each of them corresponds to an aggregation function. MOEA/D usually employs the Tchebycheff approach [15] for the decomposition of a multi-objective problem into subproblems. A coefficient vector λ^i defines each aggregation function and a set with the U coefficient vectors that are the closest to λ^i in $\{\lambda^0, \ldots, \lambda^V\}$ composes the neighborhood of λ^i [2].

The coefficient vectors should spread uniformly in the objective space. The number of vectors is $V = C_{H+o-1}^{o-1}$, where o is the number of problem objectives and $H + 1$ is the size of weight set $\{\frac{0}{H}, \frac{1}{H}, \ldots, \frac{H}{H}\}$ used to construct coefficient vectors. As a consequence, a tradeoff between o and H should be found in order to bound V, generating a number of subproblems that is computationally tractable.

In relation to MEAN, MOEA/D requires the additional parameters U and H. To work with the SR problem, we adapted MOEA/D to use NDE, which was called MOEA/D-NDE.

3.3   Evolutionary History Recombination

In this section the recombination operator for the NDE is described: Evolutionary History Recombination (EHR) proposed in [16]. As the name of the operator suggests, EHR is based on the evolutionary history of the operators PAO and CAO, that is, on the sequence of vertices (p, a), for PAO, and (p, r, a), for CAO, applied in the generation of new individuals. The history of each individual can be retrieved by using the auxiliary structures from NDE: matrix Π_x, which stores the positions of node x in each individual, and array π, which stores the ancestor of each individual.

In order to simplify the utilization of EHR, we propose a modification in array π of the NDE, called π_m, such that it can store not only the index of the ancestor but also a triple of nodes (a, r and p) that were used in the application of the operator PAO or CAO (in the case of PAO, the value in r is null). In this way, a sequence of movements to generate individual F_i from any ancestor can be accessed from π_m.

4   Discrete DE with Movements List

In [17] the authors propose an optimization approaches for the differential evolution, called Discrete Differential Evolution algorithm with List of Movements (DDELM). DDELM was applied to combinatorial problems where the operators difference, addition, and product by scalar in the differential mutation equation are redefined in the space of discrete variables.

The difference between two candidate solutions is a List of Movements in the search space defined as:

Definition 1: A list of movements M_{ij} is a list containing a sequence of valid movements m_k such that the application of these movements to a solution $s_i \in S$ leads to the solution $s_j \in S$, where S is the search space, that is, the set of all possible combinations of values for the variables.

In this way, the "difference" between two solutions is defined as being the corresponding list of movements:

$$M_{ij} = s_i \ominus s_j, \tag{4}$$

where $\ominus$ is a special binary minus operator that returns a list of movements M_{ij} that represents a path from s_i towards s_j. This list, in some sense, captures the differences between these two solutions.

The multiplication of the list of movements by a constant is defined as:

Definition 2: The multiplication of the list of movements, M_{ij}, by a constant $F \in [0,1]$, returns a list M'_{ij} with the $\lceil F \times |M_{ij}| \rceil$ movements of M_{ij}, where $|M_{ij}|$ is the size of the list.

Thus, the multiplication of the list of movements by a constant can be denoted, using the special binary multiplication operator $\otimes$, by:

$$M'_{ij} = F \otimes M_{ij}. \tag{5}$$

Finally, the application of a list of movements to a given solution is defined as follows:

Definition 3: The application of the sequence of movements in the list M'_{ij} into a solution s_k, returns a new solution s'_k:

$$s'_k = s_k \oplus M'_{ij}. \tag{6}$$

With the definition above, one can generate a mutant vector defined as:

$$v_i = x_0 \oplus F \otimes (x_1 \ominus x_2)$$
$$v_i = x_0 \oplus F \otimes M_{12}$$
$$v_i = x_0 \oplus M'_{12},$$

which is the proposed discrete version of the typical differential mutation equation.

We emphasize that in this paper we extend the ideas in [17] by proposing a list of movements based on the NDE, which is suitable for representing candidate solutions in DSR problems.

5   Proposed Approach

The proposed approach is called MEAN-DE, which consists basically of MEAN and the mutation operator of DE re-designed from the EHR operator. In other words, the list of movements [17] is obtained from the application of EHR operator. In this sense, the difference between any two individuals x_1 and x_2 is a list of movements M_{12} composed by a sequence of triples $(p, \_, a)$ and (p, r, a) obtained from π_m (Section 3.3). In fact, the list is the concatenation of two sequences: one from x_1 to x_c and another from x_c to x_2, where x_c is their common ancestor.

Thus, the EHR can be used to implement a discrete differential mutation operator that is computationally efficient ($O(\sqrt{n})$ in average). The implementation is straightforward as follows:

Be x_0, x_1, and x_2 three individuals randomly selected from the current population to participate in the differential mutation equation:

$$v_i = x_0 \oplus F \otimes (x_1 \ominus x_2) \tag{7}$$
$$v_i = x_0 \oplus F \otimes M_{12},$$

where the list of movements M_{12} is obtained from the history of applications of PAO, CAO and EHR, which are stored into the modified array π_m.

To illustrate the proposed differential mutation operator based on EHR, let us consider the DS in Fig. 2(a), consisting of 3 feeders. The NDE of each feeder is shown in Fig. 2(b). In this sense, consider the tree representation of the common ancestor x_c ((Fig. 3(a)), which is the forest shown in Fig. 2(a)) of individuals x_1 and x_2 (shown in Figs. 3(b) and 3(c), respectively) generated through the application of CAO and PAO. Individual 1 comes from ancestor x_c by the following sequence of PAO applications $(11,\_,17)$, $(7,\_,6)$ and $(24,\_,23)$. Individual 2 derives from x_c by applying CAO as follows: $(21, 20, 14)$, and $(11, 12, 13)$.

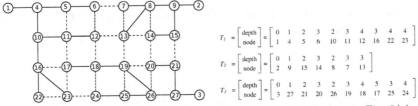

(a) DS with 3 feeders modeled by a graph with three trees

(b) NDE for the feeders in Fig. 2(a).

Fig. 2. NDE arrays for three trees of the spanning forests

Thus, with the aid of the EHR, the list of movements for each individual is written as:

$$M_{c1} = [(11, \_, 17)\,(7, \_, 6)\,(24, \_, 23)] \tag{8}$$
$$M_{c2} = [(21, 20, 14)\,(11, 12, 13)]$$

The list of movements M_{12} between individuals x_1 and x_2 is the junction of this two lists. M_{12} is built by choosing alternately one movement from each list of the individuals in order to avoid a bias of the movements list from one only individual. So, the movement list M_{12} is given by:

$$M_{12} = [(11, \_, 17), (21, 20, 14), (7, \_, 6), \tag{9}$$
$$(11, 12, 13), (24, \_, 23)]$$

Assuming that $F = 0.6$ [17], the number of movements used from M_{12} is $\lceil F \times |M_{12}| \rceil = \lceil 0.6 \times 5 \rceil = 3$. Thus M'_{12} results in:

$$M'_{12} = [(11, \_, 17), (21, 20, 14), (7, \_, 6)]. \tag{10}$$

The mutant vector v_i is obtained by applying each movement of M'_{12} into the base vector x_0:

$$v_i = x_0 \oplus M'_{12}. \tag{11}$$

Thus, using x_0 from Fig. 3(d) and M'_{12} obtained above, the resultant mutant vector is the one shown in Fig. 3(e).

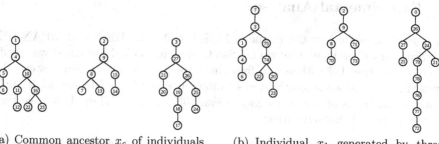

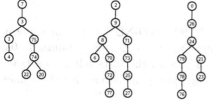

(a) Common ancestor x_c of individuals x_1 and x_2

(b) Individual x_1 generated by three applications of the PAO operator into the common ancestor of Fig.3(a)

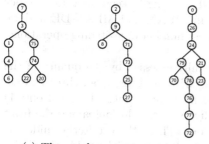

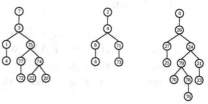

(c) Individual x_2 generated by three applications of the CAO operator into the common ancestor of Fig.3(a)

(d) Individual x_0, base vector, randomly choosing in a current population

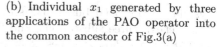

(e) The result mutant vector v_i

Fig. 3. Example of Differential mutation operator based on EHR

In summary, the proposed approach allows the implementation of the differential mutation operator for DSR problems by using the EHR operator to build the list of movements as proposed in [17]. The list of movements captures the differences between any two individuals and can be scaled and applied to the base individual in order to generate a mutant solution. MEAN-DE employs the differential mutation operator as the search engine within the MEAN framework.

6 Experimental Analyses

In order to analyze how the methods MOEA/D-NDE, MEAN and MEAN-DE performs for SR problem, the real DS Sao Carlos city (called System 1) was used to compose other DS with size of four times the original DS (called System 2). System 2 is composed of four Systems 1 interconnected by 49 NO new additional switches (the data of the two DSs are available in [18]). These DSs have the following general characteristics:

System 1 (S1): 3860 buses, 532 sectors, 632 switches (509 NC and 123 NO switches), three substations, and 23 feeders; System 2 (S2): 15 440 buses, 2128 sectors, 2577 switches (2036 NC and 541 NO switches), 12 substations, and 92 feeders.

The approaches MEAN, MEAN-DE and MOEA/D-NDE were run using a Core 2 Quad 2.4GHz, 8G RAM, with Linux Operating System Ubuntu 14.04 version, and the language compiler C gcc-4.4 and all the tests refers to a fault at the largest feeder in Systems 1 and 2, interrupting the service for the whole feeder.

Parameters of MEAN and MEAN-DE are the subpopulation table sizes, which were all setup to $S_{T_i} = 5$. MEAN, MEAN-DE and MOEA/D-NDE used dynamic probability of PAO and CAO applications. We evaluated different values of parameters U and H of MOEA/D-NDE in order to keep the total number of evaluations closest to the number used in MEAN and MEAN-DE and we chose the set that corresponded to the smallest number of switching operations, returning $U = 30$ and $H = 10$.

The performance between MOEAs is usually assessed by the quality of the approximated Pareto fronts found by the algorithms. In general, three characteristics are taken into account to evaluate an approximated Pareto front: 1) proximity to the Pareto-optimal front, 2) diversity of solutions along the front and 3) uniformity of solutions along the front. These three criteria guide the search to a high-quality and diversified set of solutions which enable the choice of the most appropriate solution in a posterior decision-making process [19].

To quantify these three characteristics in a set of non-dominated solutions, various measures have been developed, as example, Error Ratio, Generational Distance, the R_2 and R_3 [20] Hypervolume (HV) [21] and ϵ-indicator [21]. In this paper, R_2, R_3, HV and ϵ indicators are used to assess the performance of the proposed algorithm, each of them is based on different preference information, then by using them all we provide a range of comparisons intead of just one point-of-view.

In this context, the experiments with the system test evaluated the approaches according to: (a) the performance of them for the SR problem; and (b) the relative performance of those MOEAs concerning R_2, R_3, HV and ϵ-indicators.

MEAN, MEAN-DE and MOEA/D-NDE were run 50 times (with different seeds for the used random number generator). Each run evaluated 100,000 solutions. In this paper the MEAN, MEAN-DE and MOEA/D-NDE approaches will be search for SR plans which restore the entire out-of-service area (full restoration cases) respecting radiality and all the operational constraints (voltage drop, substation and network loading).

Table 1 enables a comparison of MEAN, MEAN-DE and MOEA/D-NDE for S1 and S2 according to the number of switching operations for SR plans (the most critical aspect for the SR problem). Those results concern only the feasible solutions with the smallest number of switching operations found by each approach in each run. Clearly, MEAN-DE and MEAN outperform MOEA/D-NDE according to the number of switching operations for SR plans.

Table 1. *Simulations with Single Fault in Systems 1 and 2*

		Switching Operations		
		MEAN-DE	MEAN	MOEA/D-NDE
	Minimum	7	7	9
	Average	9	13	19
S1	Maximum	11	29	73
	Standard Deviation	1.46	5.48	13.41
	Minimum	7	11	27
	Average	20	25	77
S2	Maximum	77	107	105
	Standard Deviation	14.07	20.1	22.11

Table 2 synthesize other electrical aspects of the best solutions found by MEAN, MEAN-DE and MOEA/D-NDE for each objective and constraint. Basically, they emphasize that the found solutions are all feasible and don't significantly differ from each other according to those aspects, thus, the critical aspect is the number of switching operations, as shown by Table 1.

Table 2. *Simulation Results - Single Fault in Systems 1 and 2*

		MEAN-DE		MEAN		MOEA/D-NDE	
		Avg[1]	Dev.[2]	Avg	Dev.	Avg	Dev.
	Power Losses	377.2	29.8	353.8	36.1	370.19	28.6
	Voltage Ratio(%)	4.1	0.8	3.8	0.7	3.3	0.07
S1	Network Loading (%)	77.7	7.7	74.4	8.4	86.3	6.9
	Substation Loading (%)	53.9	2.1	53.3	1.5	53.1	2.9
	Running Time	13.6	0.2	12.8	1.2	5.7	0.2
	Power Losses	1195.6	50.8	1191.6	69.5	1238.3	91.8
	Voltage Ratio(%)	3.7	0.6	3.9	0.8	3.5	0.5
S2	Network Loading (%)	80.7	10.1	83.1	9.5	88.1	7.2
	Substation Loading (%)	55.2	1.9	57.1	4.8	57.2	3.4
	Running Time	16.3	0.6	14.9	0.4	8.5	1.1

[1] Average.
[2] Standard Deviation.

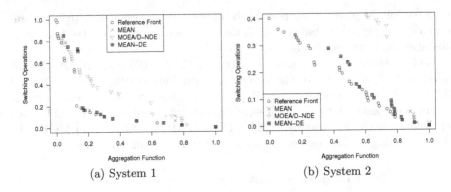

Fig. 4. Pareto fronts obtained from Systems 1 and 2.

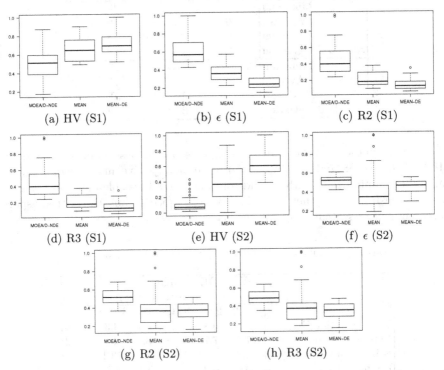

Fig. 5. Box plots for metrics R_2, R_3, HV and ϵ-indicators obtained using Systems 1 and 2.

Moreover, analyses of the results according to metrics used to compare MOEAs show that the MEAN-DE outperforms MEAN and MOEA/D-NDE for test problem (S1 and S2) in terms of approximating the Pareto optimal set while preserving a diverse, evenly-distributed set of nondominated solutions. Fig. 4 indicates that MEAN-DE is able to evolve individuals near to the Reference Front (which is composed using solutions of all found fronts obtained from 50 trials with each method)when compared with the approaches MEAN and MOEA/D-NDE.

The distribution of the performance metrics R_2, R_3, HV and ϵ-indicators for System 1 are shown in Figs 5(a), 5(b), 5(c) and, 5(d), respectively. The MEAN-DE can find in average a front that is diverse and uniformly distributed for test problem when compared with other approaches. Moreover, Figs 5(e), 5(f), 5(g) and, 5(h) corroborate such performance for System 2.

7   Conclusions

This paper presented a new MOEA using NDE with a powerful differential mutation operator to solve the SR problem in large-scale DS (i.e., DS with thousands of buses and switches).

The proposed approach, called MEAN-DE, combines the main characteristics of MEAN, EHR operator and list of movements of the DDELM proposed in [17]. MEAN and MEAN-DE are both based on the strategy of subpopulation tables to deal with multiobjective optimization involving more than two objectives. The MEAN-DE uses a new differential mutation operator structured from the EHR operator, providing computational efficiency in order to deal with relatively large DS. In addition, a MOEA using subproblem Decomposition and NDE (MOEA/D-NDE) was investigated.

In the experiments three approaches, named MEAN, MEAN-DE and MOEA/D-NDE were applied to two DS test problem. To measure the quality of obtained solutions of MOEAs using NDE, the metrics R_2, R_3, HV and ϵ-indicators were used. According to the simulation results, MEAN-DE showed a better performance in terms of R_2, R_3, HV and ϵ-indicators in relation to other approaches analyzed.

Acknowledgments. The authors would like to acknowledge CAPES, CNPq, Fundaão Araucária and FAPESP for the financial support given to this research.

References

1. Deb, K., Pratap, A., Agarwal, S., Meyarivan, T.: A fast and elitist multiobjective genetic algorithm: Nsga-ii. IEEE Transactions on Evolutionary Computation **6**(2), 182–197 (2002)
2. Zhang, Q., Li, H.: Moea/d: A multiobjective evolutionary algorithm based on decomposition. IEEE Transactions on Evolutionary Computation **11**(6), 712–731 (December 2007)

3. Carreno, E., Romero, R., Padilha-Feltrin, A.: An efficient codification to solve distribution network reconfiguration for loss reduction problem. IEEE Transactions on Power Systems **23**(4), 1542–1551 (November 2008)
4. Santos, A., Delbem, A., London, J., Bretas, N.: Node-depth encoding and multi-objective evolutionary algorithm applied to large-scale distribution system reconfiguration. IEEE Transactions on Power Systems **25**(3), 1254–1265 (August 2010)
5. Delbem, A.C.B., de Carvalho, A., Policastro, C.A., Pinto, A.K.O., Honda, K., Garcia, A.C.: Node-depth encoding for evolutionary algorithms applied to network design. In: Deb, K., Tari, Z. (eds.) GECCO 2004. LNCS, vol. 3102, pp. 678–687. Springer, Heidelberg (2004)
6. Sanches, D.S., Junior, J.B.A.L., Delbem, A.C.B.: Multi-objective evolutionary algorithm for single and multiple fault service restoration in large-scale distribution systems. Electric Power Systems Research **110**, 144–153 (2014)
7. Mansour, M., Santos, A., London, J., Delbem, A., Bretas, N.: Node-depth encoding and evolutionary algorithms applied to service restoration in distribution systems. In: Power and Energy Society General Meeting, pp. 1–8. IEEE, July 2010
8. Gois, M.M., Sanches, D.S., Martins, J., Junior, J.B.A.L., Delbem, A.C.B.: Multi-objective evolutionary algorithm with node-depth encoding and strength pareto for service restoration in large-scale distribution systems. In: Purshouse, R.C., Fleming, P.J., Fonseca, C.M., Greco, S., Shaw, J. (eds.) EMO 2013. LNCS, vol. 7811, pp. 771–786. Springer, Heidelberg (2013)
9. Diestel, R.: Graph Theory. Graduate Texts in Mathematics, vol. 173, 3rd edn. Springer, Heidelberg (2005)
10. Ahuja, R.K., Magnanti, T.L., Orlin, J.B.: Network Flows: Theory, Algorithms, and Applications. Printce Hall, Englewood Cliffs (1993)
11. Shirmohammadi, D., Hong, H., Semlyen, A., Luo, G.: A compensation-based power flow method for weakly meshed distribution and transmission networks. IEEE Transactions on Power Systems **3**(2), 753–762 (May 1988)
12. Deb, K.: Multi-objective optimization using evolutionary altorithms. Wiley, New York (2001)
13. Coello, C.A.C., Lamont, G.B., Veldhuizen, D.A.V.: Evolutionary Algorithms for Solving Multi-Objective Problems (Genetic and Evolutionary Computation). Springer, Secaucus (2006)
14. Schaffer, J.D.: Multiple objective optimization with vector evaluated genetic algorithms. In: Proceedings of the 1st International Conference on Genetic Algorithms, pp. 93–100. L. Erlbaum Associates Inc., Hillsdale (1985)
15. Miettinen, K.: Nonlinear Multiobjective Optimization. International Series in Operations Research and Management Science, vol. 12. Kluwer Academic Publishers, Dordrecht (1999)
16. Sanches, D., Lima, T., Santos, A., Delbem, A., London, J.: Node-depth encoding with recombination for multi-objective evolutionary algorithm to solve loss reduction problem in large-scale distribution systems. In: Power and Energy Society General Meeting, pp. 1–8. IEEE, July 2012
17. Prado, R.S., Silva, R.C.P., Guimar aes, F.G., Neto, O.M.: A new differential evolution based metaheuristic for discrete optimization. International Journal of Natural Computing Research **1**(2), 15–32 (2010)
18. Source Project (2009). http://lcr.icmc.usp.br/colab/browser/Projetos/MEAN

19. Coelho, G., Von Zuben, F., da Silva, A.: A multiobjective approach to phylogenetic trees: selecting the most promising solutions from the pareto front. In: Seventh International Conference on Intelligent Systems Design and Applications, ISDA 2007, pp. 837–842, October 2007
20. Hansen, M., Jaszkiewicz, A.: Evaluating the quality of approximations to the non-dominated set. Technical report, Poznan University of Technology (1998)
21. Zitzler, E., Thiele, L., Laumanns, M., Fonseca, C., da Fonseca, V.: Performance assessment of multiobjective optimizers: an analysis and review. IEEE Transactions on Evolutionary Computation 7(2), 117–132 (April 2003)

Combining Data Mining and Evolutionary Computation for Multi-Criteria Optimization of Earthworks

Manuel Parente[1(✉)], Paulo Cortez[2], and António Gomes Correia[3]

[1] ISISE Institute for Sustainability and Innovation in Structural Engineering
and ALGORITMI Research Centre, University of Minho, Guimarães, Portugal
map@civil.uminho.pt
[2] ALGORITMI Research Centre, Department of Information Systems,
University of Minho, Guimarães, Portugal
pcortez@dsi.uminho.pt
[3] ISISE Institute for Sustainability and Innovation in Structural Engineering,
University of Minho, Guimarães, Portugal
agc@civil.uminho.pt

Abstract. Earthworks tasks aim at levelling the ground surface at a target construction area and precede any kind of structural construction (e.g., road and railway construction). It is comprised of sequential tasks, such as excavation, transportation, spreading and compaction, and it is strongly based on heavy mechanical equipment and repetitive processes. Under this context, it is essential to optimize the usage of all available resources under two key criteria: the costs and duration of earthwork projects. In this paper, we present an integrated system that uses two artificial intelligence based techniques: data mining and evolutionary multi-objective optimization. The former is used to build data-driven models capable of providing realistic estimates of resource productivity, while the latter is used to optimize resource allocation considering the two main earthwork objectives (duration and cost). Experiments held using real-world data, from a construction site, have shown that the proposed system is competitive when compared with current manual earthwork design.

Keywords: Earthworks · Equipment allocation · Metaheuristics · Data mining

1 Introduction

Levelling the ground surface and preparing the required foundation conditions are necessary steps prior to the construction of most Civil Engineering structures. These steps are especially important in the construction of linear structures, as is the case of roads or railways, since they imply the levelling of large extensions of ground surface. In order to achieve this, engineers rely on heavy mechanical equipment, such as excavators, dumper trucks, bulldozers and compactors, which allow them to handle large amounts of soil or other materials. Usually these tasks include excavating material from areas that are above the target height and transporting them to the areas below target height, where they are spread into layers and compacted, forming an embankment. The tasks

© Springer International Publishing Switzerland 2015
A. Gaspar-Cunha et al. (Eds.): EMO 2015, Part II, LNCS 9019, pp. 514–528, 2015.
DOI: 10.1007/978-3-319-15892-1_35

associated with the usage of mechanical equipment to excavate, transport, spread and compact material in order to shape the ground surface in order to fulfil a specific purpose are often referred to as earthworks.

The design of earthworks tasks is often performed by a human expert. Such expert often uses her/his experience and intuition as the main criteria for the selection and allocation of resources throughout the construction process, in order to achieve a fixed trade-off between the two key earthwork design objectives, cost and duration. This is not a trivial task. Similarly to a wide range of real-word resource allocation tasks, the two-goal optimization is nonlinear and involves a large search space of design solutions for placing the available equipment in an earthworks project.

Considering such human allocation practice, there is a high potential for reducing costs and duration of earthwork projects by adopting artificial intelligence techniques, such as data mining and Metaheuristics. In effect, data mining techniques have been proposed within this domain, taking advantage of the recent increase of available construction databases to accurately predict the productivity of mechanical equipment given specific site conditions [1–6]. Moreover, several Metaheuristics, such as evolutionary computation, ant colony optimization and swarm intelligence, have been proposed for optimal allocation of resources within the earthworks domain [7–13].

This paper presents a proposal of an intelligent system that uses both data mining and evolutionary computation to tackle the multi-criteria optimization problem associated with resource allocation in earthwork construction. The optimization task addressed in this work is a particular instance of the more general job shop scheduling problem. Considering that the data mining approach has been presented in [18], a stronger emphasis is given towards the evolutionary multi-objective optimization component of the proposed system. The main contributions are associated with the architecture and methodology that comprise the presented optimization system. In terms of optimization methods, previous works either optimize a single objective, such as cost [7, 8] or duration [9], or adopt a weighted approach [10], that optimizes separately three duration-costs weight setups (i.e. 0.8/0.2; 0.7/0.3; and 0.5/0.5). In contrast with these solutions, the system discussed in this paper takes a Pareto front optimization approach, which not only optimizes both objectives simultaneously, but also outputs a set of interesting trade-off solutions. Depending on the budget and deadline restraints, the solution that best adjusts the objectives of the designer can then be selected. Furthermore, regarding productivity estimation, while existent applications lean on the experience of the designer [9, 10], resulting in rough estimation of equipment work rates, others attempt to build computer-demanding simulation models to solve this issue [7, 8, 11]. Contrariwise, the novel system uses data-driven models (fit to real data) to estimate equipment productivity, which allows for a realistic estimation. Finally, proposed the system is validated by experimenting with real-world data from a construction site and comparing the results with those obtained by conventional earthwork design.

The paper is organized as follows. Firstly, the optimization framework for the design of earthworks, where the earthwork problem is described as a series of simultaneous production lines, susceptible to optimization, is presented in Section 2. Then, a brief state of the art description of data mining and Metaheuristics applications to the earthwork domain is described in Section 3. Next, the multi-criteria optimization system is detailed in Section 4, featuring the description of the system and results that were obtained when applying such system with real-world data from a construction site. Finally, closing conclusions and perspectives of future work are presented in Section 5.

2   An Optimization Framework for the Design of Earthworks

Taking into account an optimization point of view, earthwork construction can be described as a number of production lines based on resources and dependency relations between sequential tasks. The resources are the mechanical equipment that is essential for the development of the project, namely excavators, dumper trucks, bulldozers and compactors, while the sequential tasks correspond to the associated processes, specifically excavation, transportation, spreading and compaction, respectively. The speed at which the latter can be completed depends on the amount of the former being allocated into each task. In other words, the work rate (in this case often measured in volume of handled material per hour, m³/h) in each sequential task can be manipulated by increasing or decreasing the amount of associated resources allocated to it. This means that earthworks are strongly susceptible to optimization, which is aimed at minimizing both execution cost and duration. The multi-criteria include conflicting properties: in general, one can decrease execution duration by increasing the amount of allocated resources (mechanical equipment) to a task, but such results in an increase of the associated execution costs and vice-versa. However, it should be noted that the costs related to fuel and machinery maintenance (indirect costs) are substantial. Since these increase accordingly to the duration that mechanical equipment are working, a solution with the least possible amount of allocated resources is not necessarily the least costly. Thus, the optimal balance between the criteria must be established.

The tasks that comprise earthwork projects have a set of specific characteristics in this context, of which the focal point is interdependency. Indeed, earthwork tasks are not only sequential, but also the work rate of each of them is always limited to the work rate of its preceding task. For instance, the dumper trucks cannot undergo the transportation of soil if the latter has yet to be excavated and loaded into them; and bulldozers cannot spread soil into layers so as to allow compaction if the material has not been brought to them by the dumper trucks, and so on. Furthermore, when dealing with sequential and interdependent tasks such as these, the speed at which a single production line can carry out its work is equivalent to the work rate associated with its last task. In this context, maximizing the work rate in the final task (in this case, compaction) would correspond to a solution with minimum execution time for a production line. However, it is noteworthy to emphasize such allocation is limited by the available equipment and also by the site conditions, such as space restrictions in excavation or compaction areas (usually designated as fronts). To fully take advantage of the available resources, one must guarantee that the allocated compaction equipment is fed enough material so as to allow for constant production. In other words, the work rate in all tasks prior to compaction (excavation, transportation and spreading) must be equal or similar to the work rate obtained in the associated compaction front. Should the work rate of a task fall short of the work rate of succeeding tasks, then the productivity of the whole production line will be limited to the one obtained in that task. This keeps the equipment from reaching its maximum potential in terms of work rate, i.e. by forcing it to idle while waiting for material. Therefore, it is essential to control the work rate in each task within a production line.

Naturally, an earthwork construction is not depicted in a single production line, but rather in several independent production lines working simultaneously. Each of these production lines is associated with a compaction front, since that is the final stage for handling the geomaterials. Moreover, there is one more characteristic specific to these production lines that significantly increases its complexity. As construction ensues in several simultaneous production lines, compaction work will come to completion in one production line at a time. At the point when one production line has completed its assignment, the associated equipment is no longer contributing towards the completion of the earthwork project, thus calling for its reallocation into either an existent or a new production line. However, considering that site conditions have changed since the previous allocation, this reallocation should include all available equipment once again if it is to keep its optimal status. Thus, the whole resource allocation must be reorganized in order to optimally resume the execution of the project. This enhances the problem with a dynamic nonlinear feature, which must always be taken into account in earthworks design.

3   Artificial Intelligence in Earthworks Equipment Allocation

3.1   Data Mining

The quality of an earthworks project design can only be as good as the ability to estimate the associated equipment productivity as close to reality as possible. Nowadays, this parameter estimation is often based on the experience of the designer. In most cases, designers either settle for a somewhat random distribution of equipment, just as long as it is feasible, or attempt to apply a set of standardized teams to every production line, of which an average productivity can roughly be estimated. Obviously these neither guarantee a good design, nor result in optimal executions of the projects. In this context, data mining provides an interesting alternative approach for estimating productivity parameters. Data mining [14] allows the extraction of useful knowledge (e.g. predictive models) from raw data (often based on vast databases and/or with complex relationships), searching for patterns and tendencies in the data. Guided by domain knowledge and under a semi-automated process that uses computational tools, data mining is an iterative and interactive process. Popular predictive data mining models are based on machine learning techniques such as multiple regression (MR), artificial neural networks (ANN) [15] and support vector machines (SVM) [16]. These techniques are capable of automatically analyzing complex relationships in the data, turning them into knowledge which can be used to predict future values in new environments and for a better understanding of the problem domain variable relationships. Data mining is often framed in the context of a methodology, such as CRISP-DM (Cross Industry Standard Process for Data Mining) [17], which includes six phases (Figure 1) and facilitates the execution of data mining projects in real-world applications.

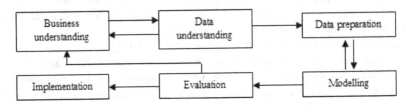

Fig. 1. The six phases of the CRISP-DM methodology (adapted from [17])

Most data mining applications to earthworks construction feature the estimation of equipment productivity, namely artificial neural network (ANN) for the estimation of excavation and transport equipment productivity rates [1, 3, 4], execution time and cost in earthwork [5] or productivity of earthwork production lines [6]. Other data mining techniques, such a multiple regression, have been successfully applied to the prediction of excavator cycle time [2]. However, a characteristic shared by all data mining techniques is that the quality of the models is highly dependent on the quality and availability the data used to fit such models. In cases for which databases are not available or do not include the necessary quality or variability to be targeted by data mining applications, it is still possible to use data stemming from technical guides or reports, which is a common practice within the geotechnical field. The practicality of this approach has been demonstrated by [18], in which several data mining models was adjusted for the purpose of estimating the productivity values for compactors in different conditions. In this work, the database consisted of the compaction tables featured in the GTR [19], a widely used empirical compaction guide. It encompassed several variables that were used as inputs for the model and can be summarized into qualitative variables (i.e., material and roller types, as well as compaction energy level) and quantitative variables (i.e., Q/S ratio, layer thickness, roller speed and number of roller passes). These inputs concern the use of soil in embankments and capping layers, representing a thorough description of the compaction conditions in regular construction cases (e.g., number of passes is one of the factors that determines the speed at which the compaction of each layer is completed, while layer thickness is directly related to the volume of material compacted after each set of roller passes). Using the R statistical tool [20] and the rminer library [21], the model that showed the best adjustment to the data was based on an ANN, achieving positive results. The ANN achieved a high coefficient of determination values (e.g., $R^2=0.99$), as well as low values for root mean squared error and mean absolute error (e.g., RMSE=0.0068, MAE=0.0039), for unseen test data using 20 runs of a 10-fold cross-validation, corresponding to a reliable prediction model. Such predictive model is capable of automatically estimating the compaction conditions, specifically the productivity of compaction equipment for any practical case with high efficiency.

3.2   Metaheuristics

Although data mining can be used for estimating parameters with a good adjustment to reality, it cannot, by itself, guarantee an optimal solution in terms of execution costs and durations. Since these criteria are a function of the allocation solution chosen by the designer, optimization becomes a complex task. Considering the non-linear characteristics of the problem and since the solution space includes a large search space (in terms of distribution combinations of equipment throughout the construction site in each phase), conventional Operational Research (e.g. linear programming) and blind search methods are not effective for solving this problem. As such, Metaheuristics are an interesting solution within this domain, since they are capable of searching interesting search space regions under a reasonable use of computational resources. Indeed, several studies have followed this approach by using optimization methods such as Genetic Algorithms (GA) [7, 12, 22] and Swarm Intelligence [9, 10, 13, 23]. Yet, the optimization carried out in most of these systems (e.g., [7, 9, 10, 12, 13, 23]) still requires an estimation of parameters, especially equipment productivity, which is still left to the experience gathered by the designer or attempted to be estimated in theoretical simulation models. Moreover, many of these applications focus on single tasks or partial processes that comprise earthworks, i.e., excavation and hauling [9, 12], in an attempt to deal with the high complexity of the problem. For this reason, these systems lack the advantages of a global optimization of execution durations and costs throughout all construction phases. In terms of optimization objectives, existent systems tend to be limited to single objective optimization, such as cost [7] or duration [9], or attempt to consider both objectives via a weight-based optimization [10]. Although these solutions are considered effective in reducing computation effort requirements, they overlook the advantages of optimizing both objectives simultaneously. Even if it can be looked at as multi-criteria optimization, the weighted-based approach used in [10] only outputs a single trade-off for a particular weight combination (e.g. 0.8 for first criteria and 0.2 for second). However, as one can easily infer in non-trivial multi-criteria optimization problems, often there is not a single optimal trade-off solution, but rather a set of trade-offs with conflicting objectives. Thus, a much natural multi-criteria optimization approach is to optimize a Pareto front of solutions, where each solution is called non-dominated, or Pareto optimal, if none of the objectives can be improved in value without worsening the other. In the context of earthwork optimization, all Pareto-optimal solutions are considered equally good and the main choice criteria for selecting one solution over the other is often decided by the project designer based on the construction final deadline and/or budget. Obviously, secondary criteria may be used to support the final decision, such as environmental aspects, which can be assessed by the determination of carbon emissions in each solution.

Taking into account that Pareto front multi-optimization requires the tracking of a population of solutions, population based Metaheuristics such as evolutionary computation, have become a natural and popular solution. Evolutionary computation is inspired in natural evolution and selection processes. Several computational variants have been proposed, such as GA [24], which are quite used within the earthwork construction domain. Evolutionary computation methods often start with a random population of possible solutions (or individuals), which are evaluated according to their fitness in a given situation. Then, the best fitted solutions are most likely to produce offspring, in the form of a new set of solutions that include characteristics from the individuals that originated then. This way, the initial set of solutions is improved in each iteration (or generation), ultimately coming to an optimal or near-optimal set of solutions.

Several evolutionary computation methods have been proposed for Pareto front optimization. In this work, we adopt the Non-dominated Sorting Genetic Algorithm-II (NSGA-II) [25] due two main reasons. Firstly, NSGA-II is a popular and standard method for multi-criteria evolutionary optimization. Secondly, NSGA-II is easily available for a computational use in the R statistical tool [20] via the package mco [26], which is the same tool adopted for the development of our integrated optimization system. The R tool was selected since the data mining models (i.e., ANNs) were also fit using this computational environment, thus allowing an easier of integration of both data mining and NSGA-II methods. Moreover, the R tool includes several conventional optimization methods, such as the Linear Programming (LP) method that is used for individual fitness calculation (see Section 4.1).

4   Multi-criteria Optimization of Earthworks

4.1   System Overview

The developed system is comprised of a data mining module integrated into the multi-objective optimization module of equipment distribution in earthworks. The first module takes care of estimating equipment productivity, while the second module carries out its optimal allocation. The system architecture falls into the framework proposed in [6]. In this work, the data mining module was applied to the GTR guide, aiming to determine compactor productivity given the material and site conditions.

The algorithmic flow for the multi-criteria evolutionary optimization method and its associated fitness function is shown in Figure 2. By interpreting the problem as a series of production lines, it becomes possible to focus the NSGA-II allocation of resources to the compaction task (last task of the production lines), which sets the work rate target value for each production line. Each solution is represents the compaction equipment for all necessary construction phases. For a particular construction phase, the solution is composed of a sequence of C integer genes: $g_1 \ g_2 \ g_3 ... g_C$, where g_i denotes the position of the i-th compactor (or roller) in terms of its compaction front and C represents the total number of compactors. The genes can take a value that ranges from 0 (not used) to the maximum number of compaction fronts F that have to be completed. The whole individual (or chromosome) includes all construction phase gene sequences, thus the total number of genes corresponds to the number of available compactors times the number of necessary construction phases: $C \times F$. For demonstration purposes, Figure 3 exemplifies a particular case where there are $C=2$ rollers and $F=2$, thus individuals are represented using four genes.

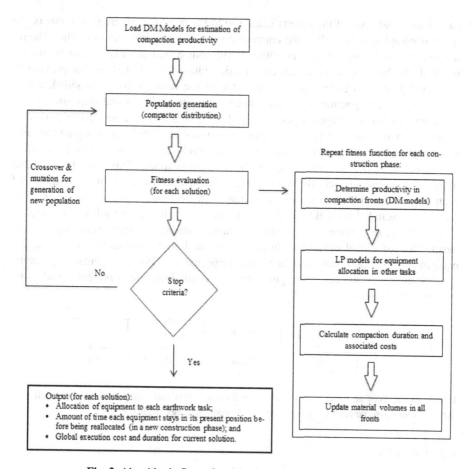

Fig. 2. Algorithmic flow of multi-criteria optimization system

At the start of the fitness evaluation procedure, only the genes that correspond to the first construction phase are selected and a repair strategy is implemented to ensure that all solutions are feasible. Given that the NSGAII implementation of the R tool only works with real values, the first step of the repair strategy is to round each chromosome to the nearest integer. Then, the work rate in each compaction front is then estimated by the data mining module. Under these conditions, the equipment for the remaining tasks (excavation, transportation, spreading) of each production line (associated with each compaction front) is then distributed by using LP optimization models. Essentially, there is one LP model per equipment type (or per task type) which is responsible for distributing the associated equipment according to the work rate obtained in the last task of the each production line. Each model targets the minimization of the total cost of the allocated equipment, while ensuring that total productivity is as close as possible to the productivity estimated in the associated compaction front.

With the allocation process completed for each production line (or compaction front), it is then possible to ascertain which one will be completed first. The information regarding the duration and costs of the current phase (up to the point the compaction in one of the fronts is completed), as well as the completed front, is saved into memory. Then, the remaining material volumes in every other active compaction and excavation fronts are updated. As the genes for the next construction phase are selected and the previously described process is executed, a second step of the repair strategy is added, which verifies which compaction fronts have been completed. The second step assures that any compactor in the current construction phase that is allocated to a completed front is: a) allocated to another front, if possible; or b) not allocated (by changing the gene to zero). For executing step a), the gene is iteratively changed according to the rule: $g_i = (g_i+1) \mod (F+1)$, until a feasible front value is found. This way the available equipment is reorganized throughout the construction fronts at the beginning of each construction phase, while assuring that work fronts that have already been completed are excluded from future allocations, as exemplified in the left of Figure 3. In each solution, this process is repeated for each construction phase, resulting in a determination of global costs and durations for the initial distribution of compaction equipment. The best solutions are then subjected to NSGA-II genetic operators, namely crossover and mutation, generating new solutions which are evaluated using the same methodology.

Fig. 3. Example of an initial chromosome for 2 compactors and 2 compaction fronts (left) and the final chromosome after the execution of the repair strategy (right)

4.2   Results

The proposed system was tested using real-word data from a construction site. A subset from a database that has been previously used during the development of the data mining models [6] was used as a reference. The available data includes the daily allocation of earthwork equipment throughout a road construction site, including information on available equipment, material volumes and types in excavation and compaction fronts and distances between fronts. The selected subset includes five production lines working simultaneously ($F=5$), to which equipment was originally allocated by conventional design methodologies. Since there is a total of C=5 available compactors, the resulting individuals in the optimization system will be comprised of 5x5=25 genes each, defining the search space for this problem.

Pareto Results. Using the methodology described in Section 4.1, the system outputs a Pareto-optimal set of solutions and their associated global costs and durations. The Pareto line represents several potential allocations of equipment throughout the construction site, in each construction phase, and this information can be accessed for each solution. Such Pareto-optimal set of solutions is useful for the designer or engineer, as she/he might want to choose different solutions depending on the available budget and the required deadlines. As earthwork construction is inherently a dynamic and unpredictable environment, different cost-duration solutions might become better adjusted to the ever-changing site conditions as construction develops.

The default parameterization of NSGA-II method, as implemented in the R tool, was adopted, namely: population size of 100, stop after 100 generations, crossover probability of 0.7 and mutation probability of 0.2. The rationale is to focus more on assessing and validating the capabilities of the proposed integrated system when compared with current human design, rather than calibrating the optimization algorithm. We note that in preliminary tests, smaller population sizes (e.g., 20) were explored, but the obtained results were worse than the default population size of 100. Also, the fitness evaluation is computationally costly, as it requires several data mining model estimations and LP optimizations (for each front), thus a population size much larger than 100 individuals would increase the computational effort. In effect, with the default NSGA-II population size, the method was executed with 3 runs on an Intel Core 2 Duo 2.66 GHz processor, and it required from 24 to 28 hours to complete each single run. The total computational time for the computational experiment (all 3 runs) was approximately 76 hours, even though this could be easily reduced using parallel computation (e.g., server with several multicore processors). The average Pareto-optimal front (over all 3 runs) is shown in Figure 4a, while a comparison between the conventional human design and those obtained by the proposed optimization system is illustrated in Figure 4b. The Pareto-optimal front (Figure 4a) is the result of a vertical averaging (i.e., according to the Duration objective) of the Pareto curves outputted by each run, using the averaging algorithm proposed by Fawcett [27] for vertical averaging of ROC curves. Since these are mean values, indication of the 95% confidence interval according to a t-student distribution was also included.

In this optimization attempt, the equipment available for the conventional design allocation was kept fixed. In other words, the presented results stem from a simple reorganization of the available equipment throughout the construction fronts, without the addition of any other piece of equipment. Bearing in mind Figure 4b, it is easy to infer how the solution obtained by conventional design is far from optimal. In fact, should this system be implemented for this construction project, a high impact could be achieved, with an estimated reduction of around 50% to 70% of both costs and durations. Still, the system does not take into account the occurrence of unpredictable events during construction (i.e., equipment malfunction). However, this could be mitigated by rerunning a new optimization procedure with new restraints, which would result in a new set of optimal solutions for current site conditions.

Allocation Analysis. From the original solution regarding equipment allocation throughout the five production lines, a general improvement could be observed. In most cases, a reasonable reduction in both durations and costs was attained by the optimization system, which is the case of the second example described in Table 1 (production line 2). One production line featured a significant increase in global work rate without increasing costs, while in other cases considerable reductions in costs were achieved without a relevant increase in total duration. Regarding the latter, a comparison between

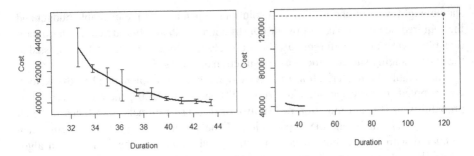

Fig. 4. Optimization results: a) vertically averaged Pareto-optimal front; b) comparison between optimized Pareto front (line) and the real-world human based allocation solution (dot); in both graphs, the Cost objective, y-axis, is presented in Euro, while the Duration objective, x-axis, is presented in hours

the resource allocation in the original human based solution and the one obtained by the proposed optimization system is shown in Table 1 (production line 1). This also corresponds to the production line where the highest amount of material volume was handled. The same methodology was used to determine costs and durations for both the original and the optimized setup.

It is easy to infer that, for both production lines depicted in Table 1, the work rates in each task of the original setup are not homogeneous, as opposed to the work rates of the optimized solution. For both cases, the whole production line is limited by the work rate of excavators in the original setup, which means that the other tasks have to wait for material to be excavated in order to allow for its transport, spreading and finally compaction. This incurs in equipment idle time while waiting for material to be ready for handling, which represents wastes in terms of resources (since these do not work at full efficiency) and fuel (contributing to unnecessary costs), as well as an increase on unnecessary carbon emissions. As a result, the total work rate of these production lines cannot be considered superior to that of the minimum work rate obtained in the production line tasks, in this case excavation (1080 m^3/h in production line 1 and 394 m^3/h in production line 2). In contrast, the work rates obtained in the proposed optimized solutions for each task that comprises the production line are as homogeneous as possible, given the available equipment. As such, a constant flow of material throughout tasks can be achieved, using the allocated resources to their full potential and efficiency. It is noteworthy to emphasize that, besides optimizing the whole allocation in terms of costs and durations, the developed system is expected to always keep the allocated equipment working at full efficiency. This is done by minimizing equipment idle time as much as possible, which will also result in minimization of unnecessary carbon emissions. This is very challenging to achieve by conventional design methodologies.

Although the total work rate of the original setup is still slightly superior to the one obtained in its optimized counterpart in production line 1, the human based allocation solution features several pieces of equipment which are not necessary for its progress, as is the case of the six 40 ton dumpers that have been originally selected, for instance.

In this case, the optimization system allocated five considerably smaller trucks (lower capacity, but lower fuel consumption and, thus, lower operation costs) to fulfil this role instead. As a result, the optimized setup for this case resulted in a decrease of 75% in total costs, while not incurring in any significant increase in duration (the actual increase in total duration is less than 1 hour of work). In the case of production line 2, besides solving the problem of work rate bottlenecks when compared to the original solution, the proposed optimized solution also features the allocation of higher productivity equipment. Consequently, a substantial decrease, over 50% in both cost and duration objectives, is obtained when comparing the integrated system optimized setup with the original human based solution. These results emphasize the importance of using intelligent computational tools for optimizing this type of construction works, also revealing how conventional human design allocation methodologies can be relatively counter-productive in some situations.

Table 1. Comparison between the conventional allocation the optimized allocation for two different production lines

Parameter	Production line 1		Production line 2	
	Original solution	Optimized solution	Original solution	Optimized solution
Average distance to excavation fronts (m)	700		175	
C - Number of Compactors	2	2	1	1
Compactor work rate (m³/h)	1831	1008	614	1055
Number of spreaders	2	2	1	2
Spreader work rate (m³/h)	1500	1088	413	1239
Number of dumper trucks	6	5	2	2
Dumper truck work rate (m³/h)	2228	1009	2960	1600
Number of excavators	2	2	1	2
Excavator work rate (m³/h)	1080	1080	394	1080

Finally, is important to note that the results associated with Figure 4 and Table 1 were obtained using an efficiency factor for mechanical equipment, k, of 0.75. This efficiency factor is related to the amount of time that the mechanical equipment spends in actual production. According to earthwork technical guides [19], actual "on-the-job" productivity is commonly influenced by factors such as operator skill, personal delays, job layout and other delays. Since one of the main focuses of this system is to maximize productivity of all the allocated mechanical equipment, it makes sense to consider the maximum value commonly suggested in earthwork technical guides (k=0.75). However, it is very hard to achieve the same efficiency factor in practice by means of conventional design, especially taking into account the fact that, as previously mentioned, it is mostly based on the experience of each designer. Additionally, unforeseen delays due to unpredictable situations that can occur in a real environment often have a significant impact on the actual efficiency factor of equipment in a construction site. In the present case, the available data indicates that the average actual efficiency factor for the mechanical equipment was just over k=0.3, which is not uncommon in this type of construction. As such, this large gap between these efficiency factors must be taken into account when analyzing the apparent discrepancy between the optimized results and the ones obtained by conventional design.

5   Conclusions and Future Work

Earthwork tasks are resource-dependant processes which aim to level target ground areas so as to allow for the construction of structures or infra-structures. Considering that these tasks represent a significant percentage of total execution durations and costs of road and railway projects, optimizing the resources involved is essential. However, the fact that conventional design methodologies lack the tools for optimal resource allocation can significantly hinder the durations and costs associated with the obtained solutions. Moreover, these methodologies are not prepared to keep up with the recent increasing demands regarding higher productivities and environmental aspects, such as minimizing carbon emissions.

In this work, a Pareto approach based on Non-dominated Sorting Genetic Algorithm-II (NSGA-II) was chosen as a basis for the development of an earthworks optimization system. The proposed system integrates several technologies, including artificial intelligence, in the form of evolutionary computation and data mining methods, and linear programming optimization, in an attempt to adjust to the complex reality associated with these types of constructions. The aim is to optimize the available resource allocation (represented by mechanical equipment) throughout the sequential tasks (namely excavation, transportation, spreading and compaction of geomaterials) that comprise the earthworks process. In this framework, the data mining technology supports the optimization techniques by providing realistic estimates to the productivity of the available equipment given site conditions.

Experiments have been carried out, using real-word data from a construction site and focusing on the assessment of the capabilities of the integrated system when compared with human allocation design. Competitive results were achieved by the proposed system, stressing the importance of using intelligent optimization tools in the design of earthworks. Also, some limitations of conventional human allocation design were shown, in particular where the production line equipment is either significantly above the required work rate requirements (incurring in unnecessary costs) or below it (resulting in idle times and low efficiency ratios). Moreover, it was possible to verify the capability of the proposed system to distribute equipment in a relatively homogeneous way (when compared to conventional design), while minimizing costs and durations, which was the goal of this research.

Future work should include the addition of features which should allow a better adjustment to reality by the system, as is the case of better grasping of space restriction conditions in the construction site. The determination of carbon emissions, either to be used as secondary criteria or a minimization objective, also fits the future work category. Furthermore, the exploration of different NSGA-II parameterization (e.g. crossover and mutation probability), as well as other multi-objective optimization methods, such as Strength Pareto Evolutionary Algorithm 2 (SPEA-2) or S-Metric Selection Evolutionary Multi-objective Optimization Algorithm (SMS-EMOA), will be addressed in future work.

Acknowledgement. The authors wish to thank FCT for the financial support under the doctoral Grant SFRH/BD/71501/2010.

References

1. Shi, J.J.: A neural network based system for predicting earthmoving production. Constr. Manag. Econ. **17**, 463–471 (1999)
2. Edwards, D.J., Griffiths, I.J.: Artificial intelligence approach to calculation of hydraulic excavator cycle time and output. Min. Technol. **109**, 23–29 (2000)
3. Tam, C.M., Tong, T., Tse, S.: Artificial neural networks model for predicting excavator productivity. J. Eng. Constr. Archit. Manag. **9**, 446–452 (2002)
4. Schabowicz, K., Hoła, B.: Application of artificial neural networks in predicting earthmoving machinery effectiveness ratios. Arch. Civ. Mech. Eng. **8**, 73–84 (2008)
5. Hola, B., Schabowicz, K.: Estimation of earthworks execution time cost by means of artificial neural networks. Autom. Constr. **19**, 570–579 (2010)
6. Parente, M., Gomes Correia, A., Cortez, P.: Artificial neural networks applied to an earthwork construction database. In: Toll, D., Zhu, H., Osman, A., Coombs, W., Li, X., Rouainia, M. (eds.) Advances in Soil Mechanics and Geotechnical Engineering, pp. 200–205. IOS Press, Durham (2014)
7. Marzouk, M., Moselhi, O.: Selecting earthmoving equipment fleets using genetic algorithms. In: Yucesan, E., Chen, C.-H., Snowdon, J.L., Charnes, J.M. (eds.) Proceedings of the 2002 Winter Simulation Conference, Montreal, Canada, pp. 1789–1796 (2002)
8. Cheng, T., Feng, C., Chen, Y.: A hybrid mechanism for optimizing construction simulation models. Autom. Constr. **14**, 85–98 (2005)
9. Kataria, S., Samdani, S.A., Singh, A.K.: Ant Colony Optimization in Earthwork Allocation. Int. Conf. Intell. Syst. 1–9 (2005)
10. Zhang, H.: Multi-objective simulation-optimization for earthmoving operations. Autom. Constr. **18**, 79–86 (2008)
11. Cheng, F., Wang, Y., Ling, X.: Multi-Objective Dynamic Simulation-Optimization for Equipment Allocation of Earthmoving Operations. Constr. Res. Congr. 328–338 (2010)
12. Xu, Y., Wang, L., Xia, G.: Research on the optimization algorithm for machinery allocation of materials transportation based on evolutionary strategy. Procedia Eng. **15**, 4205–4210 (2011)
13. Nassar, K., Hosny, O.: Solving the Least-Cost Route Cut and Fill Sequencing Problem Using Particle Swarm. J. Constr. Eng. Manag. **138**, 931–942 (2012)

14. Fayyad, U., Piatetsky-Shapiro, G., Smyth, P.: From Data Mining to Knowledge Discovery in Databases. Am. Assoc. Artif. Intell. **17**, 1–18 (1996)
15. Haykin, S.: Neural Networks – A Compreensive Foundation. Prentice Hall (1999)
16. Hearst, M.A.: Support vector machines. IEEE Intell. Syst. **13**, 18–28 (1998)
17. Chapman, P., Clinton, J., Kerber, R., Khabaza, T., Reinartz, T., Shearer, C., Wirth, R.: CRISP-DM 1.0 Step-by-step data mining guide (2000)
18. Marques, R., Gomes Correia, A., Cortez, P.: Data mining applied to compaction of geomaterials. In: Eight International Conference on the Bearing Capacity of Roads, Railways and Airfields (BCR2A 2009), Montreal, Canada, pp. 597–605. Taylor & Francis (2009)
19. SETRA, LCPC: Guide des Terrassements Routiers - Réalisation des Semblais et des Couches de Forme (2000)
20. R Development Core Team: R: A language and environment for statistical computing. R Foundation for Statistical Computing, Vienna, Austria (2011)
21. Cortez, P.: Data mining with neural networks and support vector machines using the R/rminer tool. In: Perner, P. (ed.) ICDM 2010. LNCS (LNAI), vol. 6171, pp. 572–583. Springer, Heidelberg (2010)
22. Moselhi, O., Alshibani, A.: Crew optimization in planning and control of earthmoving operations using spatial technologies. J. Inf. Technol. Constr. **12**, 1–17 (2007)
23. Miao, K., Sun, X., Li, L.: A roadbed earthwork allocation model based on ACO algorithm. Appl. Mech. Mater. (44-47), 3483–3486 (2011)
24. Holland, J.H.: Adaptation in Natural and Artificial Systems. University of Michigan Press, Ann Arbor (1975)
25. Deb, K., Pratap, A., Agarwal, S., Meyarivan, T.: A fast and elitist multiobjective genetic algorithm: NSGA-II. IEEE Trans. Evol. Comput. **6**, 182–197 (2002)
26. Mersmann, O., Trautmann, H., Steuer, D., Bischl, B., Deb, K.: Package "mco": Multiple Criteria Optimization Algorithms and Related Functions (2014). http://git.p-value.net/p/mco.git
27. Fawcett, T.: An introduction to ROC analysis. Pattern Recognit. Lett. **27**, 861–874 (2006)

Exploration of Two-Objective Scenarios on Supervised Evolutionary Feature Selection: A Survey and a Case Study (Application to Music Categorisation)

Igor Vatolkin[✉]

Department of Computer Science, TU Dortmund, Dortmund, Germany
igor.vatolkin@tu-dortmund.de

Abstract. Almost all studies which apply feature selection for supervised classification are limited to single-objective optimisation, validating feature sets with only one criterion like accuracy, classification error, correlation with the category, etc. However, this approach usually leads to a decrease of performance with respect to other relevant criteria. In this paper, we provide a summary of previous studies on supervised evolutionary multi-objective feature selection with a focus on the choice of the objectives. Further, we explore the application of EMO-FS for 28 pairs of evaluation measures in a case study predicting musical genres and styles based on the initial set of 636 features. To measure the advantage of a multi-objective approach over a single-objective one, we propose two metrics based on hypervolume and provide a statistical comparison of multi-objective performance across 14 categorisation tasks.

Keywords: Multi-objective feature selection · Evolutionary feature selection · Music classification · Genre recognition

1 Introduction and Literature Survey

1.1 Single- and Multi-objective Feature Selection

Supervised classification of data instances is based on the previously extracted and typically numerical data characteristics (features) together with labels, for example binary relationships to given categories. In real-world classification scenarios, automatic classification often faces a problem that high demands on time and/or computing resources are necessary to create a large enough set of labelled instances to learn from: the labelling may require intensive manual efforts, expensive probes, or complex simulations.

If a large number of feature candidates are used to construct classification models from a rather limited training set of data instances, the danger of overfitting increases and some irrelevant features may be identified as relevant by chance. Highly overfitted classification models perform very well for the training

A. Gaspar-Cunha et al. (Eds.): EMO 2015, Part II, LNCS 9019, pp. 529–543, 2015.
DOI: 10.1007/978-3-319-15892-1_36

data, but their generic performance is poor [22]. The identification of a "perfect" feature set instead of the training of classification models with all available features may provide further benefits. Because the number of features is reduced, it is possible to save time and storage resources of feature extraction, preprocessing, training of classification models, and classification itself. For instance, the classification of unlabelled data is done faster with small and robust models instead of complex and probably overfitted ones.

In general, feature selection can be defined as follows:

$$q^* = \arg\min_{q} \left[m\left(\boldsymbol{y}, \hat{\boldsymbol{y}}, \varPhi(\mathcal{F}, \boldsymbol{q})\right) \right], \tag{1}$$

where $\boldsymbol{q}$ is a binary vector which denotes the features to be selected (one at i-th position means that i-th feature should be selected), $\varPhi$ is the corresponding subset of the complete feature set $\mathcal{F}$, and m is a relevance function (also referred to as evaluation measure) for the evaluation of feature sets. $\boldsymbol{y}$ are the true labels, $\hat{\boldsymbol{y}}$ the predicted ones, and $\boldsymbol{q}^*$ is the optimal feature set w.r.t. the given relevance function. Note that some relevance function are not dependent on the labels (amount of correlated features, classification runtime, etc.). Measures to be maximised like accuracy can be adapted for the minimisation.

The decision to optimise a feature set with respect to a particular relevance function often leads to a decrease of performance w.r.t. other relevant criteria, consider several examples:

- Models with very small classification errors often require more features, have a lower generalisation ability, and the classification may be slower.
- For binary classification, the performance on positive and negative data instances may be in conflict.
- Models trained with small training sets may efficiently reduce efforts for data labelling but have a lower classification quality.

In the multi-objective feature selection (MO-FS), K conflicting relevance functions $m_1, ..., m_K$ are taken into account:

$$q^* = \arg\min_{q} [m_1\left(\boldsymbol{y}, \hat{\boldsymbol{y}}, \varPhi(\mathcal{F}, \boldsymbol{q})\right), ..., m_K\left(\boldsymbol{y}, \hat{\boldsymbol{y}}, \varPhi(\mathcal{F}, \boldsymbol{q})\right)]. \tag{2}$$

For a set of F features, there are exist $2^F - 1$ combinations of possible non-empty subsets, and the problems related to feature selection were described as NP-hard [16]. Because many deterministic strategies require a large number of feature subset evaluations (for an exhaustive overview of methods see [13]), heuristics like evolutionary algorithms (EAs) may help to find a (sub)optimal feature subset within an acceptable amount of evaluations. Because EAs evolve a set of (among others non-comparable) solutions at the same time, they are also well suited to solve MO-FS.

The first application of EAs for feature selection was proposed in [33]. Evolutionary multi-objective feature selection (EMO-FS) was introduced approximately a decade later [6]. In the following years and until now, a number of studies on EMO-FS was reported. Probably the most exhaustive up-to-date overview is provided in [23] - however only ten references on supervised EMO-FS are listed (in the next section we refer to several further works). This indicates a strong potential for further studies.

1.2 Evaluation Measures in Previous Works on Supervised Evolutionary Multi-Objective Feature Selection

There exist many relevance functions which may be in conflict and important for a concrete application scenario. In [39,41], we discussed several groups of evaluation measures with a specific aim to improve the reliability of music classification. However, this categorisation is also applicable for other supervised classification scenarios:

- Measures of **classification performance** include commonly applied methods for the evaluation of supervised classification and are typically based on the confusion matrix: accuracy, precision, recall, etc., for a general discussion see [1,35]. Some of these measures were constructed for highly imbalanced data sets [34].
- **Resource** measures describe demands on runtime and storage space. It is usually hard to measure runtime in a credible way because it depends on the hardware, load of the operating system, but also on the (in)efficient implementation of the method or differences between runtime environments.
- **Model complexity** measures help to identify simpler models which are more robust against overfitting. Because more complex models often have larger demands on resources, this group of measures is related to resource measures. However, the primary goals are different and sometimes a model with larger demands on storage space can be indeed more robust: consider a k-nearest-neighbours model which requires more storage place compared to, e.g., a decision tree model.
- Optimisation of **user related** measures aims at the reduction of any personal efforts necessary to create a supervised classification model (e.g., labelling of a large training set or any interaction during the training process), but also to increase personal satisfaction with classification results.
- **Specific performance** measures are designed for the evaluation of a particular task. For example, the evaluation of music segmentation (binary recognition of boundaries between song sections like intro, bridge, or verse) may be based on a non-linearly weighting of the distances between predicted and existing boundaries.

Even closely related measures may be less correlated as observed in our previous study on music genre and style recognition [40] and thus can be used as objectives for EMO-FS. Following combinations of objectives were reported in the literature:

- In a majority of works the goal is to **minimise the number of features and the classification error** (or maximise the accuracy) [6,11,12,14, 19,21,26,27,39,41,42,45,47,48]. Sometimes the definition of multi-objective feature selection is explicitly restricted to this case [14,47,48].
- **Further combinations of two objectives related to classification performance** are recall and specificity [7,10,39] and precision and recall [5]. Only a few works explore more combinations of two objectives: 3 combinations in [12] and 4 in [11], however always with the number of features as one of both criteria (together with accuracy, F-measure and two error measures).
- Optimisation of pairs of **measures beyond classification performance and the number of features** is seldom applied. The classification error and the size of decision tree classifier are minimised in [29]. Filter-based measures are included in [46] (inter-correlation between a feature set and a category against intra-correlation across features in the set[1]) and in [44] (15 combinations of 6 importance measures). The last work reports the largest number of combinations of objectives in our survey.
- **Three and more objectives** optimised at the same time are the number of features together with the classification error and F-measure [12], extended with the number of features, classification error, and recall in [11]. Further combinations are the number of features, the classification error, and the difference in error rate among classes [20], the number of features, recall, and specificity [37], the number of features, accuracy, and mutual information [45], the number of features, information gain, and mutual correlation [31], and the number of features together with 3 pairs of criteria selected from accuracy, F-measure, and Matthews correlation coefficient [21]. Simultaneous optimisation of four objectives (the number of features, error, recall, and F-measure) was applied in [11].

Not only numbers of objective combinations, but also the sizes of feature sets were typically not very large in related works. From the above mentioned studies only [10,20,29,39,41,42,47] addressed sets of features above two hundred. Though, in one of the first exhaustive studies on evolutionary feature selection it was recommended to apply EAs for "very large" sets with more than 100 features [18].

Various popular multi-objective evolutionary algorithms (MOEAs) were applied for feature selection (however, without a systematic comparison across the methods), for example, NSGA-II [3] in [5,14,21,31,44,46], SMS-SMOA [8] in [39,41,42], PSO [15] in [47], and differential evolution [30] in [48]. The design of specific operators to enhance the original methods was addressed in some works, e.g., commonality-based crossover designed to preserve "building blocks" of features with high performance [6], memetic framework to improve EA with a local search [49], asymmetric mutation to favour smaller feature sets [42], or the ensemble of multi-objective optimisers [37].

---

[1] Note that the evaluation of feature sets w.r.t. filter-based measures only does not belong any more to supervised feature selection.

1.3   Music Categorisation and Evolutionary Feature Selection

Music information retrieval (MIR) is a growing research domain which comprises subfields such as music transcription, recommendation of new music, optical music recognition, analysis of listening preferences, automatic correction of vocals, detection of plagiarism, and so on. Because the current scope of MIR goes far beyond the "retrieval of information", a probably better acronym is "music information research" as introduced in [32]. The last document provides a well structured description of current tasks and future challenges of MIR. The history of the earlier stage of MIR is discussed in [4].

Categorisation of music data plays one of the most prominent roles in MIR. The categories to predict are musical genres, styles, personal preferences, moods, instruments, harmonic properties such as key and mode, and so on. Among others, the recognition of genres is a very common application, [36] refers to several hundreds of related publications.

In contrast to some other classification scenarios, however, feature selection was until now not very often applied for music classification. The simple reason is that the number of features was not very high in the recent past. For example, in one of the pioneering works on audio genre classification [38], 30 features were used. However, this situation is changing:

- The number of new approaches for the extraction of **high-level, semantic features** relevant to music theory is growing. For example, in [41] we extracted a set of 566 high-level descriptors, partly integrated from several MIR frameworks and partly developed by ourselves.
- The analysis of "big data" gathered from **listener statistics** on the Internet can be used for categorisation. Automatically predicted user tags [2] can be a feature source for the recognition of further categories, or playlist statistics may lead to successful recognition of musical genres [43].
- In some works **automatic feature construction** was proposed [24, 25, 28]. The evolution of method chains built as combinations of various signal-related operators (e.g., Fourier transform, filters) and general operators (product, autocorrelation, etc.) leads to a theoretically unlimited number of features, so that the proper identification of relevant feature sets becomes necessary to control this process.

Therefore, music classification is a promising and praxis-related application for the comparison and analysis of feature selection methods. Evolutionary FS was applied probably the first time for music classification in [9]. A list with further studies on EA-FS in MIR is provided in [41]. EMO-FS is until now almost unexplored for MIR, and we hope that after our first studies [39, 41, 42] the interest on this topic will grow.

The choice of conflicting objectives for the optimisation of music classification depends on a concrete application. For example, resources play a more important role for the applications on mobile devices. The balance between surprise and trustworthiness for the recommendation of new music depends on the preferences of a listener. In other scenarios the limitation of human efforts may be very important (learning only from a few music tracks or even only by positives).

2   Experimental Setup

2.1   Classification Scenarios

For a case study on the exploration of combinations of different evaluation measures for supervised EMO-FS we use our music data set of 120 albums[2]. The categories to recognise are 6 genres (Classic, Electronic, Jazz, Pop, Rap, R&B) and 8 styles (AdultContemporary, AlbumRock, AlternativePopRock, etc.). A complete set of 636 audio signal features contains various characteristics of time domain, spectrum, cepstrum, phase domain, etc. For the exact list of features and references to their definitions see [41]. Feature vectors were extracted for classification windows of 4s with 2s overlap, and the classification models are created with three binary supervised methods: naive Bayes, random forest, and support vector machine with a linear kernel.

2.2   Measures for the Evaluation of Feature Subsets

SMS-EMOA [8] was applied for two-objective optimisation of all 28 pairs of 8 evaluation measures described below. 5 statistical runs for each of 3 classification methods and 14 categories led to an overall number of $28 \cdot 5 \cdot 3 \cdot 14 = 5,880$ experiments. The number of EA iterations in each statistical run was limited to 3,000. The details about our adaptation to the original SMS-EMOA are provided in [41].

Two evaluation measures to minimise were the rate of selected features m_{FR} and the balanced relative classification error m_{BRE}:

$$m_{BRE} = \frac{1}{2} \left(\frac{FN}{TP + FN} + \frac{FP}{TN + FP} \right), \tag{3}$$

where TP denotes true positives (number of classification windows labelled as belonging to a category and correctly predicted as belonging to it), TN true negatives (labelled as not belonging and predicted as not belonging), FP false positives (not belonging but predicted as belonging), and FN (belonging but predicted as not belonging). Estimation of m_{BRE} makes sense for imbalanced sets if the performance on both categories is relevant.

Evaluation measures to maximise (Equations 4-9) were precision m_{PREC}, recall (sensitivity) m_{REC}, specificity m_{SPEC}, F-measure m_{F1}, geometric mean m_{GEO}, and Spearman's correlation coefficient m_{SPEAR}.

Precision, recall, and specificity evaluate the performance for instances of one class or identified as belonging to one class.

$$m_{PREC} = \frac{TP}{TP + FP}, \tag{4}$$

$$m_{REC} = \frac{TP}{TP + FN}, \tag{5}$$

---

[2] http://ls11-www.cs.uni-dortmund.de/rudolph/mi#music_test_database

$$m_{SPEC} = \frac{TN}{FP + TN}. \tag{6}$$

F-measure and geometric mean are the combinations of these measures:

$$m_F = \frac{(\beta^2 + 1) \cdot m_{PREC} \cdot m_{REC}}{\beta^2 \cdot m_{PREC} + m_{REC}}, \tag{7}$$

and m_{F1} is a special case with $\beta = 1$ (m_{PREC} and m_{REC} are evenly balanced).

Geometric mean was proposed in [17] for imbalanced training sets:

$$m_{GEO} = \sqrt{m_{REC} \cdot m_{SPEC}}. \tag{8}$$

Finally, Spearman's correlation coefficient measures the rank-based correlation between true and predicted labels:

$$m_{SPEAR} = \frac{\sum_{i=1}^{T} (R(\hat{y}(i)) \cdot R(y(i))) - T\left(\frac{T+1}{2}\right)^2}{\sqrt{\left(\sum_{i=1}^{T} (R^2(\hat{y}(i))) - T\left(\frac{T+1}{2}\right)^2\right) \cdot \left(\sum_{i=1}^{T} (R^2(y(i))) - T\left(\frac{T+1}{2}\right)^2\right)}}, \tag{9}$$

where T is the number of all classification windows, $y(i)$ the true label of the i-th window, $\hat{y}(i)$ the predicted label of the i-th window, and $R(\cdot)$ the rank after the sorting of windows.

2.3   Evaluation of Multi-objectiveness

If two evaluation measures are strongly (anti)correlated, MOO is not necessary: minimisation or maximisation of one of both criteria is sufficient. However, if the correlation between two measures can not be explained theoretically, a large number of evaluations of classification models is necessary to decide if MOO makes sense. Further, dependencies between objectives may be stronger or weaker for certain regions of the search space. For example, an increasing number of features often leads to a larger risk that some of irrelevant features would be identified as relevant, and the general performance suffers (we observed this behaviour in our previous studies on EMO-FS). On the other side, this does not mean that a simple reduction of the number of features would increase the classification performance. To provide an automatic justification for the decision for or against MOO, we propose below two a posteriori validation measures based on hypervolume which describe the best found non-dominated front.

Dominated hypervolume, or $\mathcal{S}$-metric, evaluates both the closeness to the Pareto-front and the diversity of solutions [50]:

$$\mathcal{S}(\mathbf{a}_1, ..., \mathbf{a}_N) = vol\left(\bigcup_{i=1}^{N} [\mathbf{a}_i, \mathbf{r}]\right), \tag{10}$$

where $\mathbf{a}_1, ..., \mathbf{a}_N$ are N solutions (feature subsets in case of EMO-FS) of the front, $\mathbf{r}$ is a reference point (usually worst possible solution), and $vol(\cdot)$ is the united volume of all hypercubes between the reference point and $\mathbf{a}_1, ..., \mathbf{a}_N$.

For a multi-objective minimisation of K objectives, consider the ideal solution $\mathbf{a}_{ID}$ which components are built from the best individual values for each objective. If the minimisation of one of both objectives would be enough to (almost) achieve this ideal solution, the volume between the non-dominated front and $\mathbf{a}_{ID}$ would be small and in extreme case equal to zero, when single- and multi-objective approaches would converge to the same single optimum. Thus, we can measure the volume exquisitely dominated by the ideal solution as $\mathcal{S}(\mathbf{a}_{ID}) - \mathcal{S}(\mathbf{a}_1, ..., \mathbf{a}_N)$. The share of this volume in per cent to the overall volume dominated by $\mathbf{a}_{ID}$ is calculated as:

$$\epsilon_{ID} = \frac{\mathcal{S}(\mathbf{a}_{ID}) - \mathcal{S}(\mathbf{a}_1, ..., \mathbf{a}_N)}{\mathcal{S}(\mathbf{a}_{ID})} \cdot 100\%. \tag{11}$$

To illustrate ϵ_{ID}, Fig. 1 sketches two examples. A higher diversity of tradeoff solutions in the left subfigure leads to a larger volume exquisitely dominated by the ideal solution (shaded area). The right subfigure presents an example with only a marginal advantage of MOO over a single-objective approach.

Another possibility to measure the advantage of MOO over SOO is to find a solution with the maximum contribution to the dominated hypervolume and to estimate the hypervolume exquisitely dominated by other solutions of the non-dominated front. If this volume is very small, the search for the solution with the maximum contribution to the dominated hypervolume may be sufficient and can be achieved optimising a weighted sum of both objectives. This volume corresponds to areas marked with diagonal lines in Fig. 1. The share of this volume to the volume dominated by the complete front is defined as:

$$\epsilon_{MAX} = \frac{\mathcal{S}(\mathbf{a}_1, ..., \mathbf{a}_N) - \max\limits_{i \in \{1,...,N\}} \mathcal{S}(\mathbf{a}_i)}{\mathcal{S}(\mathbf{a}_1, ..., \mathbf{a}_N)} \cdot 100\%. \tag{12}$$

For the estimation of ϵ_{ID} and ϵ_{MAX}, we set the reference point to the Nadir point after all experiments for the related combination of two objectives.

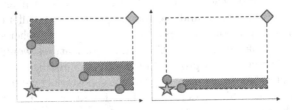

Fig. 1. Stronger (left subfigure) and weaker (right subfigure) advantage of MOO against single-objective approach. Circles: solutions from the non-dominated front; Asterisk: ideal solution; diamond: reference point.

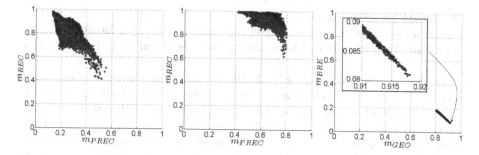

Fig. 2. Left subfigure: All solutions found during the maximisation of precision and recall for 'Electronic'; Middle subfigure: EMO with the same pair of objectives for 'Classic'; Right subfigure: Minimisation of classification error and maximisation of geometric mean for 'Rap'.

3   Discussion of Results

3.1   General Comparison of Objective Pairs

Table 1 in the Appendix lists ϵ_{ID} and ϵ_{MAX} for all combinations of objectives and classification tasks. The measures are estimated for non-dominated fronts after 15 experiments (5 statistical repetitions for each of 3 classifiers), cf. Sect. 2.2. As it can be expected, pairs of objectives are very differently suited for MOO. This is illustrated in Fig. 2. The best combination on average across 14 categories is precision and recall: mean of ϵ_{ID} is 21.27%, and mean of ϵ_{MAX} is 30.87%. The left subfigure shows all feature subsets found during EMO process for this pair of objectives and a category with the highest values of ϵ_{ID} and ϵ_{MAX} ('Electronic', $\epsilon_{ID} = 33.90\%$, $\epsilon_{MAX} = 39.52\%$) and the subfigure in the middle for a category with the lowest values of multi-objectiveness ('Classic', $\epsilon_{ID} = 4.20\%$, $\epsilon_{MAX} = 12.99\%$). Differing complexities of classification tasks explain smaller values of ϵ_{ID} and ϵ_{MAX} for 'Classic' compared to 'Electronic': the distinction of classical music against popular tracks was the simplest task across all categories.

Fig. 2, right subfigure presents an example for a pair of objectives which is at least reasonable for MOO, geometric mean and the balanced relative classification error ('Rap', $\epsilon_{ID} = 0.00\%$, $\epsilon_{MAX} = 0.14\%$). The enlargement in the rectangle shows that there are two non-dominated solutions even if SOO would be sufficient for feature selection. All pairs with zero entries in Table 1 belong to 6 possible combinations of m_{BRE}, m_{F1}, m_{GEO}, and m_{SPEAR}. Thus, it can be suggested after the experiments that only one of these measures should be selected for MOO. In particular, these measures evaluate the classification quality w.r.t. the imbalance of the data set.

3.2   Statistical Comparison of Objective Pairs Across Categories

The values of ϵ_{ID} and ϵ_{MAX} may vary for different categorisation tasks as shown above. To compare the pairs across the categories, Mann-Whitney test was applied for all vectors $\mathbf{u}(i,j)$ and $\mathbf{v}(i,j)$, where $i \in \{1, ..., 28 \cdot 28\}$ iterates through all combinations of objective pairs and $j \in \{1,2\}$ indicates if ϵ_{ID} or ϵ_{MAX} build the elements of 14-dimensional vectors $\mathbf{u}$ and $\mathbf{v}$ (14 is the number of categories).

The results are plotted in Fig. 3. Black entries in matrices indicate that the pair of objectives in the corresponding row has a significantly higher multi-objectiveness than the pair in the column and white entries that the pair in the row has a significantly lower value of multi-objectiveness than the pair in the column. Grey values indicate no statistical difference. For example, the optimisation of m_{BRE} and m_{PREC} led to larger ϵ_{ID} compared to m_{BRE} and m_{FR} or m_{BRE} and m_{REC}, but there is no statistical difference across categories between m_{BRE}, m_{PREC} and m_{BRE}, m_{SPEC}.

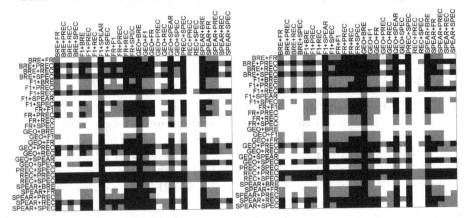

Fig. 3. Comparison of pairs of objectives against each other based on ϵ_{ID} (left subfigure) and ϵ_{MAX} (right subfigure). For further details see the text.

Another observation is that there are many entries marked with grey colour: 19.31% for the matrix with ϵ_{ID} values (28 diagonal entries were not considered for this statistic) and 24.87% for the matrix with ϵ_{MAX} values. This means, for instance, that it can not be stated that the application of EMO is more promising for m_{BRE} and m_{PREC} rather than for m_{BRE} and m_{SPEC}. On the other side, there are some pairs of objectives which are in general better suited for EMO. In particular, the $m_{REC+PREC}$ row contains black entries except for the comparison with itself and $m_{REC+SPEC}$; this holds for both ϵ_{ID} and ϵ_{MAX}. Other combinations make no sense for MOO, e.g., the $m_{F1+SPEAR}$ row contains white entries except for the comparison with itself and $m_{GEO+BRE}$ for both matrices.

3.3 Difference between ϵ_{ID} and ϵ_{MAX}

ϵ_{ID} and ϵ_{MAX} were defined having slightly different aims in mind. Despite of expected correlation between these measures we may expect also some differences for particular combinations of objectives and classification tasks. Fig. 4, left subfigure plots all pairs of ϵ_{ID}, ϵ_{MAX} from the Table 1.

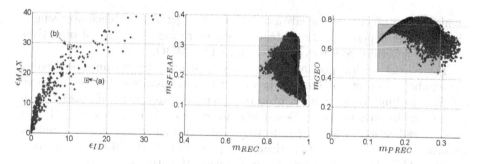

Fig. 4. Comparison of ϵ_{ID} and ϵ_{MAX}

To show the most extreme cases for imbalance between ϵ_{ID} and ϵ_{MAX}, the hypervolume was estimated for all points of the Fig. 4, left subfigure, with the goal to maximise ϵ_{ID} and minimise ϵ_{MAX}. In other words, it was searched for a non-dominated front with the highest share of the volume between the front and the ideal feature subset (better for MOO) but also the largest share of the volume exisuitely dominated by the solution with the largest hypervolume (better for SOO). This combination of ϵ_{ID} and ϵ_{MAX} is marked with (a). Feature subsets from all corresponding experiments are visualised in the middle subfigure (maximisation of m_{REC} and m_{SPEAR}, category 'ClubDance'), and the shaded rectangle marks the dominated hypervolume of the solution with the largest S from the non-dominated front.

Respectively, (b) in the left subfigure and the right subfigure (m_{PREC}, m_{GEO}, 'SoftRock') illustrate the opposite situation. Here, the share of the volume between the non-dominated front and the ideal solution is comparably smaller, but the share of the hypervolume of the solution with the largest hypervolume lower.

A possible problematic issue of ϵ_{ID} is that if the hypervolume between the ideal and Nadir points is very large, the share of the volume between the non-dominated front and the ideal solution may be smaller, although both objectives may be in conflict and of interest for a decision maker. For example, the row 'BRE,FR' in the upper half of Table 1 contains smaller values than 'BRE,FR' in the bottom half of the table. As discussed in our previous studies on the optimisation of m_{BRE} and m_{FR}, the choice of solutions from the non-dominated front depends on the application scenario. E.g., for a mobile device with limited resources it could be promising to accept a slightly worse classification error removing approximately half of features necessary to achieve the minimal error. Nevertheless, ϵ_{ID} still describes important properties of non-dominated fronts.

4   Conclusions and Outlook

In this paper, we provided an up-to-date overview of related works on evolutionary multi-objective feature selection for supervised classification. Further, we proposed two measures to evaluate the multi-objectiveness which may help to identify reasonable combinations of objectives. As a case study we optimised 28 pairs of criteria for 14 music classification tasks. The results showed that for many pairs of objectives the advantage of mutli-objective approach over single-objective one depends on the category, but there are also pairs of objectives which are generally very well or barely suited for MOO.

There exist several possibilities for further studies, for instance, extending the number of objectives, but also investigating combinations of three and more criteria. To limit their number, objectives can be sorted w.r.t. discussed measures for multi-objectiveness. We also plan to compare the impact of EMO-FS for music classification to other classification tasks with large feature sets. Another promising direction is to systematically compare the impact of different enhancements to MOEAs proposed for the optimisation of feature selection.

References

1. Alpaydin, E.: Introduction to Machine Learning. The MIT Press (2010)
2. Bertin-Mahieux, T., Eck, D., Maillet, F., Lamere, P.: Autotagger: A Model for Predicting Social Tags from Acoustic Features on Large Music Databases. Journal of New Music Research **37**(2), 115–135 (2008)
3. Deb, K., Pratap, A., Agarwal, S., Meyarivan, T.: A Fast and Elitist Multiobjective Genetic Algorithm: NSGA-II. IEEE Trans. on Evol. Comp. **6**(2), 182–197 (2002)
4. Downie, J.S.: Music Information Retrieval. Annual Review of Information Science and Technology **37**(1), 295–340 (2003)
5. Ekbal, A., Saha, S., Garbe, C.S.: Feature selection using multiobjective optimization for named entity recognition. In: Proc. of the 20th Int'l Conf. on Pattern Recognition (ICPR), pp. 1937–1940 (2010)
6. Emmanouilidis, C., Hunter, A., MacIntyre, J.: A multiobjective evolutionary setting for feature selection and a commonality-based crossover pperator. In: Proc. of the IEEE Congress on Evolutionary Computation (CEC), vol. 1, pp. 309–316 (2000)
7. Emmanouilidis, C.: Evolutionary Multi-Objective Feature Selection and ROC Analysis With Application to Industrial Machinery Fault Diagnosis. Evolutionary Methods for Optimization, Design and Control, CIMNE (2002)
8. Emmerich, M.T.M., Beume, N., Naujoks, B.: An EMO algorithm using the hypervolume measure as selection criterion. In: Coello Coello, C.A., Hernández Aguirre, A., Zitzler, E. (eds.) EMO 2005. LNCS, vol. 3410, pp. 62–76. Springer, Heidelberg (2005)
9. Fujinaga, I.: Machine recognition of timbre using steady-state tone of acoustic musical instruments. In: Proc. of the Int'l Comp. Music Conf. (ICMC), pp. 207–210 (1998)
10. García-Nieto, J., Alba, E., Jourdan, L., Talbi, E.: Sensitivity and Specificity based Multiobjective Approach for Feature Selection: Application to Cancer Diagnosis. Inf. Process. Lett. **109**(16), 887–896 (2009)
11. Gaspar-Cunha, A.: Feature selection using multi-objective evolutionary algorithms: application to cardiac SPECT diagnosis. In: Rocha, M.P., Riverola, F.F., Shatkay, H., Corchado, J.M. (eds.) IWPACBB 2010. AISC, vol. 74, pp. 85–92. Springer, Heidelberg (2010)
12. Gaspar-Cunha, A., Mendes, F., Duarte, J., Vieira, A., Ribeiro, B., Ribeiro, A., Neves, J.: Multi-objective evolutionary algorithms for feature selection: application in bankruptcy prediction. In: Deb, K., Bhattacharya, A., Chakraborti, N., Chakroborty, P., Das, S., Dutta, J., Gupta, S.K., Jain, A., Aggarwal, V., Branke, J., Louis, S.J., Tan, K.C. (eds.) SEAL 2010. LNCS, vol. 6457, pp. 319–328. Springer, Heidelberg (2010)

13. Guyon, I., Nikravesh, M., Gunn, S., Zadeh, L.A. (eds.): Feature Extraction. Foundations and Applications. Springer (2006)
14. Hamdani, T.M., Won, J.-M., Alimi, M.A.M., Karray, F.: Multi-objective feature selection with NSGA II. In: Beliczynski, B., Dzielinski, A., Iwanowski, M., Ribeiro, B. (eds.) ICANNGA 2007. LNCS, vol. 4431, pp. 240–247. Springer, Heidelberg (2007)
15. Kennedy, J., Eberhart, R.: Particle swarm optimization. In: Proc. of the IEEE International Conference on Neural Networks (ICNN), vol. 4, pp. 1942–1948 (1995)
16. Kohavi, R., John, G.H.: Wrappers for feature subset selection. Artificial Intelligence 97(1–2), 273–324 (1997)
17. Kubat, M., Matwin, S.: Addressing the curse of imbalanced training sets: one-sided selection. In: Proc. of 14th Int'l Conf. on Machine Learn. (ICML), pp. 179–186 (1997)
18. Kudo, M., Sklansky, J.: Comparison of Algorithms that Select Features for Pattern Classifiers. Pattern Recognition 33(1), 25–41 (2000)
19. Lac, H.C., Stacey, D.A.: Feature subset selection via multi-objective genetic algorithm. In: Proc. IEEE Int'l Joint Conf. on Neural Networks (IJCNN), pp. 1349–1354 (2005)
20. Liu, J., Iba, H.: Selecting informative genes using a multiobjective evolutionary algorithm. In: Proc. of the 2002 Congr. on Evolutionary Comput. (CEC), pp. 297–302 (2002)
21. Martins, M., Costa, L., Frizera, A., Ceres, R., Santos, C.: Hybridization between Multi-Objective Genetic Algorithm and Support Vector Machine for Feature Selection in Walker-Assisted Gait. Comp. Meth. and Prog. in Biom. 113, 746–748 (2014)
22. Mitchell, T.: Machine Learning. McGraw-Hill (1997)
23. Mukhopadhyay, A., Maulik, U., Bandyopadhyay, S., Coello Coello, C.A.: A Survey of Multiobjective Evolutionary Algorithms for Data Mining: Part I. IEEE Trans. on Evolutionary Computation 18(1), 4–19 (2014)
24. Mäkinen, T., Kiranyaz, S., Raitoharju, J., Gabbouj, M.: An Evolutionary Feature Synthesis Approach for Content-Based Audio Retrieval. EURASIP Journal on Audio, Speech and Music Processing 2012 (2012)
25. Mierswa, I., Morik, K.: Automatic Feature Extraction for Classifying Audio Data. Machine Learning Journal 58(2–3), 127–149 (2005)
26. Oliveira, L.S., Sabourin, R., Bortolozzi, F., Suen, C.Y.: A Methodology for Feature Selection Using Multiobjective Genetic Algorithms for Handwritten Digit String Recognition. Int'l Journal of Pattern Recogn. and Artif. Intell. 17, 903–929 (2003)
27. Oliveira, L.S., Morita, M., Sabourin, R.: Feature selection for ensembles using the multi-objective optimization approach. In: Jin, Y. (ed.) Multi-Objective Machine Learning, Studies on Computational Intelligence, vol. 16, pp. 49–74. Springer (2006)
28. Pachet, F., Roy, P.: Analytical features: A Knowledge-Based Approach to Audio Feature Generation. EURASIP J. on Audio, Speech, and Mus. Proc. 2009 (2009)
29. Pappa, G.L., Freitas, A.A., Kaestner, C.A.A.: Attribute Selection with a Multiobjective Genetic Algorithm. Adv. in Artificial Intell., 280–290. Springer (2002)
30. Robič, T., Filipič, B.: DEMO: differential evolution for multiobjective optimization. In: Coello Coello, C.A., Hernández Aguirre, A., Zitzler, E. (eds.) EMO 2005. LNCS, vol. 3410, pp. 520–533. Springer, Heidelberg (2005)
31. Saroj, J.: Multi-objective genetic algorithm approach to feature subset optimization. In: Proc. of IEEE Int'l Advance Computing Conf. (IACC), pp. 544–548 (2014)
32. Serra, S., Magas, M., Benetos, E., Chudy, M., Dixon, S., Flexer, A., Gómez, E., Gouyon, F., Herrera, P., Jorda, S., Paytuvi, O., Peeters, G., Schlüter, J., Vinet, H., Widmer, G.: Roadmap for Music Information Research. The MIReS Consortium, Tech. Rep. (2013)
33. Siedlecki, W.W., Sklansky, J.: A Note on Genetic Algorithms for Large-Scale Feature Selection. Pattern Recognition Letters 10(5), 335–347 (1989)

34. Sokolova, M., Japkowicz, N., Szpakowicz, S.: Beyond accuracy, F-score and ROC: a family of discriminant measures for performance evaluation. In: Proc. of the Adv. in Artif. Intell. Works. on Eval. Methods for Machine Learning (AI), pp. 1015–1021 (2006)
35. Sokolova, M., Lapalme, G.: A Systematic Analysis of Performance Measures for Classification Tasks. Information Processing and Management **45**(4), 427–437 (2009)
36. Sturm, B.L.: A survey of evaluation in music genre recognition. In: Nürnberger, A., Stober, S., Larsen, B., Detyniecki, M. (eds.) AMR 2012. LNCS, vol. 8382, pp. 29–66. Springer, Heidelberg (2014)
37. Tan, C.J., Lim, C.P., Cheah, Y.-N.: A Multi-Objective Evolutionary Algorithm-based Ensemble Optimizer for Feature Selection and Classification with Neural Network Models. Neurocomputing **125**, 217–228 (2014)
38. Tzanetakis, G., Cook, P.: Musical Genre Classification of Audio Signals. IEEE Transactions on Speech and Audio Processing **10**(5), 293–302 (2002)
39. Vatolkin, I., Preuß, M., Rudolph, G.: Multi-objective feature selection in music genre and style recognition tasks. In: Proc. of the 13th Genetic and Evolutionary Computation Conference (GECCO), pp. 411–418. ACM (2011)
40. Vatolkin, I.: Multi-objective evaluation of music classification. In: Gaul, W.A., Geyer-Schulz, A., Schmidt-Thieme, L., Kunze, J. (eds.) Studies in Classification, Data Analysis, and Knowledge Organization, pp. 401–410. Springer (2012)
41. Vatolkin, I.: Improving Supervised Music Classification by Means of Multi-Objective Evolutionary Feature Selection. PhD thesis, TU Dortmund (2013)
42. Vatolkin, I., Nagathil, A., Theimer, W., Martin, R.: Performance of specific vs. generic feature sets in polyphonic music instrument recognition. In: Purshouse, R.C., Fleming, P.J., Fonseca, C.M., Greco, S., Shaw, J. (eds.) EMO 2013. LNCS, vol. 7811, pp. 587–599. Springer, Heidelberg (2013)
43. Vatolkin, I., Bonnin, G., Jannach, D.: Comparing audio features and playlist statistics for music classification. In: Submitted to Post-Conference Proc. of the 2nd European Conf. on Data Analysis (ECDA) (2014)
44. Venkatadri, M., Srinivasa, R.K.: A Multiobjective Genetic Algorithm for Feature Selection in Data Mining. Int'l Journal of Computer Science and Information Technologies **1**(5), 443–448 (2010)
45. Vignolo, L.D., Milone, D.H., Scharcanski, J.: Feature Selection for Face Recognition Based on Multi-Objective Evolutionary Wrappers. Expert Systems with Applications **40**, 5077–5084 (2013)
46. Wang, C.-M., Huang, Y.-F.: Evolutionary-Based Feature Selection Approaches with New Criteria for Data Mining: A case study of credit approval data. Expert Systems with Applications **36**, 5900–5908 (2009)
47. Xue, B., Zhang, M., Browne, W.N.: Particle Swarm Optimization for Feature Selection in Classification: A Multi-Objective Approach. IEEE Trans. on Cybernetics **43**(6), 1656–1671 (2013)
48. Xue, B., Fu, W., Zhang, M.: Differential evolution (DE) for Multi-objective feature selection in classification. In: Proc. of the Companion of Genetic and Evolutionary Computation Conference (GECCO), pp. 83–84 (2014)
49. Zhu, Z., Ong, Y.-S., Kuo, J.-L.: Feature selection using single/multi-objective memetic frameworks. In: Goh, C.-K., Ong, Y.-S., Tan, K.C. (eds.) Multi-Objective Memetic Algorithms, pp. 111–131. Springer (2009)
50. Zitzler, E., Thiele, L.: Multiobjective optimization using evolutionary algorithms - a comparative case study. In: Eiben, A.E., Bäck, T., Schoenauer, M., Schwefel, H.-P. (eds.) PPSN 1998. LNCS, vol. 1498, p. 292. Springer, Heidelberg (1998)

Appendix

Table 1. Results of experiments for all combinations of objectives and categories

	Clas	Elec	Jazz	Pop	Rap	R&B	Adul	Albu	Alte	Club	Heav	Prog	Soft	Urba
						ϵ_{ID}								
BRE,FR	0.53	1.97	1.51	1.30	1.38	0.81	1.00	1.03	1.03	2.12	1.20	1.61	1.77	0.85
BRE,PREC	3.11	16.24	9.54	8.09	6.78	4.72	17.30	8.25	10.29	7.31	5.27	10.87	7.79	7.62
BRE,REC	1.09	8.37	4.52	4.96	1.98	6.64	1.82	2.91	3.29	7.56	4.48	0.52	0.82	8.73
BRE,SPEC	8.09	4.96	6.69	5.45	4.18	6.06	14.62	10.68	12.38	14.00	5.74	11.02	13.71	5.39
F1,BRE	0.00	1.97	0.98	0.00	1.19	0.20	4.80	1.43	1.70	0.00	1.77	1.65	1.80	1.80
F1,PREC	1.20	3.33	2.53	8.94	0.23	0.04	1.91	2.06	1.24	1.17	0.17	0.21	3.06	2.47
F1,REC	3.07	16.58	9.80	3.01	6.37	21.48	17.82	17.03	18.46	10.26	16.71	5.47	9.55	19.01
F1,SPEAR	0.00	0.00	0.00	0.00	0.00	0.12	0.00	0.00	0.00	0.02	0.00	0.19	0.06	0.14
F1,SPEC	4.86	1.36	1.21	4.80	0.11	3.30	2.85	4.67	2.33	6.67	1.64	3.47	4.51	1.33
FR,F1	0.91	1.67	1.31	0.86	1.02	3.35	2.28	2.07	1.45	1.91	1.56	2.37	1.09	2.19
FR,PREC	1.10	1.60	2.10	1.12	2.34	2.52	3.81	2.11	2.08	1.61	2.16	3.26	2.84	2.39
FR,REC	1.01	0.14	0.16	0.32	0.20	0.79	0.39	0.37	0.51	0.46	0.09	0.31	0.34	0.38
FR,SPEC	0.25	0.18	0.37	0.07	1.01	1.16	0.40	0.58	0.25	0.44	0.57	0.48	0.68	0.10
GEO,BRE	0.00	0.00	0.00	0.00	0.00	0.00	0.00	0.00	0.00	0.00	0.00	0.00	0.00	0.00
GEO,F1	0.28	1.56	0.03	0.00	0.13	0.24	4.89	0.25	5.52	0.06	1.90	0.61	0.10	1.09
GEO,FR	0.76	0.63	0.56	0.79	1.91	0.82	0.77	0.60	0.64	1.29	0.47	0.59	0.79	0.77
GEO,PREC	3.55	11.70	8.35	11.11	4.21	2.28	24.14	13.40	12.62	0.56	6.53	4.36	9.89	4.76
GEO,REC	1.13	7.62	3.81	7.07	1.40	7.68	1.51	4.67	3.39	7.62	3.04	0.78	1.05	7.95
GEO,SPEAR	0.11	1.17	0.06	0.00	0.40	0.05	2.03	0.16	0.45	0.00	0.32	0.00	0.03	0.05
GEO,SPEC	8.19	3.25	5.78	4.84	3.08	5.03	8.79	7.47	9.35	7.78	4.29	5.90	7.31	4.42
PREC,SPEC	0.18	0.18	0.78	0.11	0.02	0.54	0.12	0.93	0.41	0.97	0.66	1.27	0.71	0.06
REC,PREC	4.20	33.90	18.69	32.03	11.93	20.14	31.68	20.42	27.57	11.04	22.76	12.74	26.10	24.61
REC,SPEC	6.41	14.07	12.45	19.06	8.00	20.57	25.39	19.52	14.74	14.51	12.86	19.61	19.24	13.47
SPEAR,BRE	0.87	1.71	0.02	0.00	1.00	0.05	0.01	0.27	0.23	0.00	0.76	0.00	0.22	0.44
SPEAR,FR	0.89	2.07	1.50	0.67	1.77	1.59	2.25	2.81	1.11	1.91	0.99	1.55	1.35	1.54
SPEAR,PREC	1.75	1.30	3.01	2.88	0.19	0.77	3.27	4.74	0.47	0.36	0.36	2.20	5.58	1.41
SPEAR,REC	2.58	10.86	6.00	3.06	5.49	8.09	2.45	10.40	7.75	11.32	10.32	0.84	5.00	15.22
SPEAR,SPEC	5.96	0.48	3.47	1.28	0.70	4.43	4.00	4.17	3.55	10.07	1.82	6.15	5.00	1.95
	Clas	Elec	Jazz	Pop	Rap	R&B	Adul	Albu	Alte	Club	Heav	Prog	Soft	Urba
						ϵ_{MAX}								
BRE,FR	7.33	11.19	10.53	9.62	11.15	6.33	8.57	8.00	4.68	12.71	6.98	11.46	11.46	4.72
BRE,PREC	17.13	30.63	24.97	25.45	21.86	16.70	29.57	23.39	26.93	11.82	18.28	28.95	19.58	25.21
BRE,REC	4.41	12.75	11.11	13.73	9.32	15.60	8.83	12.22	15.31	11.50	15.83	4.90	6.33	15.66
BRE,SPEC	21.24	18.64	18.17	21.38	14.40	15.74	26.26	24.18	24.92	22.74	16.72	24.20	25.00	14.49
F1,BRE	0.00	12.85	3.15	0.00	8.77	2.91	14.46	9.23	9.85	0.11	12.55	11.84	9.88	10.16
F1,PREC	9.91	11.65	11.03	26.10	3.50	0.88	11.64	10.78	6.11	8.59	2.90	1.90	14.12	7.27
F1,REC	14.11	21.62	19.82	16.68	16.59	33.66	29.85	27.62	32.60	16.44	30.02	20.79	27.89	27.78
F1,SPEAR	0.00	0.00	0.34	0.32	0.00	1.35	0.00	0.00	0.00	1.20	0.00	4.34	0.62	2.47
F1,SPEC	13.52	5.46	4.08	17.44	1.82	9.45	8.18	12.35	9.87	10.72	5.46	8.93	17.53	3.51
FR,F1	8.32	9.43	6.45	7.02	4.75	7.39	9.04	4.76	6.22	9.76	7.12	11.68	7.80	9.16
FR,PREC	5.85	4.21	4.24	3.44	7.05	4.74	8.65	7.52	8.85	5.32	7.80	10.84	7.66	4.48
FR,REC	2.91	3.08	3.64	2.67	2.79	1.30	3.20	4.42	4.23	2.23	2.25	3.94	4.06	2.40
FR,SPEC	2.61	3.16	1.21	2.34	8.47	4.71	2.96	1.22	1.77	4.25	1.84	1.89	4.10	2.37
GEO,BRE	0.00	0.00	0.00	0.00	0.14	0.00	0.00	0.21	0.00	0.11	0.00	0.00	0.26	0.00
GEO,F1	3.46	10.09	1.13	0.00	1.96	3.28	19.82	4.89	14.69	0.72	11.48	8.42	1.09	6.61
GEO,FR	7.23	7.27	6.44	5.96	11.60	8.37	4.92	6.65	6.80	10.51	7.19	6.78	8.95	6.10
GEO,PREC	13.61	28.72	22.72	29.12	16.47	11.87	27.40	25.48	25.45	2.78	22.85	18.11	28.70	19.49
GEO,REC	4.94	15.70	12.21	15.96	6.62	19.63	6.74	16.59	15.56	15.55	15.39	6.76	7.25	21.67
GEO,SPEAR	2.74	10.34	2.06	0.00	3.36	2.07	8.62	2.49	5.46	0.00	4.04	0.00	1.16	2.12
GEO,SPEC	19.92	14.30	19.68	17.96	15.62	16.73	24.14	19.92	22.14	18.87	14.41	21.26	21.88	14.62
PREC,SPEC	2.15	0.89	1.96	2.20	0.71	4.38	0.52	4.28	3.21	6.20	2.76	3.45	1.72	1.31
REC,PREC	12.99	39.52	29.46	38.11	27.23	28.69	39.26	33.92	38.95	16.16	36.01	28.75	32.10	30.91
REC,SPEC	13.05	22.17	21.57	33.84	21.32	27.19	38.50	31.25	25.42	17.64	26.89	33.73	34.22	23.71
SPEAR,BRE	7.84	10.76	1.13	0.00	6.05	0.83	0.15	4.11	2.22	0.00	5.63	0.00	2.24	5.04
SPEAR,FR	7.68	7.80	5.86	6.26	10.81	7.33	12.56	8.54	4.21	12.01	7.29	5.75	7.35	6.77
SPEAR,PREC	10.28	6.42	13.55	14.39	4.08	6.09	13.19	13.22	4.58	1.52	5.21	9.88	22.05	8.08
SPEAR,REC	10.68	19.59	13.19	18.16	15.00	19.20	11.29	18.90	22.79	14.61	24.96	5.19	10.78	27.08
SPEAR,SPEC	15.74	3.27	12.84	11.62	2.76	15.03	13.60	15.87	12.78	20.13	8.27	16.98	15.23	8.48

A Multi-objective Approach
for Building Hyperspectral Remote Sensed
Image Classifier Combiners

S.L.J.L.Tinoco[1], D. Menotti[1], J.A. dos Santos[2], and G.J.P. Moreira[1(✉)]

[1] Computing Department, Universidade Federal de Ouro Preto, Ouro
Preto, MG, Brazil
gladston@iceb.ufop.br
[2] Computer Science Department, Universidade Federal de Minas Gerais,
Belo Horizonte, MG, Brazil

Abstract. Hyperspectral images are one of the most important data
source for land cover analysis. These images encode information about
the earth surface expressed in terms of spectral bands, allowing us to
precisely classify and identify materials of interest. An approach that
has been widely used is the combination of various classification methods in order to produce a more accurate thematic map based on classification of hyperspectral images. Our multi-objective remote sensed
hyperspectral image classifier combiner (MORSHICC) approach uses a
genetic algorithm-based strategy for choosing the best subset of classifiers, that is, the one which provides higher accuracy with the fewest
possible amount of classifiers. We propose to use combiners that linearly
weigh each classification approach through Genetic Algorithm (WLC-
GA) and Integer Linear Programming (WLC-ILP). For building the combiners, we used three data representations and four learning algorithms,
producing twelve classification approaches such that the multi-objective
approach can select the best subset. Experimental results on well-known
datasets show that the MORSHICC approach with WLC-GA and WLC-
IP not only produces combiners with fewer classifier approaches but also
improves the final accuracy rates. Therefore, these combiners may produce more accurate thematic maps for real and large datasets in a short
time.

1   Introduction

Remote sensing data have been used as source of information for many applications such as urban planning, agriculture, and environmental monitoring. Most
of these applications require automatic pattern analysis, which enables great
advances in the interpretation of the materials in the earth surface [3,14,18]. In
this context, the main step consists in classifying each pixel of the image [2].

This work was partially supported by CAPES, CNPq (grant 449638/2014-6), and
FAPEMIG (grant APQ-00768-14).

© Springer International Publishing Switzerland 2015
A. Gaspar-Cunha et al. (Eds.): EMO 2015, Part II, LNCS 9019, pp. 544–556, 2015.
DOI: 10.1007/978-3-319-15892-1_37

In typical pattern recognition problems, the objective is to yield the best results in terms of accuracy rates [13]. Given a scenario with a set of classifiers, the most naïve strategy is to select the classifier that achieves the best performance as the final solution for the classification problem.

However, it has been observed that among the non-selected classifiers, or even including the best ones, the sets of misclassified patterns are not always correlated. It suggests that different classifiers can provide some information to improve the final results [11]. Thus, combination of classifiers have been widely employed, with the goal of using all available information, when a single classifier can not achieve the expected results [19].

Several works have proposed effective strategies to construct good ensembles of classifiers [4,17]. They state that the key issue for achieving the highest possible accuracy rates is to exploit the diversity among the classifiers. They make errors on different instances. Hence, a combination of these classifiers can reduce the total error [17]. In [4], the authors propose the concept of "good" and "bad" diversity to the *Majority Vote* rule. The greater the "good" diversity value, the smaller the *Majority Vote* error is.

Nonetheless, there is no a widely accepted definition for diversity. It is not clear at this moment, what is the correlation between diversity and accuracy [7]. For instance, diversity is used in order to reduce the generalization error in [9]. As a conclusion, the authors pointed out that using only diversity measures is not a good strategy to reach a suitable combination of classifiers. Also, dos Santos *et al.* [7] noted that bad individual classifiers do not should be included in the final ensemble even if it has high diversity in comparison with others.

The quality of an ensemble depends on the careful selection of classifiers to be combined. One way to perform a suitable combination, *i.e.*, how many and which are the best classifiers, would be evaluate every possible combination given set of classifiers. This task would require a high computational effort even for a small number of classifiers/approaches, because there are 2^n - 1 possible combinations (for $n = 12$, 4095 combinations would be evaluated). Another option to deal with this problem, due to the combinatorial nature of the search space, would be the usage of algorithms that optimize combinatorial problems, such as Evolutionary Algorithms. It is noteworthy that it is also interesting to get a combination with a smaller set of classifiers. Therefore, the problem can be described as a search for the accuracy maximization and minimization of the number of classifiers.

Having this context in mind, we propose in this paper the *use* of a multi-objective approach for remote sensed hyperspectral image classifier combiner (MORSHICC) based on genetic algorithm to determine the Pareto's front (*i.e.*, set of non dominated individuals) which represents the set of best combiners in accord to two objectives: maximization of accuracy and minimization of the number of classifiers used in the combiner. From our previous works, here, we use linearly weighted combiners generated by Genetic Algorithms (WLC-GA)[19], and Integer Linear Programming (WLC-ILP) [21]. For building the combiners, we used three types of data representation and four well-known learning algorithms (Support Vector Machines (SVM) with linear and RBF kernels,

Backpropagation-based Multilayer Perceptron Neural Network (MLP) and K-Nearest Neighbor (KNN)) generating twelve classification approaches. For more details regarding the classification approaches, since the focus of this work is not on them, we suggest the reader to see [19–21] for more details. Experiments were carried out in well-known datasets: Indian Pines and Pavia obtained by AVIRIS and ROSIS sensors, respectively [16].

2   Background

The main goal of combining multiple classifiers is to improve the performance of the final classification in comparison with single classifiers. It comprises the selection of the most suitable classifiers. In this section, we present the methods for combination and search that were used in this work.

2.1   Combination Methods

The input for the combination methods we have employed in this paper is the output of the single classifiers. For each class, the classifiers produce a *soft* value, *i.e.*, a certain degree of support [13]. These outputs can be fuzzy, posterior probabilities, certainty, or possibility values [10]. Based on these *soft* outputs one can build a Decision Profile (DP). Formally, a DP for a given sample x can be defined as a $L \times C$ matrix, *i.e.*, $DP(x) = [D_1(x), D_2(x), ..., D_l(x), ..., D_L(x)]$ in which $D_l(x) = [d_{l,1}(x), d_{l,2}, ..., d_{l,c}, ..., d_{l,C}(x)]^T$, L is the number of classifiers, C is the number of classes, and $d_{l,c}(x)$ is the degree of support given by classifier D_l to class c [10,13], as illustrates Fig. 1. After building support degrees for each input sample, a crisp value (the final label) can be assigned by using the maximum support value in the set, for instance.

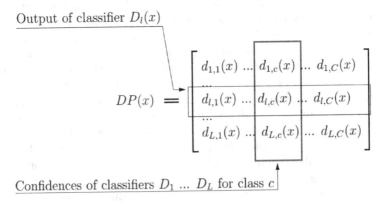

$$DP(x) = \begin{bmatrix} d_{1,1}(x) \,...\, d_{1,c}(x) \,|..\, d_{1,C}(x) \\ d_{l,1}(x) \,...\, d_{l,c}(x) \,..\, d_{l,C}(x) \\ d_{L,1}(x) \,...\, d_{L,c}(x) \,...\, d_{L,C}(x) \end{bmatrix}$$

Output of classifier $D_l(x)$

Confidences of classifiers $D_1 ... D_L$ for class c

Fig. 1. Decison Profile. Adapted and modified from [13].

According to Kuncheva [13], combiners are methods that use all predictions produced by two or more classifiers to build an accurate final decision. They can be divided into "nontrainable" and "trainable" combiners. The "nontrainable" combiners have no need of training any parameter. They perform some basic operation (for instance: average, maximum, minimum and product) in the DP to produce new support values and, hence, a final decision.

The purpose of the "trainable" combiners is to give more discriminant power to classifiers that have greater accuracy [13] when classifiers have different outputs. Weighted Average, Weighted Majority Vote (WMV), and other weighted approaches are based on this idea. In the following, we briefly describe the two linearly weighted combiners used in our proposed selection method, which, in turn, is described further.

In this work, in particular, we use weighted linear combiners, from our previous works [19,21]. Let us first define a Weighted Linear Combination (WLC). Given a sample x, let $\mu_c(x) = \sum_{l=1}^{L} w_l \times d_{l,c}(x)$ be the support for the class c, w_l be the weight of the l-th classifier and $d_{l,c}(x)$ be the support of l-th classifier for the class c.

The task of finding the best weights can be seen as an *Integer Linear Programming* (ILP) optimization problem, generated the WLC-ILP approach [21]. This problem requires the minimization (or maximization) of a linear form subject to linear inequality constraints. New supports for each class are built using the WLC and the weights found by running the simplex method. Then, a label is assigned, for a given sample x, as the index of the maximum support $\mu_c(x)$. The IBM CPLEX solver [12], a state-of-art ILP solver, is used as optimization routine.

In [19], predictions of classifiers are also combined using a weighted linear combination of the DP, as stated above. However, the weights are found using a global search performed by a GA named WLC-GA. The fitness function was built based on the accuracy produced using the WLC in the dataset. A bit string representation encode the weights in individual chromosomes. Each weight can be a non-negative integer value between 0 and 127, which means that there are 7 bits in chromosome for each weight.

2.2   Search Methods

Many works have investigated methods for selecting subsets of classifiers rather than combining all classifiers [7,8,19,22]. This selection aims at improving the performance of the combination, since it focuses on finding the subset of the most relevant classifiers. From a set of classifiers Cl, we apply search algorithms to select the subset of the best performing classifiers S, where $|S| \leq |Cl|$. We can notice two important aspects: the search algorithm and the search criterion.

Evolutionary Algorithms, such as Genetic Algorithms, attempt to find an optimal or near optimal global solution. More specifically Multi-Objective Genetic Algorithms seem to be a better option to the classifier selection problem due to the possibility of dealing with a population of solutions.

Another important aspect is the choice of the most appropriate search criterion. Although there is no consensus, the role played by diversity is emphasized in the literature. However, diversity and accuracy does not exhibit a strong relationship [4] and the estimated accuracy can not be replaced by diversity [9].

Our approach exploits a Multi-Objective Genetic Algorithm for searching. The search criteria used are the accuracy and the number of classifiers. We intend to reduce the number of classifiers, but also increase the accuracy. The final accuracy of the combination is the one obtained by either the WLC-ILP or WLC-GA combiners using the selected subsets of classifiers.

3   Multi-objective Optimization Approach

In this section, we present the multi-objective remote sensed hyperspectral image classifier combiner (MORSHICC) based on genetic algorithm. We evolve a population of classifier combiners aiming at accuracy maximization and minimization of the number of classifiers. The latter objective searches for faster and less expensive combiners to efficiently classify large datasets.

Each individual is a combiner which, in turn, is represented by a set of classifiers and its weights computed by a method, *e.g.*, WLC-ILP or WLC-GA. The set of classifiers for combination contains twelve classification approaches as shown in Tables 1 and 2. We use a binary chromosome representation, in which each position (gene) on chromosome represents the presence (or absence) of a classifier. The population successively evolves through the generations following a tournament which randomly selects two individuals and then applies crossover and mutation rules. The binary selection is run six times such that twelve (M) child individuals are generated. We use the one-point crossover, *e.g.*, a point along the chromosome is randomly selected, then the pieces to the left of that point are exchanged between the chromosomes, producing a pair of offspring chromosomes as crossover operator, and bit inversion as mutation operator.

We start with a randomly chosen initial population of size M, *e.g.*, the number of classifier combiners used, in which the number of classification approaches in each individual also randomly varies. For each individual of the initial population, a combiner (WLC-ILP or WLC-GA) is run, and based on the classifier combiner generated, the pair (classification accuracy, number of classifiers) is computed. Note that either the WLC-ILP or WLC-GA is adopted during the optimization process.

The evolution step is defined in terms of two objectives: maximizing the classification accuracy and minimizing the number of classifiers. This step relies on the concept of dominance: a point is said to be dominated if it is worse than another point in at least one objective, while not being better than that point in any other objective. The Pareto-set is the set that contains no dominated solution, thus it consists of points that are not simultaneously worse than any other point in both objectives.

More specifically, a **generation** of the MORSHICC genetic algorithm works as presented in Algorithm 1. Note that the initial population (generation zero) is first evaluated (*i.e.*, combiners' computation). Then, it is submitted to the Generation step. At the beginning of each generation, we create more M child

Algorithm 1. Generation step of our MORSHICC

1: **input**: *current: set of M evaluated individuals*
2: *child* ← bin. select. on *current* and mutation & crossover
3: **for** each *ind* ∈ *child* **do** {evaluating every child}
4:    (*acc.*, #*class.*) ←WLC-ILP (or WLC-GA) of *ind*
5: *current* ← *current* ∪ *child* {|*current*| = 2 × *M*}
6: *next* ← ∅
7: **repeat** {Pareto-set evaluation procedure}
8:    *best* ← non-dominance set from *current*
9:    *next* ← *next* ∪ *best*
10:    *current* ← *current* − *best*
11: **until** |*next*| ≤ *M*
12: remove extra individuals
13: **output**: *next: set of M evolved and evaluated individuals*

individuals using binary selection and crossover/mutation operators (lines 2). The child individuals are evaluated (lines 3-4) and joined to the parent individuals updating the so called *current* generation set to $2M$ individuals (line 5). The next generation set (**output**) is composed of individuals that survive to the recursive Pareto-set evaluation procedure (lines 6-11). The Pareto-set of the current generation set (line 8) is inserted into the next generation set (line 9), which is initially empty (line 6). The current Pareto-set (*best*) is removed from the current generation set (line 10). The process of computing a new Pareto-set of the remaining individuals is repeated (lines 8-10) until the next generation set contains M individuals. If eventually the last Pareto-set collects individuals that extrapolates the M size limit of the generation set, among the last inserted individuals, the ones with highest accuracy standard deviation[1] are removed, so that the output set reach exactly M individuals (lines 12-13).

The evolution (generation) process is repeated until a predetermined maximum number of generations is reached. This procedure is similar to the "Non-Dominated Sorting" selection operator, which is employed in the NSGA-II [6]. It is noticeable that the final subset of classifier combiners is reprocessed such that only non-dominated solutions remain.

4   Experimental Results and Discussion

The classification approaches selected for combination should constitute a diverse set and provide additional information. For such aim we used data representations such as: Pixelwise [15]; *Extend Morphological Profile* (EMP) [1]; and *Feature Extraction by Genetic Algorithms* (FEGA) [20]. For classification, we used well-known learning algorithms such as *Support Vector Machines* (SVM) [15], with *Radial Basis Function (RBF)* and *Linear* kernels, *Multilayer Perceptron Neural Network (MLP)* [1], and *k-Nearest Neighbor* (KNN) [5]. The full set of

---

[1] Each combiner is evaluated using several training/testing sets.

Table 1. IP dataset: classification approaches

Identifier	Classification Approaches		Train Accuracy (%) 10%	15%
1	*EMP*	*RBF-SVM*	88.77-($\pm$0.39)	90.88-($\pm$0.29)
2	*PixelWise*	*RBF-SVM*	80.38-($\pm$0.44)	83.51-($\pm$0.29)
3	*FEGA*	*RBF-SVM*	76.27-($\pm$0.50)	78.99-($\pm$0.34)
4	*EMP*	*MLP*	81.60-($\pm$0.76)	82.62-($\pm$0.67)
5	*PixelWise*	*MLP*	73.04-($\pm$0.75)	76.83-($\pm$0.48)
6	*FEGA*	*MLP*	72.26-($\pm$0.80)	75.14-($\pm$0.64)
7	*EMP*	*kNN*	83.94-($\pm$0.34)	86.62-($\pm$0.38)
8	*PixelWise*	*kNN*	67.28-($\pm$0.44)	69.40-($\pm$0.33)
9	*FEGA*	*kNN*	61.76-($\pm$0.62)	63.66-($\pm$0.38)
10	*EMP*	*linSVM*	79.10-($\pm$0.55)	80.14-($\pm$0.43)
11	*PixelWise*	*linSVM*	77.17-($\pm$0.50)	80.55-($\pm$0.39)
12	*FEGA*	*linSVM*	72.96-($\pm$0.59)	75.67-($\pm$0.49)

classifiers used for combination contains twelve classification approaches (Tables 1 and 2).

Experiments were carried out in two training set scenarios, *i.e.*, with 10% and 15% of samples using the well-know *Indian Pines* (IP) and *Pavia University* (PU) datasets.

In both scenarios, the testing set were adjusted to 85% of unseen samples for a fair comparison of the obtained effectiveness. During the MORSHICC evolution, each individual (classifier combiner) was run 30 times using 30 different training and testing sets randomly created. Mean and variances of these experiments were used to compute the confidence intervals of each combiner using a 0.05 confidence level. For each run of each evaluated combiner, it is important to note that 50% of the training data is used for initially train the classifiers and the remaining 50% used for weights estimation in order to avoid biased and specialized weights, and in a second training phase the classifiers are retrained with the entire training set. Note that these all subsets (training+testing) were initially created and then used in all experiments such that a fair comparison can be performed. **MORSHICC approach setup**: We set the generation number to 10 with a population of 12 individuals. We evaluated only 120 individuals during the evolution process due to the high computation cost of our approach, which took almost one week for both datasets using a personal computer with *Intel(R) Core(TM) i5-2450M* processor and 4 GB of main memory with *Ubuntu 12.04* Operating System. We used 12 bits to represent the presence/absence of

Table 2. PU dataset: classification approaches

Identifier	Classification Approaches		Train Accuracy (%) 10%	15%
1	EMP	RBF-SVM	97.20-(±0.11)	97.56-(±0.08)
2	PixelWise	RBF-SVM	93.17-(±0.13)	93.50-(±0.10)
3	FEGA	RBF-SVM	90.87-(±0.22)	91.39-(±0.11)
4	EMP	MLP	94.42-(±0.50)	94.48-(±0.74)
5	PixelWise	MLP	92.43-(±0.20)	92.91-(±0.14)
6	FEGA	MLP	89.70-(±0.27)	90.04-(±0.19)
7	EMP	kNN	95.56-(±0.12)	96.23-(±0.08)
8	PixelWise	kNN	84.99-(±0.17)	85.82-(±0.13)
9	FEGA	kNN	88.31-(±0.18)	89.06-(±0.11)
10	EMP	linSVM	90.72-(±0.16)	91.31-(±0.13)
11	PixelWise	linSVM	90.96-(±0.18)	91.10-(±0.13)
12	FEGA	linSVM	87.52-(±0.22)	87.61-(±0.13)

Table 3. Results for IP dataset using 10% training set

training set 10%	Classifiers' ID	Number of Classifiers	Accuracy (%)	Confidence Interval
WLC-ILP	1-2-3-4-5-6-7-8-9-10-11-12	12	89.57	0.80
WLC-GA	1-2-3-4-5-6-7-8-9-10-11-12	12	89.72	0.52
MORSHICC in WLC-ILP	1-2-4-5-7-11-12	7	90.15	0.46
MORSHICC in WLC-GA	1-2-4-6-7-8-10-11	8	91.12	0.29
Best Individual Classifier	1	1	88.77	0.39

each classification approach in combination, and the probabilities of crossover and mutation to 80% and 0.9%, respectively.

In the following, the analysis for claiming statistically significance takes into account the confidence intervals and mean accuracies reported in Tables 3, 4, 5, and 6. In these same tables, from the Pareto's front combiners obtained for MORSHICC in WLC-ILP and WLC-GA, we choose to report the best mean accuracies obtained by the combiners which are the ones with the largest number of classifiers.

Table 4. Results for IP dataset using 15% of training set

training set 15%	Classifiers' ID	Number of Classifiers	Accuracy (%)	Confidence Interval
WLC-ILP	1-2-3-4-5-6-7-8-9-10-11-12	12	91.55	0.62
WLC-GA	1-2-3-4-5-6-7-8-9-10-11-12	12	91.63	0.54
MORSHICC in WLC-ILP	1-2-3-4-7-10-11	7	93.00	0.24
MORSHICC in WLC-GA	1-2-4-5-7-8-11	7	93.30	0.21
Best Individual Classifier	1	1	90.88	0.29

For the IP dataset using 10% of training samples, in Table 3, it is shown that our approaches achieved significantly better results than the best individual classifier. Also observe that the MORSHICC approach in the WLC-ILP produced statistically similar accuracy, and the MORSHICC approach in the WLC-GA produced statically better accuracy when compared with their respective combiner using all classification approaches. Anyway in both cases MORSHICC approach employed fewer classifier.

For the IP dataset using 15% of training samples, our approaches have also achieved significantly better results than the best individual classifier accuracy as shown in Table 4. Moreover, in this scenario, the MORSHICC approaches produced significantly better accuracies with fewer classifiers when compared to both the WLC-ILP and WLC-GA combiners using all classifiers. Note that the classification accuracies obtained using 15% for training the classifiers are significantly higher than when using 10%.

Figure 2 shows the graphs of the Pareto's fronts produced by our MORSHICC approach training in 10% and 15% of training samples and tested in 85% of samples. It is important to claim that the individual approach 1 ($EMP+RBF$-SVM), which has higher accuracy, is present in most combinations that generated the Pareto's fronts. By observing and comparing the red points (combiners) of each graph of the Pareto's front, it is possible to see that smaller combiners (few classifiers) can also produce higher accuracies with no statistically difference to the best combiners. This information is useful when faster combiners (with low computational cost) are required.

For the PU dataset, using 10% and 15% of training samples, Tables 5 and 6, respectively, show that the MORSHICC approach in the WLC-GA combiner achieved significantly better results than the best individual classifier. However its result is not statistically better if compared with the one obtained by the WLC-GA approach using all combiners. Moreover, the MORSHICC in the WLC-ILP performed poorly than the best individual combiner and also with respect to

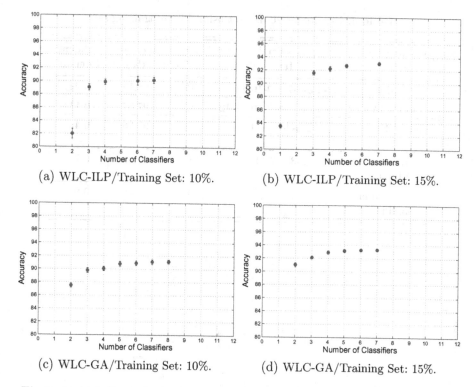

(a) WLC-ILP/Training Set: 10%.   (b) WLC-ILP/Training Set: 15%.

(c) WLC-GA/Training Set: 10%.   (d) WLC-GA/Training Set: 15%.

Fig. 2. Pareto's fronts for IP dataset. Blue bars stand for the confidence intervals.

Table 5. Results for PU dataset using 10% of training set

Approaches	Classifiers' ID	Number Classifiers	Accuracy (%)	Confidence Interval
WLC-ILP	1-2-3-4-5-6-7-8-9-10-11-12	12	97.73	0.18
WLC-GA	1-2-3-4-5-6-7-8-9-10-11-12	12	97.55	0.62
MORSHICC in WLC-ILP	1-4-7-9-11	5	97.11	0.30
MORSHICC in WLC-GA	1-2-5-7-10-11-12	7	98.00	0.09
Best Individual Classifier	1	1	97.20	0.11

the WLC-ILP approach using all combiners. Notice that the accuracy improvement obtained in this dataset is smaller if compared with the obtained results for the IP dataset since here there is few room for improvement, *i.e.*, the best individual classifier produces less than 3% of error. As a possible consequence

Table 6. Results for PU dataset using 15% of training set

training set 15%	Classifiers' ID	Number Classifiers	Accuracy (%)	Confidence Interval
WLC-ILP	1-2-3-4-5-6-7-8-9-10-11-12	12	98.12	0.24
WLC-GA	1-2-3-4-5-6-7-8-9-10-11-12	12	97.87	0.11
MORSHICC in WLC-ILP	1-2-3-4-5-7-11	7	97.28	0.40
MORSHICC in WLC-GA	1-2-3-5-7-11	6	98.44	0.79
Best Individual Classifier	1	1	97.56	0.08

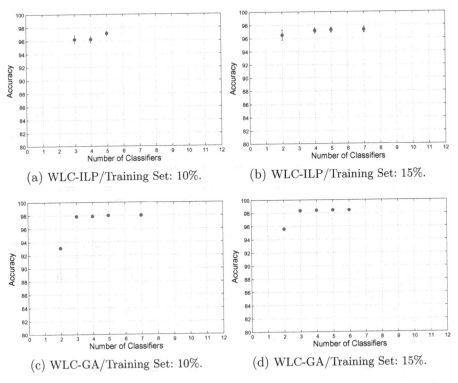

(a) WLC-ILP/Training Set: 10%.

(b) WLC-ILP/Training Set: 15%.

(c) WLC-GA/Training Set: 10%.

(d) WLC-GA/Training Set: 15%.

Fig. 3. Pareto's fronts for PU dataset. Blue bars stand for the confidence intervals.

of this fact, observe that the best results obtained for 15% (the MORSHICC in the WLC-GA) is not significantly better than the one obtained for 10% (the MORSHICC in the WLC-GA), even it is slightly higher.

Figure 3 shows the graphs of the Pareto's front produced by our approach for the PU dataset using 10% and 15% of training samples. We can observe the same conclusions we had in Figure 2 (IP dataset), however here with higher accuracies.

In general (both datasets), the MORSHICC approaches in the WLC-GA combiner produces higher accuracies than the ones MORSHICC approaches in the WLC-ILP method, although they are not necessarily better in terms of statistical significance. This negative result of the WLC-IP method can be justified by the numerical instability faced by it for solving the optimization problem. Nonetheless the WLC-ILP method is in average 20 times faster than the WLC-GA combiner for estimating the final weights as shown in [21].

5   Conclusions

In this paper, we have presented a multi-objective remote sensed hyperspectral image classifier combiner (MORSHICC) approach based on genetic algorithm to determine the Pareto's front. Our aim is to use the Pareto's front to determine the set of best combiners. We have modeled the problem according to two objectives: maximization of accuracy and minimization of the number of classifiers used in the combiner. Experimental analysis shows that MORSHICC not only produces an ensemble with a very small set of classifiers but also improves the final accuracy results. Furthermore, the obtained ensembles may achieve more accurate thematic maps for real and large datasets in a short time. Future work includes the application of the proposed techniques in real world problems, such as automatic agricultural crop recognition.

Acknowledgments. The authors thank UFOP and funding Brazilian agencies CNPq, Fapemig and CAPES.

References

1. Benediktsson, J., Palmason, J., Sveinsson, J.: Classification of hyperspectral data from urban areas based on extended morphological profiles. IEEE Trans. on Geoscience and Remote Sensing (TGARS) **43**(3), 480–491 (2005)
2. Benediktsson, J.A., Chanussot, J., Fauvel, M.: Multiple classifier systems in remote sensing: From basics to recent developments. In: Benediktsson, J.A., Chanussot, J., Fauvel, M. (eds.) Multiple Classifier Systems. LNCS, vol. 4472. Springer, Heidelberg (2007)
3. Benediktsson, J.A., Chanussot, J., Moon, W.M.: Very high-resolution remote sensing: Challenges and opportunities [point of view]. Proceedings of the IEEE **100**(6), 1907–1910 (2012)
4. Brown, G., Kuncheva, L.I.: "Good" and "Bad" diversity in majority vote ensembles. In: El Gayar, N., Kittler, J., Roli, F. (eds.) MCS 2010. LNCS, vol. 5997, pp. 124–133. Springer, Heidelberg (2010)
5. Cover, T., Hart, P.: Nearest neighbor pattern classification. IEEE Trans. on Information Theory **13**(1), 21–27 (January 1967)

6. Deb, K.: Multi-objective optimisation using evolutionary algorithms: an introduction. In: Multi-objective Evolutionary Optimisation for Product Design and Manufacturing, pp. 3–34. Wiley (2011)
7. Dos Santos, J.A., Faria, F.A., da S Torres, R., Rocha, A., Gosselin, P.H., Philipp-Foliguet, S., Falcao, A.: Descriptor correlation analysis for remote sensing image multi-scale classification. In: 2012 21st International Conference on Pattern Recognition (ICPR), pp. 3078–3081 (2012)
8. Faria, F.A., dos Santos, J.A., Rocha, A., da, S., Torres, R.: A framework for selection and fusion of pattern classifiers in multimedia recognition. Pattern Recognition Letters 39, 52–64 (2014)
9. Gabrys, B., Ruta, D.: Genetic algorithms in classifier fusion. Applied soft computing 6(4), 337–347 (2006)
10. Ghosh, A., Shankar, B., Bruzzone, L., Meher, S.: Neuro-fuzzy-combiner: an effective multiple classifier system. Int. J. of Knowledge Engineering and Soft Data Paradigms 2(2), 107–129 (2010)
11. Hadjitodorov, S.T., Kuncheva, L.I., Todorova, L.P.: Moderate diversity for better cluster ensembles. Information Fusion 7(3), 264–275 (2006)
12. ILOG S.A.: CPLEX 12.5 User's Manual (2012)
13. Kuncheva, L.: Combining Pattern Classifiers: Methods and Algorithms. Wiley-Interscience (2004)
14. Licciardi, G., Marpu, P., Chanussot, J., Benediktsson, J.: Linear versus nonlinear PCA for the classification of hyperspectral data based on the extended morphological profiles. IEEE Geoscience and Remote Sensing Letters (GRSL) 9(3), 447–451 (2012)
15. Melgani, F., Bruzzone, L.: Classification of hyperspectral remote sensing images with support vector machines. IEEE Trans. on Geoscience and Remote Sensing (TGARS) 42(8), 1778–1790 (2004)
16. Plaza, A., et al.: Recent advances in techniques for hyperspectral image processing. Remote Sensing Environment 113(1), 110–122 (2009)
17. Polikar, R.: Ensemble based systems in decision making. IEEE Circuits and Systems Magazine 6(3), 21–45 (2006)
18. Prasad, S., Bruce, L.M., Chanussot, J.: Optical Remote Sensing: Advances in Signal Processing and Exploitation Techniques, vol. 3. Springer (2011)
19. Santos, A.B., de, A., Araújo, A., Menotti, D.: Combining multiple classification methods for hyperspectral data interpretation. IEEE Journal of Selected Topics in Applied Earth Observations and Remote Sensing 6(3), 1450–1459 (2013)
20. Santos, A.B., de S. Celes, C.S.F., de A.Arajo, A., Menotti, D.: Feature selection for classification of remote sensed hyperspectral images: a filter approach using genetic algorithm and cluster validity. In: The 2012 International Conference on Image Processing, Computer Vision, and Pattern Recognition (IPCV), vol. 2, pp. 675–681 (2012)
21. Tinôco, S., Santos, A., Santos, H., dos Santos, J.A., Menotti, D.: Ensemble of classifiers for remote sensed hyperspectral land cover analysis: An approach based on linear programming and weighted linear combination. In: IEEE Int. Geoscience and Remote Sensing Symposium (IGARSS), pp. 4082–4085 (2013)
22. Zhang, L., Zhang, L., Tao, D., Huang, X.: On combining multiple features for hyperspectral remote sensing image classification. IEEE Trans. on Geoscience and Remote Sensing (TGARS) 50(3), 879–893 (2012)

Multi-objective Optimization of Barrier Coverage with Wireless Sensors

Xiao Zhang[✉], Yu Zhou, Qingfu Zhang, Victor C.S. Lee, and Minming Li

Department of Computer Science,City University of Hong Kong,
83 Tat Chee Avenue, Kowloon, Hong Kong
{xiao.zhang,yzhou57-c}@my.cityu.edu.hk
{qingfu.zhang,csvlee,minming.li}@cityu.edu.hk

Abstract. Barrier coverage focuses on detecting intruders in an attempt to cross a specific region, in which limited-power sensors in these scenarios are supposed to be distributed remotely in an indeterminate way. In this paper, we consider a scenario where sensors with adjustable ranges and a few sink nodes are deployed to form a virtual sensor barrier for monitoring a belt-shaped region and gathering incidents data. The problem takes into account three relevant objectives: minimizing power consumption while meeting the barrier coverage requirement, minimizing the number of active sensors (reliability) and minimizing the transmission distances between active sensors and the nearest sink node (efficiency of data gathering). It is shown that these three objectives are conflicting in some degree. A Problem Specific MOEA/D with local search methods is proposed for finding optimal tradeoff solutions and compared with a classical algorithm. Experimental results indicate that knee regions exist, and these knee regions may provide the best possible tradeoff for decision makers.

1 Introduction

In recent years, there has been increasing development in the field of wireless sensor networks (WSNs). One of the most important applications in WSNs is border surveillance and intrusion detection, such as detecting intruders crossing country borders or boundaries of battlefields. Many recent works have addressed such surveillance applications by using WSNs to organize the network nodes as a barrier [1]. For deterministic deployment of sensors, the high performance can be achieved sufficiently by analysis. However, surveillance tasks may involve hard-to-reach areas, in which case unmanned mission way is more desirable. Specifically, limited-power sensors and several sink nodes in these scenarios are supposed to be distributed remotely, for example, dropped from aircraft; they wake up, organize themselves as a network, and start sensing the area for intrusion. When a sensor detects an intrusion, the event is reported to the sink node so that an appropriate decision is made.

Power efficient is always a critical issue in wireless barrier coverage. The single objective optimization problem, minimizing the total power consumption while the barrier is full covered, is referred to as *General Min-Cost Linear Coverage*

© Springer International Publishing Switzerland 2015
A. Gaspar-Cunha et al. (Eds.): EMO 2015, Part II, LNCS 9019, pp. 557–572, 2015.
DOI: 10.1007/978-3-319-15892-1_38

problem (GMCLC) [2]. However, there is no efficient way to get exact optimal solutions since it is proved to be NP-hard [3]. In addition, since sensors are vulnerable to failure, it is important to minimize the number of active sensors to improve the reliability while meeting the coverage requirement. Besides, in general wireless sensor networks, a large number of sensor nodes, which are generally compact and inexpensive, are distributed in an observation area while sink nodes with comparatively sufficient power are defined as the data gathering center. Long transmission distances between sensor nodes and sink nodes cause low efficiency of data gathering and high energy consumption.

In this paper, we take the considerations above into account simultaneously and propose an algorithm to achieve the following objectives:

- *Objective 1*: Minimizing the total power consumption via activating a subset of the sensor nodes and adjusting their sensing ranges.
- *Objective 2*: Minimizing the number of active sensors to improve the reliability of coverage.
- *Objective 3*: Minimizing the active sensors' average distance from the nearest sink node to improve the efficiency of data gathering.

However, these three objectives are conflicting in nature. The sensors are failure-prone: each sensor fails independently with a certain probability. Under the condition of fully coverage, the fewer sensors activated, the higher reliability achieved. Meanwhile, the power consumption is proportional to the radii of active sensors. Next, we take a simple instance to illustrate the conflict among objectives. Fig. 1 shows two feasible solutions for the coverage problem. In the first solution (the left one), the power consumption is $Cost_1 = r_1^\kappa$ and the number of active sensors $|\mathbf{S_1}^*|$ is *one*. In the second solution (the right one), the power consumption is calculated by $Cost_2 = r_2^\kappa + r_3^\kappa$, and number of active sensors $|\mathbf{S_2}^*|$ is *two*. Since $r_1 = r_2 + r_3 = \frac{m}{2}$, we have $Cost_1 > Cost_2$. That is to say, minimizing the total power consumption may increase the number of active sensors, and require the active sensors to be distributed evenly along the barrier. In addition, minimizing the active sensors average distance from the nearest sink node may result in more active sensors close to the sink node. However, to meet the coverage requirement, either more sensors, if available, are activated to cover the region far away from the sink node or a larger sensing range is assigned to the farthest sensors, leading to higher power consumption. Thus, finding the tradeoff among them is worth exploring.

This problem can be formulated as a *Multi-objective Optimization Problem (MOP)*. Classical algorithms may not be applicable and few approaches tackle these objectives simultaneously. It is reasonable to use *Multi-Objective Evolutionary Algorithms (MOEAs)*, which have been proven efficient and effective in dealing with MOPs in wireless sensor networks [4] [5].

In this paper, we refine the barrier coverage problem to an *MOP* with three objectives, which is referred to as *Tradeoff on Barrier Coverage with Adjustable Sensing Radius Problem (TBCAP)*. Solutions are obtained through a problem specific MOEA, which adopts the framework of decomposition-based multiobjective

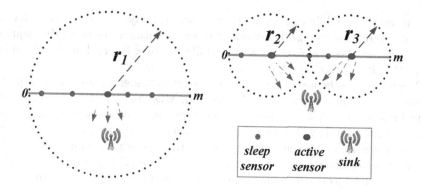

Fig. 1. Illustration of conflict among the objectives

evolutionary algorithm (MOEA/D) [6] as the baseline algorithm. We call it PS-MOEA/D. The PS-MOEA/D employs the problem-specific operators and local search methods. Besides, in order to improve the search, we incorporate a dynamic strategy of computational resource assignment. Moreover, a perturbation is involved to search for the global optimal solution.

The remainder of this paper is organized as follows. The related works is presented in Section 2. In Section 3, we define and formulate the problem, and give a naive algorithm for finding the tradeoff. Section 4 presents the details of the problem specific MOEA/D. Section 5 shows the experimental results and analysis. Finally, Section 6 outlines the conclusions and future directions.

2   Related Works

A heterogeneous WSN consists of several types of nodes with different capability, in which a large number of sensor nodes with the capabilities of sensing data, while fewer sink nodes may have larger battery and more powerful processing resource [7]. They are widely used in surveillance [8] [9]. Among them, barrier coverage problem deals with how to deploy sensor nodes to form barrier coverage for detecting intruders crossing a belt-shaped area of interest [10] [11] [12].

Optimizing the efficiency of data gathering and transmission quality between sensors and sinks have been widely studied [7] [13]. Mhatre et al. [7] studies a heterogeneous sensor network in which nodes are to be deployed over a unit area for the purpose of surveillance. They determined the optimum sensor nodes and sink nodes intensities (λ_0, λ_1).

Power consumption is always a critical issue in wireless barrier coverage. It helps to prolong the network lifetime by turning off some sensors while meeting given coverage requirements. Since it is proved to be NP-hard [3], several approximation algorithms have been proposed in recent years [2] [3] [14] .

Moreover, network failure, partial or whole, may not only be due to power exhaustion of the sensor nodes. Some sensors may stop functioning due to mechanical problems when they are working. This may result in unexpected consequences. Very few researchers focus on the reliability of the sensor networks for coverage.

To improve the reliability of coverage requirement, Sanjay et al. [15] consider an unreliable wireless sensor grid network for coverage with sensors placed in a square of unit area. In this model, all sensors are failure-prone, i.e., each node fails independently with a certain probability.

Proposing a scheme for wireless coverage considering so many aspects together is a challenging problem. To this end, MOEAs may provide a desirable model for solving such sensor network design problems. While both coverage and power consumption have been extensively studied in the past [16], few attempts however, have been made on tackling the coverage, power consumption, reliability and efficiency of data gathering simultaneously or explicitly. Martins et al. [17] presented multiobjective hybrid optimization algorithms for minimizing the power consumption and maximizing the coverage in flat WSNs subject to node failures. In [16], the problem objectives are stated as maximizing the coverage and minimizing energy consumption for maximizing the network lifetime. A sleep scheduling method is incorporated into a multiobjective optimization framework. Recently, Lanza-Gutierrez et al. [18] use MOEAs to optimize a WSN composed of a set of sensors, a sink node and relay nodes, analyzing the performance of algorithms by objectives of the average energy consumption suffered by the sensors and the average coverage provided by the network.

3   Preliminaries

3.1   Multiobjective Problem and MOEA/D

An MOP is generally formulated as follows.

$$minimize\ F(x) = (f_1(x), \ldots, f_m(x)) \tag{1}$$
$$subject\ to\ x \in \Omega$$

where Ω is the decision space and $x \in \Omega$ is a decision variable. R^m consists of m objective functions $f_1, \ldots, f_m$: R^m is the objective space. The objectives in problem (1) often conflict with each other and an improvement on one objective may lead to the deterioration of another. A Pareto optimal solution is an optimal tradeoff candidates among all objectives. The Pareto optimum terminology is described in [19], in which *Pareto dominance, Pareto optimal, Pareto Set* (PS) and *Pareto Front* (PF) are defined formally. The decision makers require an approximation to the PF for a good insight to the problem and make the decision.

Tchebycheff approach [20] is employed to decompose the MOP into a number of sub-problems. Let $\lambda^1, \lambda^2, \ldots, \lambda^n$, be a set of uniformly spread weighted vectors and z^* be an ideal point. The problem can be decomposed into scalar optimization sub-problems as follows.

$$minimize\ g^{te}(x|\lambda^j, z^*) = max_{1 \le i \le m}\{\lambda_i^j|f_i(x) - z^*|\} \tag{2}$$

Therefore, one is able to obtain different Pareto optimal solutions by solving a set of single objective optimization problems defined by the Tchebycheff approach with different weight vectors. MOEA/D minimizes all these m objective functions simultaneously in a single run. Neighborhood relations among these single objective sub-problems are defined based on the distances among their weight vectors. Each sub-problem is optimized by using information mainly from its neighboring sub-problems. The details of MOEA/D can be found in [6].

3.2 Problem Formulation

Barrier Model. Consider a WSN consisting of a set of sensor nodes and several sink nodes, in which sensor nodes form a virtual sensor barrier for monitoring a belt-shaped region to detect and send intruding events to one of the sink nodes. Fig. 2 shows an illustration of the barrier model. Intrusion is assumed to occur from top to bottom. The assumptions are as follows.

- The sensor nodes and sink nodes are assumed to be randomly deployed and static once deployed with known positions.
- Assume that each sensor has an adjustable disk sensing range r and is equipped with limited power.
- The sink nodes with sufficient energy (comparing to sensor nodes) are not failure-prone.

Mathematical Model. We define the following notations formally, which are used in the analysis in the mathematical model:

- **S**: a set **S** of N sensors $\{\mu_1, \mu_2, ..., \mu_N\}$ are randomly distributed on a belt region which needs to be monitored.
- r_i: each sensor μ_i has an adjustable sensing range r_i. The power consumption of each active sensor is proportional to r_i^κ for some positive constant $\kappa \geq 2$.
- Π: a set Π of π sink nodes $\{s_1, s_2, ..., s_\pi\}$ are distributed on a belt region, in which $\pi << N$.
- (x_i, y_i): each sensor μ_i has a coordinate to denote the location.
- (x_j^s, y_j^s): each sink node s_j has a coordinate to denote the location.
- $d^j{}_i$: the distance between each sensor $\mu_i \in \mathbf{S}^*$ to its closest sink node $s_j \in \Pi$. The distance from the sensor μ_i to the sink node s_j is $d_i^j = \sqrt{(x_j^s - x_i)^2 + (y_j^s - y_i)^2}$.

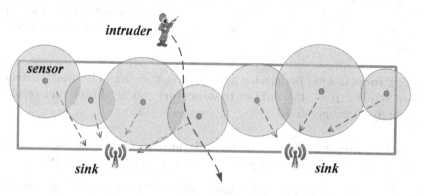

Fig. 2. Wireless barrier coverage model

- **S***: a subset $S^* \subseteq S$ of sensors are activated and assigned ranges to formulate a two-layer decision variable $\Omega = \{(u_1, r_1), (u_2, r_2), ..., (u_i, r_i), ..., (u_N, r_N)\}$, $u_i \in \{0, 1\}$.

In this way, *TBCAP* can be stated as an MOP, where we minimize the power consumption (f_1), the number of active sensors (f_2), and the active sensors' average distance from the closest sink node (f_3).

$$f_1 = \sum_{\mu_i \in S^*} r_i^\kappa$$

$$f_2 = |S^*| \qquad (3)$$

$$f_3 = \frac{\sum_{\mu_i \in S^*} d_i^j}{|S^*|}$$

3.3   Weighted-Sum Algorithm

Weighted-Sum Algorithm (WSA) as the most widely used classical method for MOP is used for comparing performance with our proposed PS-MOEA/D. It is the simplest yet efficient approach to find solutions on the entire Pareto-optimal set. The WSA in this paper is based on a genetic algorithm, which has the following procedures.

Solution Encoding. The solution is represented by a two-layer coding structure $C = \{(u_1, r_1), (u_2, r_2), ..., (u_i, r_i), ..., (u_N, r_N)\}$. The boolean u_i describes the working status of the sensor node and r_i indicates the value of its radii.

Repairing Initial Solutions. An approximation algorithm is necessary to guarantee the barrier coverage requirement, which can be found in [3].

Steady State Evolution. The Steady State Genetic Algorithm(SSGA) [21] based operators are adopted in WSA, which benefit from selecting two individuals and combining them to obtain two offsprings by crossover and mutation operators. Then, if these two new individuals are more adapted than the worst two individuals of the population, the former are included in the population by replacing the latter.

Shrink Process. After performing initialization, there could be several overlaps between sensors. If so, the radii of those sensors can be shrunk and repaired immediately after operations.

Evaluation. In each generation, the fitness is calculated by weighted-sum method after normalization. Specifically, *fitness* $= \omega_1 \times \sum_{\mu_i \in S^*} r_i^\kappa + \omega_2 \times |S^*| + \omega_3 \times \frac{\sum_{\mu_i \in S^*} d_i^j}{|S^*|}$, where ω_1, ω_2 and ω_3 are weights varying between *zero* and *one* and $\omega_1 + \omega_2 + \omega_3 = 1$.

4   Problem-Specific MOEA/D

In general, encoding representation, repair process and shrink process are identical to the WSA in Section 3.3. In order to improve the search ability of MOEA/D for *TBCAP*, some modification and improvement have been introduced. In the following part, we explain the procedure of Algorithm 1 in detail. The following sections are related to the main steps of the PS-MOEA/D.

Algorithm 1. PS-MOEA/D Framework for *TBCAP*

Input:

N_P: Population size and number of sub-problems

N_N: Size of neighborhood

M_E: Maximum number of evaluations

Output:

P: Final solutions

Step 1-Initialization: randomly generate an initial population and set parameters

Step 2-Repairing: repair the solutions to meet problem requirement

Step 3-Decomposition: decompose the *TBCAP* to N_P sub-problems

Step 4-Evaluation

While $e < M_E$

  Step 4.1-Selection: selection of sub-problems by using tournament selection based on μ_i as Sel_P or Per_P

  For $i = 1 : |Sel_P|$

    Step 4.2-Mutation: generate a new solution by mutation operator

    Step 4.3-Local Search: use of forward-LS and backward-LS

    Step 4.4-Update: update of current and neighboring solutions

  End-for

  Step 4.5-Perturbation: perturbation operator on Per_P

  $e \leftarrow e + 1$

  Step 4.6-Update Utility: calculate and update the utility

End-while

return P

End

4.1   Problem Decomposition

let $\Lambda = \lambda^1, \lambda^2, \ldots, \lambda^n$, be a set of uniformly spread weighted vectors, z^* be the ideal point and values of $f_j(x)$ in problem (3) have been normalized. Thus, the objective function of i-th sub-problem can be referred to problem (2). *TBCAP* is decomposed into scalar optimization sub-problems. A neighborhood of weight vector λ^i is defined as a set of its several weight vectors in Λ. The neighborhood of i-th sub-problem consists of all the sub-problems with the weight vectors from the neighborhood of λ^i. MOEA/D provides an easy yet efficient way to take the advantage of scalarization method and solve all subproblems simultaneously with different objective preference in a single run. In this paper, Λ is used to guide the problem specific operators for adjusting the degree of power consumption, reliability and efficiency of data gathering and therefor obtaining different preference barrier coverage.

Algorithm 2. Local Search Strategies

Input:
    χ_k (one individual of k-th sub-problem)
    N_N^k (neighborhood size of the sub-problem)
Output:
    χ_k (updated individual of k-th sub-problem)
1: Randomly choose a neighborhood j of χ_k
2: If $j < k$
3:    For $i = 1 : k - j$
       Forward-LS(χ_k,χ_j)
    End-for
4: else
5:    For $i = 1 : j - k$
       Backward-LS(χ_k,χ_j)
    End-for
6: **return** χ_k
7: End

4.2   Genetic Operators

Selection operators choose the most suitable solutions to produce offspring. In this paper, we have adopted a tournament selection operator based on utility for each sub-problem, which has been tested to be fast and effective [22]. Mutation operator randomly selects two genes within a specific range (a relatively small interval), in order to be further improved by fine-tuning the solution.

4.3   Local Search: Forward-LS and Backward-LS

Two original problem-specific local search strategies, as shown in Algorithm 2, have been developed. There are two search directions, i.e., *Forward-LS* (Fig. 3(a)) and *Backward-LS* (Fig. 3(b)).

    The idea of problem-specific local search strategies is inspired by workload balancing, which is to construct two possible search directions for an offspring whose performance is better. The search procedure is from starting point to ending point. We set the search direction based on the number of active sensors of starting point and ending point. The starting point can be randomly selected, and the ending point is the best individual of the neighborhood. When an off-spring shows improvement in terms of the objective function, it is adopted as the solution of this subproblem. The details are given in Algorithm 3 and 4.

    For example, consider the *Forward-LS* in Fig. 3(a), the active sensor j with a large sensing radius to cover a specific region B of the barrier. Then, search from the nearby sleeping sensors to check if there exists two sleeping sensors i and k, which can be assigned sensing ranges to cover B. If exists, we set sensor i from the status active to sleep, and sensor i and k from sleep to active with corresponding radii.

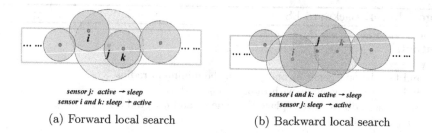

sensor j: *active → sleep*
sensor i and k: *sleep → active*

(a) Forward local search

sensor i and k: *active → sleep*
sensor j: *sleep → active*

(b) Backward local search

Fig. 3. Local search procedures description

Algorithm 3. Forward-LS

Input: χ_k, χ_j
Output:
   χ_k
1: Find the gene g with maximum radii r_g
2: Find the two nearest genes g_1 and g_2 with radii *zero* around gene g
3: Assign the radius to genes g_1 and g_2 to replace the gene g, produce χ_k'
4:   If χ_k' is better than χ_k
   $\chi_k \leftarrow \chi_k'$
   End-if
5: **return** P
6: End

4.4 Dynamical Resource Allocation

The sub-problems may have different computational difficulties, which makes it reasonable to assign different amounts of computational effort to different problems [22]. As we can see, for the *TBCAP*, the complexity fits binomial distribution with the number of active sensors. Thereby, more computational resource based on utility will be assigned to the sub-problems with higher complexity.

4.5 Utilities Update

We define and compute a utility for each sub-problem. Computational efforts are distributed to these sub-problems based on their utilities. If evaluation times is a multiplication of a certain number, then we compute the relative decrease of the objective for each sub-problem i, Δ_i. The utility of the sub-problem can be calculated as follows.

$$\mu_i = \begin{cases} 1.00 & \text{if } \Delta_i > 0.001 \\ (0.99 + 0.01\frac{\Delta_i}{0.001}) & \text{otherwise} \end{cases}$$

4.6 Perturbation

Perturbation improves the quality of solutions found by PS-MOEA/D, thereby speeding up the search for global optimal solution. Let χ_k be the current solution

---

Algorithm 4. Backward-LS
_____
Input: χ_k, χ_j
Output:
    χ_k
1: Find two disjoint genes g_1 and g_2 with the minimum radius
2: Find the nearest gene g_0 with radii 0 to g_1 and g_2
3: Assign the radii to g_0 to replace the gene g_1 and g_2, produce χ'_k
4:    If χ'_k is better than χ_k
        $\chi_k \leftarrow \chi'_k$
        End-if
5: **return** P
6: End
_____

to the k-th sub-problem, we apply a random interchange move on χ_k to produce χ'_k. It randomly selects a number of genes within a specific range (a relatively large interval), in order to jump out of local optimum.

5   Experiments and Discussions

5.1   Experimentation

This section presents the setup of the experimentation, with the purpose of validating the performance of the implemented PS-MOEA/D. The experiments are conducted on a 3.4GHz Intel PC with 4GB RAM. The programming language is MATLAB(R2013a). The proposed algorithm runs with the following parameter values: the maximum number of evaluations $M_E = 1,000$, neighborhood size $niche = 20$, mutation rate $P_m = 1.0$ and the experiments for each instance are replicated for 10 independent trials. Since an analysis of the parameter sensitivity is not a major concern of this study, we have not performed any previous analysis to fix these values.

Depending on the deployment method, the coordinates of the sensor positions may follow a particular distribution. For instance, if sensors are thrown off an aircraft that flies over the middle of a field, most sensors are expected to fall somewhere close to the central line, and several sensors are likely to end up further out. One could then argue that the sensor distribution is uniform along the axis of route. Thus, the experiments fall into two major parts, i.e., Uniform distribution and Gaussian distribution. In the experiments, the length of barrier and the default number of sink node is set as 1000 units and *one*, respectively. The offsets of the sensors are assumed to be 0.

5.2   Performance Comparison

In this section, we study the effectiveness of the proposed PS-MOEA/D on *TBCAP*. To do so, we compare the proposed method with the WSA.

Performance Measure The quality of the obtained non-dominated solutions is usually evaluated from three perspectives: (i) the closeness to the true PF, (ii) diversity and (iii) uniformity. No single metric can reflect all these aspects and often a number of metrics are used. In this study, we use the *Set Coverage* $C(X,Y)$ ($X \succeq Y$) [23] and *distance to reference set* $D_{ref}(X,R)$ [24] metrics.

$$C(X,Y) = \frac{|y \in Y | \exists x \in X : x \prec y|}{|Y|} \tag{4}$$

The $C(X,Y)$ metric compute the percentage of solutions in Y dominated by solutions in X, divided by the total number of solutions in Y. The higher the value of $C(X,Y)$ obtained, more diversely and uniformly the solution set X distributed.

$$D_{ref}(X,R) = \frac{\sum_{r \in R} \{ min_{x \in X} \{ dis(x,r) \} \}}{|R|} \tag{5}$$

The distance from reference set calculates the average distance from a solution in the reference set R to the closest solution in X. The smaller the value of $D_{ref}(X,R)$, the closer the set X is to R. In the absence of the real reference set (i.e., true PF), we calculate the average distance of each single point to the nadir point since we consider minimization objectives.

Comparison with the WSA. We validate the performance of PS-MOEA/D by conducting comparison experiments in different scale (the number of randomly deployed sensors) *TBCAP*. Fig. 4 shows that the PS-MOEA/D outperforms the WSA in terms of *set coverage* and *distance to reference set* on Uniform instances, where the horizontal axis represents the number of randomly deployed sensors and the vertical axis represents the mean values of *set coverage* and *distance to reference set*. Similar results on Gaussian instances can be found in Fig. 5. From the experimental results, it is observed that PS-MOEA/D obtains better PFs than the WSA. Specifically, in both figures, the PS-MOEA/D obtains a percentage of dominance 30% to 60%; if we check the inverse coverage relation, the fraction of non-dominated solutions achieved by the WSA that dominates the Pareto sets obtained by PS-MOEA/D, in all cases, this fraction is close to 0%. Besides, PS-MOEA/D performs better on average than WSA in terms of *distance to reference set*. In addition, it can be noticed that PS-MOEA/D shows better stability of results as the instance scale increases. Summarizing that, the PS-MOEA/D has obtained more evenly distributed PFs providing a better approximation towards the nadir point than the WSA.

5.3   Existence of Knee Regions

Knee points are made up of Pareto-optimal solutions, which provide the best possible tradeoff among the three conflicting objectives, in other words, any improvement in one objective must outweigh the aggregated deterioration of other objectives. These are probably the most interesting solutions in many real-world problems. Faced with multiple methods for finding knee points [25] [26] [27], since it is a challenging topic to find the true extreme Pareto optimal solutions for the *TBCAP*, we propose to find the knee points based on a trade-off metric designed by Rachmawati and Srinivasan [26].

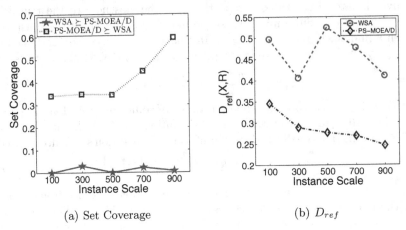

(a) Set Coverage   (b) D_{ref}

Fig. 4. Comparison between PS-MOEA/D and WSA on Uniform instance

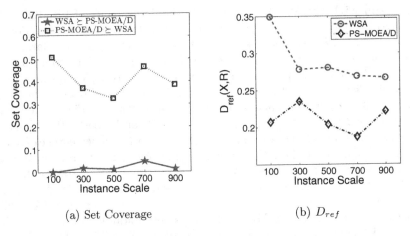

(a) Set Coverage   (b) D_{ref}

Fig. 5. Comparison between PS-MOEA/D and WSA on Gaussian instance

Following this metric, we define a notation $\rho(X_i, S)$ to represent the least amount of improvement per unit deterioration by substituting any alternative solution from non-dominated solution set S with X_i. Solutions residing in convex knee regions have the highest values in terms of $\rho(X_i, S)$. It allows us to define the strong degree of knee points by setting a threshold value θ. It can be mathematically defined as follow.

$$\rho(X_i, S) = min_{X_j \in S; i \neq j} \frac{\sum_{1 \leq m \leq M} max(0, f_m(X_j) - f_m(X_i))}{\sum_{1 \leq m \leq M} max(0, f_m(X_i) - f_m(X_j))} \tag{6}$$

$$S_{knee}^{\theta} = \{X_i | \rho(X_i, S) \geq \theta, X_i \in S\} \tag{7}$$

X_j corresponds to a member of the non-dominated solutions S that are non-dominated with respect to X_i; $f_m(X_i)$ corresponds to the m-th objective value of solution X_i, S_{knee}^{θ} denotes the set of knee points with the threshold value θ.

To demonstrate the existence of a knee region for this problem, two sets of experiments have been conducted with different θ values, namely 0.5 and 0.25. In Fig. 6, it shows that the obtained knee points obtained on the 500 sensors Uniform and Gaussian deployment. Note that in this part of simulation, we assume there is only one sink node, which is located in the middle of the barrier. It can be noticed that the fronts have clear knee points in which it is more reasonable to take a final decision about which solution should be adopted.

Besides, comparing with the Uniform deployment method, more knee points have been found by the Gaussian deployment method. The PF obtained by the Uniform deployments spreads more evenly than the Gaussian deployment. Except for the reason of sensors' positions, the biggest reason is the location of the sink node. Since in this part of simulation, the sink node is assumed to be located in the middle of the barrier. Following the Gaussian deployment, a large number of sensors are deployed closely to the middle of the barrier. Thus, in this case, the solutions tend to be high quality for the objective of average distance.

5.4   Effect of the Number of Sink Nodes

Intuitively, when more sink nodes are deployed, the estimated average distance from sensor nodes to the nearest sink node should be shorter and influence other objectives. The number of sink nodes is one of the factors that may influence the obtained PFs. We compare the obtained PFs by PS-MOEA/D with the number of sink nodes from 1 to 5. Assume that the sink nodes are uniformly located along the barrier, then we run the experiments to validate the effect of the number of sink nodes. As expected, from the results of Fig. 7, we can observe an important property that more sink nodes are uniformly deployed, better PFs can be obtained. The major reason is that the active sensors may have more sink nodes to be chosen as the nearest sink.

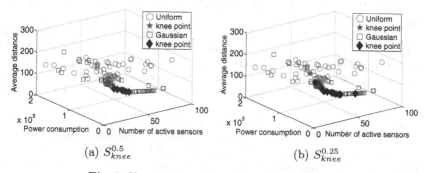

(a) $S_{knee}^{0.5}$ (b) $S_{knee}^{0.25}$

Fig. 6. Knee region of instance with 500 sensors

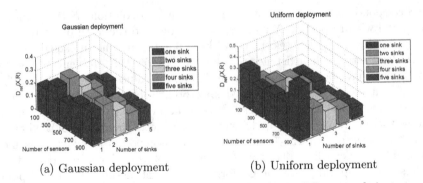

(a) Gaussian deployment   (b) Uniform deployment

Fig. 7. D_{ref} values found by varying number of sinks on different scale instances

6   Conclusion

This paper has made several contributions. Firstly, *TBCAP* is defined and formulated. Secondly, a PS-MOEA/D has been proposed for finding optimal tradeoff solutions. Thirdly, an experimental investigation has been presented, which explores the tradeoff among reliability, power consumption and average distance. A comparative study is conducted to evaluate the proposed approach. Additionally, the effect of the number of sink nodes to the PF have also been studied. Our future work will enrich the model to make it closer to reality and further improve the performance of the PS-MOEA/D.

Acknowledgments. This work was supported in part by a grant from the Research Grants Council of the Hong Kong Special Administrative Region, China [Project No. CityU 115312].

References

1. Yu, Z., Teng, J., Li, X., Xuan, D.: On wireless network coverage in bounded areas. In: Proceedings of the IEEE Conference on Computer Communications, pp. 1195–1203. IEEE (2013)
2. Fan, H., Lee, V.C.S., Li, M., Zhang, X., Zhao, Y.: Barrier coverage using sensors with offsets. In: Cai, Z., Wang, C., Cheng, S., Wang, H., Gao, H. (eds.) WASA 2014. LNCS, vol. 8491, pp. 389–400. Springer, Heidelberg (2014)
3. Fan, H., LI, M., Sun, X., Wan, P.J., Zhao, Y.: Barrier coverage by sensors with adjustable ranges. ACM Transactions on Sensor Networks **11**(1), 14(1)—14(20) (2014)
4. Konstantinidis, A., Yang, K., Zhang, Q., Zeinalipour-Yazti, D.: A multi-objective evolutionary algorithm for the deployment and power assignment problem in wireless sensor networks. Computer Networks **54**(6), 960–976 (2010)
5. Yoon, Y., Kim, Y.H.: An efficient genetic algorithm for maximum coverage deployment in wireless sensor networks. IEEE Transactions on Cybernetics **43**(5), 1473–1483 (2013)

6. Zhang, Q., Li, H.: Moea/d: A multiobjective evolutionary algorithm based on decomposition. IEEE Transactions on Evolutionary Computation 11(6), 712–731 (2007)

7. Mhatre, V.P., Rosenberg, C., Kofman, D., Mazumdar, R., Shroff, N.: A minimum cost heterogeneous sensor network with a lifetime constraint. IEEE Transactions on Mobile Computing 4(1), 4–15 (2005)

8. Onur, E., Ersoy, C., Deliç, H., Akarun, L.: Surveillance wireless sensor networks: deployment quality analysis. IEEE Network 21(6), 48–53 (2007)

9. Sun, Z., Wang, P., Vuran, M.C., Al-Rodhaan, M.A., Al-Dhelaan, A.M., Akyildiz, I.F.: Bordersense: Border patrol through advanced wireless sensor networks. Ad Hoc Networks 9(3), 468–477 (2011)

10. Kumar, S., Lai, T.H., Arora, A.: Barrier coverage with wireless sensors. In: Proceedings of Annual International Conference on Mobile Computing And Networking, pp. 284–298. ACM (2005)

11. Saipulla, A., Westphal, C., Liu, B., Wang, J.: Barrier coverage with line-based deployed mobile sensors. Ad Hoc Networks 11(4), 1381–1391 (2013)

12. He, S., Gong, X., Zhang, J., Chen, J., Sun, Y.: Curve-based deployment for barrier coverage in wireless sensor networks. IEEE Transactions on Wireless Communications 13(2), 724–735 (February 2014)

13. Lee, E., Park, S., Lee, J., Oh, S., Kim, S.H.: Novel service protocol for supporting remote and mobile users in wireless sensor networks with multiple static sinks. Wireless Networks 17(4), 861–875 (2011)

14. Chen, J., Li, J., He, S., He, T., Gu, Y., Sun, Y.: On energy-efficient trap coverage in wireless sensor networks. ACM Transactions on Sensor Networks 10(1), 2–29 (2013)

15. Shakkottai, S., Srikant, R., Shroff, N.: Unreliable sensor grids: coverage, connectivity and diameter. Proceedings of the IEEE Conference on Computer Communications 2, 1073–1083 (March 2003)

16. Sengupta, S., Das, S., Nasir, M., Vasilakos, A.V., Pedrycz, W.: An evolutionary multiobjective sleep-scheduling scheme for differentiated coverage in wireless sensor networks. IEEE Transactions on Systems, Man, and Cybernetics, Part C: Applications and Reviews 42(6), 1093–1102 (2012)

17. Martins, F.V., Carrano, E.G., Wanner, E.F., Takahashi, R.H., Mateus, G.R.: A hybrid multiobjective evolutionary approach for improving the performance of wireless sensor networks. IEEE Sensors Journal 11(3), 545–554 (2011)

18. Lanza-Gutierrez, J.M., Gomez-Pulido, J.A., Vega-Rodriguez, M.A., Sanchez-Perez, J.M.: A parallel evolutionary approach to solve the relay node placement problem in wireless sensor networks. In: Proceeding of Annual Conference on Genetic and Evolutionary Computation, pp. 1157–1164. ACM (2013)

19. Coello, C.C., Lamont, G.B., Van Veldhuizen, D.A.: Evolutionary algorithms for solving multi-objective problems. Springer (2007)

20. Miettinen, K.: Nonlinear multiobjective optimization, vol. 12. Springer (1999)

21. Smith, J., Fogarty, T.C.: Self adaptation of mutation rates in a steady state genetic algorithm. In: Proceedings of IEEE International Conference on Evolutionary Computation, pp. 318–323. IEEE (1996)

22. Zhang, Q., Liu, W., Li, H.: The performance of a new version of moea/d on cec09 unconstrained mop test instances. In: IEEE Congress on Evolutionary Computation, pp. 203–208 (2009)

23. Zitzler, E., Thiele, L.: Multiobjective evolutionary algorithms: a comparative case study and the strength pareto approach. IEEE Transactions on Evolutionary Computation 3(4), 257–271 (1999)

24. Czyzżak, P., Jaszkiewicz, A.: Pareto simulated annealing a metaheuristic technique for multiple-objective combinatorial optimization. Journal of Multi-Criteria Decision Analysis **7**(1), 34–47 (1998)

25. Deb, K., Miettinen, K., Sharma, D.: A hybrid integrated multi-objective optimization procedure for estimating nadir point. In: Ehrgott, M., Fonseca, C.M., Gandibleux, X., Hao, J.-K., Sevaux, M. (eds.) EMO 2009. LNCS, vol. 5467, pp. 569–583. Springer, Heidelberg (2009)

26. Rachmawati, L., Srinivasan, D.: Multiobjective evolutionary algorithm with controllable focus on the knees of the pareto front. IEEE Transactions on Evolutionary Computation **13**(4), 810–824 (2009)

27. Bechikh, S., Ben Said, L., Ghédira, K.: Searching for knee regions in multi-objective optimization using mobile reference points. In: Proceedings of the ACM Symposium on Applied Computing, pp. 1118–1125. ACM (2010)

Comparison of Single and Multi-objective Evolutionary Algorithms for Robust Link-State Routing

Vitor Pereira[1], Pedro Sousa[1], Paulo Cortez[2], Miguel Rio[3], and Miguel Rocha[4](✉)

[1] Centro Algoritmi/Department of Informatics,
University of Minho, Braga, Portugal
vitor.pereira@algoritmi.uminho.pt, pns@di.uminho.pt
[2] Centro Algoritmi/Department of Information Systems,
University of Minho, Braga, Portugal
pcortez@dsi.uminho.pt
[3] Department of Electronic and Electrical Engineering,
University College London, London, UK
m.rio@ee.ucl.ac.uk
[4] Centre Biological Engineering/Department of Informatics,
University of Minho, Braga, Portugal
mrocha@di.uminho.pt

Abstract. Traffic Engineering (TE) approaches are increasingly important in network management to allow an optimized configuration and resource allocation. In link-state routing, the task of setting appropriate weights to the links is both an important and a challenging optimization task. A number of different approaches has been put forward towards this aim, including the successful use of Evolutionary Algorithms (EAs). In this context, this work addresses the evaluation of three distinct EAs, a single and two multi-objective EAs, in two tasks related to weight setting optimization towards optimal intra-domain routing, knowing the network topology and aggregated traffic demands and seeking to minimize network congestion. In both tasks, the optimization considers scenarios where there is a dynamic alteration in the state of the system, in the first considering changes in the traffic demand matrices and in the latter considering the possibility of link failures. The methods will, thus, need to simultaneously optimize for both conditions, the normal and the altered one, following a preventive TE approach towards robust configurations. Since this can be formulated as a bi-objective function, the use of multi-objective EAs, such as SPEA2 and NSGA-II, came naturally, being those compared to a single-objective EA. The results show a remarkable behavior of NSGA-II in all proposed tasks scaling well for harder instances, and thus presenting itself as the most promising option for TE in these scenarios.

Keywords: Multi-objective evolutionary algorithms · Traffic engineering · NSGA · SPEA · Intra-domain routing · OSPF

© Springer International Publishing Switzerland 2015
A. Gaspar-Cunha et al. (Eds.): EMO 2015, Part II, LNCS 9019, pp. 573–587, 2015.
DOI: 10.1007/978-3-319-15892-1_39

1   Introduction

Link-State protocols, such as Intermediate-System to Intermediate-System (ISIS) [12] and Open Shortest Path First (OSPF)[11], are widely used routing protocols. Positive weights, assigned to each link in the network, are used to compute the shortest path (SP) between each source-destination pair, through which network traffic flows. The SPs are obtained using the Dijkstra algorithm [5], and minimize the total sum of link weights in the path. Thus, link weights define how traffic is accommodated onto the underlying network topology, being, in this context, the most important decision factor for the configuration of the traffic routing process. The decision making involved in link weights configuration, usually performed by a network administrator, is not an easy task when the scale of the network and the typically high volume of traffic and flows are taken into consideration. If inadequate, a configuration can cause the misallocation of traffic into the available resources, resulting in packet loss, increasing delays, and, potentially, in the unfulfillment of service level agreements (SLAs).

The Traffic Engineering (TE) problem addressed by this work arises in this context. It consists in finding a set of weights that optimize the congestion levels of the network, for which there are known aggregated traffic demands specified for each source-destination pair. This NP-hard optimization problem has been covered in previous efforts [1,8] with good results, resorting to several optimization approaches, which include, for instance, Evolutionary Algorithms (EA) in previous work by the authors [13,14]. Indeed, EA based approaches to TE have been proven to deliver near optimal solutions for the weight setting problem with several advantages when compared with other optimization techniques. Their ability to provide a set of possible solutions, with distinct trade-offs between objectives, enables network administrators to choose from a broader set of configurations, and consequently offers a conscious choice of the most adequate solution. However, distinct EAs have different merits and limitations [15] and consequently some approaches may not offer equally good solutions.

In this context, the present work offers a comparative study of three popular EAs spanning both single and multi-objective alternatives: the Non-dominated Sorting Genetic Algorithm (NSGA-II), the Strength Pareto Evolutionary Algorithm (SPEA2) and a Single-Objective Evolutionary Algorithm (SOEA) previously proposed by the authors. The experimental study allows to compare the performance of the three approaches in two extensions of the described problem, where the weights need to be set for scenarios considering the network's dynamic behavior, namely considering changes on the traffic demands over distinct time periods, in the first case, and the possibility of a single link failure, in the latter case.

The paper proceeds with section 2, describing the experimental model, the framework that sustained the experiments and EAs configuration; section 3 presents the results for the scenario with two distinct traffic demand matrices; section 4 presents the results for the scenarios with a single link failure; finally, section 5 presents the conclusions of this study.

2   Experimental Model

Changes on traffic demands and link failures are dynamic conditions that undermine the operational performance of a network. Traffic demands undergo periodic changes during specific periods of time, such as night and day, which affect the congestion levels of the network. To address effective TE under those changes, network administrators could, eventually, perform alterations on the installed weights configuration to induce the redistribution of traffic. However, weight configuration changes cause a temporary instability on the traffic flows due to the distributed nature and convergence time of the routing protocol. Furthermore, changes on traffic paths disrupt the performance of higher level protocols, such as the Transport Control Protocol (TCP) whose connections may become degraded by out of order packet delivery.

There are also similar considerations to be made when re-configuring weights in response to link failures. The majority of these faults are single link failures, and last, usually, a relatively short amount of time [9]. Frequent link weights reconfigurations are thereby not considered a good approach to the problem. A more appealing solution consists in finding a single weights setting that would allow the network to maintain a good performance level against such events. In this case, the weights configuration to seek would guarantee a good traffic distribution in normal network conditions and continue to provide a good congestion level after a link failure or in case of foreseen changes of traffic demands. The next section presents an overview of the mathematical model used to support the simulations.

2.1   Mathematical Model

Network topologies are modelled as directed graphs $G(N, A)$, where N represents a set of nodes, and A a set of arcs, with capacity constrains c_a for each $a \in A$. The amount of demand routed on the arc a, induced by a particular weight configuration, with source s and destination t, is denoted by $f_a^{(s,t)}$. We define the utilization of an arc a as $u_a = \frac{\ell_a}{c_a}$ where ℓ_a is the sum of all flows $f_a^{(s,t)}$ that travel over it. A well known piece-wise linear cost function Φ_a, proposed by Fortz and Thorup [7], is used to heavily penalize over-utilized links. The derivative of Φ_a is defined as:

$$\Phi_a' = \begin{cases} 1 & for \ 0 \leq u_a < 1/3 \\ 3 & for \ 1/3 \leq u_a < 2/3 \\ 10 & for \ 2/3 \leq u_a < 9/10 \\ 70 & for \ 9/10 \leq u_a < 1 \\ 500 & for \ 1 \leq u_a < 11/10 \\ 5000 & for \ u_a \geq 11/10 \end{cases} \tag{1}$$

The single optimization objective consists in distributing traffic demands in order to minimize the sum of all costs, as expressed in Equation 2.

$$\Phi = \sum_{a \in A} \Phi_a \tag{2}$$

A normalized congestion measure Φ^* is used to enable results comparison between distinct topologies and, for single objective EAs, to linearly combine the normal state congestion value of a network with the congestion after the occurrence of an event. It is important to note that when Φ^* equals 1, all loads are below 1/3 of the link capacity, while when all arcs are exactly full the value of Φ^* is 10 2/3. This value will be considered as a threshold that bounds the acceptable working region of the network.

It is now possible to define the general multi-objective optimization problem addressed in this work. Given a network represented by a graph $G = (N, A)$ and one or more demand matrices D_i, the aim is to find the set of weights (w) that simultaneously minimizes the objective functions Φ_1^* and Φ_2^*, that, respectively, evaluate the congestion level of the network on a normal state and the congestion level after a change on the network operational conditions. For single objective optimization, the algorithms use a linear weighting scheme where the cost of the solution is given by:

$$f(w) = \alpha \times \Phi_1^* + (1 - \alpha) \times \Phi_2^*, \ \alpha \in [0; 1] \tag{3}$$

2.2   Experimental Framework

The experimental simulations were run on a publicly available optimization framework, NetOpt [13], previously developed by the authors, in which the optimization meta-heuristic algorithms are provided by a Java-based library, JEColi [6]. An OSPF routing simulator is used to accommodate the traffic demands onto the networks topology arcs, and therefore enabling the application of the congestion evaluation function Φ^*. An overall view of the framework architecture is shown in Figure 1 that also translates the general multi-objective optimization problem defined in the previous section.

The simulations were run for two synthetic topologies with 30 nodes, named 30_2 and 30_4 (the indexes 2 and 4 stand for the average in/out degree of each node), and a real-world backbone topology, the well known Abilene topology. The synthetic topologies were generated by the Brite topology generator [10], using the Barabasi-Albert model, with a heavy-tail distribution and an incremental grow type. The link capacities uniformly vary in the interval $[1; 10]$ Gbits. The characteristics of each topology are summarized in Table 1.

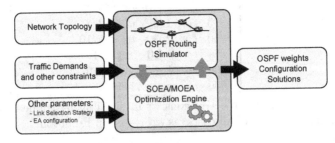

Fig. 1. General architecture of the optimization framework

Table 1. Synthetic and realistic network topologies

Name	Topology	Nodes	Edges
Abilene	backbone	12	15
30_2	random	30	55
30_4	random	30	110

Traffic demand matrices provide, for every ingress point a and every egress point b in the network, the volume of traffic from a to b over a given time interval. For each topology, three distinct levels of traffic demand D_i, $i \in \{0.3, 0.4, 0.5\}$, were used in the experiments, where i represents the expected mean of congestion in each link. Larger level values imply more difficult problems, as the volume of traffic to accommodate is greater. The set of demands D_i for the Abilene network were obtained by scaling Netflow data [3] publicly available and measured on March 1st 2004 and September 1st 2004. The set of demand matrices for the synthetic topologies were randomly generate to fulfill the requirements of the expected mean of congestion.

2.3 EAs Setup and Evaluation Metrics

Three different EAs were considered in this study, and applied to the contemplated optimization problems. The first approach was the use of a single objective EA (SOEA) based on previous work by the authors [14]. Alternative approaches were provided by two of the most popular multi-objective EAs, namely SPEA2 [16] and NSGA-II [4]. All algorithms were configured to use the same encoding and reproduction operators and their configurations was done to reduce any differences not related with the inherent differences of the optimization engines.

In all EAs, each individual encodes a solution as a vector of integer values, where each value (gene) corresponds to the weight of a link (arc) in the network, and therefore the size of the individual equals the number of links in the network. Although OSPF link weights are integers valued from 1 to 65535, only values in range [1; 20] were considered, allowing to reduce the search space and, simultaneously, increasing the probability of finding equal cost multipaths (ECMP). ECMP offers substantial increases in bandwidth by load-balancing traffic over multiple paths.

The individuals that populate the initial populations were randomly generated, with arc weights taken from a uniform distribution within the reduced range. All EAs resort to the same reproduction operators for solutions combination and genetic diversity:

- *Random mutation*, replaces a given gene by a random value, within the allowed range.
- *Incremental/decremental mutation*, replaces a given gene by the next or by the previous integer value, with equal probabilities, within the allowed range.
- *Uniform crossover*, this operator works by taking two parents as input and generating two offspring. For each position in the genome, a binary variable

is randomly generated: if its value is 1, the first offspring takes the gene from the first parent in that position, while the second offspring takes the gene from the second parent; if the random value is 0, the roles of the parents are reversed.

The single objective EAs use a *roulette wheel scheme* in the selection procedure, by converting the fitness value into a linear ranking in the population. In the experiments, a population size of 100 was considered, and for the MOEAs, an archive of the same size was used. For the SOEA experiments, the final objective value is taken as a linear combination of the two objectives, weighted by a factor (α) that defines the trade-off; three values were considered for $\alpha \in \{0.25, 0.5, 0.75\}$. Each simulation configuration was run 30 times with a stopping criteria of 1000 generations.

Three performance metrics that enable results comparison and the evaluation of the MOEA and SOEA algorithms performance were used in the experimental study:

- *C-measure*: It is based on the concept of solution dominance. Given two Pareto Fronts (PF1,PF2), the measure C(PF1; PF2) returns the fraction of solutions in PF2 that are dominated by at least one solution in PF1. A value of 1 indicates that all points in PF2 are dominated by points in PF1, so values near 1 clearly favour the method that generated PF1; values near 0 show that few solutions in PF2 are dominated by solutions in PF1.
- *Trade-off analysis* (TOA): For a pareto front PF1, and given a value of α, the solution that minimizes $\alpha \times \Phi_1^* + (1 - \alpha) \times \Phi_2^*$ is selected. Parameter α can take distinct values in the range $[0; 1]$, thus defining different trade-offs between the objectives. The values with the same α can be compared among the several multi objective optimizers (MOOs) and also with those from traditional algorithms.
- *Hypervolume*: It is the n-dimensional space that is contained by a set of points. It encapsulates in a single unary value a measure of the spread of the solutions along the Pareto front, as well as the closeness of the solutions to the Pareto-optimal front. We considered as an approximation for the Pareto-optimal front the non dominated solutions of all simulations in the same context, regardless of the algorithm.

The next two sections present more precised definitions of the two studied case problems where changes in the operational conditions of a network undermine its performance. In each case, the results produced by the three algorithms, SOEA, NSGA-II and SPEA2, are discussed and compared.

3   Optimization for Two Traffic Demands Matrices

3.1   Problem Definition

Traffic demands possess temporal properties that have a significant impact on internet traffic engineering. The diversity of services available on contemporary networks, as well as human behaviors and habits, provoke variations on traffic volumes and flow patterns not accommodated by traditional routing solutions. To acknowledge those variations, for example between two periods, such as night and day, we aim to find a link weight configuration that enables the network to sustain good functional performance in both periods. Thus, given two demand matrices, D_1 and D_2, that represent the traffic requirements of two distinct periods, we want to find a link weight configuration w that simultaneously minimizes the congestion functions Φ_1^* and Φ_2^*. Each Φ_i^* is the normalized cost function Φ (Equation 2) that evaluate the network congestion considering the traffic demands matrix D_i. The SOEA weighted-sum aggregation function for this set of experiments is defined in accordance with Equation 3. The main idea behind the optimization process is that, by compromising the congestion level in each individual scenario, it is possible to obtain a suitable configuration for both matrices. Under the SOEA algorithm, an administrator is able to fine tune adjustments, such as favouring one of the matrices and penalizing the other, by setting the α parameter accordingly in Equation 3. Under MOEA algorithms, the produced solutions feature distinct trade-offs between the objectives which enables network administrators to select the most appropriate solution.

3.2 Simulation Results

The experimental results, for each of the three algorithms (SOEA, NSGA-II and SPEA2), are summarized in Table 2 and Table 3 which respectively present the best and the mean fitness values of all runs with distinct trade-offs, organized by traffic demands levels and α values. In the experiments with the 30_4 network topology only $D0.3$ level traffic demand matrices were considered as for higher levels of demands the obtained congestion values surpass the threshold of 10 2/3, above which the network ceases to operate acceptably. As the size and degree of each node increase, the difficulty of the optimization problem also increases. It is important to mention that, in all simulations, the linear correlation between the two considered traffic demands matrices, D_1 and D_2, for which the congestion is simultaneously optimized, is approximately 0.5.

The results for the Abilene topology show that all three algorithms were able to converge to the same best solution in at least one of the 30 simulations. The average fitness values, for all levels of demands and trade-offs, Table 3, are also very similar among the three algorithms. The performance metric C-measure, given in Table 4, where the overall mean value for all the distinct instances and runs was computed, reinforces the conclusion that all performances are akin with respect to the Abilene topology. The SOEA, NSGA-II and SPEA2 algorithms were able to provide equally good solutions as all values are of the same magnitude and, consequently, no algorithm's pareto fronts are considered to dominate the others.

For larger network topologies, the performance of the three algorithms starts to diverge. The results for the synthetic topology 30_2, with 30 nodes and 55 edges, show that the NSGA-II algorithm is able to attain best fitness values for every α and demands level. This can be observed, for instance, with $D0.4$

Table 2. Best fitness comparison for two demand matrices optimization

Algorithm	First Demands	Second Demands	Abilene			30_2			30_4		
			0.25	0.50	0.75	0.25	0.50	0.75	0.25	0.50	0.75
SOEA			1.139	1.161	1.179	1.374	1.386	1.398	2.503	2.487	2.472
NSGA-II	0.3	0.3	1.139	1.161	1.179	1.338	1.349	1.357	1.907	1.968	1.961
SPEA2			1.139	1.161	1.179	1.461	1.452	1.442	4.336	5.620	6.178
SOEA			1.446	1.367	1.283	1.745	1.638	1.531	-	-	-
NSGA-II	0.3	0.4	1.446	1.367	1.283	1.659	1.559	1.453	-	-	-
SPEA2			1.446	1.367	1.283	1.878	1.718	1.559	-	-	-
SOEA			1.522	1.522	1.521	1.951	1.985	2.019	-	-	-
NSGA-II	0.4	0.4	1.522	1.522	1.521	1.841	1.882	1.916	-	-	-
SPEA2			1.522	1.522	1.521	2.139	2.184	2.214	-	-	-

Table 3. Mean fitness comparison for two demand matrices optimization

Algorithm	First Demands	Second Demands	Abilene			30_2			30_4		
			0.25	0.50	0.75	0.25	0.50	0.75	0.25	0.50	0.75
SOEA			1.218	1.218	1.218	3.103	2.977	2.785	38.017	40.592	43.167
NSGA-II	0.3	0.3	1.212	1.213	1.213	1.925	1.826	1.728	4.306	4.440	4.575
SPEA2			1.213	1.213	1.214	2.186	2.064	1.942	43.983	46.922	49.862
SOEA			1.489	1.399	1.308	7.946	6.115	4.285	-	-	-
NSGA-II	0.3	0.4	1.482	1.393	1.304	3.538	2.860	2.183	-	-	-
SPEA2			1.482	1.393	1.304	4.551	3.622	2.693	-	-	-
SOEA			1.565	1.554	1.543	10.446	10.213	9.980	-	-	-
NSGA-II	0.4	0.4	1.559	1.549	1.540	2.919	2.801	2.684	-	-	-
SPEA2			1.559	1.549	1.540	5.401	5.271	5.141	-	-	-

matrices and $\alpha = 0.5$, where the minimum and average fitness values are, respectively, 1.951 and 10.446 (SOEA), 1.841 and 2.919 (NSGA-II), 2.139 and 5.401 (SPEA2). Although NSGA-II and SOEA best values are very similar, and better than SPEA2 results, the average congestion values for NSGA-II are substantially smaller, that is, the NSGA-II solutions are globally better than those provided by the other two algorithms. The averaged C metric values for the 30_2 network topology scenarios, Table 4, show that NSGA-II solutions dominate SPEA2 ones in more than 56% on average, with a reverse C-measure of almost 0%. When compared with SOEA solutions, NSGA-II solutions dominate approximately 16% of the SOEA ones and, for the reverse case, SOEA solutions dominate NSGA-II ones in only about 6%. It is therefore possible to conclude, that, for the 30_2 topology experiment scenarios, the NSGA-II algorithm offers generally better solutions than any of the two other algorithms and that the SPEA2 algorithm had the worst performance of all.

As the size of the used topology increases, the performance of the NSGA-II algorithm detaches from the others. The experiments with the 30_4 network topology show that while the best values of the three algorithm remain acceptable, NSGA-II features the best solutions, and is the only algorithm whose mean fitness values remain within the acceptable operating limits of the network (Table 3). The C metric values for this new set of simulations are very similar to those obtained for the 30_2 topology, and again, NSGA-II solutions are globally better than those provided by SOEA and SPEA2.

Table 4. Overall C-Measure for two traffic demand matrices optimization

	Abilene			30_2			30_4		
	SOEA	NSGA-II	SPEA2	SOEA	NSGA-II	SPEA2	SOEA	NSGA-II	SPEA2
SOEA	-	0.143	0.179	-	0.062	0.553	-	0.029	0.602
NSGA-II	0.150	-	0.182	0.161	-	0.564	0.150	-	0.600
SPEA2	0.110	0.113	-	0.001	0.004	-	0.000	0.001	-

Table 5. Average hypervolume for two traffic demand matrices optimization

Algorithm	Topology		
	Abilene	30_2	30_4
SOEA	0.002	0.585	34.359
NSGA-II	0.002	0.321	4.868
SPEA2	0.002	4.592	7950.838

It is possible to identify a consistency in the performance of all three algorithms where NSGA-II is the algorithm that show comparatively best results. The hypervolume indicators, presented in Table 5, also support that NSGA-II is the best choice algorithm in the context of weights setting optimization for two traffic demand matrices. The NSGA-II pareto fronts are closer to the Pareto-optimal approximation, and better spread, than those provided by SOEA and SPEA2.

4 Single Link Failure Optimization

4.1 Problem Definition

Link failures on network topologies can occur for different reasons. At the physical layer, a fiber cut or a failure of optical equipment may cause a loss of physical connectivity. Other failures may be related to hardware, such as linecard failures. Router processor overloads, software errors, protocol implementation and misconfiguration errors may also lead to loss of connectivity between routers. Failures may also vary in nature. They can be due to scheduled network maintenance or be unplanned. Although backbone networks are usually well planned and adequately provisioned, link failures may still occur and undermine their operational performance. Several mechanisms can be used to protect an IP network against link failures, such as overlay protection or MPLS fast re-route [2], but protecting all links remains a very difficult task, or even impossible, especially for large network topologies. Thus, protection against failure continues to be link based.

The NetOpt framework supports several criteria to select the failing link, some are dynamic, that depend on the solution that is being evaluated, while others are user choices. The framework also allows to select more than one link to fail simultaneously each corresponding to an optimization objective. This study only considers two of the available single link selection criteria:

- *Highest Load*: The selected link, for each solution being evaluated, is the one that has the highest load for the traffic demands given as parameter. Therefore, distinct solutions may have a different failing link.
- *User Selected*: A network administrator identifies the link against whose failure the network should be protected.

For a given network topology with n links and a traffic demands matrix D, the aim is to find a set of weights w that simultaneously minimize the function Φ_n^*, representing the congestion cost of the network in the normal state, and Φ_{n-1}^*, representing the congestion cost of the network when foreseeing that a selected link from the topology will fail. The SOEA weighted-sum aggregation model is described in Equation 4:

$$f(w) = \alpha \times \Phi_n^* + (1 - \alpha) \times \Phi_{n-1}^*, \ \alpha \in [0; 1] \tag{4}$$

An administrator is able to define a trade-off between the objectives by tuning the value of the α weight. When $\alpha = 1$, the optimization is only performed for the normal state topology, without any link failure, whereas when using $\alpha = 0.5$ the same level of importance is given to the two topology states. However, as the link failure optimization can compromise the network congestion level in a normal state, a network administrator may wish to focus on the performance of the normal state network, e.g. using a α value between 0.5 and 1, at the expense of the congestion level in a failed state, that may not occur. Although this feature offers a good tuning tool for administrator, it requires several distinct runs in order to assert the best compromised solution. MOEA algorithms, on the other hand, are able to deliver such knowledge base and choice selection after a single run, and therefore, being more appealing in this context.

4.2 Highest Load Link Failure Optimization

The failure of the network link that carries the highest traffic load is one of the worst case scenarios for the failure of a single link in a network. Its failure would translate into the re-routing of the higher amount of traffic and potentially the worst case for out of order TCP packet delivery. Distinct levels of traffic demands, $D0.3$, $D0.4$ and $D0.5$ were used to compare the algorithms in problems with increasing difficulty. For comparison purpose, Table 6 that shows the obtained minimum weighted-sum aggregation fitness values, also includes the optimized congestion values for the networks without link failure optimization, and the respective congestion level after the failure of the link with higher load.

The simulation results show that, for the smallest topology, Abilene, all three algorithm behave alike producing equally good solutions. But, as the topology size increases, or with the escalation of traffic requirements, NSGA-II is able to obtain solutions which translate into lower congestion values before and after the link failure. In the 30_4 network topology scenario, with $D0.3$ traffic demands and $\alpha = .5$, the fitness values before and after the link failure are, respectively, 2.19 and 2.29 for NSGA-II; 5.81 and 33.01 for SOEA; 204.29 and 219.19 for SPEA2. These results are even more relevant when comparing with the congestion values

Table 6. Best fitness values for single link failure weights setting optimization - Highest Load Link

Topology	Demand	Without Link Failure Optimization		With Link Failure Optimization						Algorithm
				$\alpha = 0.25$		$\alpha = 0.5$		$\alpha = 0.75$		
		Before	After	Before	After	Before	After	Before	After	
Abilene	0.3	1.20	1.76	1.29	1.23	1.29	1.23	1.23	1.33	NSGA-II
				1.34	1.21	1.33	1.22	1.24	1.35	SOEA
				1.29	1.23	1.29	1.23	1.23	1.34	SPEA2
	0.4	1.53	32.22	1.63	1.58	1.63	1.58	1.55	1.70	NSGA-II
				1.69	1.58	1.69	1.58	1.55	1.73	SOEA
				1.64	1.58	1.64	1.58	1.55	1.70	SPEA2
	0.5	1.91	309.48	2.14	1.91	2.14	1.91	2.05	2.17	NSGA-II
				2.26	1.93	2.26	1.93	2.26	1.93	SOEA
				2.14	1.91	2.14	1.91	2.04	2.14	SPEA2
30_2	0.3	1.49	14.20	1.55	1.42	1.44	1.48	1.44	1.48	NSGA-II
				1.56	1.58	1.56	1.58	1.54	1.61	SOEA
				1.57	1.50	1.56	1.51	1.49	1.64	SPEA2
	0.4	1.79	41.44	1.83	1.76	1.75	1.80	1.75	1.80	NSGA-II
				2.07	2.09	1.85	2.22	1.85	2.22	SOEA
				1.95	1.93	1.91	1.96	1.91	1.96	SPEA2
	0.5	5.49	180.94	4.99	3.70	4.99	3.70	4.11	5.31	NSGA-II
				12.61	17.58	12.61	17.58	12.61	17.58	SOEA
				8.23	8.41	8.15	8.48	7.86	9.03	SPEA2
30_4	0.3	3.67	73.69	2.38	2.20	2.30	2.25	2.10	2.59	NSGA-II
				11.14	7.91	11.14	7.91	6.04	13.64	SOEA
				59.48	29.64	28.95	47.39	28.95	47.39	SPEA2
	0.4	33.93	223.04	18.66	10.13	18.66	10.13	10.07	28.42	NSGA-II
				77.09	88.80	77.09	88.80	58.81	140.07	SOEA
				355.03	139.57	205.65	190.92	159.03	325.12	SPEA2
	0.5	126.90	158.44	157.19	95.15	97.19	132.37	97.19	132.37	NSGA-II
				310.85	180.52	310.85	180.52	224.07	277.66	SOEA
				490.70	466.66	490.70	466.66	467.31	504.91	SPEA2

when only the congestion of the network in the normal state is optimized by resourcing to a single objective algorithm ($\alpha = 1$). The NSGA-II algorithm was able to provide a better solution while optimizing two objectives than a SOEA algorithm that optimizes a single objective, the congestion of the network before the link failure. This result is observed in all scenarios that are more demanding, allowing to conclude that NSGA-II performs better in these more difficult optimization tasks than SOEA even considering two objectives rather than a single one.

Although congestion values above 10 2/3 are not acceptable within an operational network, the results allow to observe that the more difficult the problem, the greater the difference between the quality of the solutions produced by each of the three EAs. NSGA-II is able to outperform SOEA and SPEA2 in all scenarios. The lack of performance of the SOEA algorithm in more demanding scenarios can be explained by its requirement of a higher number of generations to properly converge. It is also important to acknowledge that even small changes on a single weight can provoke drastic changes on shortest paths and therefore on the congestion value. The crowding distance used in the selection operator of NSGA-II, that keeps a diverse front by making sure each member stays a crowding distance apart, seems to positively influence the algorithm performance.

Table 7. C-measure of the highest Load link failure optimisation

	Abilene			30_2			30_4		
	SOEA	NSGA-II	SPEA2	SOEA	NSGA-II	SPEA2	SOEA	NSGA-II	SPEA2
SOEA	-	0.173	0.242	-	0.071	0.702	-	0.056	0.887
NSGA-II	0.007	-	0.098	0.143	-	0.709	0.143	-	0.887
SPEA2	0.010	0.108	-	0.003	0.005	-	0.000	0.000	-

Table 8. Best fitness values for single link failure weights setting optimization - User Select Link

Topology	Demand	Without Link Failure Optimization		With Link Failure Optimization						Algorithm
				$\alpha = 0.25$		$\alpha = 0.5$		$\alpha = 0.75$		
		Before	After	Before	After	Before	After	Before	After	
				1.23	1.73	1.20	1.74	1.20	1.74	NSGA-II
	0.3	1.20	1.76	1.22	1.71	1.21	1.72	1.20	1.74	SOEA
				1.24	1.72	1.20	1.74	1.20	1.74	SPEA2
				1.58	33.44	1.58	33.44	1.53	33.52	NSGA-II
Abilene	0.4	1.53	25.57	1.56	5.26	1.56	5.26	1.56	5.26	SOEA
				1.58	33.44	1.58	33.44	1.53	33.52	SPEA2
				1.97	119.30	1.95	119.32	1.93	119.34	NSGA-II
	0.5	1.91	309.48	1.98	281.63	1.98	281.63	1.98	281.63	SOEA
				2.01	119.29	1.95	119.32	1.93	119.34	SPEA2
				1.40	1.50	1.40	1.50	1.40	1.50	NSGA-II
	0.3	1.49	8.17	1.54	4.55	1.54	4.55	1.54	4.55	SOEA
				1.61	1.74	1.60	1.74	1.60	1.74	SPEA2
				1.76	1.93	1.76	1.93	1.75	1.93	NSGA-II
30_2	0.4	1.79	58.65	2.10	3.40	2.10	3.40	2.10	3.40	SOEA
				2.18	2.50	2.18	2.50	2.18	2.50	SPEA2
				5.79	41.19	5.44	41.51	5.44	41.51	NSGA-II
	0.5	5.49	193.16	22.45	87.53	17.23	91.56	8.07	117.40	SOEA
				28.13	47.86	12.86	55.66	12.30	56.43	SPEA2
				2.20	2.29	2.19	2.29	2.18	2.33	NSGA-II
	0.3	3.67	117.13	5.81	33.01	5.81	33.01	5.81	33.01	SOEA
				204.29	219.19	204.29	219.19	204.29	219.19	SPEA2
				10.43	9.67	9.90	10.05	9.85	10.11	NSGA-II
30_4	0.4	33.93	98.98	49.27	111.08	49.27	111.08	49.27	111.08	SOEA
				456.36	509.50	456.36	509.50	440.71	544.54	SPEA2
				89.91	88.98	62.21	112.74	62.21	112.74	NSGA-II
	0.5	126.90	421.07	165.46	319.34	165.46	319.34	165.46	319.34	SOEA
				557.49	600.71	557.49	600.71	557.49	600.71	SPEA2

The C-measure values in Table 7 show that, despite being able to offer solutions with equivalent best fitness for the Abilene topology, the SO algorithm produces more solutions that are neither dominated by NSGA-II or SPEA2 solutions. In contrast, for the more demanding topologies, 30_2 and 30_4, NSGA-II solutions dominate approximately 14% of the SOEA solutions, when the reverse is 7% or less. When compared against SPEA2, both NSGA-II and SOEA present better values.

4.3   User Choice Link Failure Optimization

A network administrator can consider that a particular link is more crucial than others, because of its capacity or for other reasons. It is therefore important to enable an administrator to select the link that needs to be protected against failure. For this set of simulations, the selected link in each topology is such that it occurs in the largest number of shortest paths when assigning to each link a weight inversely proportional to its capacity.

The minimal congestion values before and after the failure of the selected link, for distinct trade-offs ($\alpha = 0.25, 0.5, 0.75$), are presented in Table 8.

The results of this new test suite consolidate previous observations, that is, for simpler problems, with smaller topologies and lower traffic demand levels, the SOEA and MOEAs algorithms provide equally good solutions, but, as the number of nodes and links increases, or with the growth of traffic demands, NSGA-II is able to deliver better solutions, in the large majority of scenarios, both before and after the link failure. The C metric values, Table 9, are also similar to those observed for the higher load link failure optimization. In average and in the context of simpler problems, SOEA continues to have more solutions that are not dominated by any of the non-dominated sets of solutions resulting from NSGA-II and SPEA2 based optimizations. As the difficulty of the problem increases, NSGA-II stands out, providing better sets of non-dominated solutions.

Table 9. C-measure of User Choice link failure optimization

	Abilene			30_2			30_4		
	SOEA	NSGA-II	SPEA2	SOEA	NSGA-II	SPEA2	SOEA	NSGA-II	SPEA2
SOEA	-	0.162	0.203	-	0.074	0.443	-	0.065	0.659
NSGA-II	0.008	-	0.063	0.122	-	0.468	0.163	-	0.657
SPEA2	0.024	0.125	-	0.014	0.019	-	0.000	0.001	-

5   Conclusion

The simplicity of link-state protocols, and their reliability proven over the last two decades, continues to justify the use of such routing algorithms in the context of IP backbone networks. However, the dynamic conditions of IP networks, such as changes on traffic demands and disruptions on the underlying topology need to be addressed so that the network continues to ensure a good operational performance even if such events take place. An administrator could react to such changes by re-configuring the link weights but with a temporary negative impact on traffic flows. Other approaches, such as preventive optimization, can effectively take into consideration foreseen changes to compute weight configurations that allow the network to ensure a continues good levels of performance even in dynamic conditions. In this context, two multi objective problems were addressed, that consider changes on traffic demands and single link failure, resourcing to three popular EAs spanning both single and multi-objective: NSGA-II, SPEA2 and single objective EA using weighted-sum aggregation.

The results showed that for simpler problems the single objective optimization approaches provide solutions with best fitness values as good as the MOEA algorithms but, as the difficulty of the problems increases, for more complex network topologies and for more demanding traffic requirements, NSGA-II provides better solutions. By comparing the obtained results with previous work by the authors, it can be observed that SOEA algorithms require a greater number of generations for more demanding problems than MOEA algorithms. The two MOEAs, NSGA-II and SPEA2, rely heavily on their density estimator mechanisms, where the NSGA-II ability to provide a broader spread seems to influence more positively the optimization process than a better solution distribution attained by SPEA2.

Apart from the quality of the solutions other more practical aspects help determine the most appropriate algorithm to the problem. The single objective approaches have an important limitation. They assume, in each individual optimization process, that there is a single optimum trade-off between the objectives. A network administrator needs to guess which value of the weighting trade-off parameter better fits the needs of a network on the addressed operational conditions. In contrast, MOEA algorithms are able to calculate a set of solutions with distinct trade-offs between the two objectives, and let the network administrator decide which solution to implement. Moreover, NSGA-II is able to offer this broader set of solutions within a shorter time than the SOEA using weight-aggregation, or SPEA2 in the same conditions.

References

1. Altin, A., Fortz, B., Thorup, M., Ümit, H.: Intra-domain traffic engineering with shortest path routing protocols. Annals of Operations Research **204**(1), 56–95 (2013)
2. Awduche, D., Malcolm, J., Agogbua, J., O'Dell, M., McManus, J.: Requirements for Traffic Engineering Over MPLS. RFC 2702 (Informational), September 1999
3. Claise, B.: RFC 3954 - Cisco Systems NetFlow Services Export Version 9, October 2004
4. Deb, K., Agrawal, S., Pratap, A., Meyarivan, T.: A fast and elitist multiobjective genetic algorithm: NSGA-II. IEEE Trans. Evolutionary Computation **6**(2), 182–197 (2002)
5. Dijkstra, E.: A note on two problems in connexion with graphs. Numerische Mathematik **1**(1), 269–271 (1959)
6. Evangelista, P., Maia, P., Rocha, M.: Implementing metaheuristic optimization algorithms with jecoli. In: Proceedings of the 2009 Ninth International Conference on Intelligent Systems Design and Applications, ISDA '09, pp. 505–510. IEEE Computer Society, Washington, DC, USA (2009)
7. Fortz, B.: Internet traffic engineering by optimizing ospf weights. In: Proceedings of IEEE INFOCOM, pp. 519–528 (2000)
8. Fortz, B., Thorup, M.: Optimizing ospf/is-is weights in a changing world. IEEE Journal on Selected Areas in Communications **20**(4), 756–767 (2002)

9. Iannaccone, G., Chuah, C., Mortier, R., Bhattacharyya, S., Diot, C.: Analysis of link failures in an ip backbone. In: Proceedings of the 2Nd ACM SIGCOMM Workshop on Internet Measurment, IMW '02, pp. 237–242. ACM, New York, NY, USA (2002)

10. Medina, A., Lakhina, A., Matta, I., Byers, J.: Brite: Universal topology generation from a users perspective. Technical report, Boston, MA, USA (2001)

11. Moy, J.: OSPF Version 2. RFC 2328 (Standard), April 1998. Updated by RFC 5709

12. Oran, D.: OSI IS-IS Intra-domain Routing Protocol. Technical report, IETF, February 1990

13. Pereira, Vitor, Rocha, Miguel, Cortez, Paulo, Rio, Miguel, Sousa, Pedro: A Framework for Robust Traffic Engineering Using Evolutionary Computation. In: Doyen, Guillaume, Waldburger, Martin, Čeleda, Pavel, Sperotto, Anna, Stiller, Burkhard (eds.) AIMS 2013. LNCS, vol. 7943, pp. 1–12. Springer, Heidelberg (2013)

14. Rocha, M., Sousa, P., Cortez, P., Rio, M.: Quality of Service Constrained Routing Optimization Using Evolutionary Computation. Applied Soft Computing **11**(1), 356–364 (2011)

15. Tan, K.C., Lee, T.H., Khor, E.F.: Evolutionary algorithms for multi-objective optimization: Performance assessments and comparisons. Artif. Intell. Rev. **17**(4), pp. 251–290, June 2002

16. Zitzler, E., Laumanns, M., Thiele, L.: Spea 2: Improving the strength pareto evolutionary algorithm. Technical report (2001)

Author Index